797,885 Books
are available to read at

Forgotten Books

www.ForgottenBooks.com

Forgotten Books' App
Available for mobile, tablet & eReader

ISBN 978-1-331-52780-0
PIBN 10201495

This book is a reproduction of an important historical work. Forgotten Books uses state-of-the-art technology to digitally reconstruct the work, preserving the original format whilst repairing imperfections present in the aged copy. In rare cases, an imperfection in the original, such as a blemish or missing page, may be replicated in our edition. We do, however, repair the vast majority of imperfections successfully; any imperfections that remain are intentionally left to preserve the state of such historical works.

Forgotten Books is a registered trademark of FB &c Ltd.
Copyright © 2015 FB &c Ltd.
FB &c Ltd, Dalton House, 60 Windsor Avenue, London, SW19 2RR.
Company number 08720141. Registered in England and Wales.

For support please visit www.forgottenbooks.com

1 MONTH OF FREE READING

at

www.ForgottenBooks.com

By purchasing this book you are eligible for one month membership to ForgottenBooks.com, giving you unlimited access to our entire collection of over 700,000 titles via our web site and mobile apps.

To claim your free month visit:
www.forgottenbooks.com/free201495

* Offer is valid for 45 days from date of purchase. Terms and conditions apply.

Similar Books Are Available from
www.forgottenbooks.com

Zoology of the Invertebrata
A Text-Book for Students, by A. E. Shipley

Text Book of Vertebrate Zoology
by J. S. Kingsley

Zoology for High Schools and Colleges
by A. S. Packard

Practical Zoology
by Robert William Hegner

Handbook of Invertebrate Zoology
For Laboratories and Seaside Work, by William Keith Brooks

The Descent of Man, and Selection in Relation to Sex, Vol. 1 of 2
by Charles Darwin

Zoology
by A. S. Packard

The Aquarium Its Inhabitants, Structure, and Management
by J. E. Taylor

Agricultural Zoology
by Jan Ritzema Bos

Fauna of Baluchistan and N. W. Frontier of India, Vol. 1
The Jurassic Fauna, by Fritz Noetling

A Guide for Laboratory and Field Work in Zoology
by Henry R. Linville and Henry A. Kelly

Dogs, Jackals, Wolves, and Foxes
A Monograph of the Canidæ, by George Mivart

A Text-Book of Field Zoology
Insects and Their Near Relatives and Birds, by Lottie E. Crary

Animal Life
A First Book of Zoölogy, by David Starr Jordan

Elementary Textbook of Economic Zoology and Entomology
by Vernon L. Kellogg

Elements of Medical Zoology
by Alfred Moquin-Tandon

British Zoology, Vol. 1
by Thomas Pennant

Savage Survivals
by J. Howard Moore

The Fauna of the Deep Sea
by Sydney J. Hickson

Elements of Comparative Zoology
by J. S. Kingsley

QL
444
.D3
R18
c.3

SMITHSONIAN INSTITUTION
UNITED STATES NATIONAL MUSEUM
Bulletin 97

THE GRAPSOID CRABS OF AMERICA

BY

MARY J. RATHBUN

Associate in Zoology, United States National Museum

WASHINGTON
GOVERNMENT PRINTING OFFICE
1918.

BULLETIN OF THE UNITED STATES NATIONAL MUSEUM.

Issued JAN 25 1918

ADVERTISEMENT.

The scientific publications of the United States National Museum consist of two series, the *Proceedings* and the *Bulletins*.

The *Proceedings*, the first volume of which was issued in 1878, are intended primarily as a medium for the publication of original, and usually brief, papers based on the collections of the National Museum, presenting newly acquired facts in zoology, geology, and anthropology, including descriptions of new forms of animals, and revisions of limited groups. One or two volumes are issued annually and distributed to libraries and scientific organizations. A limited number of copies of each paper, in pamphlet form, is distributed to specialists and others interested in the different subjects, as soon as printed. The date of publication is printed on each paper, and these dates are also recorded in the table of contents of the volumes.

The *Bulletins*, the first of which was issued in 1875, consist of a series of separate publications comprising chiefly monographs of large zoological groups and other general systematic treatises (occasionally in several volumes), faunal works, reports of expeditions, and catalogues of type-specimens, special collections, etc. The majority of the volumes are octavos, but a quarto size has been adopted in a few instances in which large plates were regarded as indispensable.

Since 1902 a series of octavo volumes containing papers relating to the botanical collections of the Museum, and known as the *Contributions from the National Herbarium*, has been published as bulletins.

The present work forms No. 97 of the *Bulletin* series.

RICHARD RATHBUN,
Assistant Secretary, Smithsonian Institution,
In charge of the United States National Museum.
WASHINGTON, D. C., *December 10, 1917.*

TABLE OF CONTENTS.

	Page.
Introduction	1
Sources of material	1
Special researches	2
Acknowledgments	3
Explanation of terms and abbreviations used	6
The grapsoid or catometopous crabs of America	9
Systematic discussion	11
Family Goneplacidae	15
Subfamily Carcinoplacinae	17
Genus Trizocarcinus	17
Trizocarcinus dentatus	18
Genus Bathyplax	19
Bathyplax typhla	19
Genus Pilumnoplax	21
Pilumnoplax americana	21
elata	23
Subfamily Goneplacinae	24
Genus Goneplax	25
Goneplax barbata	26
sigsbei	26
rosaea	27
hirsuta	28
tridentata	29
Subfamily Prionoplacinae	29
Genus Prionoplax	29
Prionoplax atlantica	30
ciliata	31
Genus Tetraplax	32
Tetraplax quadridentata	32
Genus Euryplax	34
Euryplax nitida	34
polita	36
Genus Chasmophora	37
Chasmophora macrophthalma	37
Genus Speocarcinus	38
Speocarcinus carolinensis	39
granulimanus	40
ostrearicola	41
californiensis	42
Genus Pseudorhombila	42
Pseudorhombila octodentata	43
Genus Oediplax	44
Oediplax granulata	44
Genus Cyrtoplax	45
Cyrtoplax spinidentata	46

TABLE OF CONTENTS.

	Page.
Family Goneplacidae—Continued.	
Subfamily Prionoplacinae—Continued.	
Genus Panoplax	47
Panoplax depressa	47
Genus Glyptoplax	48
Glyptoplax pugnax	50
smithii	51
Genus Eucratopsis	52
Eucratopsis crassimanus	52
Subfamily Rhizopinae	54
Genus Chasmocarcinus	54
Chasmocarcinus typicus	55
latipes	57
obliquus	58
cylindricus	59
Family Pinnotheridae	61
Subfamily Pinnotherinae	61
Genus Pinnotheres	62
Pinnotheres ostreum	66
holmesi	68
geddesi	70
politus	71
angelicus	72
lithodomi	73
maculatus	74
bipunctatus	78
depressus	79
muliniarum	81
pugettensis	82
nudus	83
serrei	84
concharum	86
pubescens	87
barbatus	88
strombi	90
silvestrii	91
margarita	91
reticulatus	93
moseri	94
shoemakeri	95
taylori	97
orcutti	98
hemphilli	99
guerini	101
hirtimanus	101
Genus Fabia	101
Fabia subquadrata	102
lowei	104
byssomiae	105
canfieldi	106
Genus Parapinnixa	107
Parapinnixa nitida	107
hendersoni	109
affinis	111

TABLE OF CONTENTS.

Family Pinnotheridae—Continued.
  Subfamily Pinnotherinae—Continued.
    Genus Parapinnixa—Continued.

	Page.
Parapinnixa bouvieri	111
beaufortensis	112
Genus Dissodactylus	114
Dissodactylus nitidus	116
mellitae	117
encopei	119
borradailei	121
stebbingi	123
alcocki	124
calmani	125
Subfamily Pinnothereliinae	127
Genus Pinnixa	128
Pinnixa transversalis	131
faxoni	133
cristata	134
patagoniensis	135
monodactyla	136
longipes	137
floridana	138
retinens	139
tomentosa	141
faba	142
littoralis	145
barnharti	149
minuta	150
chaetopterana	151
valdiviensis	154
occidentalis	155
sayana	156
cylindrica	159
franciscana	161
schmitti	162
hiatus	164
tubicola	165
weymouthi	166
affinis	168
brevipollex	169
Genus Scleroplax	170
Scleroplax granulata	171
Genus Opisthopus	172
Opisthopus transversus	173
Genus Pinnaxodes	174
Pinnaxodes chilensis	175
meinerti	177
tomentosus	178
Genus Tetrias	179
Tetrias scabripes	179
Genus Pinnotherelia	180
Pinnotherelia laevigata	181

TABLE OF CONTENTS.

	Page.
Family Cymopoliidae	182
Genus Cymopolia	183
Cymopolia cristatipes	186
alternata	188
zonata	190
lucasii	193
faxoni	194
affinis	196
rathbuni	198
bahamensis	200
isthmia	201
dentata	202
obesa	205
tuberculata	207
sica	208
angusta	210
depressa	212
fragilis	213
cursor	215
gracilis	218
floridana	220
gracilipes	221
acutifrons	223
Family Grapsidae	224
Subfamily Grapsinae	226
Genus Grapsus	226
Grapsus grapsus	227
Genus Geograpsus	231
Geograpsus lividus	232
Genus Leptograpsus	234
Leptograpsus variegatus	234
Genus Goniopsis	236
Goniopsis cruentata	237
pulchra	239
Genus Pachygrapsus	240
Pachygrapsus crassipes	241
maurus	244
transversus	244
gracilis	249
marmoratus	250
pubescens	252
corrugatus	252
Genus Planes	253
Planes minutus	253
marinus	258
Genus Grapsodius	259
Grapsodius eximius	259
Subfamily Varuninae	260
Genus Cyrtograpsus	260
Cyrtograpsus angulatus	261
altimanus	262

Family Grapsidae—Continued.
Subfamily Varuninae—Continued.

	Page.
Genus Hemigrapsus	264
Hemigrapsus affinis	264
crenulatus	266
nudus	267
oregonensis	270
Genus Tetragrapsus	273
Tetragrapsus jouyi	273
Genus Glyptograpsus	275
Glyptograpsus impressus	275
jamaicensis	277
Genus Platychirograpsus	278
Platychirograpsus typicus	278
Genus Euchirograpsus	281
Euchirograpsus americanus	282
Subfamily Sesarminae	283
Genus Sesarma	284
Subgenus Chiromantes	287
Sesarma africanum	287
Subgenus Sesarma	288
Sesarma verleyi	288
sulcatum	289
reticulatum	290
aequatoriale	292
curacaoense	293
rhizophorae	294
crassipes	294
bidentatum	295
jarvisi	296
ophioderma	297
barbimanum	298
Subgenus Holometopus	298
Sesarma rectum	298
occidentale	299
cinereum	300
miersii	303
miersii iheringi	304
magdalenense	305
biolleyi	306
tampicense	307
ricordi	308
angustipes	311
roberti	312
festae	313
angustum	314
hanseni	315
benedicti	316
Genus Metopaulias	317
Metopaulias depressus	318
Genus Metasesarma	319
Metasesarma rubripes	319
Genus Sarmatium	321
Sarmatium curvatum	321

TABLE OF CONTENTS.

	Page.
Family Grapsidae—Continued.	
Subfamily Sesarminae—Continued.	
Genus Aratus	322
Aratus pisonii	323
Genus Cyclograpsus	325
Cyclograpsus integer	326
cinereus	327
punctatus	328
Genus Chasmagnathus	329
Chasmagnathus granulata	329
Subfamily Plagusiinae	331
Genus Plagusia	331
Plagusia depressa	332
depressa tuberculata	334
immaculata	335
chabrus	336
Genus Percnon	337
Percnon gibbesi	337
Family Gecarcinidae	339
Genus Cardisoma	340
Cardisoma guanhumi	341
crassum	345
Genus Ucides	346
Ucides cordatus	347
occidentalis	350
Genus Gecarcinus	351
Gecarcinus ruricola	352
lateralis	355
quadratus	358
planatus	359
lagostoma	361
Genus Gecarcoidea	362
?Gecarcoidea lalandii	364
Family Ocypodidae	365
Subfamily Ocypodinae	366
Genus Ocypode	366
Ocypode albicans	367
occidentalis	372
gaudichaudii	373
Genus Uca	374
Uca maracoani	378
monilifera	380
heterochelos	381
princeps	382
stylifera	383
heteropleura	385
insignis	385
tangeri	387
minax	389
mordax	391
brevifrons	393
pugnax	395
pugnax rapax	397
pugilator	400
galapagensis	403
macrodactylus	404

Family Ocypodidae—Continued.
  Subfamily Ocypodinae—Continued.
    Genus Uca—Continued.

	Page.
Uca rectilatus	
thayeri	405
speciosa	406
crenulata	408
coloradensis	409
spinicarpa	410
panamensis	411
uruguayensis	412
oerstedi	413
helleri	414
stenodactylus	415
musica	416
subcylindrica	417
festae	419
leptodactyla	420
latimanus	420
Subfamily Macrophthalminae	422
Genus Euplax	423
Euplax leptophthalma	423
Explanation of plates	424
Index	425
	447

LIST OF ILLUSTRATIONS.

TEXT FIGURES.

	Page.
1. Diagrammatic dorsal view of a grapsoid crab, showing the terms used in description	1
2. Diagrammatic ventral view of a grapsoid crab, showing the terms used in description	5
3. *Trizocarcinus dentatus*, ventral view of left side of carapace of male holotype, showing stridulating ridge, × 4	18
4. *Bathyplax typhla*, male (9729). *a*, Outer maxillipeds, showing approximation, × 4; *b*, abdomen and part of sternum, × 3; *c*, left chela, × 2	20
5. *Pilumnoplax americana*, male, station 62, dorsal view, × 1½	22
6. *Pilumnoplax americana*, male, station 62, anterior portion of carapace, viewed from before, × 3½	22
7. *Goneplax hirsuta*, cotype, × 2⅔. *a*, Dorsal view; *b*, right chela and carpus, outer view. (After Borradaile.)	28
8. *Prionoplax atlantica*, male, holotype. *a*, Abdomen and part of sternum, × 8; *b*, outer maxilliped, dotted line marking median axis, × 12	30
9. *Tetraplax quadridentata*, male (24564), outer maxillipeds, showing approximation, × 8	33
10. *Tetraplax quadridentata*, male (24564), abdomen, part of sternum, and coxa of last leg, × 6¾	33
11. *Euryplax nitida*, male (15012). *a*, Abdomen and part of sternum, × 3½; *b*, outer maxilliped, dotted line marking median axis, × 5½	34
12. *Chasmophora macrophthalma*, female holotype. *a*, Carapace, eyes, and antennae, dorsal view, × 5⅔; *b*, right chela, outer view, × 7⅓	38
13. *Chasmophora macrophthalma*, anterior view of orbit of female holotype, showing hiatus in which lies the antennal flagellum, × about 13	38
14. *Speocarcinus granulimanus*, male holotype. *a*, Abdomen and part of sternum × 3; *b*, outer maxillipeds, showing approximation, × 5½	40
15. *Speocarcinus granulimanus*, male holotype, right chela, outer view, × 2⅓	40
16. *Speocarcinus californiensis*, male (32966). *a*, Outer maxilliped, dotted line marking median axis, × 4½; *b*, abdomen and part of sternum, × 2⅔	42
17. *Pseudorhombila octodentata*, male holotype. *a*, Right chela, outer view, × 1; *b*, abdomen, part of sternum, and coxa of last leg, × 1¾	43
18. *Pseudorhombila octodentata*, male holotype, outer maxillipeds, showing approximation, × 3	43
19. *Oediplax granulata*, female holotype. *a*, Antennal and buccal region, × 2⅓; *b*, right chela, outer view, × 1½	44
20. *Cyrtoplax spinidentata*, male holotype, outer maxillipeds, showing approximation, × 6	46
21. *Panoplax depressa*, male (25624). *a*, Outer maxilliped, × 9; *b*, abdomen and part of sternum, × 6	48
22. *Eucratopsis crassimanus*, female (17801), outer maxilliped, × 8	53
23. *Chasmocarcinus typicus*, male holotype. *a*, Dorsal view, × 2; *b*, abdomen and sternum, flattened, × 2	55

LIST OF ILLUSTRATIONS.

	Page.
24. *Chasmocarcinus typicus*, male (22073), outer maxillipeds, showing approximation, × 2½	56
25. *Chasmocarcinus latipes*, female holotype, dorsal view, × 1⅜	57
26. *Chasmocarcinus latipes*, female holotype. *a*, Outer maxillipeds, in place, × 5; *b*, abdomen and part of sternum, × 3	57
27. *Chasmocarcinus obliquus*, male holotype. *a*, Right chela, outer view, × 8; *b*, abdomen and part of sternum, × 14	58
28. *Chasmocarcinus cylindricus*, female (24551), dorsal view, × 2⅓.	60
29. *Chasmocarcinus cylindricus*. *a*, Right chela of male holotype, × 6; *b*, abdomen and sternum of male holotype, × 6½; *c*, abdomen of female (24551), × 6½	60
30. *Pinnotheres ostreum*, outer maxilliped of female (2542), × 15½.	67
31. *Pinnotheres holmesi*, outer maxilliped of female holotype, × 15½	69
32. *Pinnotheres geddesi*, female (23767). *a*, Chela, × 16; *b*, outer maxilliped, × 16	70
33. *Pinnotheres politus*, outer maxilliped of female (40448), × 50	71
34. *Pinnotheres angelicus*, outer maxilliped of female (17467), × 15½	72
35. *Pinnotheres maculatus*, male, × 8. (After Smith.)	75
36. *Pinnotheres maculatus*, outer maxilliped of female (36782), × 10	75
37. *Pinnotheres depressus*, outer maxilliped of male (48594), × 137½	80
38. *Pinnotheres muliniarum*, outer maxilliped of male holotype, × 90	81
39. *Pinnotheres pugettensis*, endognath of outer maxilliped of female (40396), × 21	82
40. *Pinnotheres nudus*, female holotype. *a*, Carapace, slightly enlarged; *b*, abdomen; *c*, outer maxilliped; *d*, left cheliped; *e*, first leg. (After Holmes.)	83
41. *Pinnotheres serrei*, outer maxilliped of female (48571), × 20	84
42. *Pinnotheres concharum*, endognath of outer maxilliped of male (23929), × 20	86
43. *Pinnotheres pubescens*, female holotype. *a*, General outline, × 2 1/16; *b*, buccal area. (After Holmes.)	88
44. *Pinnotheres barbatus* (23435). *a*, Endognath of outer maxilliped of female, × 20; *b*, abdomen of male, × 7½	89
45. *Pinnotheres strombi*, endognath of outer maxilliped of female holotype, much enlarged	90
46. *Pinnotheres reticulatus*, outer maxilliped of female holotype, × 15½.	93
47. *Pinnotheres moseri*, endognath of outer maxilliped of female holotype, × 17.	94
48. *Pinnotheres shoemakeri*, endognath of outer maxilliped of male holotype, × 97	96
49. *Pinnotheres taylori*, endognath of outer maxilliped of male (40397), × 40	97
50. *Pinnotheres orcutti*, endognath of outer maxilliped of male holotype, × 67	99
51. *Pinnotheres hemphilli*, outer maxilliped of male holotype, × 110	100
52. *Pinnotheres guerini*, outer maxilliped of holotype, enlarged. (After Milne Edwards.)	101
53. *Fabia subquadrata*, female. *a*, Outer maxilliped (17480), × 14; *b*, dorsal view of body and last two legs, × 1½; *c*, right chela, × 1½; *b* and *c* from Monterey Bay, after Weymouth	102
54. *Fabia subquadrata*. *a*, Ventral view of front and mouth, enlarged; *b*, end of a leg, enlarged. (After Dana.)	103
55. *Fabia lowei*, female (23437). *a*, Endognath of outer maxilliped, × 10; *b*, left chela, × 4	104
56. *Fabia byssomiae*, outer maxilliped of female (25648), × 21	105
57. *Fabia canfieldi*, female holotype, outer maxilliped, × 21	106
58. *Parapinnixa nitida*, female holotype. *a*, General outline, × 2¾; *b*, outer maxilliped, enlarged. (After Holmes.)	108

LIST OF ILLUSTRATIONS.

	Page.
59. *Parapinnixa hendersoni*, female (48711), outer maxilliped, × 61..	109
60. *Parapinnixa bouvieri*, female holotype, outer maxilliped, × 100..	112
61. *Parapinnixa beaufortensis*, male holotype, dorsal view, with right cheliped detached, × 20.	113
62. *Parapinnixa beaufortensis*, outer maxilliped of male holotype, × 176	113
63. *Parapinnixa beaufortensis*, male holotype, ventral view, × 20.	113
64. *Dissodactylus nitidus*, outer maxilliped of female (22113), × 50	116
65. *Dissodactylus nitidus*, third leg of female (22113), × 11.	117
66. *Dissodactylus mellitae*. *a*, Outer maxilliped of female (40272), × 23; *b*, leg of paratype (23434), × 15.	118
67. *Dissodactylus encopei*, male (23430). *a*, Outline of carapace, × 6; *b*, third leg, × 16; *c*, endognath of outer maxilliped, × 46; *d*, fourth leg, × 16; *e*, chela, × 16.	120
68. *Dissodactylus borradailei*, female (49231). *a*, Outer maxilliped, × 23; *b*, leg × 11.	122
69. *Dissodactylus stebbingi*, outer maxilliped of male holotype, × 63	123
70. *Dissodactylus alcocki*, outer maxilliped of female holotype, × 73	124
71. *Dissodactylus alcocki*, leg of male paratype, × 16.	125
72. *Dissodactylus calmani*, outer maxilliped of female (48570), × 50	126
73. *Dissodactylus calmani*. *a*, Last right leg of female holotype, × 14; *b*, a'domen of male (50168), × 12.	126
74. *Pinnixa transversalis*, outer maxilliped of female (40446), × 16½	132
75. *Pinnixa transversalis*. *a*, A¹domen of male, enlarged; *b*, right cheliped, enlarged. (After Faxon.)	132
76. *Pinnixa transversalis*, a'domen and sternum of male, enlarged. (After Milne Edwards and Lucas.)	132
77. *Pinnixa faxoni* (7639) × 13¾. *a*, A¹domen of male; *b*, outer maxilliped..	133
78. *Pinnixa cristata*, outer maxilliped of female holotype, × 15	134
79. *Pinnixa patagoniensis*, male holotype. *a*, Outer maxilliped, × 18¾; abdomen, × 7¼	135
80. *Pinnixa longipes*, cotype, general outline, enlarged. (After Holmes.)	137
81. *Pinnixa longipes*, cotype, outer maxilliped, enlarged. (After Holmes.)	137
82. *Pinnixa floridana*. *a*, A¹domen of male (49249), × 19; *b*, outer maxilliped of female paratype (6696), × 19.	138
83. *Pinnixa retinens*, endognath of outer maxilliped of male holotype, × 33½...	140
84. *Pinnixa retinens*, male holotype, × 12. *a*, A¹domen; *b*, third leg, lower side; *c*, left chela.	140
85. *Pinnixa tomentosa*, outer maxilliped of female (29948), × 15½.	141
86. *Pinnixa tomentosa*, female holotype, enlarged. *a*, Chela; *b*, third leg; *c*, first leg. (After Holmes.)	142
87. *Pinnixa faba* (31599), × 6. *a*, Third leg of female; *b*, third leg of male; *c*, right chela of male; *d*, left chela of female; *e*, eye in or it, front view.	143
88. *Pinnixa faba* (31599). *a*, Outer maxilliped of female, × 16; *b*, a'domen of male, × 6½.	143
89. *Pinnixa littoralis* (31600), × 6. *a*, Left chela of female; *b*, front, and eye in or it, fro*n*t view; *c*, third leg of female; *d*, left chela of male; *d*, third leg of male.	146
90. *Pinnixa littoralis* (31600). *a*, Outer maxilliped of female, × 14; *b*, a'domen of male, × 5⅛.	147
91. *Pinnixa barnharti*, outer maxilliped of female (31510), × 15½..	149
92. *Pinnixa minuta*, male holotype. *a*, Carapace. antennae and eyes, × 18½; *b*, last three a' dominal segments and adjacent sternum, × 18; *c*, right chela, × 27; *d*, third leg, × 18; *e*, endognath of outer maxilliped, much enlarged.	151

Page.

93. *Pinnixa chaetopterana*, Buzzards Bay. *a*, Chela of adult female, × 6; *b*, chela of adult male, × 4. (After Faxon.) 152
94. *Pinnixa chaetopterana* (5043). *a*, Outer maxilliped of female, × 17; *b*, a' domen of male, × $6\frac{2}{3}$ 152
95. *Pinnixa valdiviensis*. *a*, A' domen of male (5740), × $5\frac{1}{2}$; *b*, left chela of male, (5740) × 5; outer maxilliped of cotype, much enlarged .. 154
96. *Pinnixa occidentalis*, left chela of male (17470), × $3\frac{1}{2}$ 155
97. *Pinnixa occidentalis*, male holotype. *a*, A' domen, × $5\frac{1}{5}$; *b*, outer maxilliped, × $13\frac{3}{5}$ 156
98. *Pinnixa sayana*, male (18211). *a*, Outer maxilliped, × 20; *b*, a' domen × $11\frac{4}{5}$ 158
99. *Pinnixa cylindrica*, male (17952). *a*, Outer maxilliped, × 19; *b*, a' domen, × $7\frac{3}{4}$ 160
100. *Pinnixa franciscana*. *a*, Al domen of male (48445), × $13\frac{3}{4}$; *b*, outer maxilliped of female holotype, × $13\frac{3}{4}$ 161
101. *Pinnixa schmitti*, male (25850). *a*, Outer maxilliped, × $15\frac{1}{2}$; *b*, a' domen, × $7\frac{3}{4}$; *c*, left or larger chela, × 7 163
102. *Pinnixa hiatus*, outer maxilliped of female holotype, × $19\frac{3}{5}$. 164
103. *Pinnixa tubicola*. *a*, Outer maxilliped of female (20860), × $22\frac{4}{5}$; *b*, a' domen of male, Trinidad, × $15\frac{1}{2}$; *c*, left chela of male, × 6, after Weymouth. 165
104. *Pinnixa weymouthi*, male holotype. *a*, Outer maxilliped, × 20; *b*, left chela, × $7\frac{3}{4}$, after Weymouth; *c*, abdomen, × 16 167
105 *Pinnixa affinis*, female holotype. *a*, Dorsal view, × 3; *b*, endognath of outer maxilliped, × $22\frac{3}{5}$ 168
106. *Pinnixa affinis*, right chela of female holotype, × 13 168
107. *Pinnixa brevipollex*, female holotype, × $2\frac{3}{4}$ 169
108. *Pinnixa brevipollex*, *a*, Outer maxilliped of female holotype, × $18\frac{3}{5}$; *b*, abdomen of male (21593), × $7\frac{1}{8}$ 170
109. *Scleroplax granulata*. *a*, Abdomen of male (49247), × $15\frac{1}{2}$; *b*, endognath of outer maxilliped of female paratype, × $15\frac{1}{2}$ 171
110. *Opisthopus transversus*, female (23927). *a*, Dorsal view, × $1\frac{4}{5}$; *b*, endognath of outer maxilliped, × 8 173
111. *Pinnaxodes chilensis*, male (22112). *a*, Endognath of outer maxilliped, much enlarged; *b*, abdomen, × $4\frac{1}{8}$ 176
112. *Pinnaxodes meinerti*. *a*, Outer maxilliped of female (5760), × $15\frac{1}{2}$; *b*, abdomen of male holotype, × 8 178
113. *Pinnaxodes tomentosus*, female, cotype. *a*, Outer maxilliped, × about 10; *b*, general outline, nat. size. (After Ortmann.) 178
114. *Tetrias scabripes*, female holotype. *a*, Right chela, × 5; *b*, outer maxillipeds, × $5\frac{1}{2}$; *c*, dorsal aspect, × $2\frac{3}{10}$ 180
115. *Pinnotherelia laevigata*, male holotype. *a*, Dorsal view, × $\frac{3}{5}$; *b*, antennal and buccal regions; *c*, first maxilliped; *d*, outer maxilliped; *e*, second maxilliped; *f*, extremity of a leg; *b–f* are enlarged. (After Milne Edwards and Lucas.) 182
116. *Cymopolia cristatipes*, male holotype, after A. Milne Edwards and Bouvier. *a*, Dorsal view, × $3\frac{1}{4}$; *b*, front, dorsal view, × 8; *c*, ventral view of anterior half of body, × 8; *d*, abdomen, × 4; *e*, endognath of left outer maxilliped, × $7\frac{1}{2}$ 187
117. *Cymopolia alternata*, outline of front and orbits of male (19840), × 6 188
118. *Cymopolia zonata*, outline of front and orbits of male (22071), × 6. 192
119. *Cymopolia lucasii*, male holotype, dorsal view, × nearly 2 .. 193
120. *Cymopolia faxoni*, outline of front and orbits of female holotype, × 6.... 194
121. *Cymopolia affinis*, outline of front and orbits of male (24515), × 6.. 196

Page.

122. *Cymopolia rathbuni*, female holotype, after A. Milne Edwards and Bouvier. *a*, Right oculo-antennal region, ventral view, × 20; *b*, right antero-lateral portion of carapace, dorsal view, × 14; *c*, second left leg, upper face, × 11½; *d*, merus and carpus of third right leg, upper face, × 11½; *e*, first left leg, lower face, × 11½; *f*, right last leg, × 11½; *g*, posterior border of carapace, × 14 199

123. *Cymopolia isthmia*, outline of front and orbits, of female holotype, × 10.... 202

124. *Cymopolia dentata*, after A. Milne Edwards and Bouvier. *a*, Dorsal view of female holotype, × 3 about; *b*, frontal border of the same, × 7½; *c*, carapace of a male, dorsal view, × 3½; *d*, right outer maxilliped, × 7½; *e*, abdomen of male (same as *c*), × 7½; *f*, left oculo-antennal region of same male, × 7½; *g*, merus and carpus of first left leg of female holotype, upper face, × 5; *h*, left cheliped of male, outer face, × 4; *i*, right cheliped of male, outer face, × 4 203

125. *Cymopolia obesa*, immature female (not male) holotype, after A. Milne Edwards and Bouvier. *a*, Dorsal view, × 3 about; *b*, endognath of right outer maxilliped, × 6½; *c*, first left leg, upper side, × 5; *d*, last left leg, × 5; *e*, left frontal and orbital border, from above, × 7½; *f*, left oculo-antennal region, ventral view, × 7½ 205

126. *Cymopolia tuberculata*, male, × 1⅘, after Faxon. *a*, Dorsal view; *b*, ventral view 207

127. *Cymopolia sica*, after A. Milne Edwards and Bouvier. *a*, Male, dorsal view, × 3; *b*, carapace of same, dorsal view, × 4½; *c*, right oculo-antennal region of same, lower view, × 7½; *d*, merus and carpus of first right leg of same, upper side, × 5; *e*, left outer maxilliped, × 8; *f*, sternum and abdomen of male, × 5¾ 209

128. *Cymopolia depressa*, young female (6505, M. C. Z.), after A. Milne Edwards and Bouvier. *a*, Left oculo-antennal region, ventral view, × 15; *b*, carapace, dorsal view, × 8½; *c*, merus and carpus of second right leg, upper side, × 11; *d*, endognath of left outer maxilliped, × 18; *e*, posterior border of carapace, enlarged 212

129. *Cymopolia fragilis*, male, enlarged, after Faxon. *a*, Dorsal view; *b*, ventral view 214

130. *Cymopolia cursor*, female, after A. Milne Edwards and Bouvier. *a*, Dorsal view, × 2½; *b*, right oculo-antennal region, ventral view, × 8; *c*, left half of frontal and lateral border, × 5; *d*, merus and carpus of third left leg, upper side, × 5; *e*, merus and carpus of first left leg, upper side, × 5; *f*, left cheliped, outer side, × 8 216

131. *Cymopolia cursor*, after A. Milne Edwards and Bouvier. *a*, Dorsal view of body of female shown in fig. 130, × 3½; *b*, same view of female (6496), type of *C. dilatata*, × 4 217

132. *Cymopolia gracilis*, outline of front and orbits of female (11411), × 6...... 219

133. *Cymopolia gracilipes*, after A. Milne Edwards and Bouvier. *a*, Dorsal view of female holotype, × 5; *b*, endognath of the left outer maxilliped, much enlarged; *c*, merus and carpus of second left leg, upper side, × 8; *d*, left oculo-antennal region, lower view, × 10; *e*, abdomen of male, × 8½.... 222

134. *Cymopolia acutifrons*, female holotype, after A. Milne Edwards and Bouvier. *a*, Right frontal border, × 15 (edge between median teeth and orbit broken); *b*, right outer orbital angle and adjoining tooth of infra-orbital border, × 15; *c*, median part of frontal border, × 20; *d*, left pterygostomian lobe and contiguous antennal peduncle, ventral view, × 15; *e*, right eye, from above, × 15 224

LIST OF ILLUSTRATIONS.

	Page.
135. *Grapsus grapsus*, outer maxilliped, enlarged. (After Milne Edwards.)	228
136. *Goniopsis cruentata*. *a*, Outer maxilliped, enlarged (after Milne Edwards); *b*, left first appendage of a' domen of male (7677), lower side, × 2	237
137. *Goniopsis pulchra*, a' domen of male (12467) × 1½	240
138. *Pachygrapsus marmoratus*, outer maxilliped, enlarged. (After Milne Edwards.)	251
139. *Tetragrapsus jouyi*, male (17496). *a*, Buccal cavity, × 8¼; *b*, orbital and antennal region, × 6¾; *c*, abdomen, × 6¾	274
140. *Glyptograpsus jamaicensis*, male. *a*, Abdomen, holotype, × 1½; *b*, first right a' dominal appendage (42881), inner side, × 2⅔; *c*, outer maxilliped of holotype, × 3⅔; *d*, top view of right chela of same, slightly enlarged; *e*, outer view of same, slightly enlarged	277
141. *Platychirograpsus typicus*, male (19863), dorsal view, slightly reduced	279
142. *Platychirograpsus typicus*, male, holotype. *a*, Outer maxilliped (dotted line showing median line), × 4; *b*, abdomen, × 1⅛; *c*, right orbital and antennal region, ventral view, × 2¼	279
143. *Platychirograpsus typicus*, right or major cheliped. *a*, Anterior view of specimen in Halifax Museum, natural size; *b*, upper view of same, natural size; *c*, anterior view of 19863, × 1¼	280
144. *Euchirograpsus americanus*, outer maxilliped, enlarged. (After A. Milne Edwards and Bouvier.)	282
145. *Sesarma (Sesarma) sulcatum*, abdomen of male (4631), × 1⅓	290
146. *Sesarma (Sesarma) aequatoriale*, natural size. *a*, Right chela of male, outer view; *b*, dorsal view; *c*, a' domen of male. (After Ortmann.)	292
147. *Sesarma (Sesarma) curacaoense*, left appendage of first segment of male a'-domen (17678), lower view, × 4½	293
148. *Sesarma (Holometopus) occidentale*, male holotype. *a*, Anterior view of front, × 1⅜; *b*, carapace and eyes, dorsal view, × 1⅜; *c*, third leg, × 1⅜; *d*, left chela, outer view, × 1⅜; *e*, last five segments of a' domen, × 1¾; *f*, appendage of first segment of a'domen, × 7	300
149. *Sesarma (Holometopus) cinereum*, male (15072). *a*, Abdomen, × 3½; *b*, right appendage of first segment of a' domen, ventral view, × 7½	301
150. *Sesarma (Holometopus) biolleyi*, abdomen of male holotype, × 3½	306
151. *Sesarma (Holometopus) tampicense*, left appendage of first segment of a' domen of male holotype, ventral view, × 8	307
152. *Sesarma (Holometopus) hanseni*, male holotype. *a*, Abdomen, × 4⅔; *b*, appendage of first segment of a' domen, ventral view, × 5¼	316
153. *Cyclograpsus punctatus*, outer maxilliped, enlarged. (After Milne Edwards.)	328
154. Coxa of third right leg of male of, *a*, *Plagusia depressa* (40609), × 1⅔; *b*, *P. depressa tuberculata* (18826), × 1½	333
155. *Cardisoma guanhumi*, abdomen of male (17987), natural size	343
156. *Cardisoma crassum*, front view of orbital and antennal region of male (2137), × 1¼	345
157. *Cardisoma crassum*, abdomen of male (48809), natural size	346
158. *Ucides cordatus*, abdomen of male (17595), natural size	348
159. *Ucides occidentalis*, abdomen of male (40490), natural size	350
160. *Gecarcinus ruricola*, abdomen of male (7343), natural size	353
161. *Gecarcinus lateralis*, male (11368), × 2. *a*, Outer maxillipeds, in place; *b*, abdomen of male	356
162. *Gecarcinus quadratus*, male, × 2. *a*, Outer maxillipeds, in place, of specimen in Copenhagen Museum; *b*, abdomen (32311)	359
163. *Gecarcinus planatus*, abdomen of male (20650), natural size	360

LIST OF ILLUSTRATIONS.

	Page.
164. *Gecarcinus lagostoma*, outer maxilliped, Ascension Island, slightly reduced. *a*, Upper view; *b*, lower view. (After Miers.)	361
165. *Gecarcinus lagostoma*, abdomen of male (14869), natural size	362
166. *Uca mordax*, inner surface of larger chela of male from Para, slightly reduced. (After Smith.)	392
167. *Uca galapagensis*, male (22319), × 1½. *a*, Dorsal view; *b*, inner surface of larger chela	403
168. *Uca rectilatus*, cotype, enlarged, after Holmes. *a*, Dorsal view of carapace, eyes and legs; *b*, ischium and merus of right outer maxilliped; *c*, carpus and chela of smaller cheliped; *d*, merus of right or larger cheliped, inner face; *e*, larger chela, inner face	406
169. *Uca thayeri*, male (23753), natural size. *a*, Dorsal view; *b*, inner surface of larger chela	407
170. *Uca helleri*, male, × 3. *a*, Dorsal view of holotype; *b*, inner surface of larger chela (25666)	415
171. *Uca musica*, male holotype, × 3⅓. *a*, Lower view of larger, left chela, showing stridulating ridge; *b*, anterior (lower) view of portion of first left leg, showing granules which play against stridulating ridge	418
172. Prehistoric bowl unearthed in Costa Rica, the base representing a female land crab, *Cardisoma guanhumi*, with the claws folded under the eyes, × about ⅔. Original in possession of Mrs. Zeledon, cast in U. S. National Museum	424

PLATES.

1. *Trizocarcinus dentatus.*
2. *Bathyplax typhla.*
3. *Pilumnoplax elata.*
4. *Goneplax barbata* and *sigsbei.*
5. *Goneplax barbata.*
6. *Prionoplax atlantica* and *Tetraplax quadridentata.*
7. *Euryplax nitida.*
8. *Speocarcinus carolinensis.*
9. *Speocarcinus granulimanus.*
10. *Speocarcinus ostrearicola* and *californiensis.*
11. *Cyrtoplax spinidentata.*
12. *Panoplax depressa* and *Eucratopsis crassimanus.*
13. *Oediplax granulata* and *Glyptoplax smithii.*
14. *Chasmocarcinus obliquus* and *Pseudorhombila octodentata.*
15. *Pinnotheres holmesi* and *ostreum.*
16. *Pinnotheres geddesi* and *angelicus.*
17. *Pinnotheres depressus, maculatus,* and *pugettensis.*
18. *Pinnotheres pugettensis* and *muliniarum.*
19. *Pinnotheres serrei* and *barbatus.*
20. *Pinnotheres strombi* and *concharum.*
21. *Pinnotheres reticulatus, moseri,* and *taylori.*
22. *Pinnotheres shoemakeri* and *orcutti.*
23. *Pinnotheres hemphilli.*
24. *Fabia subquadrata, lowei, canfieldi,* and *byssomiae.*
25. *Pinnaxodes meinerti* and *Parapinnixa bouvieri.*
26. *Parapinnixa hendersoni* and *Dissodactylus nitidus.*
27. *Dissodactylus encopei* and *borradailei.*
28. *Dissodactylus stebbingi, alcocki, calmani,* and *mellitae.*

LIST OF ILLUSTRATIONS.

29. *Pinnixa transversalis, faxoni,* and *cristata.*
30. *Pinnixa patagoniensis, floridana,* and *tomentosa.*
31. *Pinnixa faba* and *littoralis.*
32. *Pinnixa barnharti.*
33. *Pinnixa valdiviensis* and *chaetopterana.*
34. *Pinnixa occidentalis, sayana,* and *valdiviensis.*
35. *Pinnixa franciscana, cylindrica,* and *schmitti.*
36. *Pinnixa hiatus, tubicola,* and *weymouthi.*
37. *Scleroplax granulata* and *Opisthopus transversus.*
38. *Pinnaxodes chilensis.*
39. *Pinnotherelia laevigata* and *Tetrias scabripes.*
40. *Pinnotherelia laevigata* and *Cymopolia sica.*
41. *Pinnixa retinens* and *Cymopolia floridana.*
42. *Cymopolia alternata.*
43. *Cymopolia alternata.*
44. *Cymopolia lucasii* and *zonata.*
45. *Cymopolia zonata* and *faxoni.*
46. *Cymopolia affinis.*
47. *Cymopolia bahamensis* and *affinis.*
48. *Cymopolia rathbuni* and *isthmia.*
49. *Cymopolia obesa.*
50. *Cymopolia gracilis.*
51. *Cymopolia gracilis* and *fragilis.*
52. *Cymopolia cursor* and *gracilipes.*
53. *Grapsus grapsus.*
54. *Grapsus grapsus.*
55. *Geograpsus lividus.*
56. *Leptograpsus variegatus.*
57. *Goniopsis cruentata.*
58. *Goniopsis pulchra.*
59. *Pachygrapsus crassipes.*
60. *Pachygrapsus maurus* and *gracilis.*
61. *Pachygrapsus gracilis* and *transversus.*
62. *Pachygrapsus marmoratus.*
63. *Planes minutus.*
64. *Planes marinus.*
65. *Cyrtograpsus angulatus.*
66. *Cyrtograpsus altimanus.*
67. *Hemigrapsus affinis.*
68. *Hemigrapsus crenulatus.*
69. *Hemigrapsus nudus.*
70. *Hemigrapsus oregonensis.*
71. *Tetragrapsus jouyi.*
72. *Glyptograpsus impressus* and *jamaicensis.*
73. *Platychirograpsus typicus.*
74. *Euchirograpsus americanus.*
75. *Sesarma (Chiromantes) africanum.*
76. *Sesarma (Sesarma) verleyi.*
77. *Sesarma (Sesarma) reticulatum.*
78. *Sesarma (Sesarma) curacaoense* and *sulcatum.*
79. *Sesarma (Sesarma) rhizophorae.*
80. *Sesarma (Sesarma) bidentatum.*
81. *Sesarma (Sesarma) jarvisi.*

LIST OF ILLUSTRATIONS.

82. *Sesarma (Holometopus) rectum.*
83. *Sesarma (Holometopus) cinereum.*
84. *Sesarma (Holometopus) miersii.*
85. *Sesarma (Holometopus) miersii iheringi.*
86. *Sesarma (Holometopus) magdalenense.*
87. *Sesarma (Holometopus) hanseni* and *biolleyi.*
88. *Sesarma (Holometopus) tampicense.*
89. *Sesarma (Holometopus) ricordi.*
90. *Sesarma (Holometopus) angustipes.*
91. *Sesarma (Holometopus) roberti.*
92. *Sesarma (Holometopus) angustum.*
93. *Sesarma (Holometopus) benedicti.*
94. *Metasesarma rubripes.*
95. *Sarmatium curvatum.*
96. *Aratus pisonii.*
97. *Cyclograpsus integer* and *Metopaulias depressus.*
98. *Cyclograpsus cinereus.*
99. *Cyclograpsus punctatus.*
100. *Chasmagnathus granulata.*
101. *Plagusia depressa.*
102. *Plagusia depressa tuberculata.*
103. *Plagusia immaculata.*
104. *Plagusia chabrus.*
105. *Percnon gibbesi.*
106. *Cardisoma guanhumi*, male, dorsal view.
107. *Cardisoma guanhumi*, female, ventral view.
108. *Cardisoma crassum*, male, dorsal view.
109. *Cardisoma crassum*, male, ventral view.
110. *Ucides cordatus*, male, dorsal view.
111. *Ucides cordatus*, male, ventral view.
112. *Ucides cordatus*, female, dorsal view.
113. *Ucides cordatus*, female, ventral view.
114. *Ucides occidentalis*, male, dorsal view.
115. *Ucides occidentalis*, male, ventral view.
116. *Ucides occidentalis*, male, antero-dorsal view.
117. *Gecarcinus ruricola*, female, dorsal view.
118. *Gecarcinus ruricola*, female, ventral view.
119. *Gecarcinus lateralis*, male, dorsal view.
120. *Gecarcinus lateralis*, male, ventral view.
121. *Gecarcinus quadratus*, male, dorsal view.
122. *Gecarcinus quadratus*, male, ventral views.
123. *Gecarcinus planatus*, male, dorsal view.
124. *Gecarcinus planatus*, male, ventral view.
125. *Gecarcinus lagostoma*, male, dorsal view.
126. *Gecarcinus lagostoma*, male, ventral view.
127. *Ocypode albicans*, male, dorsal view.
128. *Ocypode albicans*, male, ventral view.
129. *Ocypode gaudichaudii* and *occidentalis.*
130. *Ocypode gaudichaudii* and *Uca maracoani.*
131. *Uca heterochelos* and *maracoani.*
132. *Uca monilifera.*
133. *Uca princeps.*
134. *Uca stylifera* and *mordax*

LIST OF ILLUSTRATIONS.

135. *Uca tangeri.*
136. *Uca tangeri.*
137. *Uca minax.*
138. *Uca brevifrons.*
139. *Uca pugnax.*
140. *Uca pugnax rapax.*
141. *Uca pugilator.*
142. *Uca galapagensis.*
143. *Uca macrodactylus.*
144. *Uca thayeri.*
145. *Uca speciosa.*
146. *Uca crenulata.*
147. *Uca coloradensis.*
148. *Uca spinicarpa.*
149. *Uca panamensis.*
150. *Uca uruguayensis.*
151. *Uca helleri.*
152. *Uca oerstedi* and *stenodactylus.*
153. *Uca stenodactylus.*
154. *Uca musica.*
155. *Uca subcylindrica.*
156. *Uca leptodactyla.*
157. *Uca latimanus.*
158. *Glyptoplax pugnax* and *smithii.*
159. *Eucratopsis crassimanus, Ucides cordatus, Pinnotheres politus* and *bipunctatus, Speocarcinus carolinensis, Cyrtograpsus angulatus,* and *Chasmagnathus granulata.*
160. *Pachygrapsus pubescens* and *corrugatus, Uca pugilator, subcylindrica* and *princeps, Sesarma (Sesarma) curacaoense,* and *Gecarcoides lalandii.*
161. *Uca heteropleura* and *insignis.*

THE GRAPSOID CRABS OF AMERICA.

By MARY J. RATHBUN,

Associate in Zoology, United States National Museum.

INTRODUCTION.

This volume is part of a work projected many years ago as a handbook for the study of American crabs, the main purpose being to give a brief description with figures of each species. In the meantime the character of the work has been somewhat changed so as to include a detailed catalogue of the specimens in the United States National Museum. Unavoidable delay in preparation has led to an accumulation of material which has greatly augmented the number of known species. As a result the work has been expanded into four volumes, of which the present one, dealing with the Grapsoids or Catometopes, is the first.

SOURCES OF MATERIAL.

The collections in the United States National Museum form the basis of this bulletin. They consist chiefly of material obtained by United States Government bureaus, such as the Bureau of Fisheries (known previous to July 1, 1903, as the United States Fish Commission). Through the activities of its various vessels and laboratories the Bureau of Fisheries has been able to transfer to the National Museum vast accumulations from nearly all the coasts of America. The amount of work accomplished by the steamers *Albatross* and *Fish Hawk* and the schooner *Grampus*, as well as by other vessels of the commission in earlier years, is indicated in the detailed lists of specimens.

Other Government explorations that have yielded considerable results are those constantly carried on by the Bureau of Biological Survey of the Department of Agriculture and those occasional expeditions under the auspices of the National Museum itself and the Smithsonian Institution. Of the Crustacea obtained by the United States exploring expedition in 1838–1842 and by the North Pacific exploring expedition in 1853–1856 very little remains, owing to the inadequate housing of the former collection before the existence of a National Museum building, and to the destruction of the latter collection in the Chicago fire of 1871 while it was in the custody of Dr. William Stimpson.[1]

Among private institutions contributing largely are the Carnegie Institution, of Washington, District of Columbia; the Venice Marine Biological Station, Venice, California; and various universities, such as the Stanford University and the University of California.

Through a system of exchanges reciprocal benefit has been derived from many of the larger museums, such as the Museum of Comparative Zoology, Cambridge, Massachusetts; the Peabody Museum of Yale University, New Haven, Connecticut; the Museu Paulista, São Paulo, Brazil; the Museo Nacional, at Valparaiso, Chile; and in Europe, the Muséum d'Histoire Naturelle, Paris, France; the Zoological Museum, Copenhagen, Denmark; and others.

SPECIAL RESEARCHES.

In the eighties, the early days of the United States National Museum, the collection of crustaceans, as well as of other invertebrates, was in charge of Dr. Richard Rathbun, at that time an assistant on the United States Fish Commission. The collection was of so moderate a size that it was possible for a single curator to classify all the larger and more conspicuous forms in various groups. At that time Prof. Sidney I. Smith, of the United States Fish Commission, had charge of the Decapods of the northeast Atlantic coast of America which were obtained by the commission, and the material forming the basis of his reports was subsequently transferred to the museum. Later Dr. James E. Benedict was made an assistant curator of marine invertebrates, and the then rapidly increasing numbers of decapod crustaceans occupied a large share of his attention. When the present writer took up the subject the commoner and more abundant forms were already worked over. The nomenclature, however, proved to be in a very unsettled condition, on account of our ignorance of the true status of type-specimens in some of the European museums, the misinterpretation of the descriptions of those types by others, and the consequent repetition of errors in literature.

In 1896 therefore the writer visited six European museums to examine certain types, those of J. C. Fabricius in the museums in Copenhagen and Kiel, of Herbst in Berlin, of Saussure in Geneva, of the two Milne Edwards' in Paris, and of Miers and others in the British Museum. The results were of great value in revising the

---

[1] For a full account of the sources of the invertebrate collection in the National Museum up to 1883 see "Great International Fisheries Exhibition, London, 1883, Section G. Descriptive Catalogue of the Collection illustrating the scientific investigation of the sea and fresh waters." By Richard Rathbun. Washington: Government Printing Office, 1883.

United States collection of crabs. Not only was a series of photographs obtained, but arrangements were made for an exchange of specimens whereby many cotypes and specimens directly compared with types were secured for this museum.

ACKNOWLEDGMENTS.

My thanks are due to all those who assisted me during this European trip—to Prof. E. L. Bouvier, Dr. F. Meinert, Dr. H. J. Hansen, Dr. K. Brandt, Prof. F. Jeffrey Bell, and Dr. R. I. Pocock; and also to those who in the meantime have passed away—Prof. A. Milne Edwards, Dr. C. Lütken, Dr. F. Hilgendorf, and Prof. Henri de Saussure.

Were I to name all of the correspondents who have contributed by advice, loans, gifts, notes, or otherwise toward the completion of these volumes on American crabs, the list would include nearly all carcinologists, museum curators, professors of zoology, and collectors.

At the moment, however, I am especially indebted to those who have obtained for my use additional material in the Pinnotheridae, a group difficult to understand without well-preserved and abundant material. They are Dr. C. McLean Fraser, who recently collected several hundred specimens in the neighborhood of Vancouver Island; Dr. F. W. Weymouth, who forwarded the collection belonging to Stanford University; Dr. Walter Faxon, who loaned the collection in the Museum of Comparative Zoology; and Dr. W. T. Calman, of the British Museum, who arranged the loan of a valuable type-specimen.

Permission has been freely granted by Dr. H. M. Smith, Commissioner of Fisheries, and by Dr. C. H. Townsend, of the New York Zoological Society, to use data which form part of special reports as yet unpublished.

The classification of the higher groups adopted in this report is that of Borradaile, and the keys to the same have been borrowed from his summary published in 1907. Likewise the definitions of families and subfamilies are copied or adapted from those given by Alcock in his work on the Catometopa of India.[1]

In the immediate preparation of this report I am indebted to my colleagues in the United States National Museum, and above all to Mr. Waldo L. Schmitt, assistant curator of marine invertebrates, who has made most of the drawings of microscopic mounts of the Pinnotheridae with the use of the Edinger drawing apparatus, and has rendered important assistance in the preparation of manuscript and plates.

---

[1] Materials for a Carcinological Fauna of India, No. 6. The Brachyura Catometopa or Grapsoidea. Journ. Asiat. Soc. Bengal, vol. 69, Calcutta, 1900, pp. 279–456 [621–798].

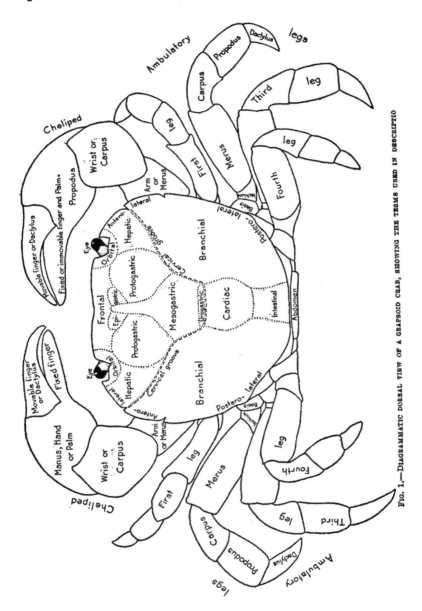

Fig. 1.—Diagrammatic dorsal view of a grapsoid crab, showing the terms used in description.

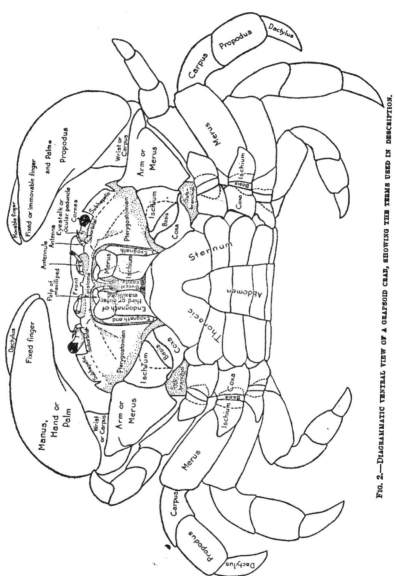

FIG. 2.—DIAGRAMMATIC VENTRAL VIEW OF A GRAPSOID CRAB, SHOWING THE TERMS USED IN DESCRIPTION.

BULLETIN 97, UNITED STATES NATIONAL MUSEUM.

EXPLANATION OF TERMS AND ABBREVIATIONS USED.

Glossary of terms used in classification and description.

Abdomen, or tail, the jointed or hinder part of the body which is reduced in size and is bent forward under the thorax. In this work the abdomen is considered as consisting of seven segments, inclusive of the telson. It is much wider in the female than in the male.

Afferent channels, the openings through which water passes to the branchiae. In the Brachyuran crabs they usually open behind the pterygostomian regions and in front of the chelipeds, save in certain of the Oxystomata where they open at the antero-lateral angles of the palate or endostome.

Ambulatory, or walking legs, usually four pairs, are behind the chelipeds. They may be only 3 in number, as in the genus *Cymopolia* where the hind pair of thoracic legs are delicate and tendril-like, as if used for attachment to foreign objects.

Antennae, or second pair of antennae, that pair of antennae situated between the antennules and the orbits.

Antennules, or antennae of the first pair, those antennae lying near together, either side of the median line.

Arm, the merus of a cheliped.

Basis, or basipodite, the second segment (from the body) of a leg or maxilliped.

Branchial region (paired), the very large lateral area of the carapace, behind the cervical suture.

Buccal cavity, or buccal cavern, the cavity on the ventral surface of the body, in which are situated the mouth-parts; it is bounded anteriorly by the epistome, laterally by the free edges of the carapace.

Carapace, or cephalothorax, the shell which covers the dorsal surface, and the lateral portions of the ventral surface, of the body.

Cardiac region, the median area of the carapace, behind the cervical suture.

Carpus, or carpopodite, the fifth segment (from the body) of a leg or a maxilliped.

Cervical groove, the complex groove, or series of grooves, running across the carapace, the groove being transverse at the middle, then turning on each side obliquely forward and outward to the lateral margin. It separates the hepatic and gastric regions from the branchial and cardiac regions.

Chela, or claw, the two last segments of a cheliped.

Cheliped, the pair of thoracic legs immediately behind the maxillipeds or jaw-feet., They bear the chelae or pincer-claws and are usually stouter, sometimes much stouter than the succeeding or walking legs.

Coxa, or coxopodite, the first or proximal segment of a leg or a maxilliped.

Dactylus, or dactylopodite, the seventh or terminal segment of a leg or maxilliped. The dactylus is the movable finger of a cheliped.

Distal, farthest from the center of the body; opposed to proximal.

Efferent channels, the channels through which water passes from the branchiae; they open at the sides of the endostome, except in the subtribe Oxystomata where they open at the middle of the endostome.

Endognath, the inner or principal branch of the maxilliped.

Endostome, or palate, the roof of the buccal cavity.

Epigastric lobes, the anterior lobes of the gastric region.

Epistome, the transverse plate forming the anterior border of the buccal cavity, its sides being fused with the carapace.

Exognath, the outer or secondary branch of the maxilliped.

Eyebrow, a term here applied to the narrow surface in the fiddler crabs between the upper border of the orbit and the orbital cavity itself.

Fingers, or digits, the narrow scissorlike blades of the claw end of a cheliped, the movable finger being the dactylus, the immovable finger the terminal part of the propodus.

Flagellum of the antennae and antennules, the long, narrow, terminal portion, composed of numerous short segments.

Gastric region, the large median area, bounded behind by the cervical suture, outside by hepatic regions, and anteriorly by the fronto-orbital regions. It is divisible into the following subregions or lobes: Mesogastric, protogastric, epigastric, metagastric, and urogastric.

Hepatic region (paired), a small subtriangular, antero-lateral region, wedged between the branchial and gastric regions and either the margin of the carapace or the margin of the orbit.

Interantennular septum, the plate which separates the two antennular cavities from each other.

Intestinal region, a short transverse area behind the cardiac region. Sometimes called the posterior cardiac lobe.

Ischium, or ischiopodite, the third segment (from the body) of a leg or maxilliped. It is usually the first large segment of the maxilliped.

Jugal region. See *Pterygostomian region*.

Labial border, the anterior border of the buccal, or mouth cavity.

Manus, or palm, the broad, proximal part of the propodus of a cheliped.

Maxillipeds, the three outermost pairs of jaw-feet, the third or outer pair forming more or less of an operculum to the buccal or mouth cavity.

Merus, or meropodite, the fourth segment (from the body) of a leg or maxilliped. It is usually the first long segment of a cheliped, or walking leg.

Mesogastric lobe or subregion, the median division of the gastric region, pentagonal in form and with a long, narrow, anterior prolongation.

Metagastric lobes, the postero-lateral lobes of the gastric region; often not defined.

Orbital hiatus, the gap in the orbital margin at its lower, inner angle.

Orbital region, the narrow space bordering the upper margin of the orbit; not always distinguishable.

Palp, or palpus, of maxilliped, consists of the last two or three segments following the merus-joint.

Pereiopods, a term applied to the chelipeds and four pairs of legs.

Pollex, the immovable finger of the cheliped. The term "pollex" has been used, however, by some writers, for the movable finger, a use not without justification.

Propodus, or propodite, the sixth or penultimate segment of a leg or maxilliped. In a cheliped, the propodus consists of the palmar portion or manus, and a narrower, immovable finger.

Protogastric lobes or subregions, the antero-lateral lobes of the gastric region.

Proximal, nearest the center of the body; opposed to distal.

Pterygostomian region, the triangular space on the ventral surface of the carapace, on either side of the buccal cavity. Sometimes called the jugal region.

Rostrum, or front, that part of the carapace which projects forward from between the bases of the eyestalks.

Sternum, or sternal plastron, the ventral, segmented wall of the thorax.

Stridulating ridge or organ, a ridge or surface made up of a close series of cross ridges or tubercles, so placed as to rub against another surface and thus produce a noise.

Subequal, nearly equal.

Subhepatic region, that area below the hepatic region and below the lateral border of the carapace.
Suborbital region, the narrow space bordering the lower margin of the orbit; not always distinguishable.
Telson, the terminal segment of the abdomen. See Abdomen.
Thumb, the immovable finger of the cheliped.
Urogastric lobe, the postero-median lobe of the gastric region. Sometimes called the genital region.
Wrist, the carpus of a cheliped.

Explanation of measurements.

The length of the carapace, unless otherwise stated, is measured on the median line, from the anterior to the posterior margin.

The width of the carapace is measured at the widest part.

The fronto-orbital width or exorbital width is measured from the outer angle of one orbit to the outer angle of the other.

The length of the rostrum is measured from the tip to the posterior line of the upper margins of the orbits.

The width of the rostrum is measured at its posterior end.

The length of the segments of the chelipeds and legs is measured on the upper or anterior margin. The length of the whole cheliped or leg is measured on the lower margin, from the articulation of the coxa with the sternum to the tip of the dactylus.

The width of the segments of the chelipeds and legs is measured at the widest part.

The length of the immovable finger is measured from the tip to the extremity of the sinus between the fingers.

Character of bottom.

Under "Material examined," the abbreviations indicating the character of the bottom, are those employed by the Bureau of Fisheries. Nouns begin with a capital, adjectives with a small letter.

bk_____black	gy_____gray	S_____sand
br_____brown	hrd_____hard	sctrd_____scattered
brk_____broken	lge_____large	sft_____soft
bu_____blue	lt_____light	Sh_____shells
Co_____coral	M_____mud	sm_____small
crs_____coarse	Oz_____ooze	Sp_____specks
dk_____dark	P_____pebbles	St_____stones
fne_____fine	Ptr_____Pteropod	stky_____sticky
For_____foraminifera	R_____rock	vol_____volcanic
G_____gravel	rd_____red	W_____seaweed
Glob_____globigerina	Rf_____reef	wh_____white
gn_____green	rky_____rocky	yl_____yellow

Additional abbreviations and notes.

In the synonymy an attempt has been made to give all the different names or combinations which have been used, but not all the references to a species.

In the lists under "Material examined" a number in parenthesis following an indication of a specimen or specimens denotes a catalogue number of the United States National Museum unless otherwise indicated. M. C. Z. = Museum of Comparative Zoology; P. M. Y. U. = Peabody Museum of Yale University; Mus. S. U. I. = Museum of the State University of Iowa; y. = young. The words "U. S. Fisheries" should be understood before "Str. *Albatross*," "Str. *Fish Hawk*," or "Sch. *Grampus*."

In the same lists there have been entered, besides specimens in the National Museum, many types examined elsewhere, as well as such specimens from other collections as increase our knowledge of the range of the species, but for lack of space no attempt has been made to record all of the many specimens examined in museum and private collections.

THE GRAPSOID OR CATOMETOPOUS CRABS OF AMERICA.

The term Catometopa or "square-fronted" was early[1] applied to a group of crabs which was contrasted with the Cyclometopa or "round-fronted" crabs. These terms were abandoned[2] because the one group was found to merge gradually into the other. The name Brachyrhyncha was given to the whole. Aside from intergrading forms, the so-called Catometopa contain many types that are not "square-fronted." In 1851[3] Dana first used the word Grapsoidea[4] for this group, *Grapsus* being the typical genus. The name Grapsoid is therefore used in the title of the present paper as a short and convenient term to indicate the content of the volume.

The key on pages 13–15 gives the relation of the families here dealt with to the remainder of the Brachyura, or short-tailed crabs.

1. The family Goneplacidae links the Catometopes to the Cyclometopes and is most closely allied to the family Xanthidae, some of its genera having the form of the Panopeids or mud crabs, and differing from them chiefly in the characters of abdomen and sternum and the shape of the chelae. The Goneplacids, being bottom-dwelling forms, are taken almost exclusively in the dredge or tangle.

2. The Pinnotheridae is a large family of small crabs, chiefly commensal, sometimes free-swimming and occasionally parasitic. While in most crabs the male is stronger than the female, especially as regards the chelipeds, and often attains a greater maximum size, the reverse is true in many Pinnotherids. The female may not only be much larger than the male, but have a very different shape and appearance, due, no doubt, to the difference in mode of life of the two sexes. The female may be commensal or parasitic, while the male is free-swimming. Much is yet to be learned regarding the habitat of

---

[1] Catométopes Milne Edwards (part), Hist. Nat. Crust., vol. 2, 1837, p. 1.
[2] See Borradaile, Ann. Mag. Nat. Hist., ser. 7, vol. 19, 1907, p. 466.
[3] Amer. Journ. Sci., ser. 2, vol. 12, 1851, pp. 283 and 285.
[4] The Grapsoïdiens of Milne Edwards is equivalent only to the Family Grapsidae.

the various species. One hundred years ago Thomas Say, the pioneer American naturalist, remarked, "It is a curious fact that, although the female of this species is so very often found occupying the oyster, the male is absolutely unknown."[1] The same might almost be said to-day, as up to this time not a single male of that species has been recorded, and only one immature specimen has been seen by the present writer or is known to exist in any collection.[2]

In order to emphasize our lack of knowledge of this important and interesting little family I have, in dealing with the genus *Pinnotheres*, made a list of the American species with an indication of the known sex or sexes of each. It is hoped that this will lead to greater interest among workers at the laboratories along our coasts in seeking and preserving examples of Pinnotheridae. They are to be sought for not in the usual haunts of crabs, but in the shells of bivalve mollusks, such as oysters, clams, mussels, and scallops, in the tubes of annelid and sipunculid worms, in ascidians, in the intestinal tract of certain globular sea-urchins, and on the outer surface of sand dollars and other flat urchins, with all of which animals they may be commensal. In some cases the males, however, may be free-swimming and should be looked for in plankton or tow-net hauls.

3. The Cymopoliidae are unique in structure, related in most features to the Catometopes, but in the small filiform legs of the last pair resembling the Dorippidae, near which they are grouped by some authors. They come from water of considerable depth and nothing is yet known of their habits or the function of their delicate hind legs, whether used to support a protective covering of sponge or ascidian, as in the Dromiidae, or serving as tendrils to cling to the branches of alcyonarians, hydroids, and algae.

4. The Grapsidae embrace large numbers of shore and shallow-water crabs, as well as a few which inhabit fresh water or are semi-terrestrial, living at considerable distance from the sea.

5. The Gecarcinidae are land crabs, often of large size, with smooth, thick carapaces and more or less spinous legs.

6. The family Ocypodidae is represented in this hemisphere by the sand crabs or ghost crabs and the fiddler crabs (genus *Uca*, formerly *Gelasimus*). In this group the length of the eyestalks and the corresponding narrowness of the frontal projection or rostrum are carried to an extreme, as is also the difference between the chelipeds of the two sexes. Both chelipeds in the female and one (either right or left) in the male are small, feeble, and similar. The other male cheliped is stout and of enormous length, especially the arm and fingers, the chela (palm and digits) usually much surpassing in length the width of the carapace.

---

[1] *Pinnotheres ostreum*, Journ. Acad. Nat. Sci. Philadelphia, vol. 1, 1817, p. 68.
[2] In November, 1917, a second male was found in an oyster by W. P. Hay.

Order DECAPODA.
Suborder REPTANTIA.
Tribe BRACHYURA.

Definition.—The Brachyura are the short-tailed crabs, in contradistinction to the Anomura, which include the hermits, the hippas, and others. The Brachyura are characterized by having the carapace fused with the epistome at the sides and nearly always in the middle; the last of the thoracic sternal somites fused with the rest, its legs usually like the others; the basis and ischium of cheliped and legs immovably united; the abdomen brachyurous (small, straight, symmetrical, bent under the thorax, showing no traces of other function than reproduction, and without biramous limbs on the sixth segment); by lacking a movable antennal scale; and by having the third maxillipeds broad.

Development.—Crabs, as a rule, pass through two or more free-swimming stages of development after leaving the egg and before attaining the form of the adult. The first stage is known as the zoea; its carapace is relatively stout and usually spined, the eyes are conspicuous, the thoracic legs are undeveloped, the abdomen is long and slender. The zoea may molt several times, with slight and very gradual changes, but eventually in shedding its skin it suddenly develops into a very different larval stage, the megalops, which has more in common with the adult than does the zoea. In the megalops the long spines of the carapace have disappeared; the eyes are at the ends of movable stalks; the five thoracic feet are developed and similar to those of the adult; and the maxillipeds are no longer used for locomotion; the telson or abdomen may be partially bent forward under the ventral surface of the body.[1]

Some crabs are exceptions to the above rule of postembryonic larval metamorphoses. Such are the river-crabs of the family Potamon-

---

[1] For details of the metamorphosis of a crab, see:

Faxon, W., On some Young Stages in the Development of Hippa, Porcellana, and Pinnixa. Bull. Mus. Comp. Zoöl. at Harvard College, vol. 5, No. 11, Cambridge, June, 1879, pp. 253-268, pls. 1-5.

Czerniavsky, Voldemar, Megalopidea s. Larvae Anomuriformes Crustaceorum Brachyurorum. Proc. Entom. Soc. St. Petersburg, vol. 11, 1880, pp. 51-90, pls. 2 and 3. With bibliography and dichotomous table of genera. (In Latin and Russian.)

Brooks, W. K., Handbook of Invertebrate Zoology. Boston: S. E. Cassino, 1882, Chap. 21, pp. 207-223.

Faxon, Walter, Selections from Embryological Monographs. Compiled by Alexander Agassiz, Walter Faxon, and E. L. Mark. I. Crustacea. By Walter Faxon. Mem. Mus. Comp. Zoöl. at Harvard College, vol. 9, No. 1, Cambridge, July, 1882, pls. 1-14 with explanations (not paged).

Ortmann, A. E., Bronn's Klassen und Ordnungen des Thier-Reichs, vol. 5, Abth. 2, Lief. 47-49, Leipzig, 1898, pp. 1078-1105, pls. 110-112.

Calman, W. T., The Life of Crustacea. London, 1911, Chap. 4, pp. 66-87.

idae (not included in this volume), in which the young issues from the egg as a tiny crab substantially like its parent.

Surprisingly little is known of the development of different species, as the life history of not more than a dozen of our American species of Brachyura has been worked out. This would prove a fruitful field for investigators in our seaside laboratories where facilities are offered for studying transformations in the aquarium.

It is unwise to draw sweeping conclusions from a few cases, for recently the discovery has been made of the direct development of the young crab from the egg in a species of the Oxyrhyncha or spider-crabs.[1]

ANALOGOUS SPECIES ON OPPOSITE SIDES OF THE CONTINENT.

Throughout this volume, after the keys to species, I have indicated in parallel columns those closely related species occurring on opposite sides of the continent. The resemblance between the units of some of these pairs is much greater than between others. All are here assembled in a single list, followed by a list of those species which occur on both sides of the continent; those of world-wide range being indicated by an asterisk (*).

Family GONEPLACIDAE.

Atlantic.	Pacific.
Prionoplax atlantica.	*Prionoplax ciliata.*
Euryplax nitida.	*Euryplax polita.*
Speocarcinus carolinensis.	*Speocarcinus granulimanus.*
Glyptoplax smithii.	*Glyptoplax pugnax.*
Chasmocarcinus typicus.	*Chasmocarcinus latipes.*

Family PINNOTHERIDAE.

Pinnotheres ostreum.	*Pinnotheres holmesi.*
Fabia byssomiae.	*Fabia canfieldi.*
Parapinnixa hendersoni.	*Parapinnixa nitida.*
Parapinnixa bouvieri.	*Parapinnixa affinis.*
Pinnixa brevipollex.	*Pinnixa affinis.*
Pinnixa faxoni.	*Pinnixa transversalis.*
Pinnixa cylindrica.	*Pinnixa franciscana.*
Pinnixa sayana.	*Pinnixa occidentalis.*
Pinnixa floridana.	*Pinnixa longipes.*

Family CYMOPOLIIDAE.

Cymopolia alternata.	*Cymopolia zonata.*
Cymopolia faxoni.	*Cymopolia lucasii.*
Cymopolia obesa.	*Cymopolia tuberculata.*

---

[1] See Stalk-eyed Crustaceans collected at the Monte Bello Islands. By Mary J. Rathbun. Proc. Zool. Soc. London, 1914, pp. 653 and 662.

Family GRAPSIDAE.

Atlantic.
Goniopsis cruentata.
Pachygrapsus maurus.
Pachygrapsus marmoratus.
Glyptograpsus jamaicensis.
Sesarma reticulatum.
Sesarma curacaoense.
Sesarma cinereum.
Sesarma miersii iheringi.
Sesarma roberti.
Cyclograpsus integer.

Pacific.
Goniopsis pulchra.
Pachygrapsus crassipes.
Pachygrapsus pubescens.
Glyptograpsus impressus.
Sesarma aequatoriale.
Sesarma rhizophorae.
Sesarma occidentale.
Sesarma biolleyi.
Sesarma angustum.
Cyclograpsus cinereus.

Family GECARCINIDAE.

Cardisoma guanhumi.
Ucides cordatus.
Gecarcinus ruricola. }
Gecarcinus lateralis. }

Cardisoma crassum.
Ucides occidentalis.
{*Gecarcinus quadratus.* (Also on the
 Atlantic side of the Isthmus.)

Family OCYPODIDAE.

Ocypode albicans.
Uca maracoani.
Uca heterochelos.
Uca speciosa.
Uca subcylindrica.

Ocypode occidentalis.
Uca monilifera.
Uca princeps.
Uca crenulata.
Uca stenodactylus.

SPECIES ON BOTH SIDES OF THE CONTINENT.

Family GRAPSIDAE.

Grapsus grapsus.*
Geograpsus lividus.
Pachygrapsus transversus.*
Planes minutus.*
Cyrtograpsus angulatus.
Sesarma (Holometopus) angustipes.
Aratus pisonii.

Family GECARCINIDAE.

Gecarcinus quadratus.

Family OCYPODIDAE.

Uca mordax.

KEY TO THE SUBTRIBES OF THE TRIBE BRACHYURA.[1]

A^1. Mouth field (endostome) prolonged forward to form a gutter. Last pair of legs normal or abnormal. Female openings generally sternal. First abdominal limbs of female wanting. Gills few____Subtribe *Oxystomata*.

---

[1] The keys are taken almost bodily from Borradaile's On the Classification of the Decapod Crustaceans, Ann. Mag. Nat. Hist., ser. 7, vol. 19, 1907, pp. 477-483. The names in the right hand margin which are printed in capitals indicate the families and higher divisions treated of in this volume.

A². Mouth field roughly square.
  B¹. Last pair of legs abnormal, dorsal. Female openings coxal. First abdominal limbs of female present. Gills usually many.
  Subtribe *Dromiacea.*
  B². Last pair of legs normal, rarely reduced, not dorsal, except in *Cymopolia* and *Retropluma.* Female openings sternal. First abdominal limbs of female wanting. Gills few _____Subtribe BRACHYGNATHA.

KEY TO THE SUPERFAMILIES OF THE SUBTRIBE BRACHYGNATHA.

A¹. Fore part of body narrow, usually forming a distinct rostrum. Body more or less triangular. Orbits generally incomplete.
  Superfamily *Oxyrhyncha.*
A². Fore part of body broad. Rostrum usually reduced or wanting. Body oval, round, or square. Orbits nearly always well inclosed.
  Superfamily BRACHYRHYNCHA.

KEY TO THE FAMILIES OF THE SUPERFAMILY BRACHYRHYNCHA.

A¹. Orbits formed, but more or less incomplete. Second antennal flagella, when present, long and hairy. Rostrum present. Body elongate-oval. Fore edge of mouth indistinct_____Family *Euryalidae=Corystidae.*
A². Orbits complete (though fissures may remain), except in the Mictyrinae, where the eyes are almost or quite unprotected. Body rarely elongate-oval. Rostrum often wanting. Second antennal flagella usually short, not hairy.
  B¹. Carpus of third maxillipeds articulates at or near antero-internal angle of the merus. Body usually round or transversely oval. Male openings nearly always coxal. In many species the right chela is always larger than the left.
    C¹. Legs more or less distinctly adapted for swimming. Usually a small lobe on the inner angle of the endopodite in the first maxillipeds. The first antennae fold slanting or transverse_____Family *Portunidae.*
    C². Legs not adapted for swimming, or if so modified, then the male genital duct opens sternally or runs in a sternal groove. Inner lobe on the endopodite in the first maxillipeds wanting.
      D¹. Fresh-water crabs with the branchial region much developed and swollen. Body often squarish, but male openings coxal.
        Family *Potamonidae.*
      D². Marine crabs with the branchial region not greatly swollen.
        E¹. First antennae fold lengthwise.
          F¹. Carapace subcircular. Second antennal flagella either long and hairy or wanting_____Family *Atelecyclidae.*
          F². Carapace broadly oval or hexagonal. Second antennal flagella present, short, not hairy_____Family *Cancridae.*
        E². First antennae fold slanting or transversely.
          F¹. Body usually transversely oval. Male openings rarely sternal. Not sharply separated from the following family.
            Family *Xanthidae.*
          F². Body usually square or squarish. Male ducts open on the sternum, or, if coxal, pass along a groove in the sternum. Not sharply separated from the foregoing family.
            Family GONEPLACIDAE, p. 15.

B². Carpus of third maxillipeds does not articulate at or near the inner angle of the merus. Body usually square or squarish. Male openings sternal except in *Retropluma*, where the duct passes along a sternal groove to the coxopodite. In no species (*Cymopoliidae* excepted) is the right chela always larger than the left.
  C¹. Small, usually commensal crabs, with very small eyes and orbits. Body usually more or less rounded_____Family PINNOTHERIDAE, p. 61.
  C². Free-living crabs, with eyes not specially reduced and usually a square body.
    D¹. Last pair of legs dorsally placed and weaker than the others. Interantennular septum very thin. No distinct epistome. Exopodites of third maxillipeds not hidden.
      E¹. Front narrow. Female openings in normal position. Third maxillipeds subpediform, not covering the mouth.
        Family *Retroplumidae=Ptenoplacidae*.
      E². Front moderately broad. Female openings on the sternal segment corresponding to the first pair of walking legs. Third maxillipeds cover the mouth to a large extent and have very small meropodites_____Family CYMOPOLIIDAE=*Palicidae*, p. 182.
    D². Last pair of legs not dorsally placed nor markedly weaker than the rest. Interantennular septum not very thin.
      E¹. A gap of greater or less size is left between the third maxillipeds. Front very or moderately broad.
        F¹. Sides of the body either straight or very slightly arched. Shape squarish. Front broad. Rarely true land crabs.
          Family GRAPSIDAE, p. 224.
        F². Sides of the body strongly arched. Shape transversely oval. Front narrower. Land-crabs__Family GECARCINIDAE, p. 339.
      E². The third maxillipeds almost or quite close the mouth. Front moderately or very narrow_____Family OCYPODIDAE, p. 365.
B³. Merus of third maxillipeds small, bearing terminally a carpus of nearly its own width. Ischium very broad. Body somewhat oblong. First antennae not retractile into sockets. Parasitic on corals.
        Family *Hapalocarcinidae*.

Family GONEPLACIDAE (Dana).

Goneplacidae DANA, Amer. Journ. Sci., ser. 2, vol. 12, 1851, p. 285; U. S. Expl. Exped., vol. 13, Crust., pt. 1, 1852, pp. 308 and 310.—ALCOCK, Journ. Asiat. Soc. Bengal, vol. 69, 1900, pp. 283, 286, 292, 297, and synonymy.

The palp of the external maxillipeds articulates at or near the antero-internal angle of the merus; the exognath is of normal size and is not concealed. The interantennular septum is a thin plate. The division of the orbit into two fossae is usually not indicated. The genital ducts of the male usually perforate the base of the last pair of legs, often passing forward through a groove in the sternum.

KEY TO THE AMERICAN SUBFAMILIES AND GENERA OF THE FAMILY GONEPLACIDAE.

A¹. The base of the third segment of the male abdomen covers the whole space between the last pair of legs.

B^1. Carapace xanthoid, widest behind the postorbital angles. Orbits of normal size and form------------------Subfamily *Carcinoplacinae*, p. 17.
 C^1. Eyes movable.
  D^1. Carapace subquadrate; postero-lateral margins nearly parallel. Maxillipeds gaping---------------------------------*Trizocarcinus*, p. 17.
  D^2. Carapace hexagonal; postero-lateral margins distinctly converging posteriorly. Maxillipeds not gaping-----------*Pilumnoplax*, p. 21.
 C^2. Eyes immovable--*Bathyplax*, p. 19.
B^2. Carapace subquadrate, anterior border entirely occupied by square-cut front and orbits, the latter being long, narrow trenches. Carapace widest between the postorbital angles---Subfamily *Goneplacinae*, p. 24.
                         Goneplax, p. 25.

A^2. The base of the third segment of the male abdomen does not cover the whole space between the last pair of legs.
 B^1. Width of male abdomen at third segment more than half width of sternum in same line---------------- ------Subfamily *Prionoplacinae*, p. 29.
  C^1. Fronto-orbital width almost as great as the total width of the carapace. Eyestalks long. Carapace of ocypodine form, subquadrate; postero-lateral margins converging.
   D^1. Antennae excluded from the orbit------------------*Euryplax*, p. 34.
   D^2. Antennae entering the orbit.
    E^1. Carapace wide. Merus of outer maxilliped broader than long.
     F^1. Carapace quadrilateral-----------------------*Prionoplax*, p. 29.
     F^2. Carapace hexagonal------------------------*Chasmophora*, p. 37.
    E^2. Carapace narrow. Merus of outer maxilliped as long as broad.
                        Tetraplax, p. 32.
  C^2. Fronto-orbital width from one-half to three-fourths the total width of the carapace. Eyestalks short. Antero-lateral margin arcuate.
   D^1. Eyestalks diminishing to the tip.
    E^1. Postero-lateral margins converging posteriorly------*Cyrtoplax*, p. 45.
    E^2. Postero-lateral margins subparallel-------------*Speocarcinus*, p. 38.
   D^2. Eyestalks enlarged at the corneal end. Postero-lateral margins distinctly converging posteriorly.
    E^1. Fronto-orbital border about half total width of carapace.
     F^1. Upper surface of carpus of cheliped subrectangular.
                       Pseudorhombila, p. 42.
     F^2. Upper surface of carpus not subrectangular-------*Oediplax*, p. 44.
    E^2. Fronto-orbital border from three-fifths to three-fourths of the total width of the carapace.
     F^1. Carapace broad, width $1\frac{1}{2}$ times length----------*Panoplax*, p. 47.
     F^2. Carapace narrow, width about $1\frac{1}{3}$ times length.
      G^1. Merus of outer maxillipeds with antero-external angle prominent, acutangular-----------------------*Glyptoplax*, p. 48.
      G^2. Merus of outer maxillipeds with antero-external angle not prominent nor acutangular-------------*Eucratopsis*, p. 52.
 B^2. Width of male abdomen at third segment less than half width of sternum in same line----------------------------Subfamily *Rhizopinae*, p. 54.
                       Chasmocarcinus, p. 54.[1]

---

[1] *Ceratoplax ciliata* Stimpson (Proc. Acad. Nat. Sci. Philadelphia, vol. 10, 1858, p. 96 [42]).—Miers, *Challenger* Rept., Zool., vol. 17, 1886, p. 234, pl. 19, figs. 3–3b), an Indo-Pacific species, is recorded by Cano (Boll. Soc. Nat. Napoli, ser. 1, vol. 3, 1889, p. 229) from the coast of Ecuador. He regards his identification as doubtful (p. 91) and on p. 101 says simply " *Ceratoplax* sp." No description is given. If the locality cited be correct for the specimen, it is more likely to have been a species of *Chasmocarcinus*.

Subfamily CARCINOPLACINAE Miers.

Carcinoplacinae MIERS, Challenger Rept., Zool., vol. 17, 1886, p. 222.
Pseudorhombilinae ALCOCK, Journ. Asiat. Soc. Bengal, vol. 69, 1900, pp. 286, 292, and 297.

Carapace xanthoid, the regions seldom well defined; front usually of good breadth and square cut, often little deflexed; eyes and orbits usually of normal size and form, the eyes well pigmented and the eyestalks normally movable except in certain deep-sea genera; the antennules fold transversely; antennal flagella of medium length. Epistome well defined; buccal cavern square-cut and usually completely closed by the external maxillipeds, which have a subquadrate merus. The base of the third segment of the male abdomen covers the whole space between the last pair of legs. Male openings not sternal.

Represented in America by three genera.

Genus TRIZOCARCINUS Rathbun.

Trizocarcinus RATHBUN, Proc. U. S. Nat. Mus., vol. 47, 1914, p. 117; type, *T. dentatus* (Rathbun).

Carapace deep, subquadrilateral, somewhat broader than long, with little distinction of regions, convex in both directions. Fronto-orbital border about three-fourths of the greatest breadth of the carapace; antero-lateral borders arched, dentate.

Front square-cut, straight, faintly notched in the middle, distinctly separated from the supra-orbital angles, between one-third and one-fourth the width of the carapace. Upper margin of orbit with 2 distinct notches. Basal antennal joint short, the flagellum standing loosely in the orbital hiatus. The antennules fold transversely.

Buccal cavity widening distally, not completely closed by the maxillipeds, the merus of which has a concave anterior margin, and the antero-external lobe projects forward not outward. Efferent branchial channels well defined. A stridulating ridge formed of parallel striae runs obliquely backward from the antero-external angle of the buccal cavity; and is played upon by a short ridge on the merus of the chelipeds.

Chelipeds equal, much more massive than the legs. In both sexes all seven abdominal segments are distinct, and in the male the third segment covers the whole width of the sternum between the bases of the last pair of legs.

This genus is closely related to the Indo-Pacific genus *Carcinoplax*, but differs in the form of the merus of the maxillipeds, in the notch separating front and orbital angle, the superior notches of the orbit, and the stridulating ridges.

Only one species known.

TRIZOCARCINUS DENTATUS (Rathbun).[1]

Plate 1.

Carcinoplax dentatus RATHBUN, Proc. U. S. Nat. Mus., vol. 16, 1893, p. 243, (type-locality, Gulf of California, lat. 29° 40′ 00″ N.; long. 112° 57′ 00″ W., 76 fathoms; holotype, male, Cat. No. 17462, U.S.N.M.).

Trizocarcinus dentatus RATHBUN, Proc. U. S. Nat. Mus., vol. 47, 1914, p. 117, text-fig. 1, pl. 1.

Diagnosis.—A pterygostomian stridulating ridge. Merus of maxillipeds not broader than long. Segments of male abdomen free; third segment reaching coxae.

Description.—Carapace about four-fifths as long as broad; posterior median portion almost level and bordered by a blunt ridge from which the anterior and lateral portions slope downward. Cardiac and mesogastric regions in part demarcated. Surface short-hairy, unequally granulate.

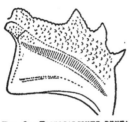

FIG. 3.—TRIZOCARCINUS DENTATUS, VENTRAL VIEW OF LEFT SIDE OF CARAPACE OF MALE HOLOTYPE, SHOWING STRIDULATING RIDGE, × 4.

Front bimarginate, deeply grooved; inner angle of orbit a subacute tooth. Three flattened, upturned, pointed and subequal antero-lateral teeth. Margins of teeth and orbit denticulate. Postero-lateral margins subparallel. Pterygostomian ridge crossed by about 70 fine striae. A short complementary ridge exists on the lower proximal margin of the inner surface of the arm, and is crossed obliquely by 10 or 11 striae.

Chelipeds granulate, outer face of carpus and chela clothed with long hair except tips of fingers. An upper marginal spine on merus at distal third; a lower subterminal spine. Inner spine of carpus curved upward. Chelae flattened, margins serrulate; fingers irregularly toothed, not gaping. Inner surface of propodus with a deeply concave surface below.

Legs long and narrow; last three joints long, hairy, margins granulate.

Third segment of male abdomen the widest, first and second segments subequal in width, exposing the sternum.

Measurements.—Male holotype, length of carapace 14, width of same 18 mm.

Range.—Gulf of California; 30 to 76 fathoms.

Material examined.—

Gulf of California, off Cape Lobos, Mexico: Lat. 29° 40′ 00″ N.; long. 112° 57′ 00″ W.; 76 fathoms; gn. M.; temp. 59° F.; Mar. 24,

---

[1] The parenthesis about an author's name after a species name indicates that the species was originally placed in a different genus.

1889; station 3016, *Albatross;* 2 males (1 is holotype) (17462). Lat. 29° 54′ 30″ N.; long. 113° 01′ 00″ W.; 58 fathoms; gn. M.; temp. 61.8° F.; Mar. 24, 1889; station 3017, *Albatross;* 1 female (17463).

Gulf of California, off Point San Fermin; lat. 30° 21′ 00″ N.; long. 114° 25′ 15″ W.; 30 fathoms; gy. M.; temp. 62° F.; Mar. 27, 1889; station 3035, *Albatross;* 1 young female (17464).

Genus BATHYPLAX A. Milne Edwards.

Bathyplax A. MILNE EDWARDS, Bull. Mus. Comp. Zoöl., vol. 8, 1880, p. 16; type, *B. typhla* A. Milne Edwards.

Carapace hexagonal; antero-lateral margins arcuate and armed, postero-lateral margins converging.

Front straight, about two-fifths width of carapace. Orbits small, shallow, ill-defined; eyes small, immovable, deficient in pigment. Antennae standing in the orbital hiatus; broad joint reaching the downward prolongation of the front; flagellum long. Antennules transverse.

Buccal cavity widening rapidly anteriorly. Palatal ridge strong. Merus of outer maxilliped much wider than the ischium, its antero-external angle expanded.

Chelipeds dissimilar. Ambulatory legs slender.

Abdomen widest at third segment which reaches the coxae of the last pair of legs; no segments fused.

Only one species known.

BATHYPLAX TYPHLA A. Milne Edwards.

Plate 2.

Bathyplax typhlus A. MILNE EDWARDS, Bull. Mus. Comp. Zoöl., vol. 8, 1880, p. 16 (type-locality, Frederickstadt. St. Croix Island, 451 fathoms; type in M. C. Z.).

1886, p. 230, pl. 20, fig. 3 (type-locality, south of Pernambuco, 30 to 400 fathoms; type in Brit. Mus.).

Bathyplax typhlus, var. *oculiferus* MIERS, Challenger Rept., Zool., vol. 17,

Diagnosis.—Orbits rudimentary; eyes immovable, almost blind. Buccal cavity very wide in front. Chelipeds dissimilar; stridulating ridge on arm.

Description.—Carapace flat transversely, convex longitudinally; posteriorly uneven, the cardiac and posterior mesogastric region deeply delimited. Surface pubescent and for the most part granulate.

Fronto-orbital width about three-fifths that of the carapace. Edge of front margined, bent down a little at the middle. Orbit without a superior inner or outer angle; too small to hold the small spherical light-brown cornea set on a short stalk; inferior inner angle dentiform. Two conical distant antero-lateral spines, the first one removed from the orbit.

The outer maxillipeds do not fit close together nor against the epistome.

Chelipeds subequal, the chelae dissimilar. Merus with a spine or tubercle near the distal end of the outer margin; on the inner surface near the distal end there is a transverse stridulating organ capable of scraping against the granules of the pterygostomian region. Carpus elongate, the right one with a sharp inner spine, the left one with a small inner tubercle or spine. Right chela with palm expanded below, both fingers narrow and gaping, more in the male than in the female. Left chela shorter, very thin, and expanded below at the middle of the propodus, immovable finger in consequence wider at base than in the right chela, fingers not gaping; upper surface of palm produced inward in a blunt tooth.

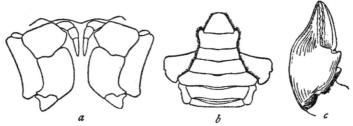

FIG. 4.—BATHYPLAX TYPHLA, MALE (9729). *a*, OUTER MAXILLIPEDS, SHOWING APPROXIMATION, × 4; *b*, ABDOMEN AND PART OF STERNUM, × 3; *c*, LEFT CHELA, × 2.

Legs long and very slender, margins granulous, surface downy, last three segments with a thin fringe of long hair.

Measurements.—Male (9724), length of carapace 16.5, width of same 21.5 mm.

Range.—From Gulf of Mexico to Pernambuco, Brazil, 280 to 463 fathoms.

Material examined.

Southwest of Cape San Blas, Florida: Lat. 28° 36' 15" N.; long. 86° 50' 00" W.; 347 fathoms; gy. M.; temp. 44.1° F.; Mar. 13, 1885; station 2395, *Albatross;* 1 male (9724). Lat. 28° 34' 00" N.; long. 86° 48' 00" W.; 335 fathoms; gy. M.; Mar. 13, 1885; station 2396, *Albatross;* 1 male, 1 female (9729). Lat. 28° 42' 00" N.; long. 86° 36' 00" W.; 280 fathoms; gy. M.; Mar. 14, 1885; station 2397, *Albatross;* 1 female (9734).

West of Cuba: Lat. 22° 35' 00" N.; long. 84° 23' 00" W.; 463 fathoms; wh. Co.; temp. 45° F.; Jan. 21, 1885; station 2352, *Albatross;* 1 female (9541).

Localities recorded.—

Off Frederickstadt, St. Croix Island, 451 fathoms (A. M. E.).

Off St. Lucia, 423 fathoms (A. M. E.).

South of Pernambuco, lat. 9° 5' to 10' S.; long. 34° 49' to 53' W., between 300 and 400 fathoms (Miers).

Genus PILUMNOPLAX Stimpson.

Pilumnoplax STIMPSON, Proc. Acad. Nat. Sci. Philadelphia. vol. 10, 1858, p. 93; type, *P. sulcatifrons* Stimpson.—ALCOCK, Journ. Asiat. Soc. Bengal, vol. 69, 1900, p. 311.

Carapace hexagonal, more or less depressed, a little broader than long, regions often faintly indicated. Fronto-orbital border two-thirds, or more, the greatest breadth of the carapace; the toothed antero-lateral borders are slightly arched or oblique.

Front straight, rather prominent, more or less confluent with the supra-orbital angles, often notched or grooved in the middle line. Supra-orbital border often with two fissures. The antennal flagellum, which is of good length, stands in the orbital hiatus. The antennules fold transversely or nearly so.

Buccal cavity widening a little anteriorly, almost closed by the maxillipeds.

Chelipeds more massive than the legs, which are slender, their dactyli compressed.

Abdomen in the female and commonly in the male seven-jointed; in the male the third segment covers the whole width of the sternum.

Tropical and South Atlantic (deep sea), Arabian Sea (deep), Japan, Hawaiian Islands, Fiji.

KEY TO THE AMERICAN SPECIES OF THE GENUS PILUMNOPLAX.

A^1. Carapace flat, regions scarcely indicated. No segments of male abdomen fused_____*americana*, p. 21.
A^2. Carapace convex in an antero-posterior direction, regions defined. Third, fourth, and fifth segments of male abdomen fused_____ _*elata*, p. 23.

PILUMNOPLAX AMERICANA Rathbun.

Pilumnoplax americanus RATHBUN, Bull. Lab. Nat. Hist. State Univ. Iowa, vol. 4, 1898, p. 283, pl. 7, figs. 1 and 2 (type-locality, off Georgia, 440 fathoms; male holotype, Cat. No. 19652, U.S.N.M.).

Pilumnoplax sinclairi ALCOCK and ANDERSON, Ann. Mag. Nat. Hist., ser. 7, vol. 3, 1899, p. 11 (type-locality, Travancore coast, 430 fathoms; type in Indian Mus.).—ALCOCK, Deep-Sea Brachyura *Investigator*, 1899, p. 74, pl. 3, figs. 1–1*a*.

Pilumnoplax americana ALCOCK, Deep-Sea Brachyura *Investigator*, 1899, p. 82; Journ. Asiat. Soc. Bengal, vol. 69, 1900, p. 311.

Diagnosis.—Carapace flat, regions scarcely indicated; edge of front bimarginate. Wrist armed with 2 teeth. Segments of male abdomen free.

Description.—Carapace a little more than three-fourths as long as broad, very finely granulate, naked, regions fairly indicated. Front as seen from above faintly notched, the truncate lobes sloping slightly backward from the middle; edge turned vertically down and deeply grooved from side to side.

FIG. 5.—PILUMNOPLAX AMERICANA, MALE, STATION G2, DORSAL VIEW, × 1½.

Antero-lateral border much shorter than the postero-lateral, thin and sharp and cut into four teeth, the first of which is lobiform and fused with the obtuse orbital tooth, the next two acute, the last one minute.

Eyes stout. Margin of orbit fissured above near the middle, deeply excavate above and below near the outer angle; inner lower angle dentiform and moderately prominent.

Chelipeds very unequal in both sexes, surface finely frosted, inner angle of wrist strongly prominent, with two acute teeth.

Legs narrow, unarmed, granulate above, very sparingly hairy; third pair the longest.

Measurements.—Male (46184), length of carapace 12, width of same 15.7 mm. Female, length of carapace 13, width of same 16 mm. (Alcock).

FIG. 6.—PILUMNOPLAX AMERICANA, MALE, STATION G2, ANTERIOR PORTION OF CARAPACE, VIEWED FROM BEFORE, × 3½.

Range.—Off Georgia and the Florida Keys; Arabian Sea. 70 to 440 fathoms.

Material examined.—

Off Georgia; lat. 30° 44′ 00″ N.; long. 79° 26′ 00″ W.; 440 fathoms; Co. crs. S. Sh. For.; temp. 45.6° F.; Apr. 1, 1885; station 2415, *Albatross;* 1 male holotype (19652); 2 males, 3 females (19653).

Off Sand Key, Florida; Sand Key Light bearing N. about 6 miles; 116 fathoms; June 19, 1893; station 28, State Univ. Iowa Exped.; 1 male (Mus. S. U. I.).

Off American Shoal, Florida; State Univ. Iowa Exped.: American Shoal Light bearing N. by W., 10 miles; about 100 fathoms; June 27,

1893; station 51; 1 male (Mus. S. U. I.). American Shoal Light bearing N. by W. ¼ W. about 10 miles; 105–110 fathoms; June 27, 1893; station 52; 1 male (20027); 2 males (Mus. S. U. I.). American Shoal Light bearing NE. by N., 8 miles; 70–80 fathoms; June 29, 1893; station 62; 1 male (Mus. S. U. I.).

Pourtales Plateau; lat. 24° 16′ 00″ N.; long. 81° 22′ 00″ W.; about 200 fathoms; June 27, 1893; station 56, State Univ. Iowa Exped.; 1 male (Mus. S. U. I.).

Gulf Stream, off Key West, Florida; lat. 24° 18′ 37″ N.; long. 81° 36′ 50″ W.; 127 fathoms; rky.; temp. 58° F.; Mar. 4, 1902; station 7301, *Fish Hawk;* 2 males (46184).

PILUMNOPLAX ELATA (A. Milne Edwards).

Plate 3.

Eucratoplax elata A. MILNE EDWARDS, Bull. Mus. Comp. Zoöl., vol. 8, 1880, p. 18 (type-locality, West Florida, 13 fathoms; holotype female in Paris Mus.).

Eucratoplax elata ? RATHBUN, Bull. Lab. Nat. Hist. State Univ. Iowa, vol. 4, 1898, p. 281.

Diagnosis.—Carapace convex, regions defined; edge of front thin. Wrist armed with one spine or tooth. Third, fourth, and fifth segments of male abdomen fused.

Description of male.—Carapace subquadrate, convex, punctate, granulate on the branchial, hepatic, and intestinal regions; regions well marked. Front deflexed, about two-fifths the width of the carapace; margin thin, divided into two very slightly convex, entire lobes. Orbit nearly as wide as half the front, with two short, closed fissures above and a large outer notch; edge granulate; outer angle obtuse, not prominent; upper margin nearly transverse. Antero-lateral margin shorter than the postero-lateral. Lateral teeth four besides the orbital; the first small, triangular; the second the broadest; the third the longest and most prominent, acuminate; the fourth postero-lateral and minute.

Chelipeds unequal, granulate; merus trigonal, thick, upper margin with a spine one-third the distance from the proximal end. Carpus with an anterior groove and sharp inner spine. Granulation of the smaller propodus coarser than that of the larger; in both there is a tendency to form longitudinal ridges. There is a broad truncate tooth on the propodus at the base of the dactylus in both chelae; in the larger, the dactylus has a large basal tooth; both fingers deeply grooved; immovable finger with a granulate ridge above the lower margin; dactylus with proximal portion of upper surface granulate. The brown color of the dactylus does not extend quite to the manus; the color of the pollex is even less extensive. Fingers not gaping.

Ambulatory legs slender; superior margin of meral, carpal, and propodal joints minutely spinulous.

Second segment of abdomen much narrower than the first and third, exposing the sternum; third segment with angular margins; third, fourth, and fifth segments coalesced; terminal segment with extremity rounded; proximal margin concave. Surface of abdomen smooth, of sternum granulate.

Description of female.—The postero-lateral margins are less convergent and the lateral teeth are three in number besides the postorbital tooth. The tooth corresponding to the first one in the male is obsolete; the next two are large and less outstanding than in the male; last tooth larger in female than in male. Chelipeds less unequal in female than in male.

Growth variation.—In the young, the first and last of the four lateral teeth of the carapace are obsolete. The merus of the cheliped bears a denticulated elevation instead of a spine, as in the adult.

Measurements.—Male (11407), length of carapace 7.6, width of same 10.5, width of front 4.1 mm. Female (19880), length of carapace 8.1, width of same 10.2, width of front 3.7 mm.

Range.—East and west Florida; 13 to 193 fathoms.

Material examined.—

Southwest of Cape San Blas, Florida; *Albatross:* Lat. 28° 41′ 00″ N.; long. 86° 07′ 00″ W.; 169 fathoms; gy. M.; Mar. 14, 1885; station 2400, 1 female (19879). Lat. 28° 38′ 30″ N.; long. 85° 52′ 30″ W.; 142 fathoms; gn. M. brk. Sh.; Mar. 14, 1885; station 2401; 1 female (19880). Lat. 28° 36′ 00″ N.; long. 85° 33′ 30″ W.; 111 fathoms; gy. M.; Mar. 14, 1885; station 2402; 1 y. (19881).

Off Sand Key, Florida; Sand Key Light bearing NW. by N., Key West Light N. ¼ E.; 50–60 fathoms; June 19, 1893; station 27, State Univ. Iowa Exped.; 1 male y. (Mus. S. U. I.).

Off Cape Florida; lat. 25° 40′ 00″ N.; long. 80° 00′ 00″ W.; 193 fathoms; gy. S.; temp. 43.4° F.; Apr. 9, 1886; station 2644, *Albatross;* 1 male (11407).

Subfamily GONEPLACINAE Miers.

Gonoplacinae MIERS, *Challenger* Rept., Zool., vol. 17, 1886, p. 237.—ALCOCK, Journ. Asiat. Soc. Bengal, vol. 69, 1900, pp. 286, 293, and 316.

The anterior border of the subquadrate carapace is entirely occupied by the square-cut front and orbits, the front being either narrow or of fair breadth, and the orbits being long, narrow trenches for the elongate eyestalks. In other respects similar to the Carcinoplacinae.

Represented in America by the genus *Goneplax*.

Genus GONEPLAX Leach.

Goneplax LEACH, Edinburgh Encyc., vol. 7, 1814. pp. 393 and 430; type, *Ocypode bispinosa* Lamarck, 1801=*Cancer rhomboides* Linnaeus, 1758 (on p. 393, spelled "*Goneplat*" by typographical error); Trans. Linn. Soc. London, vol. 11, 1815, pp. 309 and 323; Nouv. Dict. Sci. Nat., vol. 13, 1817, p. 295.

Gonoplax LEACH, Encyc. Brit., Suppl. to 4th, 5th, and 6th ed., vol. 1, 1816, p. 413; Malac. Pod. Brit., 1816, pl. 13, and explanation of plate.

Frevillea A. MILNE EDWARDS, Bull. Mus. Comp. Zoöl., vol. 8, 1880, p. 15; type, *F. barbata* A. Milne Edwards.

Carapace subquadrilateral, with the antero-lateral angles acute and the lateral borders posteriorly convergent, a good deal broader than long, moderately convex, regions faintly indicated.

The front and orbits occupy the whole anterior border of the carapace; front square cut, laminar, obliquely deflexed, occupying from a fourth to a third of the anterior border of the carapace, the rest being occupied by the trenchlike orbits.

Eyestalks typically long and slender, but sometimes strongly enlarged at the cornea; the antennules fold transversely; the antennae have a short basal joint and a slender flagellum of good length, standing in the orbital hiatus.

Buccal cavity square or anteriorly widened, well separated from the prominent epistome; efferent branchial channels not well defined. The merus of the outer maxillipeds is square and bears the palp at the antero-internal angle.

Chelipeds in both sexes much more massive than the legs, which are long and slender.

The abdomen in both sexes consists of 7 separate segments; in the male the third segment either covers or nearly covers the sternum between the last pair of legs.

Distributed along the east coast of America from Gulf of Mexico to Rio de Janeiro; in the northeast Atlantic and Mediterranean; and in the Indo-Pacific from the Persian Gulf to Japan.

KEY TO THE AMERICAN SPECIES OF THE GENUS GONEPLAX.

A^1. Carapace with 2 lateral teeth or spines.
  B^1. Sides of carapace strongly convergent posteriorly.
    C^1. Notch in upper margin of orbit. Oblique ridge on branchial region.
                                                              barbata, p. 26.
    C^2. No notch in upper margin of orbit. No oblique ridge on branchial region_____*sigsbei*, p. 26.
  B^2. Sides of carapace almost parallel.
    C^1. A tuberculiform swelling on margin of carapace between exorbital tooth and next lateral tooth_____*rosaea*, p. 27.
    C^2. No tuberculiform swelling on margin of carapace between exorbital tooth and next lateral tooth_____*hirsuta*, p. 28.
A^2. Carapace with 3 lateral teeth or spines_____*tridentata*, p. 29.

GONEPLAX BARBATA (A. Milne Edwards).

Plate 4, figs. 1 and 3; plate 5.

Frevillea barbata A. MILNE EDWARDS, Bull. Mus. Comp. Zoöl., vol. 8, 1880, p. 15 (type-locality, lat. 23° 13′ N.; long. 89° 16′ W., 84 fathoms, station 36, *Blake;* type in Paris Mus.).

Frevillea barbata ? RATHBUN, Bull. Lab. Nat. Hist. State Univ. Iowa, vol. 4, 1898, p. 287.

Diagnosis.—Two lateral spines. Orbital spine long. A spine on arm and wrist. A patch of hair on distal part of wrist and proximal part of palm.

Description.—Carapace somewhat uneven, smooth, an oblique ridge on the branchial region opposite the lateral spine. Orbital spine long, projecting laterally; next spine very small. Margin of front sinuous. a slight point at the middle. A notch in the upper margin of the orbit at its inner angle.

On the posterior border of the merus of the cheliped and distal to the middle there is a curved spine, which when the arm is flexed fits directly under the small lateral spine. A prominent curved spine at the inner angle of the carpus. Fingers nearly equal in size. A patch of very fine, thick, light yellow hair on the distal part of the wrist and the proximal part of the palm.

The first segment of the male abdomen is very short, scarcely visible; second very wide reaching the coxal joints and covering the sternum; third segment at its base equal in width to second.

Measurements.—Male, Grenada, length of carapace, 7; fronto-orbital width, 10.7; posterior width, 5.8; lateral margin 5.5 mm.

Range.—Gulf of Mexico to Grenada, West Indies.

Material examined.—

Gulf of Mexico: S. of Apalachicola, Florida; lat. 28° 45′ 00″ N.; long. 85° 02′ 00″ W.; 30 fathoms; gy. S. brk. Co.; Mar. 15, 1885; station 2405, *Albatross;* 1 male (46309).

Off Havana, Cuba; 1893; State Univ. Iowa Exped.; 1 young male (Mus. S. U. I.).

Off Grenada; lat. 12° 01′ 45″ N.; long. 61° 47′ 25″ W.; 92 fathoms; fne. S.; Mar. 1, 1879; station 262, *Blake;* 1 male (Cat. No. 4116, M. C. Z.).

Other records.—Yucatan Channel; lat. 23° 13′ 00″ N.; long. 89° 16′ 00″ W.; 84 fathoms; temp. 60° F.; station 36, U. S. C. S. Str. *Blake;* male holotype (Paris Mus.).

GONEPLAX SIGSBEI (A. Milne Edwards).

Plate 4, figs. 2 and 4.

Frevillea sigsbei A. MILNE EDWARDS, Bull. Mus. Comp. Zoöl., vol. 8, 1880. p. 16 (type-locality, Grenada, 92 fathoms, station 253, *Blake;* holotype in Paris Mus.).

Diagnosis.—Two lateral spines. Orbital spine not long. A tooth on arm and wrist.

Description.—Carapace less uneven than in *G. barbata*, a slight depression behind the orbit. Front and orbits with a distinct marginal line. No notch on upper margin of orbit; outer spine shorter than in *barbata*.

Arm projecting well beyond the carapace, a small tooth at about the middle of its outer margin. A small blunt tooth at inner angle of wrist. Propodus with a longitudinal groove on outer surface just above lower margin, and a more oblique and shorter groove on inner surface. Pollex much broader than dactylus. Fingers with shallow, irregular teeth.

Dactyli of first three pairs of legs slender and longer than their propodi. Last pair of legs with the last three joints broader than in the other legs, and broader than in *barbata*.

Measurements.—Adult female, station 254, length of carapace 8.2, fronto-orbital width 13, posterior width 6.3, length of lateral margin 7 mm.

Material examined.—Off Grenada; lat. 11° 27′ 00″ N.; long. 62° 11′ 00″ W.; 164 fathoms; S. Sh.; temp. 57° F.; Feb. 27, 1879; station 254, Str. *Blake;* 1 female (Cat. No. 4117, M. C. Z.).

Other records.—Off Grenada; lat. 11° 25′ 00″ N.; long. 62° 04′ 15″ W.; 92 fathoms; Co. brk. Sh.; temp. 58.5° F.; Feb. 27, 1879; station 253, Str. *Blake;* 1 female ovig., holotype (Paris Mus.).

GONEPLAX ROSAEA (A. Milne Edwards).

Frevillea rosœa A. MILNE EDWARDS, Bull. Mus. Comp. Zoöl., vol. 8, 1880, p. 15 (type-locality, St. Vincent, 88, correctly 87, fathoms, station 232, *Blake;* holotype in Paris Mus.).

Diagnosis.—Sides almost parallel. Two lateral spines or teeth.

Description.—Carapace thicker and narrower anteriorly than in *G. barbata*, the lateral margins being almost parallel. Front wider, its margin straighter. Eyestalks shorter and stouter. The postorbital angle is formed by a sharp tooth, behind which there is a small tuberculiform swelling followed by a short but sharp hepatic spine. Chelipeds and legs as in *G. barbata*.

Measurements.—Female holotype, length of carapace 15, width of same 20 mm.

Type-locality.—Off St. Vincent; lat. 13° 06′ 45″ N.; long. 61° 06′ 55″ W.; 87 fathoms; Co.; temp. 62° F.; Feb. 21, 1879; station 232, Str. *Blake;* female holotype in Paris Mus. (After A. Milne Edwards.)

GONEPLAX HIRSUTA Borradaile.

Goneplax hirsutus BORRADAILE, Brit. Antarctic ("*Terra Nova*") Exped., 1910, Zool., vol. 3, No. 2, 1916, p. 99, text-fig. 11 (type-locality, off Rio de Janeiro, 40 fathoms; types in Brit. Mus.).

Diagnosis.—A long and dense tuft of hair on distal half of wrist and base of palm. Chelipeds of male not much, if any, longer than legs. Carapace widest between tips of postorbital spines.

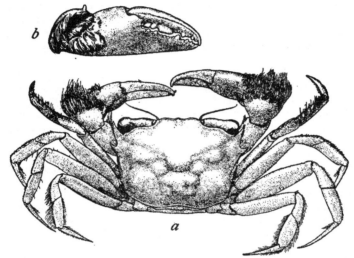

FIG. 7.—GONEPLAX HIRSUTA, COTYPE, × 2⅓. *a*, DORSAL VIEW; *b*, RIGHT CHELA AND CARPUS, OUTER VIEW. (AFTER BORRADAILE.)

Description.—Carapace about two-thirds as long as broad; its greatest width between tips of postorbital spines; its regions faintly marked except for a pronounced **H**-shaped depression in the middle; its sides converging backward from the sharp extraorbital spines, behind each of which, and nearer to it than in *G. rhomboides*, stands a smaller, very sharp spine. Front almost straight, with a shallow median notch, in which stands a rostral prominence. Orbital margin sinuous, sloping backward; width of orbit about equal to that of front.

Chelipeds almost equal, the right very slightly the larger; arm in female and (? young) male about two-thirds length of carapace, deep, with a spine a little beyond middle of upper edge; wrist about two-thirds length of arm, rather broader than long; hand longer than rest of limb; fingers about equal to palm, irregularly toothed, not gaping; a long and dense tuft of hair on outside of distal half of wrist and base of palm and a fringe of similar hairs along inner side of arm.

Legs slender, simple, fringed with hair, much like those of *G. rhomboides*, but without spine or merus. Abdomen of (? young) male narrow, like that of *G. maldivensis* Rathbun.[1] (After Borradaile.)

Size.—Length of carapace, 13 mm. (Borradaile).

Locality.—Off Rio de Janeiro, Brazil; lat. 22° 56′ S.; long. 41° 34′ W.; 73 meters (40 fathoms); May 2, 1913; *Terra Nova;* 1 male, 1 female (Brit. Mus.).

GONEPLAX TRIDENTATA (A. Milne Edwards).

Frevillea tridentata A. MILNE EDWARDS, Bull. Mus. Comp. Zoöl., vol. 8, 1880, p. 16 (type-locality, Barbados, "7½ to 50" fathoms, station 287, *Blake;* holotype in Paris Mus.).

Diagnosis.—Three lateral teeth. Two spines on wrist.

Description.—Antero-lateral teeth 3. Wrist armed with 2 spines, one inside, the other outside. No tufts of hair on the chelipeds. Dactyli of last pair of feet styliform.

Dimensions.—Female holotype, length of carapace 5, width of same 8 mm.

Type-locality.—Barbados; lat. 13° 11′ 25″ N.; long. 59° 38′ 20″ W.; 7 fathoms; Co. S. brk. Sh.; Mar. 8, 1879; station 287; Str. *Blake;* female holotype in Paris Mus.

Subfamily PRIONOPLACINAE Alcock.

Prionoplacinae ALCOCK, Journ. Asiat. Soc. Bengal, vol. 69, 1900, pp. 286 and 292.

Differs from the Carcinoplacinae in the form of the male abdomen, which is not broad enough at the third segment to cover all the space between the last pair of legs. Some of the genera approach the Goneplacinae in the form of the carapace which, however, does not attain its greatest breadth at the orbital angles.

Most of the American Goneplacidae are referable to this subfamily.

Genus PRIONOPLAX Milne Edwards.

Prionoplax MILNE EDWARDS, Ann. Sci. Nat., ser. 3, Zool., vol. 18, 1852, p. 163 [127]; type, *P. spinicarpus* Milne Edwards.

Carapace quadrilateral, transverse, lateral margins converging posteriorly, dentate. Front about one-third width of carapace. The orbits are deep trenches, which occupy the remaining width of the carapace. Eyestalks stout. Antennary flagellum elongate, situated in the orbital hiatus. Antennules transverse. Merus of outer maxilliped much shorter than the ischium, broader than long, antero-external angle prominent, a shallow notch at antero-internal angle. Chelipeds subequal, fingers flattened, pointed. Third segment of

---

[1] Bull. Mus. Comp. Zoöl., vol. 39, 1902, p. 124, plate, fig. 5.

abdomen of male not covering the space between the coxae of the last pair of legs. Third to fifth segments fused.

The type-species is said to have come from China. The other species are American and are analogous species on opposite sides of the continent: *atlantica* (Atlantic); *ciliata* (Pacific).

KEY TO THE AMERICAN SPECIES OF THE GENUS PRIONOPLAX.

A¹. Carapace with three lateral teeth_____*atlantica*, p. 30.
A². Carapace with four lateral teeth_____*ciliata*, p. 31.

PRIONOPLAX ATLANTICA Kendall.

Plate 6, figs. 1 and 2.

Prionoplax atlanticus KENDALL, Bull. U. S. Fish Comm., vol. 9, for 1889 (1891), p. 305 (type-locality, lat. 25° 23′ 00″ N.; long. 82° 43′ 00″ W., 23 fathoms; holotype male, Cat. No. 15272, U.S.N.M.).

Diagnosis.—Nearly naked. Three side teeth. Front, seen in front view, not projecting below the lower margin of the orbit. Spine at outer angle of wrist.

Description.—Carapace smooth and shining, very finely granulate along the sides, regions scarcely indicated; almost flat from side to side, convex fore and aft; sides strongly converging.

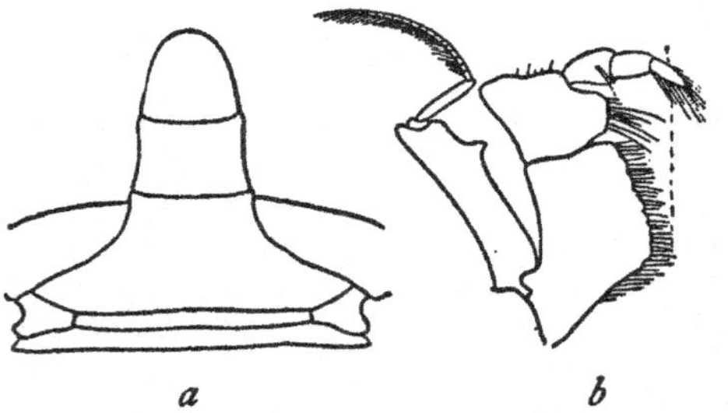

Front one-third of width of carapace, anterior margin bilobed, lobes convex, side margins slightly concave. Orbits sloping obliquely backward, upper margin slightly sinuous, lower margin deeply hollowed out below the cornea, a notch in the hollow. Eyestalks increasing in size distally, corneae very large with a ventral aspect.

FIG. 8.—PRIONOPLAX ATLANTICA, MALE HOLOTYPE. *a*, ABDOMEN AND PART OF STERNUM, × 8; *b*, OUTER MAXILLIPED, DOTTED LINE MARKING MEDIAN AXIS, × 12.

Antero-lateral margins armed with three sharp teeth, the first one forming the angle of the orbit, the second the largest and nearer the first than it is to the third.

Posterior border of the epistome very prominent.

Buccal cavity much wider in front than behind, maxillipeds with a rhomboidal gape. Merus joints granulate, ischial joints sparingly punctate.

Chelipeds not very unequal, and of moderate length, finely granulate. Merus with a tooth on upper margin near the middle. Carpus with an inner spine of good size and a small spine at outer angle. Propodus widening distally, upper and lower edges margined, upper one convex, fingers broad, grooved, fitting close together, prehensile teeth thin, broad, irregular, tips crossing, dactylus of larger chela with a strong basal tooth.

Ambulatory legs long and slender, slightly hairy, merus joints with low spinules above.

First segment of abdomen of male very wide, covering the sternum; second segment much shorter; third not reaching the coxae and completely fused with the fourth and fifth.

Measurements.—Male holotype, length of carapace 6, width of same 9.5, width of front 3.2 mm. Female, Tortugas, length of carapace 9, width of same 13.5 mm.

Range.—West and south coasts of Florida; 10 to 23 fathoms.

Material examined.—

West coast of Florida; lat. 25° 23′ 00″ N.; long. 82° 43′ 00″ W.; 23 fathoms; sft. gy. M.; temp. 66° F.; Feb. 26, 1889; station 5069, *Grampus;* 1 male holotype (15272).

Dry Tortugas, Florida; dredged; 1 female (Mus. S. U. I.).

Ship Channel, Key West, Florida; 10 fathoms; 1916; J. B. Henderson; 1 female (49649).

PRIONOPLAX CILIATA Smith.

Prionoplax spinicarpus STIMPSON, Ann. Lyc. Nat. Hist. New York, vol. 7, 1859, p. 59. Not *P. spinicarpus* Milne Edwards.

Prionoplax ciliatus SMITH, Trans. Connecticut Acad. Arts and Sci., vol. 2, 1870, p. 160 (type-locality, Panama; type in P.M.Y.U.).[1]

Diagnosis.—Margin hairy. Four side-teeth. Front, seen in front view, projecting below the lower margin of the orbit. No spine at outer angle of wrist.

Description.—Carapace very convex longitudinally, scarcely at all transversely. Surface thickly beset with small tuberculiform granules, space between the granules smooth and shining. Cervical suture indicated by a very distinct smooth sulcus, which is sharp and deep in the longitudinal portions in the middle of the carapace; branchial regions only indistinctly separated from the cardiac.

Front lamellar, very strongly deflexed and its edge divided into two prominent, rounded lobes, which, when seen in a front view, project below the inferior margins of the orbits.

Antero-lateral margin thin and divided by deep rounded sinuses into four slightly upturned lobes or teeth, of which the anterior, the hepatic, and the epibranchial are broad and truncate and their

---

[1] Unfortunately it is impossible to give illustrations of this and several other species, the types of which are said to be in the Peabody Museum, as a diligent search in 1916–1917 failed to locate the specimens.

truncated edges finely denticulated, while the posterior, or mesobranchial, is acutely pointed. Inferior lateral regions granulous, and along the lateral borders are clothed with long cilia which project beyond the margins. There are also some hairs along the lateral margins of the dorsal surface, easily removed.

Outer surface of external maxillipeds minutely granulous.

Chelipeds stout, slightly unequal. Merus armed with a spine on the posterior angle, near the distal extremity. Carpus roughened and with a long inner spine. Hands stout, compressed, perfectly smooth; upper edge angular, not crested; fingers compressed, deflexed, incurved, coarsely and irregularly toothed, not gaping. Legs slender, thickly hairy along the edges, especially on the dactyli, which are long, very slender, and cylindrical.

Sternum granulous. Abdomen of male smooth, first and third segments much wider than the second, penultimate much broader than long, its sides deeply concave.

Habits.—Probably live in holes, as the crabs have been found covered with ferrugineous mud. (Condensed from Smith.)

Measurements.—Length of carapace of male cotype 15.5, width of same 23.9 mm.

Range.—Panama (Stimpson, Smith). Guayaquil, Ecuador (Cano).

Genus TETRAPLAX Rathbun.

Tetraplax RATHBUN, Bull. U. S. Fish Comm., vol. 20, for 1900, pt. 2, 1901, p. 9; type, *T. quadridentata* (Rathbun).

Carapace quadrilateral; lateral margins dentate, somewhat converging posteriorly. Front about one-third the width of the carapace, margin nearly straight. Orbital margins nearly transverse. Eyestalks elongate and of moderate thickness. Antennae entering the orbit. Buccal cavity wider in front than behind; maxillipeds not completely filling it, outer angle not produced, inner angle slightly notched.

Chelipeds unequal, heavy, angular; fingers pointed. Ambulatory legs long, slender, compressed; dactyls of last pair concave upward and outward.

Abdominal segments in male narrower than the sternum; third to fifth segments fused.

Contains only one species.

TETRAPLAX QUADRIDENTATA (Rathbun).

Plate 6, figs. 3 and 4.

Frevillea quadridentata RATHBUN, Bull. Lab. Nat. Hist. State Univ. Iowa, vol. 4, 1898, p. 287, pl. 8, fig. 1 (type-locality, Curaçao; holotype female, Cat. No. 19974, U.S.N.M.)

Tetraplax quadridentata RATHBUN, Bull. U. S. Fish Comm., vol. 20, for 1900, pt. 2, 1901, p. 9.

Diagnosis.—Quadrilateral. Orbits elongate. Merus of maxillipeds not broader than long. First segment of male abdomen not covering the sternum; third segment not reaching the coxae.

Description.—Carapace thick, convex in both directions, about three-fourths as long as wide, covered with a short, dark-colored pubescence; when this is removed, the regions can be made out. Front about one-third width of carapace, deflexed, edge thin, from above appearing slightly emarginate. Superior margin of orbit sloping outward and slightly backward to outer orbital tooth; a notch toward inner end, and another next outer tooth; inferior margin with a notch next the outer tooth and a tooth at inner angle between which and the angle of the front there is a triangular sinus the inner side of which is formed by the outer margin of the peduncle of the antenna. Antero-lateral margin with 4 teeth, their outer edges finely denticulate; the last three teeth are very nearly in a line parallel to the median line; orbital tooth less projecting.

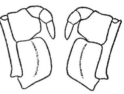

FIG. 9.—TETRAPLAX QUADRIDENTATA, MALE (24564), OUTER MAXILLIPEDS, SHOWING APPROXIMATION, × 8.

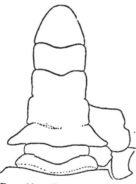

FIG. 10.—TETRAPLAX QUADRIDENTATA, MALE (24564), ABDOMEN, PART OF STERNUM, AND COXA OF LAST LEG, × 6].

Postero-lateral margins longer than antero-lateral, and moderately converging. Chelipeds subequal. Merus with a spine on superior margin at distal third. Carpus with an inner spine and an anterior fringe of hair. Hand smooth and shining, finely granulate above, upper margin acute, very finely granulate and fringed with hair. Lower outer margin of pollex with a granulated costa. Dactylus with two superior granulated costae and a superior fringe of hair. First segment of male abdomen narrower than the third. The sutures between the third and fourth, and fourth and fifth segments are partially indicated. Terminal segment rounded at extremity.

Measurements.—Male (24564), length of carapace 8.3, width of same 10.3 mm.

Range.—Cuba; Porto Rico; Curaçao. Four and one-half to 12 fathoms.

Material examined.—

Cabañas, Cuba; 2–25 fathoms; S. Sh. Grass to M.; June 8–9, 1914· Henderson and Bartsch, *Tomas Barrera* Exped; 1 female (49165).

Porto Rico: San Juan Harbor; NW. angle Morro Castle ⅜ mile; 4½–5¼ fathoms; S. M.; temp. 25.2° C.; Jan. 16, 1899; station 6054; *Fish Hawk;* 2 males, 1 female (24564). Mayaguez Harbor; Custom House E. by S. ¼ S. 2¼ miles; 12 fathoms; stky. M.; temp. 27° C.; Jan. 19, 1899; station 6060, *Fish Hawk;* 1 male (24565).

Curaçao; *Albatross;* 1 female holotype (19974).

Genus EURYPLAX Stimpson.

Euryplax STIMPSON, Ann. Lyc. Nat. Hist. New York, vol. 7, 1859, p. 60; type, *E. nitida* Stimpson.

Carapace transverse, broad, hexagonal; antero-lateral margin very short, dentate. Front nearly half as broad as the carapace. Ocular peduncles of moderate length. Antennal flagellum excluded from the orbit by an internal suborbital lobe which joins the front.

Chelipeds heavy, not very unequal.

Sternum partially exposed in the sinus of the abdomen between its second and third segments. All the segments of the abdomen distinct.

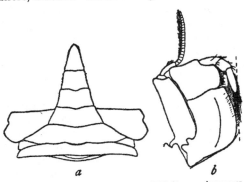

FIG. 11.—EURYPLAX NITIDA, MALE (15012). *a*, ABDOMEN AND PART OF STERNUM, × 3½; *b*, OUTER MAXILLIPED, DOTTED LINE MARKING MEDIAN AXIS, × 5½.

First segment narrow and little developed; second segment widest, covering the sternum.

An American genus. Analogous species on opposite sides of the continent: *nitida* (Atlantic); *polita* (Pacific).

KEY TO THE SPECIES OF THE GENUS EURYPLAX.

A¹. Antero-lateral margins converging anteriorly_____*nitida,* p. 34.
A². Antero-lateral margins parallel_____*polita,* p. 36.

EURYPLAX NITIDA Stimpson.

Plate 7.

Euryplax nitida STIMPSON, Ann. Lyc. Nat. Hist. New York, vol. 7, 1859, p. 60 (type-locality, Florida Keys; type not extant).—RATHBUN, Bull. U. S. Fish Comm., vol. 20, for 1900, pt. 2, 1901, p. 8.

Diagnosis.—Antero-lateral margins converging anteriorly. Arm of male with a deep pit. Fingers white at tips.

Material examined of Euryplax id.

Locality.	Lat. N. ° ' "	Long. W. ° ' "	Fath.	Bottom.	Temp.	Date.	Sta.	Collector.	Specimens.	Cat. M.
New Orleans, Louisiana								H. Hemphill	1 ♂	M.C.Z. 15012
Off Keys, Florida								Fish Hawk	1	45820
Off Key Section, Florida	28 59 15	83 32 30	9¼	dk. grassy	17° C.	Dec. 1863	7175	...do...	1 ♂	45822
Do	28 57 30	82 58 00	3	r-ky	15.5° C.	Nov. 27, 1901	7207	...do...	1 ♀	45821
Off Key Section, Florida	28 50 15	83 23 15	10	R. Co. Sh.	17° C.	Dec. 9, 1901	7187	...do...	1 ♂	45823
Do	28 19 45	83 06 30	8¼	cy. G.	13° C.	Nov. 28, 1901	7210	...do...	1 ♀	45823
Highland Section, Florida	28 08 00	82 57 00	5¾	sdy. brk. Sh.	13° C.	Jan. 21, 1902	7229	...do...	1 ♂	43948
Off Charlotte Harbor, Flori d	27 43 30	82 52 30	7½	hd. brk. Sh.	15° C.	Jan. 23, 1'02	7262	...do...	1 ♀	45824
	26 47 30	83 25 15	28	(fn. st. S. bk. (Sp. brk. Sh.		Jan. 29, 1902				
id, Florida						Mar. 18, 1885	2410	*Albatross*	1 ♂	17897
Florida Bay	25 09 52	81 21 53	3½	ry. S. Sh.	23.5° C.	Dec. 17, 1902	7352	H. Hemphill	1 ♀ ovig.	17523
Off W. Ch n nl, Florida	24 44 50	81 53 38	10	hrd. S.	19.5° C.	Feb. 24, 1902	7289	*Fish Hawk*	1 ♂ 1 ♀ y.	487
Key M, Florida								...do...	1 ♀	45825
Pine Key, Florida						Dec. 1883		H. Hemp Ill.	1 ♂	6367
Hawk Ch st, lBrida	½ m. SF. by S. of S.E. end of hk Key.	14 feet	rky			Jan. 27, 1903	7429	*Fish Hawk*	1 ♀	15099
Do	1½ m. E. of Tea Table Key.	2½	barry		Feb. 19, 1903	7466	...do...	1 ♂	45915
Off rd, Florida	Pt. Mula I. thse SSW. ½ W. 5¾ mi.	30						J. B. Henderson.	1 ♀	466
Off Vieques Id, Porto Rico		14	Co. S. Sh.	25.6° C.	Feb. 8, 1899	6085	*Fish Hawk*	1 ♂	47117
									2	63

¹Last hal tah not developed.

Description.—Carapace smooth and shining, convex fore and aft, as well as from side to side; antero-lateral margin less than half as long as the postero-lateral, and armed with three strong teeth, including the angle of the orbit; carapace widest at the third tooth. Front deeply notched on each side at the insertion of the antennæ; interantennal margin nearly straight.

Chelipeds in the male with a deep round pit at anterior distal corner of lower surface of merus; this pit is surrounded by a fringe of long hair; a sharp curved spine near the distal end of the upper surface of the merus. Carpus with a sharp inner spine; inner surface pilose. Ambulatory legs slender.

Carapace of female narrower. Chelipeds more nearly equal. Merus without pit and surrounding hair.

Color.—Distal half of fingers white.

Measurements.—Male (15012), length of carapace 15, width of same 24.9 mm.

Range.—Florida Keys, Gulf of Mexico and West Indies. Shallow water to 49 fathoms.

Material examined.—See page 35.

Localities recorded.—Key West, 2 to 5 fathoms, and off Elbow Reef, 49 fathoms (Stimpson); Sarasota Bay (Kingsley); Egmont Key (Smith); New Orleans (M. C. Z.); St. Thomas (Stimpson).

EURYPLAX POLITA Smith.

Euryplax politus SMITH, Trans. Connecticut Acad. Arts and Sci., vol. 2, 1870, p. 163 (type-locality, Panama; type in P.M.Y.U.).

Diagnosis.—Antero-lateral margins parallel. Arm of male without a deep pit. Fingers brown at tips.

Description (after Smith).—Carapace glabrous, convex longitudinally and very slightly transversely. Dorsal surface not distinctly areolated, although the cervical suture can be traced by a slight depression. Front nearly straight, a distinct marginal groove on the upper edge, and each side deeply notched at the insertion of the antennae. Antero-lateral margins parallel, very short, armed with three acute teeth. Postero-lateral margins slightly incurved. Posterior margin slightly concave in the middle.

Chelipeds nearly equal, stout, smooth, and glabrous. Merus armed with a small spiniform tooth; carpus with a small tooth within. Hands slightly swollen, superior margins high, but smooth and rounded; fingers slender, slightly deflexed. Legs smooth, nearly naked, very slender.

Appendages of first segment of male abdomen not so strongly curved at the tips as in *E. nitida*, and the terminal portion brown instead of black.

Color.—Fingers brown at tip.

Measurements.—Male holotype, length of carapace 6.9, width of same 11.2, width of front 4.4 mm.

Range.—Known only from Panama.

Genus CHASMOPHORA Rathbun.

Chasmophora RATHBUN, Proc. U. S. Nat. Mus., vol. 47, 1914, p. 119; type, *C. macrophthalma* (Rathbun).

Carapace very broad, subcylindrical, very convex longitudinally, much less so transversely; antero-lateral margin short, dentate. Fronto-orbital border about four-fifths the width of the carapace. Front separated from the orbital angle by a furrow. Eyes stout, filling the orbit; lower margin of orbit with a large outer sinus. Basal joint of antennae not reaching the front, flagellum standing in the orbital hiatus. Buccal cavity widening a little anteriorly; merus of maxillipeds with outer distal angle prominent.

Right cheliped of medium size (left not known). Legs slender. First segment of abdomen of female very broad, but not covering the whole width of the sternum; third segment narrower. It is probable that in the male these segments have a similar relation to the sternum.

Near *Euryplax*, from which it is separated by the antennae entering the orbit.

Contains only one species.

CHASMOPHORA MACROPHTHALMA (Rathbun).

Eucratopsis macrophthalma RATHBUN, Proc. U. S. Nat. Mus., vol. 21, 1898, p. 601, pl. 43, figs. 3 and 4 (type-locality, Panama Bay, 51½ fathoms; holotype female, Cat. No. 21591, U.S.N.M.).

Chasmophora macrophthalma RATHBUN, Proc. U. S. Nat. Mus., vol. 47, 1914, p. 119, text-fig. 2.

Diagnosis.—Antennae entering the orbit. Buccal cavity anteriorly widened. Eyes large, filling orbit. Carapace broad, subcylindrical.

Description.—Length three-fifths of breadth. Regions plainly marked. Surface nearly level transversely for the middle two-thirds, deflexed toward the margins; teeth directed obliquely upward. Front about one-third the width of the carapace, lobes slightly convex. Superior margin of orbit sloping backward and outward. Three large subequal antero-lateral teeth, including the orbital tooth, acute and curved slightly forward; behind the last a minute tooth or notch.

The arm is granulated toward the margins, which are unarmed. Carpus finely granulated, with an anterior submarginal sulcus and a short blunt inner tooth which is continued inferiorly as a blunt prominence. Lower margin of propodus of cheliped a little sinuous. The palmar portion is slightly margined above and below; fingers broad and flat, not gaping, with a narrow granulate border on the outer edges; prehensile margins crenate, with a slightly larger lobe near the base of the dactylus, and a three-lobed prominence at the base of the pollex.

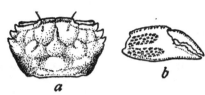

Fig. 12.—CHASMOPHORA MACROPHTHALMA, FEMALE HOLOTYPE. a, CARAPACE, EYES AND ANTENNAE, DORSAL VIEW, × 5¼; b, RIGHT CHELA, OUTER VIEW, × 7¼.

Color.—In alcohol, the carapace is marked with patches of dark blue.

Measurements.— Female holotype, length of carapace 3, width of same 5.1 mm.

Material examined.—Panama Bay; lat. 7° 56′ 00″ N.; long. 79° 41′ 30″ W.; 51½ fathoms; gn. M.; March 30, 1888; station 2805, *Albatross;* 1 female holotype (21591).

Genus SPEOCARCINUS Stimpson.

Speocarcinus STIMPSON, Ann. Lyc. Nat. Hist. New York, vol. 7, 1859, p. 58; type, *S. carolinensis* Stimpson.

Carapace subcylindrical. Fronto-orbital width three-fifths to two-thirds of the entire width. Antero-lateral margin arcuate, dentate, teeth small; postero-lateral margins parallel or nearly so. Eyestalks small, filling the orbits. Antennules transverse. Basal joint of antenna scarcely reaching the front, the next joint standing in the orbital hiatus. Epistome narrow except at the middle. Buccal cavity strongly widened anteriorly, maxillipeds gaping, the merus as long as broad, its antero-external angle prominent.

Fig. 13.—CHASMOPHORA MACROPHTHALMA, ANTERIOR VIEW OF ORBIT OF FEMALE HOLOTYPE, SHOWING HIATUS IN WHICH LIES THE ANTENNAL FLAGELLUM, × ABOUT 13.

Chelipeds moderately unequal, much stronger than the legs

Third pair of legs the longest. Dactyli sharp, those of last pair curved upward.

Abdomen of male much narrower at its base than the sternum; third to fifth segments more or less fused.

Restricted to America.

Analogous species on opposite sides of the continent: *carolinensis* (Atlantic); *granulimanus* (Pacific).

KEY TO THE SPECIES OF THE GENUS SPEOCARCINUS.

A^1. Antero-lateral teeth 5, including the orbital angle. Chelipeds nearly smooth.
carolinensis, p. 39.
A^2. Antero-lateral teeth less than 5.
  B^1. Antero-lateral teeth 4.
    C^1. Antero-lateral teeth small, separated by slight notches. Lobes of front obliquely truncate. Chelae coarsely granulate___*granulimanus*. p. 40.
    C^2. Antero-lateral teeth of good size and well separated. Lobes of front arcuate. Chelae almost smooth_____*ostrearicola*, p. 41.
  B^2. Antero-lateral teeth 3. Chelipeds nearly smooth_____*californiensis*, p. 42.

SPEOCARCINUS CAROLINENSIS Stimpson.

Plate 8; plate 159, fig. 6.

Speocarcinus carolinensis STIMPSON, Ann. Lyc. Nat. Hist. New York, vol. 7, 1859, p. 59, pl. 1, figs. 1-3 (type-locality, Charleston Harbor, S. C.; type not extant).—RATHBUN, Bull. U. S. Fish Comm., vol. 20, for 1900, pt. 2, 1901, p. 11, text-fig. 2.

Diagnosis.—Eyestalks constricted. Antero-lateral teeth five. Chelipeds nearly smooth.

Description.—Carapace nearly smooth, punctate, obscurely granulate near the margins, pubescent. Gastric region and its subdivisions well defined. Antero-lateral margin 5-toothed, including the angle of the orbit; second tooth rounded and not always distinctly separated from the first; last three teeth sharp and moderately prominent. Postero-lateral margins parallel. Front about one-fourth width of carapace, edge nearly straight, a little sinuous, and with a median emargination. Eyestalk constricted next the cornea. Maxilliped with antero-external angle of merus produced outwardly.

Chelipeds robust, nearly smooth, margins hairy. A strong, sharp spine or tooth near summit of merus. Inner margin of carpus granulated, with a blunt tooth at inner angle. Outer surface of hand glabrous, microscopically granulated. Dactylus of larger chela with a stout tooth at base. Legs with hairy margins.

Measurements.—Male, Charleston, length of carapace 23.2, width of same 28.5 mm.

Habits.—" This crab lives in the subterranean galleries excavated in the mud at low-water mark by the *Squilla, Callianassa,* and other Crustacea, or by large worms." (Stimpson.)

Range.—South Carolina to the West Indies. To a depth of 76 fathoms.

Material examined.—

Charleston Harbor, South Carolina (M. C. Z.).

Dry Tortugas, Florida; dredged; 1893; State Univ. Iowa Exped.; 1 male (22300).

West Channel entrance to Key West, Florida: 7.25 fathoms; co. S. brk. Sh.; temp. 20.2° C.; Feb. 13, 1902; station 7272, *Fish Hawk;* 1 female (45951).

Mayaguez Harbor, Porto Rico: Black buoy entrance, N. by W. ½ W. ¼ mile; 12–18 fathoms; S. M.; temp. 26° C.; Jan. 20, 1899; station 6061, *Fish Hawk;* 1 male (24560). Point del Algarrobo E. 2¾ miles; 75–76 fathoms; rky. S. Co.; temp. 68.5° F.; Jan. 20, 1899; station 6063, *Fish Hawk;* 1 male, 2 young (24561).

Off Porto Real, Porto Rico: Point Guaniquilla S. ¼ E. 2 miles; 8½ fathoms; Co. S.; temp. 26° C.; station 6074, *Fish Hawk;* 1 male, 1 female (23766).

St. Thomas; 1 male (Copenhagen Mus.).

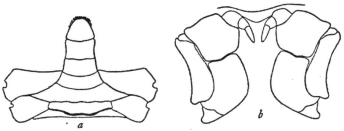

Fig. 14.—SPEOCARCINUS GRANULIMANUS, MALE HOLOTYPE. *a*, ABDOMEN AND PART OF STERNUM, × 3; *b*, OUTER MAXILLIPEDS, SHOWING APPROXIMATION, × 5½.

SPEOCARCINUS GRANULIMANUS Rathbun.

Plate 9.

Speocarcinus granulimanus RATHBUN, Proc. U. S. Nat. Mus., vol. 16, 1893, p. 242 (type-locality, lat. 30° 21′ 00″ N.; long. 114° 25′ 15″ W., 30 fathoms; holotype male, Cat. No. 17461, U.S.N.M.).

Diagnosis.—Eyestalks tapering. Antero-lateral teeth four. Chelipeds granulate.

Description.—Carapace with large punctae, distinctly granulate near the margins, regions well marked. Antero-lateral margin four-toothed, second tooth indicated by a faint notch, third more distinct, fourth acute. Front more than one-fourth width of carapace, lobes sloping a little backward toward the middle, a well-defined median notch. Eyestalks flat above, tapering to the tip.

Fig. 15.—SPEOCARCINUS GRANULIMANUS, MALE HOLOTYPE, RIGHT CHELA, OUTER VIEW, × 2½.

Cheliped s granulate; a sharp spine near summit of merus; a blunt tooth on carpus. The coarser granulation of the hand is arranged in longitudinal rows. No large basal tooth on dactylus of larger hand.
Otherwise as in *S. carolinensis*.

Measurements. — Male holotype, length of carapace 17, width of same 21 mm.

Range.—Pacific and Gulf coasts of Lower California; 23–33 fathoms.

SPEOCARCINUS OSTREARICOLA Rathbun.

Plate 10, fig. 1.

Speocarcinus ostrearicola RATHBUN, Proc. U. S. Nat. Mus., vol. 38, 1910, p. 545, pl. 48, fig. 2 Capon), Peru; holotype male, Cat. No. 40469, (type-locality, oyster beds of Matapalo (near U.S.N.M.).

Diagnosis.—Eyestalks constricted. Antero-lateral teeth four, well-marked. Chelipeds nearly smooth.

Description.—Body and legs coarsely hairy; carapace very broad, coarsely granulate; regions fairly well marked. Antero-lateral margin cut by broad V-shaped notches into four well-marked teeth. Front one-fourth as wide as carapace, having a deep median furrow; edge bilobed, lobes arcuate. Eyestalks distally slender, corneae somewhat enlarged.

Chelipeds unequal, broad, nearly smooth; a tooth on upper edge of arm and inner edge of wrist; palm high, upper margin granulate; fingers narrowly gaping, inner margins irregularly dentate; upper margin of dactylus granulate, immovable finger with a raised, granulate ridge just above lower edge.

Measurements.—Male holotype, length of carapace 12, width of same 17.6 mm.

Range.—Known only from the type-specimen, from the oyster beds of Matapalo (near Capon), Peru; Jan. 23, 1908; R. E. Coker; 1 male, received from the Peruvian Government (40469).

SPEOCARCINUS CALIFORNIENSIS (Lockington).

Plate 10, figs. 2 and 3.

Eucrate? californiensis LOCKINGTON, Proc. California Acad. Sci., vol. 7, 1876 (1877), p. 33 (type-locality, San Diego; type not extant).
Speocarcinus californiensis RATHBUN, Harriman Alaska Exped., vol. 10, 1904, p. 190, pl. 9, fig. 1.

Diagnosis.—Eyestalks tapering. Antero-lateral teeth 3. Chelipeds nearly smooth.

Description.—Regions deeply limited, surface smooth and punctate, lateral margins granulate. Antero-lateral margin upturned, tridentate, last tooth acute. Postero-lateral margins slightly converging behind. Front with straight edge, slightly emarginate, more than one-fourth width of carapace. Eyestalk flat above, and tapering to the tip, longer than in *granulimanus*. Antero-external angle of merus of maxilliped not produced outwardly, but the outer margin is thickened, somewhat revolute.

Chelipeds nearly smooth except on the margins, which are granulous. Merus and carpus each with a stout tooth. Dactylus of larger chela without a strong basal tooth. Legs fringed with hair.

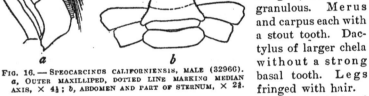

FIG. 16.—SPEOCARCINUS CALIFORNIENSIS, MALE (32966). *a*, OUTER MAXILLIPED, DOTTED LINE MARKING MEDIAN AXIS, × 4½; *b*, ABDOMEN AND PART OF STERNUM, × 2¾.

Measurements.—Male (45581), length of carapace 16, width of same 22.6 mm.

Range.—San Pedro to San Diego, California.

Material examined.—

San Pedro: H. N. Lowe; 1 male (19731); 1 male (32966). West Basin; Venice Marine Biol. Station; 1 male (45581).

Alamitos Bay; H. N. Lowe; 1 male (19731).

Genus PSEUDORHOMBILA Milne Edwards.

Pseudorhombila MILNE EDWARDS, Hist. Nat. Crust., vol. 2, 1837, p. 58; type, *P. quadridentata* (Latreille).

Carapace thick, much broader than long, convex fore and aft, very slightly so from side to side; regions partially indicated. Fronto-orbital border about half the width of the carapace. Antero-lateral borders arched, dentate. Front rather square-cut, notched in the middle, between one-third and one-fourth the width of the carapace. Orbit with two sutures above. Basal antennal joint short, the flagellum standing in the orbital hiatus. The antennules fold transversely. Buccal cavern widening anteriorly; the outer angle of the merus of the maxillipeds is prominent.

Chelipeds unequal, much more massive than the legs. Legs narrow, dactyli styliform. In the male the third to fifth segments of the abdomen are fused, and the third segment does not reach the coxae of the last pair of legs.

Locality of type-species unknown.

PSEUDORHOMBILA OCTODENTATA Rathbun.

Plate 14, fig. 3.

Pseudorhombila octodentata RATHBUN, Proc. Biol. Soc. Washington, vol. 19, 1906, p. 91 (type-locality, Dominica; holotype male, Cat. No. 32690, U.S.N.M.).

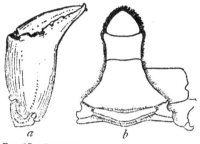

FIG. 17.—PSEUDORHOMBILA OCTODENTATA, MALE HOLOTYPE. *a*, RIGHT CHELA, OUTER VIEW, × 1; *b*, ABDOMEN, PART OF STERNUM, AND COXA OF LAST LEG, × 1½.

Diagnosis.— Antero-lateral margins arcuate. Wrist subrectangular. Abdomen of male with third to fifth segments fused and not covering the sternum at the third segment.

Description.—Carapace very convex fore and aft, regions indistinctly defined, surface closely set with flattened granules. Front subtruncate, a V-shaped median notch, a rounded lobe at outer angle. Antero-lateral teeth four (orbital angle excluded); the first small, separated from the orbit by a long, straight interval; second tooth widest; third and fourth most acute, the third the larger; the fourth the most projecting.

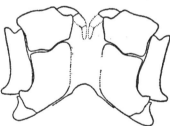

FIG. 18.—PSEUDORHOMBILA OCTODENTATA, MALE HOLOTYPE, OUTER MAXILLIPEDS, SHOWING APPROXIMATION, × 3.

Chelipeds (the right one at least) strong, covered with fine reticulated granulation; merus projecting little beyond the body, a strong subterminal tooth above; carpus subquadrate, with a conspicuous tooth at inner angle, and the outermost portion tuberculate; palm nearly twice as long as high, widening distally; dactylus as long as palm; both fingers strongly deflexed, not gaping, tips curved and overlapping.

Ambulatory legs long and narrow; meral joints granulate above and below, carpal joints above; some small superior spines on the merus. Dactyli with two fringes of long hair.

The second segment of the abdomen leaves exposed a large piece of the sternum on either side; third to fifth segments fused.

In the female there are only two antero-lateral teeth, and the merus of the legs is not spined above but granulous.

Measurements.—Male holotype, length of carapace, 33.3; width of same, 46.1; fronto-orbital width, 24.7; width of front, 12.9; length of propodus of right cheliped 39; length of merus of third ambulatory leg, 26.5 mm.

Range.—Dominica; Martinique. To 100 fathoms.

Material examined.—

Soufriere Bay, Dominica; 100 fathoms; A. H. Verrill; 1 male holotype (32690).

Martinique; 1 male, 2 females (Paris Mus.).

Genus OEDIPLAX Rathbun.

Œdiplax RATHBUN, Proc. U. S. Nat Mus., vol. 16, 1893, p. 241; type, *O. granulatus* Rathbun.

Nearly related to *Pseudorhombila*, but more xanthoid in aspect, the antero-lateral margin being longer and the posterior convergence

 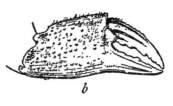

FIG. 10.—OEDIPLAX GRANULATA, FEMALE HOLOTYPE. *a*, ANTENNAL AND BUCCAL REGION. × 2½; *b*, RIGHT CHELA, OUTER VIEW, × 1½.

greater, and the wrist not obliquely quadrilateral. The regions are well separated, the front is less than one-fourth as wide as the carapace, and the buccal cavern does not widen anteriorly. The male is not known, but it is probable that, as in the female, the third segment of the abdomen does not cover the sternum between the last pair of legs.

Only the type species is known.

OEDIPLAX GRANULATA Rathbun.

Plate 13, figs. 1 and 2.

Œdiplax granulatus RATHBUN, Proc. U. S. Nat. Mus., vol. 16, 1893, p. 242 (type-locality, Gulf of California, 33 fathoms; holotype female, Cat. No. 17465, U.S.N.M.).

Diagnosis.—Antero-lateral margins strongly dentate. Wrist not subrectangular. Abdomen of male not covering sternum at third segment.

Description.—Surface coarsely granulate, regions well marked. The fronto-orbital distance is a little less than half the width of the carapace. The front has a large median notch, the two lobes are sinuous, with a distinct but nonprojecting outer tooth. The orbits trend distinctly forward and outward, their inner upper angle is separated by a furrow from the front, the lower angle is acute and prominent, there are two upper fissures and a large V-shaped sinus below the dentiform outer angle. The arcuate antero-lateral margin is upturned and is armed with four strong denticulated teeth, the first of which is the smallest and rather distant from the orbit; below this sinus is a stout subhepatic spinule. Carapace widest at the last tooth.

Chelipeds of female moderately unequal, rough, with large sharp granules arranged partly in rows. A subdistal spine on upper border of merus; a stout inner carpal spine, with a small spine below it. Dactyl granulate above for half its length, fingers narrowly gaping in the large claw, a low, stout tooth at base of dactylus. Legs hairy; merus joints spinulous above.

Dimensions.—Female holotype, length of carapace 32.5, width of same 46 mm.

Material examined.—Gulf of California; off Consag Rock, Lower California; lat. 31° 06′ 45″ N.; long. 114° 28′ 15″ W.; 33 fathoms; bn. M.; temp. 63.8° F.; Mar. 27, 1889; station 3031, *Albatross;* 2 females (1 is holotype) (17465).

Genus CYRTOPLAX Rathbun.

Cyrtoplax RATHBUN, Proc. U. S. Nat. Mus., vol. 47, 1914, p. 118; type, *C. spinidentata* (Benedict).

Carapace much broader than long, convex longitudinally and transversely. Regions well marked. Antero-lateral margins arcuate, dentate; postero-lateral converging. Fronto-orbital width three-fifths of width of carapace. Front advanced, lobes convex, separated only by a furrow from the orbital margin. Eyes rather slender. Basal antennal joint rather broad, inner angle just touching the front; flagellum standing in the orbital hiatus. Buccal cavity widening distally. Maxillipeds gaping; merus broader than long.

Chelipeds stout, unequal; wrists subtriangular in dorsal aspect, bispinose; palms high, finger strongly deflexed. Legs long and slender; dactylus of last pair recurved.

Neither the first nor the third abdominal segment in the male covers the sternum; the third to the fifth segments are fused.

Contains only one species.

CYRTOPLAX SPINIDENTATA (Benedict).

Plate 11.

Eucratoplax spinidentata BENEDICT, Johns Hopkins Univ. Circ., No. 11, 1892, p. 77 (type-locality, Jamaica; holotype male, Cat. No. 17219, U.S.N.M.).

Eucratopsis spinidentata RATHBUN, Bull. U. S. Fish Comm., vol. 20, for 1900, pt. 2, 1901, p. 11.

Cyrtoplax spinidentata RATHBUN, Proc. U. S. Nat. Mus., vol. 47, 1914, p. 119, pl. 2.

Diagnosis.—Hexagonal, very broad. Eyes narrowing distally. Wrist two-spined. Buccal cavity widening anteriorly. Legs slender. Third abdominal segment narrower than the sternum.

Description.—Carapace more convex longitudinally than transversely; lateral teeth upturned. Front between a third and a fourth of the width of the carapace, deflexed, lobes slightly convex, median notch V-shaped. The orbits trend forward and outward, and there is a shallow lobe between the two upper fissures; below the blunt outer angle, a large sinus; the feeble inner tooth is situated below the middle of the eyestalk which is narrow except at the base and fits snugly in the orbit. Of the five antero-lateral teeth, the second is lobiform and separated from the orbital angle by a shallow sinus; third tooth more or less truncate; fourth and fifth spiniform, subequal; carapace widest at the last tooth. Postero-lateral margins moderately convergent.

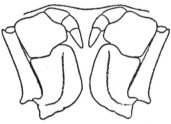

FIG. 20.—CYRTOPLAX SPINIDENTATA, MALE HOLOTYPE, OUTER MAXILLIPEDS, SHOWING APPROXIMATION, ×6.

Arm provided with a strong superior subterminal tooth. Wrist with a deep sulcus parallel to the margin next the palm, and two stout spines at the inner angle, one below the other, the upper one the longer. The palms are convex above and below, and the upper margin is granulate. The fingers are slender, much deflexed and curved inward; the dactyli are granulate above at the proximal end; the fingers are irregularly toothed along their prehensile edges which gape narrowly. The meral and carpal joints of the legs are hairy along the anterior margins, the propodi along both margins; the dactyli have four fringes of hair.

The first segment of the male abdomen is wider at its proximal than at its distal end, where it is the same width as the second segment; the third segment forms narrow but obtuse angles at its widest part.

Dimensions.—Male holotype, length of carapace, 14.6; width of same, 21 mm.

Range.—West Indies.

Material examined.

Jamaica: Kingston Harbor; T. H. Morgan; 1 male holotype (17219). Montego Bay; from sponges and algae in brackish pond; July 2, 1910; C. B. Wilson; 1 female ovig. (42935).

Porto Rico; *Fish Hawk:* Porto Real; Jan. 26, 1899; 1 male, 3 females (24554). Boqueron Bay; Jan. 27, 1899; 1 male, 2 females (24555).

St. Thomas; 1872; Krebs, coll.; 1 specimen (Copenhagen Mus.).

Trinidad; Jan. 30 to Feb. 2, 1884; *Albatross;* 2 males (7637).

Monos Island, Trinidad; Jan. 30 to Feb. 2, 1884; *Albatross;* 1 male (19651).

Genus PANOPLAX Stimpson.

Panoplax STIMPSON, Bull. Mus. Comp. Zoöl., vol. 2, 1871, p. 151; type, *P. depressa* Stimpson.

Carapace arcuate anteriorly, quadrate posteriorly; depressed, much broader than long, regions distinct. Fronto-orbital width about two-thirds of the entire width. Frontal lobes convex, well separated from the orbital angles. Antero-lateral margin dentate. Eyes of moderate size. Basal antennal joint just touching the front, flagellum standing in the orbital hiatus. Antennules transverse. Buccal cavity with sides parallel. Merus of maxillipeds broader than long. Chelipeds stout, unequal; wrists obliquely subquadrate; palms high.

The third to fifth segments of the abdomen of the male are fused, the third does not reach the coxae of the last pair of legs, while the first covers the whole width of the sternum.

Inclines toward those xanthoid genera of which *Panopeus* is the type. Contains but one species.

PANOPLAX DEPRESSA Stimpson.

Plate 12, figs. 1 and 2.

Panoplax depressa STIMPSON, Bull. Mus. Comp. Zoöl., vol. 2, 1871, p. 151 (type-localities, East and Middle Keys, Tortugas, 5 to 7 fathoms; type not extant).

Diagnosis.—Carapace depressed. Fronto-orbital width about two-thirds of entire width. Abdomen of male with third to fifth segments fused and not covering sternum at third segment.

Description.—Carapace depressed, anterior third inclined downward, regions fairly well marked, surface finely punctate, granulate along the lateral teeth; two oblique epigastric lobes. Front with two rounded lobes. Superior orbital margin with two notches; orbit as wide as each frontal lobe. Lateral teeth five, including orbital; the second shallow, separated from the first by a shallow sinus; the third large, blunt, outer margin very arcuate; fourth triangular, with a spiniform tip; carapace widest at this point; fifth tooth very small, not projecting beyond the general outline. Postero-lateral margins moderately converging behind.

Merus and carpus of chelipeds granulate toward the margins; merus dentate on upper margin; carpus oblong, with a stout blunt tooth at the inner angle and a few tubercles below it; an anterior transverse groove. Hands smooth and rounded, punctate; fingers dark brown, color not extending to palm; prehensile teeth broad, low, a large one at base of dactylus; no gape. In the female the

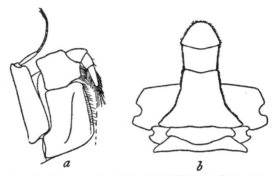

FIG. 21.—PANOPLAX DEPRESSA, MALE (25624). *a*, OUTER MAXILLIPED, ×9; *b*, ABDOMEN AND PART OF STERNUM, × 6.

upper surface of the manus is flattened, with traces of two longitudinal carinae and an intermediate furrow. Merus joints of legs roughened on anterior margins, following joints hairy.

Measurements.—Male (25624), length of carapace 8.6, width of same 13.5 mm.

Range.—West coast of Florida to Yucatan Channel and Porto Rico. To a depth of 28 fathoms.

Material examined.—See page 49.

Genus GLYPTOPLAX Smith.

Glyptoplax SMITH, Trans. Connecticut Acad. Arts and Sci., vol. 2, 1870, p. 164; type, *G. pugnax* Smith.

Carapace narrow, hexagonal, xanthoid in form, deeply areolated. Front prominent, well separated from the orbital angle. Lateral margin 4- to 5-dentate. Orbits and antennae much as in *Panoplax*. Buccal cavity widening anteriorly.

THE GRAPSOID CRABS OF AMERICA.

Material examined of Panoplax *

Locality.	Lat. N.	Long. W.	Fath.	Bottom.	Temp.	Date.	Sta.	Collector.	Specimens.	Cat. No.
Off Charlotte Harbor, Florida	26 33 00	83 10 00	28	sdy...	66° F	Apr. 2, 1901	7123	Fish Hawk	4 ♂	25624.
West Florida	26 33 30	83 15 30	27	fa. s. bk. Sp.		Mar. 18, 1885	2111	Albatross	1 ♀	263.
Do	26 18 30	83 08 45	27	fne. gy. S. bk. Sp. brk. Sh.		Mar. 19, 1885	2412	...do	2 ♂	63.
Do	26 00 00	82 57 30	24	nd. S. bk. Sp. brk. Sh.		...do...	2413	...do	1 ♀	263.
D	25 04 30	82 59 15	26	fn. wh. S. brk. Sh.		...do...	2414	...do	1 ♂ 1 ♀	92.
Off Cape San Juan, ...	22 18 00	87 04 00	24	wh. R. Co.		Jan. 30, 1885	2365	...do	1 ♂ y.	1 263.
San Antonio, Cuba		Gem House, E. ½	5					...do	1 ♂	1 268.
Mez Harbor, Porto Rico		N., 4½ m.	4-6	Co.		Jan. 20, 1899	6065	Fish Hawk	1 ♂	63.
Off Pt. de Molinas, Porto Rico		Cabo Rojo Lt. House SSE., 5¼ m.	7½	Co. S. shelly	26° C	Jan. 25, 1899	6072	...do	1 ♀ y.	21557.
Off Mio, Porto Rico		Hucares NW., ¾ W., ½ m.	9½	Co.	...do	Feb. 14, 1899	6099	...do	4 ♂ 3 y.	21558.
Ensenada Honda, Culebra, Porto Rico						Feb. 10, 1899		...do	1 y.	339.

Chelipeds short and stout. Dorsal aspect of carpus subtriangular. Manus more or less crested above, the upper proximal angle often very prominent.

Legs slender and smooth.

First segment of abdomen covers the sternum; the second is much shorter, exposing the sternum; the third does not reach the coxae of the fifth pair of legs; third, fourth and fifth segments coalesced.

Restricted to America. Analogous species on opposite sides of the continent: *smithii* (Atlantic); *pugnax* (Pacific).

KEY TO THE SPECIES OF THE GENUS GLYPTOPLAX

A^1. Lateral margin with five teeth, orbital angle included_____*pugnax*, p. 50.
A^2. Lateral margin with four teeth, orbital angle included_____*smithii*, p. 51.[1]

GLYPTOPLAX PUGNAX Smith.

Plate 158, figs. 1–6.

Glyptoplax pugnax SMITH, Trans. Connecticut Acad. Arts and Sci., vol. 2, 1870, p. 165 (type-locality, Panama; type in. P.M.Y.U.).—A. MILNE EDWARDS, Crust. Rég. Mex., 1880, p. 335, pl. 61, figs. 5–5*f*.

Diagnosis.—Five side teeth, the last one large. No lobe on upper edge of palm.

Description.—Carapace slightly convex longitudinally, not at all transversely; thickly granulous. Mesogastric lobe separated from the prominent protogastric lobes. The epigastric lobes are slight elevations separated by a marked median sulcus. Hepatic region prominent, set off by deep sulci. Mesobranchial and metabranchial lobes separated by a very slight sulcus, while the anterior portion of the branchial region is divided into three lobules, one at the base of the epibranchial tooth, a large one just within this, and a small indistinct one next the gastro-cardiac sulcus.

Front thin, horizontal, edge slightly convex, a small median notch. Inner angle of orbit a prominent tooth; two deep notches on upper border. Antero-lateral margins arcuate; the outer angle of the orbit projects only slightly beyond the second tooth and is separated from it by a slight sinus. In addition, there are three prominent triangular teeth, of which the middle one is most prominent.

Chelipeds slightly unequal, hands very large. Merus not projecting beyond carapace. Carpus short, outer face granulous, a slight groove along the margin next the propodus, a tooth on inner margin, and a small tubercle near articulation of propodus. Hand compressed, very broad, and nearly smooth, lower edge rounded, upper edges lightly crested. Pollex very broad at base, much deflexed, lower

---

[1] *Glyptoplax pusilla* (Rathbun), 1899.—*Micropanope pusilla* A. Milne Edwards, 1880, I have placed in the Family Xanthidae.

edge slightly marginate outside, tip slender, upturned. *D*actylus long and slender, upper edge slightly crested, tip hooked by the tip of the propodus. Prehensile edges sharp, slightly dentate, not gaping, or very slightly so. Fingers black.

Legs slender, minutely granulous. Dactyli clothed with very short hair, posterior pair considerably shorter than the others, propodi slightly hairy behind.

Sexual variation.—The females are more convex than the males and the front is less prominent and slightly deflexed.

Measurements.—Male, Panama, length of carapace 8.6, width of same 12.1 mm. (After Smith).

Range.—Costa Rica; Panama (Smith).

Material examined.—Punta Arenas, Costa Rica; specimens in Copenhagen Museum.

GLYPTOPLAX SMITHII A. Milne Edwards.

Plate 13, figs. 3 and 4; plate 158, figs. 7–10.

Glyptoplax smithii A. MILNE EDWARDS, Crust. Rég. Mex., 1880, p. 336, pl. 61, figs. 4–4*d* (type-locality, reefs west of Florida, 13 fathoms; type in M.C.Z.)

Diagnosis.—Four side teeth, the last one small. A lobe on upper edge of palm.

Description.—Closely related to the preceding. The carapace is slightly convex transversely, the lateral teeth upturned. The lobes of the front are truncated and slope a little backward and outward, except at the prominent and blunt outer angle. Lateral teeth four, the normal second tooth of the antero-lateral border being obsolete; the two teeth next the orbital are large, the first the larger, the second the most prominent; the last tooth is very small.

The wrists are somewhat nodulous as well as granulous. On the proximal half of the upper surface of the palm there is a lobe projecting inward. The immovable finger of the smaller propodus is slightly deflexed, but not

Material examined of Glyptoplax smithii.

Locality.	Lat. N.	Long. W.	Fath.	Bottom.	Date.	Station.	Collector.	Specimens.	Cat. No.
Off Cape Hatteras, North Carolina	° ′ 35 20	° ′ 74 45	27	crs. gy. S.	Oct. 20, 1884	2296	Albatross	2 ♂	18445
Off Cape San Blas, Florida	29 15	85 29	27	Co.	Feb. 7, 1885	2372	do	1 ♂	32721
Do	30 14	85 29	25	Co.	do	2373	do	1 ♂	18270
Do	24 45	85 02	30	gy. S. br. Co.	Mar. 15, 1885	2405	do	2 ♂	19196
Off Cape Catoche, Yucatan	22 18	87 04	24	wh. R. Co.	Jan. 30, 1885	2365	do	1 ♂ 4 ♀	18209

that of the larger; the latter has a large tooth at its base. Fingers light brown.

Otherwise much as in *G. pugnax*.

Measurements.—Male (18269), length of carapace 4.3, width of same 5.7 mm.

Range.—From Cape Hatteras, North Carolina, to Gulf of Mexico and Yucatan Channel; 13 to 30 fathoms.

Material examined.—See page 51.

Genus EUCRATOPSIS Smith.

Eucratopsis SMITH, Amer. Jour. Sci., vol. 48. Nov., 1869, p. 391; type, *E. crassimanus* (Dana); Trans. Connecticut Acad. Arts and Sci., vol. 2, 1869, p. 35 (dated "August," but issue destroyed by fire before distribution, and not reprinted until November).

Eucratoplax A. MILNE EDWARDS, Bull. Mus. Comp. Zoöl., vol. 8, 1880, p. 17; type, *E. guttata* A. MILNE EDWARDS=*crassimanus* (Dana).

Carapace narrow, panopeoid; regions plainly marked. Fronto-orbital distance from three-fifths to two-thirds of the width of the carapace. Front prominent, separated by a furrow from the slight orbital angle. Eyes stout; the three fissures of the orbit distinct. Antero-lateral margin five-toothed, orbital tooth included; postero-lateral margins moderately convergent. The basal antennal joint reaches a prolongation of the front; the flagellum stands in the orbital hiatus. The buccal cavity widens somewhat anteriorly; the merus of the maxillipeds is not much broader than long.

Chelipeds very heavy, subequal; palms very high, their upper proximal angle protuberant. Legs narrow, the first and last pairs noticeably shorter than the others. The first segment of the male abdomen reaches across the sternum, the third segment does not; the latter is fused with the two following segments.

Only the type-species included.

EUCRATOPSIS CRASSIMANUS (Dana).

Plate 12, fig. 3; plate 159, figs. 1 and 2.

Eucrate crassimanus DANA, Proc. Acad. Nat. Sci. Philadelphia, vol. 5, 1851 (1852), p. 248 (type-locality, Rio Janeiro?; Cat. No. 2332, U.S.N.M.); U. S. Expl. Exped., vol. 13, Crust., pt. 1, 1852, p. 311; atlas, 1855, pl. 19, figs. 2a–2d.

Eucratopsis crassimanus SMITH, Amer. Jour. Sci., vol. 48, 1869, p. 391; Trans. Connecticut Acad. Arts and Sci., vol. 2, 1869, p. 35.

Eucratoplax guttata A. MILNE EDWARDS, Bull. Mus. Comp. Zoöl., vol. 8, 1880, p. 17 (type-locality, Sombrero; type in Paris Mus.).

Diagnosis.—Carapace narrow, hexagonal; five antero-lateral teeth, including orbital tooth. Chelipeds very heavy. Abdomen of male not reaching the coxae of the last pair of legs.

Description.—Carapace about three-fourths as long as broad, convex in both directions, antero-lateral teeth upturned. Regions deeply separated; a transverse granulate ridge on the hepatic and on the epibranchial regions; some granular prominences on the posterolateral regions. Front prominent, two-sevenths as wide as the carapace, convex, a narrow median notch. The second lateral tooth is low and rounded and fused with the first tooth; the three last teeth are strong and prominent, the third the broadest, the fifth the most acute.

The chelipeds are very stout; the merus is as high as it is long and has a stout subdistal spine above; the upper surface of the wrist is as broad as it is long, with a blunt, stumpy inner tooth, and a groove along the anterior margin; anterior outer angle pronounced. Palms a little unequal, the lower portion of the outer surface flattened, upper aspect broad, sloping, the proximal angle of the upper margin forming a prominent lobe; the lower margin of the propodus is lightly sinuous, the pollex very broad at its base and tapering rapidly to the tip; dactylus much bent down; both digits irregularly toothed, leaving a narrow gape when closed.

FIG. 22.—EUCRATOPSIS CRASSIMANUS, FEMALE, (17801), OUTER MAXILLIPED, × 8.

Measurements.—Male holotype, length of carapace 21.4, width of same 27.9 mm.

Variations.—The type is much larger than any other specimens examined; half-grown specimens have a wider and less prominent front, and a smoother carapace which is wider in its posterior part.

Range.—From west coast of Florida to Rio de Janeiro, Brazil. Shallow water to 7¼ fathoms.

Material examined.—

Marco, Florida; H. Hemphill; 1 female ovig. (17801); 1 female (18273).

West Channel, entrance to Key West; Feb. 13, 1902; *Fish Hawk*. Beacon 2 to Sand Key Light, 48° 20′ 00″ to Key West Light, 77° 07′ 00″; 7¼ fathoms; Co. S.; temp. 20.2° C.; station 7272; 2 males, 1 female ovig. (45952). Beacon 2 to Sand Key Light, 40° 11′ 00″ to Key West Light, 119° 47′ 00″; 7½ fathoms; Co. S. br. Sh.; temp. 20.2° C.; station 7273; 4 males, 6 females (3 ovig.) (45950). Key West Light to N. W. Passage Light, 67° 35′ 00″ to Beacon 3, 103° 20′ 00″; 6¾ fathoms; Co. S. brk. Sh.; temp. 19.8° C.; station 7274; 1 male, 1 female ovig. (45949).

Jamaica (P. W. Jarvis collection).

Paqueta, Bay of Bahia, Brazil; 3–4 fathoms; 1876–1877; R. Rathbun; 1 female ovig. (16267).

Rio de Janeiro?; U. S. Exploring Expedition; 2 males (1 is holotype) (2332).

Jurujuba Bay, Rio de Janeiro; 1876–1877; R. Rathbun; 1 male (16268).

Other records.—Sombrero Key, Florida (A. M. Edw.); Port of Silam, Yucatan (Ives).

Subfamily RHIZOPINAE Miers.

Rhizopidae STIMPSON, Proc. Acad. Nat. Sci. Phila., vol. 10, 1858, p. 95 [41].
Rhizopinae MIERS, Challenger Rept., Zool., vol. 17, 1886, p. 223.—ALCOCK, Journ. Asiat. Soc. Bengal, vol. 69, 1900, pp. 287, 293, and 318.

Eyestalks often fixed, corneae small or obsolete; the lower border of the orbit has a tendency to run downwards toward the epistome. The carapace usually has its antero-lateral corners cut away and rounded off; the front may be square-cut and broad, or narrow and more or less distinctly bilobed and deflexed. The antennules may be of fair size and transversely folded, but more often, owing to the narrowness of the front, they are cramped and fold obliquely; sometimes they can not be folded in their fossae at all. Antennal flagella usually short. The epistome may either be well defined and prominent, or ill defined and sunken. The buccal cavern may be squarish, but often diminished in breadth anteriorly; the external maxillipeds have a square or suboval merus and may completely close the buccal cavern, or there may be a gap between them. The male abdomen does not nearly cover the space between the last pair of legs. Male openings sternal.

Only one genus, and that atypical, is found in American waters.

Genus CHASMOCARCINUS Rathbun.

Chasmocarcinus RATHBUN, Bull. Lab. Nat. Hist. State Univ. Iowa, vol. 4, 1898, p. 284; type, *C. typicus* Rathbun.

Carapace thick, broadest posteriorly, tapering anteriorly to the orbit and without lateral teeth. Fronto-orbital margin about half the width of the carapace. Front narrow, bifid. Orbits small, deep, outer angle not prominent. Eyestalks small, tapering, movable. The antennular cavities are entirely filled by the large basal joint. The basal joint of the antennae is short and does not nearly reach the front; flagellum of fair length. Inner orbital hiatus very broad. Epistome projecting below the maxillipeds. Buccal cavity with parallel sides. Maxillipeds standing wide apart, the merus as long as broad, suboval, shorter than the ischium. Chelipeds unequal, fingers long and slender, third pair of legs longest, dactylus of last pair recurved. Abdomen of male much narrower at base than the sternum; third to fifth segments fused.

Restricted to America and the Philippine Islands. Analagous species on opposite sides of the continent: *typicus* (Atlantic); *latipes* (Pacific).

KEY TO THE AMERICAN SPECIES OF THE GENUS CHASMOCARCINUS.

A^1. Orbits transverse in dorsal view. Carapace strongly narrowed anteriorly.
typicus, p. 55.
A^2. Orbits oblique in dorsal view.
   B^1. Orbits directed obliquely forward and outward. Merus joints of legs broad_____*latipes*, p. 57.
   B^2. Orbits directed obliquely backward and outward. Merus joints of legs narrow.
      C^1. Carapace narrow, not much broader than long_____*obliquus*. p. 58.
      C^2. Carapace broad, much broader than long_____*cylindricus*, p. 59.

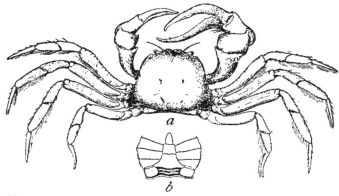

FIG. 23.—CHASMOCARCINUS TYPICUS, MALE HOLOTYPE. *a*, DORSAL VIEW, × 2; *b*, ABDOMEN AND STERNUM, FLATTENED, × 2.

CHASMOCARCINUS TYPICUS Rathbun.

Chasmocarcinus typicus RATHBUN, Bull. Lab. Nat. Hist. State Univ. Iowa, vol. 4, 1898, p. 285, pl. 7, figs. 3–5 (type-locality, north of Trinidad, 31 to 34 fathoms, stations 2121 and 2122, *Albatross;* male holotype, Cat. No. 6901, U.S.N.M.).

Diagnosis.—Orbits transverse. An antero-lateral marginal line. Sternum and abdomen granulate.

Description.—Carapace a little more than three-fourths as long as broad. Upper surface convex, rounding gradually downward into the nearly perpendicular lateral surfaces. Anterior third strongly deflexed. Lateral outline as seen from above convex and slightly emarginate at its middle where a sulcus arises which extends obliquely downward and forward joining the inferior horizontal suture. The anterior half of the side margin is marked by a row of fine granules. Cardiac region and posterior part of gastric region well marked. Surface finely granulate. On the front there is a short shallow median sulcus, margin transverse, faintly bilobed, outer corners rounded, depressed. Orbits transverse; the orbital margin incloses two areas, a deep socket in which the eye fits closely, and a shallow outer and inferior area. Second and third joints of the antennules very long. Epistome rather deep.

Chelipeds punctate, for the most part smooth. In the male the carpus has a tooth at the inner angle, blunt in the larger or right cheliped, acute in the smaller. Palms short and broad, lower margin very convex; fingers very long and slender, strongly bent downward; tips acute and bent toward each other. Dactylus of larger cheliped shorter than the pollex and very thick at base, as seen from above; prehensile margin tuberculate and very hairy. Prehensile margin of pollex denticulate and with a sinus at its base forming a slight gape. Fingers of smaller cheliped slightly bent, not at all gaping, prehensile edges very finely denticulate, with a larger tooth at the base of the dactylus. The chelipeds of the female are more nearly equal than those of the male, the right or larger resembling the left in character, the carpus having a sharp spine and the fingers very slightly bent and not gaping.

FIG. 24.—CHASMOCARCINUS TYPICUS, MALE (22073), OUTER MAXILLIPEDS, SHOWING APPROXIMATION, × 2½.

Ambulatory legs somewhat flattened, hairy; dactyli with a fringe of hair on opposite sides.

Abdomen and sternum granulate. In the male the penultimate sternal segment has a supplementary overlying plate along its posterior margin; in the female the posterior segment overlaps considerably the preceding.

Measurements.—Male holotype, length of carapace 7.5, width of same 9.8 mm.

Range.—Off Trinidad and Cape Frio, Brazil. 31 to 59 fathoms.

Material examined.—

Off Trinidad; lat. 10° 37′ 40″ to 10° 37′ 00″ N.; long. 61° 42′ 40″ to 61° 44′ 22″ W.; 31–34 fathoms; dk. slate-col. M.; temp. 67–73° F.; Feb. 3, 1884; stations 2121–2122, *Albatross;* 25 males and females (6901).

Off Cape Frio, Brazil; lat. 23° 08′ 00″ S.; long. 41° 34′ 00″ W.; 59 fathoms; bu. M.; temp. 57.1° F.; Dec. 30, 1887; station 2762, *Albatross;* 3 males, 1 female (22073).

CHASMOCARCINUS LATIPES Rathbun.

Chasmocarcinus latipes RATHBUN, Proc. U. S. Nat. Mus., vol. 21, 1898, p. 602, pl. 43, fig. 5 (type-locality, Magdalena Bay, Lower California, Mexico, 51 fathoms, station 2833, *Albatross;* holotype female, Cat. No. 21592, U.S.N.M.).

Diagnosis.—Orbits directed obliquely forward and outward. An antero-lateral marginal line. Sternum and abdomen of female smooth and punctate. Ambulatory legs wide.

FIG. 25.—CHASMOCARCINUS LATIPES, FEMALE HOLOTYPE, DORSAL VIEW, × 1⅛.

Description.—Carapace two-thirds as long as broad. Fronto-orbital distance less than one-half the width of the carapace. Surface covered with large punctae which tend to coalesce; granulate on the posterior half and toward the margins. Two very deep longitudinal impressed lines in the center of the carapace. A distinct,

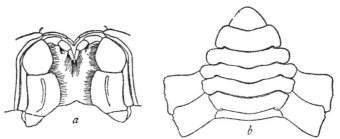

FIG. 26.—CHASMOCARCINUS LATIPES, FEMALE HOLOTYPE. *a*, OUTER MAXILLIPEDS, IN PLACE, × 5; *b*, ABDOMEN AND PART OF STERNUM, × 3.

though blunt antero-lateral margin. Front very faintly emarginate. Eye not fitting closely in the orbit; upper margin of orbit concave, directed obliquely forward and outward.

Chelipeds similar to those of *C. typicus.* Carpus, however, nearly square, propodus less arched than in *typicus,* fingers tapering regularly to the tips. A line of granules on lower margin of propodus. Margins of chelipeds fringed with hair.

Legs shorter and broader than in *C. typicus*, hairy.

Sternum and abdomen of female smooth, punctate.

Color.—In alcohol the carapace is a bluish-gray, chelipeds pale pink, ventral side of crab and ambulatory legs rust-colored.

Measurements.—Female holotype, length of carapace 12.5, width of same 17.8 mm.

Material examined.—Known only from the type-specimen from Magdalena Bay, Lower California; lat. 24° 38' 00" N.; long. 112° 17' 30" W.; 51 fathoms; gn. M.; May 2, 1888; station 2833, Str. *Albatross;* 1 female (21592).

CHASMOCARCINUS OBLIQUUS Rathbun.

Plate 14, figs. 1 and 2.

Chasmocarcinus obliquus RATHBUN, Bull. Lab. Nat. Hist. State Univ. Iowa, vol. 4, 1898, p. 286, pl. 7, fig. 6 (type-locality, southeast of Andros Island, Bahamas, 97 fathoms, station 2651, *Albatross;* holotype male, Cat. No. 20309, U.S.N.M.).

Diagnosis.—Orbits directed obliquely backward and outward. Carapace very narrow; no antero-lateral marginal line. Sternum and abdomen smooth.

Description.—Carapace narrow, nearly as long as broad, without distinct lateral margin. Surface smooth, finely and obscurely punctate. Median notch of front shallow. Superior margin of orbit

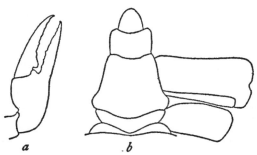

FIG. 27.—CHASMOCARCINUS OBLIQUUS, MALE HOLOTYPE. *a*, RIGHT CHELA, OUTER VIEW, × 8; *b*, ABDOMEN AND PART OF STERNUM, × 14.

directed obliquely backward and outward, outer end rounded, eye filling the orbit.

Chelipeds nearly equal in the immature male. Lower margin of merus armed with spines; a long slender spine on carpus; fingers slightly deflexed, not gaping, prehensile edges finely denticulate; dactylus of right chela with a strong basal tooth.

Legs slenderer, more cylindrical than in *C. typicus;* last two joints fringed with long hair.

Sternum and abdomen smooth; an overlying plate along posterior margin of fourth sternal segment.

Measurements.—Male holotype, length of carapace 5.3, width of same 5.9 mm.

Material examined.—Known only from the type-specimen from southeast of Andros Island, Bahamas, in Tongue of Ocean; lat. 24° 02′ 00″ N.; long. 77° 12′ 45″ W.; 97 fathoms; wh. Oz.; temp. 73.4° F.; Apr. 13, 1886; station 2651, Str. *Albatross;* 1 male (20509).

CHASMOCARCINUS CYLINDRICUS Rathbun.

Chasmocarcinus cylindricus RATHBUN, Bull. U. S. Fish Comm., vol. 20, for 1900. pt. 2, 1901, p. 10, text-fig. 1 (type-locality, Mayagüez Harbor, Porto Rico, 12 to 18 fathoms, station 6061, *Fish Hawk;* holotype male, Cat. No. 23765, U.S.N.M.).

Diagnosis.—Orbits directed obliquely forward and outward. Carapace broad, little narrowed in front; an antero-lateral marginal line. Sternum and abdomen granulate.

Description.—Carapace about two-thirds as long as broad, but wider anteriorly than in *C. latipes.* Fronto-orbital width about half width of carapace. Body subcylindrical, almost level from side to side, the frontal portion more strongly deflexed than in the other species. Anterior half of lateral margin marked by a sharp granulated line. Sutures visible about cardiac region. Surface pubescent and granulate. Front a little wider than the orbit; margin subtruncate, upper surface with a median depression, which in a dorsal view makes the margin appear bifid. Upper margin of orbit concave, directed obliquely outward and a little forward.

Material examined of Chasmocarcinus cylindricus.

Locality.	Position.	Fath.	Bottom.	Temp.	Date.	Sta.	Collector.	Specimens.	Cat. M.
Montego Bay Harbor, Porto Rico	Custom House, E. ¼ M.	7	stky. M.	27° C.	Jan. 19, 1899	6059	F. A. Andrews. *Fish Hawk.*	1 ♂ 1 ♀	42942. 21551.
Do	N. Black dy entrance, N by W. ¼ W. ¼ m.	12-18	S. M.	26° C.	Jan. 20, 1899	6061	...do...	4 ♂ 2 ♀	23765.
Do	Pt. del Algarrobo, E., 23 m.	75-76	ky. S. Co.	68.5° F.	...do...	6063	...do...	2 ♂ 2 ♀	24552.
Do	Pt. del Algarrobo, E., 4⅓ m.	161-172	M. S.	23° C.	...do...	6066	...do...	2 ♂ 2 ♀	21553.

Right cheliped larger than left. Carpus with inner angle rounded, and without a tooth. Manus of larger cheliped about one-half wider than that of the smaller cheliped in the male, and much more

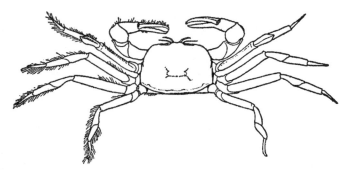

FIG. 28.—CHASMOCARCINUS CYLINDRICUS, FEMALE (24551), DORSAL VIEW, × 2½.

swollen, its fingers gaping at base; in the female the hands are more nearly equal and the fingers do not gape.

Legs narrow, fringed with hair.

Penultimate sternal segment of male without a posterior supplementary plate.

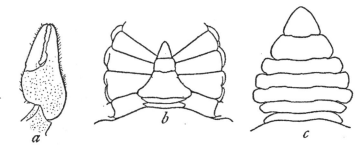

FIG. 29.—CHASMOCARCINUS CYLINDRICUS. a, RIGHT CHELA OF MALE HOLOTYPE, × 6; b, ABDOMEN AND STERNUM OF MALE HOLOTYPE, × 6½; c, ABDOMEN OF FEMALE (24551), × 6½.

Measurements.—Male holotype, length of carapace 4.6, width of same 6.5 mm. Female (24551), length of carapace 6.7, width of same 9.5 mm.

Range.—Jamaica and Porto Rico; 7 to 172 fathoms.

Material examined.—See page 59.

Family PINNOTHERIDAE Dana.

Pinnotheridea DE HAAN, Fauna Japon., Crust., 1833, p. 5; 1835, p. 34 (part).
Pinnotheridae DANA, Amer. Journ. Sci., ser. 2, vol. 2, 1851, p. 289; U. S. Expl. Exped., vol. 13, Crust., pt. 1, 1852, p. 378 (part).
Pinnotherinae MILNE EDWARDS, Ann. Sci. Nat., Zool., ser. 3, vol. 18, 1852, p. 139 [103]; vol. 20, 1853, p. 216 [182].

Carapace often more or less membranaceous, antero-lateral margins entire or very slightly dentate. Front, orbits and eyestalks very small, corneae sometimes obsolescent. Buccal cavity usually very wide, often semicircular in outline. The merus of external maxillipeds, though often very large, is never quadrilateral, and never carries the palp distinctly at the antero-internal angle; ischium usually small, sometimes absent or indistinguishably fused with the merus, in which case the merus lies with its long axis directed obliquely or almost transversely inwards; exognath small and more or less concealed. The interantennular septum, when distinguishable, is a thin plate. Male abdomen very narrow. Male openings sternal.

Small crabs, often living as commensals in bivalve mollusks, ascidians and worm tubes or on the outer surface of echinoids, and occasionally in gasteropod mollusks. Some are true parasites within the tests of echinoids, many are free-swimming, especially males.

The American genera belong to the two largest subfamilies, the Pinnotherinae and the Pinnothereliinae (p. 127).

Subfamily PINNOTHERINAE Milne Edwards.

Pinnotherinae MILNE EDWARDS, Ann. Sci. Nat., Zool., ser. 3, vol. 18, 1852, p. 139 [103], restricted.

Ischium of external maxillipeds either rudimentary or indistinguishably fused with the merus to form a single piece which is usually oblique, sometimes almost transverse. Palpus not so large as merus-ischium. Carapace usually not markedly transverse.

KEY TO THE AMERICAN GENERA OF THE SUBFAMILY PINNOTHERINAE.

A^1. Dactyls of first 3 ambulatory legs simple, not bifurcate. Palate with longitudinal ridges.
  B^1. Carapace suborbicular or subquadrate, not strikingly wider than long. Legs not successively diminishing in length from first to fourth.
    C^1. Carapace without two longitudinal, impressed lines leading back from middle of upper margin of orbit. Female larger than male and usually ill-calcified_____*Pinnotheres*, p. 62.
    C^2. Carapace with 2 longitudinal, impressed lines leading back from middle of upper margin of orbit. Orbits ventral. Margin of front flush with antennular surface and not forming a projecting hood. Male unknown _____*Fabia*, p. 101.
  B^2. Carapace much broader than long, anterior margin nearly straight. Legs diminishing in length from first to last, last very small.
Parapinnixa, p. 107.

A². Dactyls of first 3 ambulatory legs bifurcate. Palate without longitudinal ridges. Carapace subpentagonal_____*Dissodactylus*, p. 114.

Genus PINNOTHERES Latreille.

Pinnotheres LATREILLE, Hist. Nat. Crust., vol. 3, an X [1801–2], p. 25; type, *P. pisum* (Linnaeus).
Pinnothera DANA, U. S. Expl. Exped., vol. 13, Crust., pt. 1, 1852. p. 380.
Cryptophrys RATHBUN, Proc. U. S. Nat. Mus., vol. 16, 1893, p. 250; type, *C. concharum* Rathbun.
Arcotheres NAUCK, in Bürger, Zool. Jahrb., Syst., vol. 8, 1895, p. 361; type, *A. palaensis* Bürger.
Zaops RATHBUN, Amer. Nat., vol. 34, 1900, p. 588; type, *Z. depressa* (Say).
Pinnoteres ALCOCK, Journ. Asiat. Soc. Bengal, vol. 69, 1900, p. 337.

Females larger than males. Young females similar to males.

Carapace usually ill calcified in female, often calcified in male, generally convex with ill-defined edges; in shape transversely oval or circular or subquadrangular to suboctagonal, with rounded angles, sometimes longer than wide; generally surface smooth and regions not defined, occasionally surface uneven and regions indicated.

Front narrow, generally deflexed in female, more often advanced in male. Orbits small in female, larger in male, circular or oval; eyestalks short. Antennules oblique. Antennæ small, the minute flagellum standing in inner angle of orbit.

Epistome well defined. Buccal cavity crescentic, arched and very broad from side to side, but very short fore and aft. The external maxillipeds completely close the cavity; they consist chiefly of the merus which is fused with the ischium to form a single large obliquely-directed joint carrying the flagellum at its inner end; palpus typically small, though sometimes large, even half as large as the merus; it consists usually of three segments, the dactylus most often inserted on the inner or flexor border of the propodite but sometimes attached to the distal end of the propodite; very exceptionally the palpus consists of only two segments, the dactylus being absent; exognath for the most part concealed.

Chelipeds short, equal, and generally stouter than legs; legs of moderate length. Abdomen consisting usually of seven separate segments; in male narrow, in female often larger than the sternum.

Habitat.—Commensal in lamellibranch and sometimes in gasteropod mollusks, in tunicates, annelids, and sea-urchins. Males usually free-swimming.

Range.—Widely distributed in the shallow waters of tropical and warm-temperate seas.

Remarks.—Several species formerly placed in other genera are now ranged in *Pinnotheres*. *Ostracotheres politus* is found to have three segments to the palp of the outer maxilliped instead of two, which is the chief distinguishing character of *Ostracotheres*, even though it be an insufficient one. At least one of the species of *Cryptophrys* in now known to have a three-jointed palp, while *Zaops depressa*, of which only the male is recorded, is perhaps no more atypical than some other males of *Pinnotheres*.

KEY TO FEMALES OF AMERICAN SPECIES OF THE GENUS PINNOTHERES.

A^1. Palpus of outer maxilliped small, not nearly half as large as merus.
  B^1. Dactylus of third leg not longer than that of other legs.
    C^1. Carapace wider than long.
      D^1. Legs of each pair (right and left) similar.
        E^1. Dactyli of legs unequal, either the second or the fourth longer than the others.
          F^1. Dactylus of second leg longest.
            G^1. Manus widest a little behind distal end. Carapace soft and yielding.
              H^1. Dactylus of second leg nearly (about seven-eighths) as long as propodus _____ _____ *ostreum*, p. 66.
              H^2. Dactylus of second leg shorter, about two-thirds as long as propodus. Legs slenderer than in *ostreum__holmesi*, p. 68.
            G^2. Manus widest at distal end.
              H^1. Carapace about one-fifth wider than long. Palm less than twice as long as wide_____ _____*geddesi*, p. 70.
              H^2. Carapace wider, one-third wider than long. Palm more than twice as long as wide_____*angelicus*, p. 72.
          F^2. Dactylus of fourth leg longest.
            G^1. Carapace suborbicular, front a little produced. Fourth leg longest, not notably slenderer than the others.
                                                  pugettensis, p. 82.
            G^2. Carapace transversely oblong, or subpentagonal. Fourth leg not longest, but notably slenderer than the others.
                                                    politus, p. 71.
        E^2. Dactyli of legs subequal.
          F^1. Carapace thin, soft. Dactyli of legs falcate.
            G^1. Carapace considerably less than half again as wide as long.
              H^1. Fourth leg strikingly narrower, feebler than the others, its propodus with subparallel margins. Legs thinly bearded on margins_____*serrei*, p. 84.
              H^2. Fourth leg narrower than the others, but not strikingly so. Its propodus strongly narrowed distally.
                J^1. A large pit either side of middle. Posterior margin straight_____*barbatus*, p. 88.
                J^2. A longitudinal sulcus either side of middle. Posterior margin convex_____*pubescens*, p. 87.
            G^2. Carapace half again as wide as long. Fourth leg much narrower than others_____*strombi*, p. 90.
          F^2. Carapace completely calcified, hard_____*silvestrii*, p. 91.
      D^2. Legs of second pair dissimilar; right propodus and dactylus much longer than left_____*margarita*, p. 91.

C². Carapace as long as, or longer than, wide.
  D¹. Dactylus of fourth leg of different shape from the others; that is, almost straight, except for a slender, curved, horny tip___*moseri*, p. 94.
  D². Dactyli of all legs similar, falcate. Size small.
    E¹. No dorsal median groove on frontal region. Legs 1 and 4 subequal, shorter than 2 and 3, which are subequal. Two cardiac tubercles side by side_____*taylori*, p. 97.
    E². A dorsal median groove on frontal region. Legs diminishing from 1 to 4_____*shoemakeri*, p. 95.
B². Dactylus of third leg considerably longer than that of other legs.
lithodomi, p. 73.
A². Palpus of outer maxilliped large, nearly or quite half as large as merus.
  B¹. Dactylus of outer maxilliped either oblong or subspatulate, less than two-thirds as long as propodus.
    C¹. Carapace wider than long.
      D¹. Carapace suborbicular, pubescent; front a little produced.
maculatus, p. 74.
      D². Carapace nearly square, with rounded corners, naked; front not produced _____*nudus*, p. 83.
    C². Carapace as long as, or longer than, wide.
      D¹. Dactylus of fourth leg much longer than that of the other legs.
reticulatus, p. 93.
      D². Dactyli of legs similar, falcate_____*concharum*, p. 86.
  B². Dactylus of outer maxilliped spatulate, two-thirds as long as propodus.
    C¹. Hands short, bare. Carapace nearly one and a half times broader than long _____*guerini*, p. 101.
    C². Hands elongate, strongly ciliated on lower margin____*hirtimanus*, p. 101.
A³. Females not known_____*bipunctatus*, p. 78; *depressus*, p. 79; *orcutti*, p. 98; *muliniarum*, p. 81; *hemphilli*, p. 99.

KEY TO MALES OF AMERICAN SPECIES OF THE GENUS PINNOTHERES.

A¹. Carapace wider than long.
  B¹. Dactyl of last leg unlike the other three, being long and straight instead of short and curved_____*margarita*, p. 91.
  B². Dactyl of last leg not very unlike the other three.
    C¹. Carapace octagonal. Sternum sharply cristate_____ *hemphilli*, p. 99.
    C². Carapace suborbicular. Sternum not sharply cristate.
      D¹. Anterior half of carapace thickly margined with hair. Posterior margin transverse, rimmed.
        E¹. Carapace convex. Propodus of fourth leg not remarkably dilated. Eyes moderate_____*concharum*, p. 86.
        E². Carapace flat. Propodus of fourth leg broadly oval. Eyes very large _____*depressus*, p. 79.
      D². Anterior half of carapace not thickly margined with hair.
        E¹. Carapace widest at middle.
          F¹. Front strongly produced. Two large punctae near middle of carapace_____*bipunctatus*, p. 78.
          F². Front slightly produced. No large punctae near middle of carapace_____*muliniarum*, p. 81.

E². Carapace widest either in front of or behind middle.
    F¹. Carapace widest behind middle_____ostreum, p. 66.
    F². Carapace widest in front of middle.
        G¹. Of small size and nearly naked, covered everywhere with small, persistent, dark spots _____serrei, p. 84.
        G². Of good size, immaculate; chelipeds, legs and under surface pubescent; margins of merus joints of legs bearded.
                                                  barbatus, p. 88.
A². Carapace as long as, or longer than, wide.
    B¹. Postlateral portion of branchial region inclined abruptly in a steep plane, oblique to dorsal surface of carapace, in which it forms a reentering angle.
        C¹. Cheliped much wider at distal end than elsewhere. Terminal segment of abdomen transversely oblong_____orcutti, p. 98.
        C². Cheliped swollen in middle, not much if any wider at distal end than at middle.
            D¹. Carapace deeply areolated. Manus longitudinally ridged. Terminal segment of abdomen subrectangular with rounded tip.
                                                  shoemakeri, p. 95.
            D². Carapace moderately areolated. Manus scarcely ridged. Terminal segment of abdomen equilaterally triangular, blunt-pointed.
                                                  taylori, p. 97.
    B². Branchial region gradually inclined downward toward margin. Carapace with 4 large, persistent, white spots_____maculatus, p. 74.
A³. Males not known_____holmesi. p. 68; geddesi, p. 69; politus, p. 71; angelicus, p. 72; lithodomi, p. 73; pugettensis, p. 82; nudus, p. 83; pubescens, p. 87; strombi, p. 90; silvestrii, p. 91; reticulatus, p. 93; moseri, p. 94; guerini, p. 101; hirtimanus, p. 101.[1]

LIST OF KNOWN MALES AND FEMALES OF THE AMERICAN SPECIES OF PINNOTHERES.

EAST COAST SPECIES.

ostreum Say	Salem, Massachusetts, to Micco, Florida; Guadeloupe	♂	♀
geddesi Miers	Vera Cruz; Cuba; Porto Rico		♀
maculatus Say	Cape Cod to Texas; Jamaica; St. Thomas	♂	♀
depressus Say	New Jersey to Cuba	♂	
serrei Rathbun	Cuba; Jamaica; Porto Rico	♂	♀
barbatus Desbonne	St. Thomas; Guadeloupe	♂	♀
strombi Rathbun	West Florida		♀
moseri, new species	West Florida; Jamaica		♀
shoemakeri, new species	West Florida; St. Thomas	♂	♀
hemphilli, new species	West Florida	♂	
guerini M. Edwards	Cuba; Porto Rico		♀
hirtimanus M. Edwards	Cuba		♀

[1] *Pinnotheres globosum* Jacquinot (Voy. au Pole Sud sur *l'Astrolabe* et la *Zélée*, Zool., vol. 3, 1843 (?), p. 58; atlas, pl. 5, fig. 21), type-locality, Singapore, is recorded by Cano (Boll. Soc. Nat. Napoli, ser. 1, vol. 3, 1889, pp. 98 and 247) as occurring at Porto Lagunas, Patagonia. As no description of the American specimen is given, its identity with *P. globosum* needs confirmation.

WEST COAST SPECIES

holmesi, new species	?Monterey, California	♀
politus Smith	Peru; Chile	♀
angelicus Lockington	Gulf of California	♀
lithodomi Smith	Panama	♀
bipunctatus Nicolet	Chiloe	♂
muliniarum, new species	Lower California	♂
pugettensis Holmes	Puget Sound; British Columbia	♀
nudus Holmes	Santa Cruz and Monterey, California	♀
concharum (Rathbun)	British Columbia to San Diego	♂ ♀
pubescens (Holmes)	Gulf of California	♀
silvestrii Nobili	Chile	♀
margarita Smith	La Paz; Panama	♂ ♀
reticulatus, new species	Gulf of California	♀
taylori, new species	British Columbia	♂ ♀
orcutti, new species	Manzanillo	♂

Analogous species of *Pinnotheres* on opposite sides of the continent: *ostreum* (Atlantic); *holmesi* (Pacific).

PINNOTHERES OSTREUM Say.

OYSTER CRAB.

Plate 15, figs. 3–6.

? *Pinnotheres pinnophylax* Bosc, Hist. Nat. Crust., vol. 1, an X [1801–2], p. 243 (part), not pl. 6, fig. 3 (after Herbst) nor *Cancer pinnophylax* Herbst, 1783; coasts of America in *Chama lazarus* (according to Bosc). As this is a European species of *Chama*, *C. macerophylla* Gmelin is doubtless the species indicated.

Pinnotheres ostreum Say, Journ. Acad. Nat. Sci. Philadelphia, vol. 1, 1817, p. 67, pl. 4, fig. 5, female (type-locality, inhabits the common oyster; type not extant).—De Kay, Nat. Hist. New York, pt. 6, Crust., 1844, p. 12, pl. 7, fig. 16, female.—Smith, Rept. U. S. Commr. of Fish and Fisheries, pt. 1, for 1871–72 (1873), p. 546 [252]; not pl. 1, fig. 2, male.

? *Pinnotheres crassipes* Desbonne, in Desbonne and Schramm, Crust. Guadeloupe, pt. 1, 1867, p. 43 (type-locality, Guadeloupe, in *Ostrea parasitica*; type probably not extant).

Diagnosis.—First leg of female stoutest, propodus distally widened, dactylus curved, other legs similar to one another, dactylus longer, straighter than in first leg. Carapace thin. Palm very wide, just behind distal end. Propodi of all legs in male wide, dactyli longer and more curved than in female.

Description of female.—Surface glabrous for the most part, smooth, shining; carapace subcircular, with the posterior margin very broad; thin, membranaceous, yielding to the touch, convex from before backward, gastro-cardiac area separated by broad depressions from the branchio-hepatic area; lateral margins thick and bluntly rounded; front about one-seventh as wide as carapace, truncate and scarcely advanced in dorsal view, its margin deflexed and forming a rounded lobe, visible from before. Orbits small, subcircular, anteriorly placed.

Antennules large, antennae small; flagellum not as long as diameter of orbit. Carpus or first segment of palp of outer maxilliped, short, oblong; propodus elongate, end rounded; dactyl inserted behind middle of propodus, minute, and very slender.

Chelae moderate, merus and carpus rather slender; palm flattish inside, swollen outside, strongly widened from the proximal toward the distal end, then somewhat narrowed, the width across base of fingers considerably less than greatest width of palm, lower margin of propodus, as well as upper, convex; fingers, especially the immovable one, stout, not gaping, tips hooked past each other, several small teeth on prehensile edges; fixed finger horizontal.

Legs slender, subcylindrical, last 2 joints with thin fringe of hair first leg stoutest, propodus widening distally, dactylus about half as long, curved; last 3 legs similar to one another, dactyli longer and straighter; second leg longest, third next, first and fourth subequal.

Mature abdomen very large, extending beyond carapace in all directions.

Description of male.—Relatively narrower and less swollen than female; front more advanced, truncate. Chelipeds similar to those of female but palm shorter, more swollen. First leg as in female, except for the dactylus which is longer, two-thirds as long as propodus, and more curved than in female; in the remaining legs the carpus-propodus is wide as in the first leg, not slender as in female, and the dactyli are longer than in female, the fourth one quite as long as propodus; they are nearly straight proximally but toward the end are falcate. Sides of abdomen nearly straight and convergent from the third to the seventh segment, which is arcuate at the extremity.

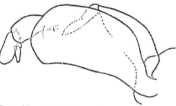

FIG. 30. — PINNOTHERES OSTREUM, OUTER MAXILLIPED OF FEMALE (2542), × 15½.

Color.—Female. Translucent whitish generally. Upper surface of carapace and median region of post-abdomen red to pinkish. Minute eyes brownish. All limbs whitish (Fowler).

Measurements.—Female (2542), length of carapace 12, width of same 15 mm. Male (49209), length of carapace 1.3, width of same 1.4 mm.

Habitat.—Commensal with oysters. Males free-swimming.

On the coast of the Carolinas *P. ostreum* is abundant, according to Dr. Bashford *D*ean;[1] however, not more than one female was noted

---

[1] Bull. U. S. Fish Comm., vol. 10 for 1890 (1892), p. 351.

in a single oyster. In most instances the crab was found well thrust in between the palps, usually between the middle ones. This is evidently annoying to the oyster, for the palps sometimes show thickened outgrowths, or are malformed and stunted in size. To what extent the crab is of value to the oyster, as commonly supposed, is debatable. The oyster is said to feed on clusters of bell-animalcules (*Zoothamnium*), which are attached to the crab;[1] on the other hand, the food of the crab, as evidenced by its stomach contents, consists in great part of such minute organisms as are sought by the host, in addition to many small crustaceans not normally the oyster's prey.

Range.—Salem, Massachusetts, to Micco, Florida; Guadeloupe (Desbonne).

Material examined.—

New York Bay; from oysters; E. G. Blackford; 74 females (2542)

New York market; from oysters; E. G. Blackford; 26 females (4991).

West coast of Chesapeake Bay, Northumberland County, Virginia; *D*ecember 23, 1914; P. L. Boone; 4 females (49208).

Rappahannock River, Virginia; in oysters; A. C. Weed; 6 females (42512).

Out of Lynn Haven oysters, Virginia; November 29, 1905; Bureau of Fisheries; 2 females (49210).

Beaufort, North Carolina; Union College Coll.; 1 male, young, depos. in U.S.N.M. (49209).

Winyah Bay, South Carolina: *D*ecember, 1890; station 54; steamer *Fish Hawk;* 5 females (18214). December 31, 1890; station 56; steamer *Fish Hawk;* 6 females (18215). One-half mile north of wharf on South Island; dredged; Sh.; temp. 49.5° F., January 3, 1891; stations 1641, 1642; steamer *Fish Hawk;* 1 female (18213).

Clam Bank Creek, South Carolina; from oysters; *D*ecember 30, 1890; steamer *Fish Hawk;* 3 females (18216).

Charleston, South Carolina; from a starfish; Louis Agassiz; 1 female, 3 mm. long, mature (5730, M.C.Z.).

Colleton River, South Carolina; 1891; steamer *Fish Hawk;* 7 females (26108).

East peninsula opposite Micco, Florida; O. Bangs; 8 females (3 ovig.) (18740).

PINNOTHERES HOLMESI,[2] new species.

Plate 15, figs. 1–2.

? *Pinnotheres nudus* WEYMOUTH, Leland Stanford Jr. Univ. Publ., Univ. Ser. No. 4, 1910, p. 53, fig. 1 (?); not *P. nudus* Holmes, 1895.

Type-locality.—? Pacific Grove, California; 1 female ovig., holotype (Mus. Stanford Univ.).

---

[1] Ryder, Rept. Maryland Fish Commission for 1880 (1881), p. 24.
[2] Named for Dr. S. J. Holmes, of the University of California.

Diagnosis.—Carapace of female soft, broadest behind, gastric region defined. Manus widest a little way behind fingers. Leg 1 stout, widest distal to middle; legs 2, 3, and 4 very slender and diminishing in order named.

Description of female.—Carapace very soft and yielding, broader than long, the front and long antero-lateral margins forming a single arch; postero-lateral margins short, oblique, concave; posterior margin long, concave; greatest width in posterior half; carapace convex, thick, rounding gradually downward, except posteriorly; gastric region defined by a depression, which is deeper and narrower than in *ostreum*, and suggests that in *Fabia*. Front between orbits truncate, medially faintly emarginate, orbits partly visible from above.

Carpus of outer maxilliped suboblong; propodus no longer than carpus, and narrower, end obliquely rounded; dactylus linear, of nearly even width throughout, rounded at extremity, inserted near proximal end of propodus and not reaching its distal end.

Chelipeds stouter than legs, but not very large; lower margin of propodus nearly straight, palmar portion widening rapidly to a point behind base of fingers; fingers stout, hooked at tips, prehensile edges uneven, an acute tooth near base of dactyl.

Relative length of legs represented by 2.3.1.4, the second longest; 2, 3, and 4 very slender; dactyl 2 much longer than 3 or 4 and curved on

FIG. 31.— PINNOTHERES HOLMESI, OUTER MAXILLIPED OF FEMALE HOLOTYPE, × 15½.

both margins, dactyl 3 less curved on anterior and nearly straight on posterior margin, dactyl 4 a trifle longer than 3, margins nearly straight; horny tips of all four very slender and hooked. First leg stout, especially propodus, not reaching middle of propodus of second; propodus widest distal to the middle and narrowest at distal extremity; dactyl shortest of all, conical except for tip.

Abdomen voluminous, much longer and wider than carapace.

Measurements.—Female holotype, length of carapace 7.2, width of same 8.7 mm.

Range.—Known only from the type-locality, which is probably Pacific Grove, California.

Affinity.—In shape resembles *P. ostreum* of the Atlantic coast, which has also a similarly enlarged propodus of first leg; first 3 legs of *holmesi* much slenderer than of *ostreum*.

PINNOTHERES GEDDESI Miers.

Plate 16, figs. 1-4.

? *Cancer* 1. The Oyster-Crab, BROWNE, Civil and Nat. Hist. Jamaica, 1756, p. 420.

Pinnotheres angelicus MIERS, Journ. Linn. Soc. London, Zool., vol. 15, 1880, p. 86; not *P. angelicus* Lockington, 1877.

Pinnotheres geddesi MIERS, Journ. Linn. Soc. London, Zool., vol. 15, 1880, p. 86 (type-locality, Vera Cruz; cotypes in British Mus.).

Pinnotheres ostrearius RATHBUN, Bull. U. S. Fish Comm., vol. 20, for 1900, pt. 2 (1901), p. 20, text-fig. 3 (type-locality, Mayaguez, Porto Rico, in an oyster from near Cabo Rojo; holotype, Cat. No. 23767, U.S.N.M.).

Diagnosis.—Near *ostreum*. Propodus of first leg of female of nearly same width throughout. Palm widest at distal end. Carapace very thin.

Description of female.—Carapace very thin and yielding, transversely suborbicular, broad behind. Gastric region distinctly outlined by a furrow, cardiac region less distinct. Front rounding downward, slightly projecting, truncate in dorsal view. Orbits circular, eyes partly visible in dorsal view.

Merus of outer maxillipeds robust, outer margin regularly convex, inner margin with bluntly rounded angle near distal extremity; carpus and propodus robust, the latter rounded and ciliated at distal end; dactyl very slender, styliform, reaching about to extremity of propodus.

FIG. 32.—PINNOTHERES GEDDESI, FEMALE (23767). *a*, CHELA, × 16; *b*, OUTER MAXILLIPED, × 16.

Chelipeds smooth; palm rapidly increasing in width from proximal to near distal end, which articulates very obliquely with the dactylus; upper margin convex at widest part, lower margin of entire propodus convex; fingers subconical, somewhat hairy, edges meeting and tips crossing when closed.

Legs slender; first one stouter than the others, propodus scarcely widening distally, dactylus about half as long, stout, anterior margin convex, posterior straight; second leg longest, third subequal to first; fourth reaches about to middle of propodus of third; dactylus of second leg long, curved, about two-thirds as long as propodus; dactylus of third leg shorter, of fourth still shorter, straighter and more hairy; third dactylus nearly straight behind, fourth quite straight behind.

Measurements.—Female cotype, length of carapace 9, width of same 10.8 mm.

Habitat.—Commensal with oysters.

Range.—Vera Cruz, Mexico; Cuba; Jamaica?; Porto Rico.

Material examined.—

Vera Cruz, Mexico; P. Geddes, collector; 1 female with ova, cotype (out of 7 in Brit. Mus.; loaned to U.S.N.M. through the courtesy of Dr. Calman).

Cuba; in oysters; Andréa, collector; received from Copenhagen Mus.; 2 females ovig. (23439).

Sagua la Grande, Santa Clara Province, Cuba; M. S. Roig, coll.; 1 female ovig. (49252).

Near Cabo Rojo, Mayaguez, Porto Rico; in oysters; Jan. 24, 1899; steamer *Fish Hawk;* 2 females (1 ovig. is holotype of *ostrearius*) (23767).

PINNOTHERES POLITUS (Smith).

Plate 159, fig. 5.

Ostracotheres politus SMITH, Trans. Connecticut Acad. Arts and Sci., vol. 2, 1870, p. 169 (type-locality, Callao, Peru; type in Peabody Mus. Yale Univ.).—LENZ, Zool. Jahrb., Suppl., vol. 5, 1902, p. 765, pl. 23, figs. 9, 9a.—PORTER, Revista Chilena Hist. Nat., vol. 13, 1909, p. 249.—RATHBUN, Proc. U. S. Nat. Mus., vol. 38, 1910, p. 545, text-fig. 3 (after Lenz).

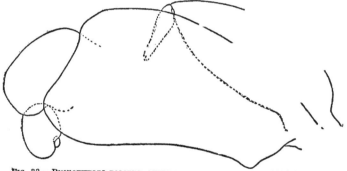

FIG. 33.—PINNOTHERES POLITUS, OUTER MAXILLIPED OF FEMALE (40448), × 50.

Diagnosis.—Carapace wide, thin, flat. Last leg very slender, with very long dactylus. Propodus of second and third legs distally expanded.

Description of female.—Carapace thin and yielding, naked, smooth and shining, transversely oblong, a little wider posteriorly than anteriorly, dorsal surface flat, borders smoothly rounded; a short median sulcus on front, a slight U-shaped one from orbits to middle of carapace. Front arcuate as seen from above, scarcely protuberant; eyes small, partly visible in dorsal view.

The second joint of the palp of the outer maxilliped is short and broad, the terminal joint is minute and has not hitherto been discovered. Because of its presence, the species is placed in *Pinnotheres.*

Chelipeds equal, rounded, smooth, and glabrous; chelae small, compressed, lower margin sinuous, upper margin of manus a little convex; fingers shorter than palm, fitting together, dactylus somewhat curved and armed near base with a small tooth which fits a slight excavation in propodal finger.

Legs short, subcylindrical and smooth; first leg shorter and narrower than second; both have the dactyli short, curved, and closing against the expanded end of propodus which is clothed at that point with a little tuft of short, stiff hair; third leg about as long as second, dactylus short and curved, but distal end of propodus not expanded; last leg shorter than third, and slenderer than the rest, dactylus slightly curved and very long and slender, equal in length to propodus.

Abdomen very wide, exceeding sternum.

Measurements.—Female (40448), length of carapace 5.7, width of same 8 mm.

Habitat.—In shells of mollusks: probably *Mytilus algosus* Gould (Smith), *Calyptraea*, species (Lenz), *Crepidula dilatata* (Coker).

Range.—Peru; Chile.

Material examined.—

Ancon Bay, Peru; found with *Crepidula dilatata* Lamarck on mussels; R. E. Coker, coll.; received from Peruvian Government; 1 female ovig. (40448).

Talcahuano, Chile; *Hassler* Exped.; received from Mus. Comp. Zoöl.; 1 female (22848).

PINNOTHERES ANGELICUS Lockington.

Plate 16, figs. 5 and 6.

Pinnotheres angelica LOCKINGTON, Proc. California Acad. Sci., vol. 7, Dec. 1, 1876 (1877), p. 155 [10] (type-locality, Angeles Bay, Gulf of California, in oysters; types not extant). Not *P. angelicus* Miers, Journ. Linn. Soc. London, Zool., vol. 15, 1880, p. 86.

Diagnosis.—Female transverse, smooth, shining. Dactylus of second leg much the longest. Prominent tubercle on basal joint of antennae. Dactylus of endognath attached to end of propodus.

Description of female.—Smooth, shining. Carapace thin, easily wrinkled, transverse, with anterior margin strongly arcuate, posterior margin long, slightly concave, sides rounded; gastric region well defined; a large pit on branchial region near inner angle. Front advanced, edge

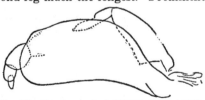

FIG. 34.—PINNOTHERES ANGELICUS, OUTER MAXILLIPED OF FEMALE (17467), × 15½.

rounded. Orbits and eyes oval, hidden from dorsal view. A large prominent tubercle at posterior end of basal joint of antenna.

Propodus of endognath distally rounded, dactylus small, attached on inner portion of extremity of propodus.

Chelipeds elongate, manus slightly compressed and increasing distally; immovable finger slightly deflexed, swollen in basal half; fingers fitting together when closed, tips curved and crossing each other.

Legs slender, second longest, third next, fourth shortest; dactyli nearly straight, except that of second leg, which is twice as long as of other legs, and nearly equaling its propodus.

Abdomen unusually large.

Measurements.—Female (17467), length of carapace 9, width of same 12.2 mm.

Habitat.—In oysters (Lockington) and mussels.

Range.—Gulf of California at Angeles Bay (type-locality) and San Josef Island.

Material examined.—

San Josef Island, Gulf of California, Mexico; Mar. 16, 1889; 5 females (3 ovig.) (17467).

Angeles Bay, Gulf of California; in *Modiola copax;* 4 females (3 ovig.) (Univ. Cal.).

PINNOTHERES LITHODOMI Smith.

Pinnotheres lithodomi SMITH, Trans. Connecticut Acad. Arts and Sci., vol. 2, 1870, p. 169 (type-locality, Pearl Islands, Bay of Panama, in *Lithodomus aristatus* Forbes and Hanley; type in Peabody Mus. Yale Univ.).

Diagnosis.—Merus of maxillipeds broadest at distal extremity, sides nearly straight. Legs very slender; dactyli slightly curved, third longest, second next, first and fourth subequal.

Description of female.—Dorsal surface of carapace smooth and naked. Merus of outer maxillipeds broadest at distal extremity, both margins nearly straight. Chelipeds equal, smooth and naked; hands cylindrical; fingers short, nearly straight, tips slightly hooked by each other, prehensile edge of dactylus armed, near base, with a small tooth which fits a slight excavation in propodal finger.

Legs very slender and wholly naked except dactyli; dactylus of first leg very short and only slightly curved, of second leg considerably longer than in first and nearly straight, of third leg very long, nearly as long as propodi, slender and slightly curved; of last leg about as long as second and ciliated along posterior edges. (After Smith.)

Measurement.—Width of carapace about 4 mm.

Habitat.—In *Lithophaga aristatus,* a boring mollusk.

Material examined by Prof. S. I. Smith.—Pearl Islands, Bay of Panama, in *Lithodomus* (now *Lithophaga*) *aristatus* Forbes and Hanley which was in its usual habitat, an excavation in the shell of a *Spondylus;* F. H. Bradley coll.; 1 female ovig., holotype (P.M.Y.U.).

PINNOTHERES MACULATUS Say.

MUSSEL CRAB.

Plate 17, figs. 3–6.

Pinnotheres maculatum SAY, Journ. Acad. Nat. Sci. Philadelphia, vol. 1, 1818, p. 450 (type-locality, "Inhabits the muricated Pinna of our coast"; cotypes, male and female, in British Mus.).

Pinnotheres maculatus VERRILL, Rept. U. S. Commr. of Fish and Fisheries, vol. 1, 1871–1872 (1873), pp. 309 [15], 434 [140], 459 [165].—SMITH, Rept. U. S. Commr. of Fish and Fisheries, vol. 1, 1871–1872 (1873), p. 546 [252].—FAXON, Bull. Mus. Comp. Zoöl., vol. 5, 1879, p. 265, footnote (Zoea).—FOWLER, Ann. Rept. New Jersey State Mus., 1911 (1912), p. 435, pls. 136 and 137; not (?) *P. byssomiae* De Kay and Gibbes.

Pinnotheres ostreum SMITH. Rept. U. S. Commr. of Fish and Fisheries, vol. 1, 1871–1872 (1873), p. 546 [252], part: pl. 1, fig. 2 (male); not *P. ostreum* Say.

Diagnosis.—Dactyli of first 3 legs hooked, of fourth nearly straight. Males usually, and the young always dark with light spots. Female with shell thick, not yielding.

Description of adult female.—Surface covered with a short, dense, deciduous tomentum. Carapace subcircular, little broader than long, thick and firm but not hard, convex, uneven, smooth, the gastrocardiac area higher and separated by depressions from the branchiohepatic area; antero-lateral angles a little prominent; front slightly advanced, about one-fifth width of carapace, in dorsal view subtruncate, bilobed by a shallow sinus, edge bent down to form a triangular lobe in front view. Orbits small, subcircular; eyes spherical. Antennae as long as width of orbit. Antennulae large, obliquely transverse.

Propodus of outer maxilliped larger than carpus and about twice as long as wide; dactylus narrow-spatulate, curved, attached near middle of propodus and reaching to extremity of the latter.

Chelae of moderate size, smooth, hairy inside; carpus elongate; palm thick, blunt-edged, increasing distally; fingers stout, fitting close together, tips hooked; propodal finger nearly horizontal: dactylus with a tooth near base fitting into a sinus of the fixed finger, sinus with a small tooth at either end.

Legs slender, smooth, hairy below, second pair longest, shorter than cheliped, first three dactyli falcate, shorter than their respective propodi, last dactylus long, equaling the propodus, and slightly curved.

Description of male.—Much smaller, diameter of carapace about half that of female, a little longer than wide, harder than female, and with a striking, light color-pattern of bare spots on a dark ground of pubescence, consisting of a median stripe, constricted in the middle and behind, a subtriangular spot each side before the middle and a linear spot each side behind.

Chelipeds shorter than in female, palms and fingers stouter. Legs wider, especially the propodal segments of the first three legs, the posterior surface of which is overlaid with a thin fringe of hairs attached near upper margin; last leg relatively shorter than in female, not reaching middle of propodus of third; dactylus more nearly like third than in female.

Abdomen at its middle about one-third width of sternum, gradually narrowing from third to seventh segment, sides of third convex,

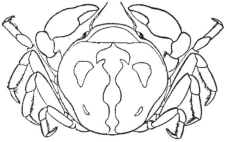

FIG. 35.—PINNOTHERES MACULATUS, MALE, × 8. (AFTER SMITH.)

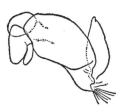

FIG. 36.—PINNOTHERES MACULATUS, OUTER MAXILLIPED OF FEMALE (36782), × 10.

seventh obtusely rounded. Sutures between segments of abdomen and sternum marked with a narrow line of dark pubescence.

Variations.—Young females resemble the male except in the shape of the abdomen and the character of its appendages; they have the coloration, pubescence and long hair on the legs, all male characters. These masculine-looking females are free-swimming like the young and sometimes attain a size of 5.2 mm. long. Normal females, colorless and commensal in habit, are 3.3 mm. (16042) and upward in length. The long hair persists on the legs in medium-sized specimens. Some males, perhaps always commensal individuals, resemble adult females in their consistency and absence of color; such range from 4 mm. (14567) long upward. The same is true of some young females, except as to size.

Individuals vary (1) in the stoutness of the chelae, the lot from Jamaica having very stout chelae, while in that from Louisiana the chelae are more slender than usual; (2) in the curvature and length of the dactyl of the second leg, normally strongly curved like the first and third but sometimes less curved and correspondingly elongate. This straightening and elongation may occur on one or both sides of a specimen and on one or several specimens from a single gathering.

Color.—Female, obscure brownish, immaculate; male, above black with whitish spots, beneath yellowish-white. For exceptions see above under variations.

Measurements.—Female (42850), length of carapace 13.7, width of same 14.3 mm. Male (18014), length of carapace 8.4, width of same 8.1 mm.

Habitat.—Males often and young of both sexes usually free-swimming. Mature males and females commensal in mussel (*Mytilus edulis*), horse mussels (*Modiola modiolus* and *M. tulipa*), "long" clam (*Mya arenaria*), scallops (*Pecten irradians* and *P. tenuicostatus*) oyster (*teste* Blackford), *Pinna muricata*, and in tubes of annelid (*Chaetopterus pergamentaceus*). On muddy bottoms of bays and sounds (Verrill).

Range.—Cape Cod to Texas; Cuba; Jamaica; St. Thomas. Surface to 21 fathoms.

Material examined.—

Off Marthas Vineyard, Massachusetts; Katama Point, E. $\frac{1}{4}$ S. $1\frac{1}{2}$ miles; 5 fathoms; S.; temp. 62° F.; Sept. 6, 1883; station 1172; steamer *Fish Hawk;* 1 female (36900).

Woods Hole, Massachusetts; U. S. Fish Comm.: 1876; V. N. Edwards; 1 y. (4082). Surface; Oct. 21, 1881; V. N. Edwards; 26 males and females y. (15022). Aug. 9, 1882; 1 male (40805). Aug. 12, 1882; 1 female y. (40804). Sept. 15, 1882; 1 male, 1 female y. (40806). Surface; Sept. 17, 1882, evening; 1 male (12561). Surface; Oct. 2, 1882, 8 p. m.; 1 female y. (12560). Surface, by electric light; Aug. 5, 1885; 24 males and females y. (10917). Surface, by electric light; August, 1885; steamer *Albatross;* 1 male y., 4 females y. (11184). 1885; 1 female ovig. (34919). From *Mya arenaria;* Aug. 26, 1886; 1 y. (11840). Mar. 28, 1888; V. N. Edwards; 1 female (26106). V. N. Edwards; 1 female ovig. (3818). From mussel; 5 females (26140). Surface; Oct. 13, 1892; V. N. Edwards; 3 males, 5 females y. (49211).

Woods Hole, Massachusetts; from tubes of *Chaetopterus;* G. M. Gray; 3 females (42148).

Vineyard Sound, Massachusetts: Surface; Aug. 29, 1882, evening; U. S. Fish Comm.; 1 female y. (40808). Gay Head Light, W. $\frac{3}{4}$ S., $2\frac{1}{4}$ miles; 10 fathoms; S.; temp. 60° F.; July 20, 1881; station 928; steamer *Fish Hawk;* 22 males (of 2 sorts), 33 females (9 ovig.) (19024, 41032).

Menemsha Bight, Marthas Vineyard, Massachusetts; $7\frac{1}{2}$ to $9\frac{1}{2}$ fathoms; S. M.; temp. 64° F.; September 6, 1882; stations 1134, 1135; steamer *Fish Hawk;* 23 males (of 2 sorts), 48 females (22 ovig.) (14567).

Off Gay Head, Marthas Vineyard, Massachusetts; from *Mytilus;* 1882; U. S. Fish Comm.; 4 females (2 ovig.) (41033).

Buzzards Bay, Massachusetts; between West Island and N. end of Woods Hole; $6-8\frac{1}{2}$ fathoms; bk. M. Sh. G.; temp. 71.1° F.; August 15, 1887; station 1215; steamer *Fish Hawk;* 1 female ovig. (12819).

Mattapoisett Harbor, Massachusetts; November, 1882; W. Nye, jr.; 2 females (6691).

Acushnet River, Massachusetts; from scallop; October 9, 1890; W. Nye, jr.; 9 males (of 2 sorts), 11 females (18014).

Off Newport, Rhode Island: Surface; July 30, 1880, evening; U. S. Fish Comm.; 3 males, 4 females y. (34082). Near Fort Adams, Rhode Island; surface; August 22, 1880; U. S. Fish Comm.; 1 male, 1 female y. (36317). Surface; September 2, 1880; U. S. Fish Comm.; 2 males (34002). Brentons Reef Lightship, NNW. $\frac{3}{4}$ W., $5\frac{1}{4}$ miles; parasitic on *Pecten;* $17\frac{1}{2}$ fathoms; S. Scallops; temp. 54° F.; August 13, 1880; station 789; steamer *Fish Hawk;* 2 males, 1 female y. (36881, 41035). Point Judith W. $\frac{1}{2}$ S., $4\frac{3}{4}$ miles; in *Modiola modiolus;* 20 fathoms; fne. S. brk. Sh.; temp. 53.5° F.; August 12, 1880; station 784; steamer *Fish Hawk;* 1 female ovig. (36741). August 24, 1880; U. S. Fish Comm.; 1 female ovig. (36786).

Narragansett Bay, Rhode Island: Beaver Tail Light, S. by E., $1\frac{1}{4}$ miles; 8 fathoms; S. Sh.; temp. 67° F.; August 6, 1880; station 772; steamer *Fish Hawk;* 1 female (34018). North End Dutch Island, S., 1 mile; 12 fathoms; G. S. M.; temp. 68° F.; August 6, 1880; station 775; steamer *Fish Hawk;* 3 males, 3 females (1 y.) (36307). In *Pecten tenuicostatus;* $16-27\frac{1}{2}$ fathoms; S. Sh. brk. Sh. Gr.; temp. 57°-58.5° F.; August 7, 1880; stations 776-781; steamer *Fish Hawk;* 10 females (8 ovig., 1 y.) (36782, 40803). Off Brentons Reef Lightship; $8\frac{1}{2}$-10 fathoms; S. brk. Sh.; temp. 63°-66° F.; August 23, 1880; stations 816-818; steamer *Fish Hawk;* 1 female y. (40802).

Off Block Island, Rhode Island: In *Modiola;* September, 1874; U. S. Fish Comm.; 1 female ovig. (36882). August 13, 1874; U. S. Fish Comm.; 1 female immat. (37856).

Off Watch Hill, Rhode Island; 17-21 fathoms; August 21, 1874; U. S. Fish Comm.; 2 females immat. (40807).

Southeast of Long Island, New York; lat. 40° 51′ N.; long. 71° 58′ W.; from scallops; 20 fathoms; August 25, 1913; station 10118; schooner *Grampus;* 5 males, 3 females (1 y.) (49212).

Gardiners Bay, Long Island, New York; from *Pecten irradians;* 1890; schooner *Grampus;* 19 males (of 2 sorts), 19 females (of 2 sorts) (16042).

Long Island Sound, New York: 1874; U. S. Fish Comm.; 2 males (colorless), 5 females (1 ovig.) (36207). 1874; U. S. Fish Comm.; 3 males (colorless), 1 female ovig. (36268). 1874; U. S. Fish Comm.; 6 females (41034).

New York market; from oysters; E. G. Blackford; 2 females ovig. (40811).

Potomac River, off Point Lookout, Virginia; surface, above oyster beds; November 16, 1880; station 916, steamer *Fish Hawk;* 1 y., about .7 mm. long (34224).

Chesapeake Bay; off New Point Comfort Light, Virginia; 5½ fathoms; hrd.; temp. 50.3° F.; Apr. 22, 1916; station 8506, *Fish Hawk;* 1 male, 1 female, both young (49644).

Beaufort, North Carolina: In tube of *Chaetopterus;* May 9, 1905; Bureau of Fisheries; 1 male (Beaufort Lab.). In shell of live *Pinna;* zoeae hatched the night of August 9, 1908; R. Binford; 2 females and zoeae (42850). Exploration of the eastern coast of United States; 1 female (2472).

Off Cape Fear, North Carolina; from *Modiola tulipa;* 15 fathoms; gy. S. brk. Co.; October 20, 1885; steamer *Albatross;* 1 female ovig. (13736).

Atlantic Ocean, 16 miles off Sapelo Island Lighthouse, Georgia; May 3, 1915; station 8259; *Fish Hawk* (Danglade); 3 males (49252).

Marco, Florida; from *Pinna muricata;* May, 1884; Henry Hemphill; 1 female ovig. (6971).

Chandeleur Islands, Louisiana; L. R. Cary; 2 females (33112).

Matagorda Bay, Texas; J. D. Mitchell; 1 female (18415).

Cuba, at Cabañas; 2–25 fathoms; taken by submarine light on sand, shell, grass to mud bottom; June 8–9, 1914; station 16, Henderson and Bartsch, *Tomas Barrera* Exped.; 2 males (48717).

Jamaica, West Indies; March 1–11, 1884; steamer *Albatross;* 3 males, 19 females (13 ovig.) (18015).

Kingston, Jamaica; taken by electric light; 1884; W. Nye, jr., steamer *Albatross;* 48 males and females y. (7737).

St. Thomas, West Indies; January 17–24, 1884; steamer *Albatross;* 2 females ovig. (7666).

From *Pecten irradians;* 1 male (20858).

PINNOTHERES BIPUNCTATUS Nicolet.

Plate 159, figs. 10–12.

Pinnotheres bipunctatum NICOLET, in Gay, Hist. Chile, Zool., vol. 3, 1849, p. 155, pl. 1, figs. 2–2c (type-locality, San Carlos de Chiloe; type not extant).

Diagnosis.—Front of male produced, truncate, with median sinus. Legs stout, ciliate. Carapace suborbicular, flattened, bipunctate.

Description of male (after Nicolet).—Carapace rounded from the front along the sides, narrowing slightly toward the back, posterior border long and straight, two rather large punctae in the middle. Front quadrilateral, transverse, prominent beyond curve of antero-lateral borders, its anterior margin broad and slightly hollowed out, its middle occupied by a longitudinal depression bordered on each side by a raised, rounded, and forward-pointing projection. Orbits small but deep.

Chelipeds and legs robust, long, and compressed; chelipeds much shorter than legs; hand short, wide, and nearly quadrate, movable finger wider than immovable and very much curved. Legs covered with very short, coarse hair, scarcely visible; dactyls strong, curved, and with a sharp claw.

Abdomen narrow, elongate; terminal segment triangular, longer than the preceding. Posterior part of body rough with spinelike hairs, of which there are some small ones on the inner margin of the outer maxillipeds.

Color.—Translucent yellow.

Measurements.—Length of male carapace, 1 to 2 lines (Nicolet); according to figure 2a cited, 3.5 mm.

Habitat.—Probably in sea-urchins (Nicolet).

Locality.—San Carlos de Chiloe, Chile.

PINNOTHERES DEPRESSUS Say.

Plate 17, figs. 1 and 2.

Pinnotheres depressum SAY, Journ. Acad. Nat. Sci. Philadelphia, vol. 1, 1817, p. 68 (type-locality, Egg Harbour; type not extant).—FOWLER, Ann. Rept. New Jersey State Mus. for 1911 (1912), p. 433, and synonymy.

Zaops depressa RATHBUN, Amer. Nat., vol. 34, 1900, p. 590.

Diagnosis.—Carapace of male flat, bordered by a rim of short hair. Eyes very large. Chelipeds very stout. Legs similar, dactyli long and curved, last propodus much expanded, oval.

Description of male.—Carapace flat, suborbicular, with the front advanced and rounding downward and the hind margin very little convex; upper sinus of orbit almost rectangular; behind it a shallow, longitudinal depression; sides of front slightly convergent to a subtruncate margin which is feebly bilobed. Margin of carapace marked by a raised line of very short, dense hair which is broader behind, straightened and narrowed over the eyes and continued around the front. Eyes very large, globular; cornea large.

End of merus of outer maxilliped longitudinally truncate; carpus not much longer than wide; propodus longer than carpus, suboblong, abruptly narrowed at middle, where the dactylus is articulated; the latter is small and narrow, and does not nearly reach end of propodus.

Chelipeds stout, margins hairy; chelae subovate; palm wider than its superior length, its margins convex, outer surface with a longitudinal elevated line bordered with a mat of coarse hair; fingers stout, especially the immovable one, which is broadly triangular except for the slender, curved tip; it has a right-angled sinus on the prehensile edge which corresponds to a tooth behind the middle of the strongly curved dactylus, remainder of dactylus crenate; when the fingers are closed, the tips cross each other.

Of the ambulatory legs the second and third are about the same length, the first a little shorter, fourth shortest; propodites convex

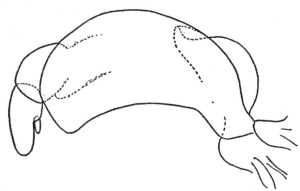

FIG. 37.—PINNOTHERES DEPRESSUS, OUTER MAXILLIPED OF MALE (48594), × 137½.

on anterior margin, faintly crenulate on posterior margin, the first very thick, its posterior margin straight, of second and third concave, of fourth convex, the fourth propodus being greatly dilated and oval in form; dactyls similar, elongate, with long, slender, curved tips.

Abdomen gradually narrowed, terminal joint no longer than wide, end rounded.

Measurements.—Male (48594), length of carapace 1.6, width of same 1.8 mm.

Range.—Egg Harbor, New Jersey, to northwest Cuba.

Material examined.—

Beaufort, North Carolina; Union College Collection, No. 71 c; 1 young male (deposit in U.S.N.M., Cat. No. 42821).

Cabañas, northwest Cuba; on sand, shell, grass to mud bottom; June 8, 1914; station 16, *Tomas Barrera* Exped.; received from Henderson and Bartsch; 1 young male (48594).

PINNOTHERES MULINIARUM, new species.

Plate 18, figs. 2 and 3.

Type-locality.—Lower California; in *Mulinia*, sp.; 1 male holotype (23443).

Diagnosis.—Male, carapace suborbicular, smooth. Fourth leg much the smallest. Dactyli of legs curved, fourth one less strongly falcate. Abdomen very long.

Description of male.—Carapace transversely suborbicular, posterior margin nearly straight; front strongly deflexed, outline slightly arched as seen from above.

Palp of outer maxilliped of large size, short and very wide; carpus short; propodus about as long as carpus, suboblong, end obliquely

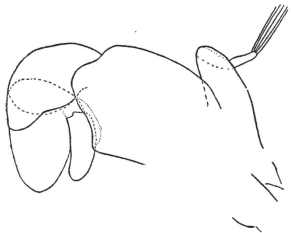

FIG. 38.—PINNOTHERES MULINIARUM, OUTER MAXILLIPED OF MALE HOLOTYPE, × 90.

rounded; dactylus broad, lunate, attached near base of propodus and following its curve, and reaches nearly to its extremity.

Chelipeds and lower surface of body pubescent. Manus widening a little distally, lower margin straight, upper convex; both fingers have tips curved and crossing each other.

Of the legs the fourth is much reduced and does not reach beyond carpus of third; margins of legs hairy, especially on merus joints; anterior margin of propodi convex, posterior margin straight in legs 1, 2, and 3, concave in leg 4; dactyli 1, 2, and 3 curved, stout except tip, which is slender and long; dactylus 4 straighter in basal half but still curved.

Abdomen very long, narrowing rapidly at fourth segment; after that, sides straight, little convergent; terminal segment subquadrate.

Measurements.—Male holotype, length of carapace 3, width of same 3.6 mm.

PINNOTHERES PUGETTENSIS Holmes.

Plate 17, figs. 7 and 8; plate 18, fig. 1.

Pinnotheres pugettensis HOLMES, Occas. Papers California Acad. Sci., vol. 7, 1900, p. 86 (type-locality, Puget Sound, in branchial cavity of *Cynthia;* type in Mus. Univ. California).

Diagnosis.—Female, carapace subpentagonal, widest anteriorly. Chela widest immediately behind fingers. Legs increasing in length from 1 to 4; dactylus of fourth pair much the longest.

Description.—Carapace soft and yielding, smooth, subpentagonal, widest between antero-lateral angles. Front broadly rounded in dorsal view, projecting little beyond general outline of carapace. Orbits nearly circular, partially visible from above.

FIG. 39.—PINNOTHERES PUGETTENSIS, ENDOGNATH OF OUTER MAXILLIPED OF FEMALE (40396), × 21.

Maxillipeds very oblique, strongly pubescent; merus of large size, somewhat pentagonal; last two segments together quadrate, distally oblique, true outer margin (inner, when folded) straight; dactylus small, attached distal to middle of propodus, the end of which it does not reach.

Merus of chelipeds furnished with long hairs on upper margin; hands elongate, rounded, widest immediately behind fingers; fingers and inner side of palm short-pubescent; fingers subcylindrical, nearly straight, a little shorter than palm, tips strongly hooked; dactyl with a low tooth near base of prehensile margin; immovable finger produced at middle in a low, broad lobe.

Legs slender, increasing slightly in length from first to fourth pair; propodi hairy above and below; dactyls narrow, compressed, pubescent, convex above, abruptly contracted near tip into a short, curved claw; in the first three legs the dactyls are shorter than propodi and have the lower margin nearly straight; in the last leg the dactylus is longer than propodus, much longer than preceding dactyli, and has the lower margin concave and long-hairy. (After Holmes chiefly.)

Variations.—In two smaller females which are, however, mature, the carapace is just as wide as long, the manus is relatively shorter, wider, and more swollen.

Color.—Outer surface of palm brownish with light-colored reticulations.

Measurements.—Female holotype, length of carapace 10, width of same 10.5; length of first leg 9.5, of last leg 10.5 mm. (Holmes).

Habitat.—Live in tunicates.

Range.—Departure Bay, British Columbia; Puget Sound.

Material examined.—Has all been dried. Departure Bay, British Columbia: George W. Taylor, collector; two females, larger one in tunicate, smaller one free (39131). 1908; Geological Survey of Canada, donor; two females, in hairy tunicate (40396).

PINNOTHERES NUDUS Holmes.

Pinnotheres nudus HOLMES, Proc. California Acad. Sci., ser. 2, vol. 4, 1894, p. 563, pl. 20, figs. 1–5 (type-locality, Santa Cruz; type destroyed in San Francisco fire); Occas. Papers California Acad. Sci., vol. 7, 1900, p. 86. Not *P. nudus* Weymouth. Leland Stanford Jr. Univ. Publ., Univ. Ser. No. 4, 1910, p. 53, text-fig. 1, except synonymy.

Diagnosis.—Carapace subquadrate with corners rounded; no indication of regions. Chela widest immediately behind fingers. Dactyli of legs nearly straight, longest and most slender in fourth pair.

Description.—Carapace a little broader than long, subquadrate with corners rounded, the anterior half nearly same shape and size

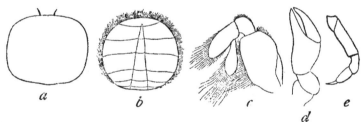

FIG. 40.—PINNOTHERES NUDUS, FEMALE HOLOTYPE. *a*, CARAPACE, SLIGHTLY ENLARGED; *b*, ABDOMEN; *c*, OUTER MAXILLIPED; *d*, LEFT CHELIPED; *e*, FIRST LEG. (AFTER HOLMES.)

as posterior half; surface curving downward toward all margins, smooth and naked; regions not defined. Front not protruding. Orbits ovate.

Chelipeds smooth, hands rather thick, widest immediately behind articulation of dactyl; fingers nearly or quite as long as palm, subconical, not conspicuously dentate on inner margins, partly covered by a very short, dense pubescence.

Three anterior legs subequal, fourth smaller; all are smooth, little compressed, and have acute, nearly straight tarsi, those of fourth pair being relatively longer and more slender than those preceding.

Abdomen of female nearly circular, covering entire sternal surface, composed of seven separate segments of which the fourth, fifth, and sixth are subequal and larger than the others.

Measurements.—Female holotype, length of carapace 20, width of same 24 mm.

Range.—Santa Cruz, California (type-locality); two females belonging to the California Academy of Sciences were destroyed in the fire following the San Francisco earthquake, 1906. Monterey (Holmes).

PINNOTHERES SERREI Rathbun.

Plate 19, figs. 1–7.

Pinnotheres serrei RATHBUN, Bull. Mus. Hist. Nat., Paris, 1909, No. 2, p. 68, 1 text-fig. (type-locality, Porto Rico; type in Paris Mus.).

Diagnosis.—Dactyli falcate. Palp of maxilliped with three joints end to end. Male spotted all over; sternum cristate. Orbits of female not visible from above; ten pits on anterior half of carapace.

Description of male.—Carapace slightly broader than long, suborbicular, widest in front of middle, posterior margin straight, anterior margin arcuate; surface punctate, flattened but with a slight convexity, a pit either side of cardiac region and a slight trace of cervical suture; no marginal line to dorsum. Front in anterior view curves downward and backward to interantennular plate, its width nearly one-third that of carapace; while width of front and orbits together slightly exceeds half width of carapace. Eye-stalks stout, diminishing from base to tip, visible from above, corneae large. Antennae project slightly beyond outer corners of front; antennules very large, transversely-obliquely placed.

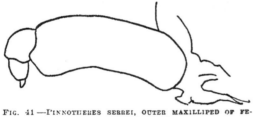

FIG. 41.—PINNOTHERES SERREI, OUTER MAXILLIPED OF FEMALE (48571), × 20.

Outer maxillipeds small, occupying but little space antero-posteriorly; palpus stout, its three segments end to end, dactylus short and stout.

Chelae stout, oblong, lower margin of propodus straight for nearly its length, upper margin convex, article widest at distal end; fingers broad at base, meeting along the edges when closed; tips curved toward each other, a basal tooth on movable finger fitting in corresponding notch in thumb.

First three legs nearly of a length, second longest, fourth shortest and narrower; stout except for the dactyls which are slender and falcate; second and third legs with an overlying fringe of long hair along distal end of carpus and continued along upper edge of posterior surface of propodus; lower margin of carpus of first three pairs short-hairy.

First sternal segment very large, anterior margin concave at middle and with a large V-shaped emargination each side behind insertion of maxilliped; anterior edge of fifth sternal segment on a much lower level than fourth segment, being somewhat cristiform; next segment similar but in lesser degree; adjacent margins of coxae of first and second legs cristiform.

Description of female (48571).—Carapace much wider than long, oblong with corners rounded, front and orbits wholly invisible in dorsal view, anterior margin as seen from above bilobed by a broad median gastric depression; on anterior half of carapace five pits each side of middle, four of which form a square, obliquely placed, the fifth pit a little outside the outermost pit of the square. Orbits oval, longer than eyes.

Chelipeds weaker than in male, propodus sinuous below, palm of subequal height throughout, thumb slightly deflexed. Merus of first three legs fringed with long hair on upper margin, this fringe continued on inner surface of carpus of first pair.

Legs less stout than in male, dactyli similar.

Variation.—A smaller female from Montego Bay has the front between the eyes, but not the orbits, visible in dorsal view so that anterior margin is not bilobed by median depression. Carapace widened a little anteriorly, legs slightly wider.

Color.—Entire surface of male above and below covered with dark pigment spots arranged with tolerable symmetry on the carapace; a few spots on anterior surface of eye-stalks.

Measurements.—Male holotype, length of carapace 3, width of same 3.3 mm.; female (48571), length of carapace 8.7, width of same 11.8 mm.

Habitat.—Female commensal in *Strombus*, male free-swimming at surface.

Range.—Cuba; Jamaica; Porto Rico.

Material examined.—

Cuba, on reef flat between Cayo Hutia and Little Cayo, northeast of light; 1914; Henderson and Bartsch, *Tomas Barrera* Exped.; 1 female (48571).

Jamaica: From mantle cavity of small *Strombus* towed in Montego Bay; Aug. 2, 1910; E. A. Andrews; 1 female (49214). Kingston; surface, by electric light; 1884; W. Nye, jr., Str. *Albatross;* 1 male (49213).

Porto Rico; 1907; P. Serre coll.; 1 male holotype (Paris Mus.).

PINNOTHERES CONCHARUM (Rathbun).

Plate 20, figs. 3–6.

Cryptophrys concharum RATHBUN, Proc. U. S. Nat. Mus., Vol. 16, 1893, p. 250 (type-localities, False Bay, San Diego County, California, from mantle of *Mya arenaria*, Cat. No. 17498, and Puget Sound, from *Cardita borealis*); Harriman Alaska Exped., vol. 10, 1904, p. 188, text-fig. 94, pl. 7, fig. 6.—WEYMOUTH, Leland' Stanford Jr. Univ. Publ., Univ. Ser. No. 4, 1910, p. 60.

Diagnosis.—Carapace almost hard, posterior margin transverse. Anterior half of carapace, as well as chelipeds and legs, thickly margined with hair. Dactyli of legs falcate, similar.

Description.—Carapace subpentagonal, a little wider than long, almost hard, smooth, a faint sulcus behind gastric region; anterior and antero-lateral margins defined by a rim of coarse setae, thickest and longest at antero-lateral angles; sides rounding gradually downward; posterior margin transverse, rimmed; front arcuate, advanced. Eyes stout, in circular orbits. Antennae equal in length to half width of front.

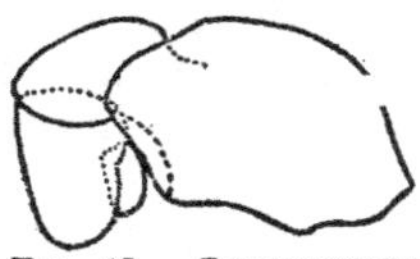

Terminal joint of palp of maxilliped small, attached to inner edge of propodus and, when the maxillipeds are in place, it is hidden under the thin edge of the merus.

FIG. 42.—PINNOTHERES CONCHARUM, ENDOGNATH OF OUTER MAXILLIPED OF MALE (23929). × 20.

Chelipeds stout, margined with a band of coarse setae, as are also the ambulatory legs. Inner surface of palm swollen, hairy; palm widening much to distal end, lower margin convex.

Length of legs 2, 3, 1, 4, the second leg longest; segments broad and flat, anterior margin of propodi and posterior margin of first propodus convex; posterior margin of second propodus concave, of third and fourth straight; dactyli similar, falcate, slender, horny tips long and very slender, making nearly half length of segment.

Sternum and abdomen smooth, the latter bordered with hair; abdomen of adult male and immature female tapering rather regularly from third to last segment, third, fourth and fifth partially fused, seventh subquadrate with distal margin arcuate. Since this paper was sent to the printer a large number of small specimens (50443) have been received, of which the males have the abdomen widened at the end, the terminal segment being wider than long, end truncate, sides rounded, while the females have the tapering abdomen characteristic of the large male. I am unable to detect other differences in the two forms of the male. All the females known have a narrow abdomen and are probably immature.

Color.—In a male and a female, preserved in alcohol, a persistent, dark spot either side of middle and a dark line in gastro-cardiac sulcus. A female (49628) in formalin is a mixture of grayish-white and tawny ochraceous, the latter color forming a pattern chiefly on the median and antero-lateral parts of the carapace, and covering the fingers of the cheliped and the last 3 segments and the distal end of the merus of the legs.

Measurements.—Male (23929), length of carapace 7, width of same 7.3 mm. Female holotype, 4.7 by 5.2 mm.

Habitat.—Commensal in bivalve mollusks and ascidians; sometimes free-swimming.

Range.—British Columbia to San Diego Bay, California.

Material examined.—

Hammond Bay, Vancouver Island, British Columbia; in *Mya arenaria;* May 20, 1916; C. McLean Fraser; 1 female (49628)

Departure Bay, Vancouver Island, British Columbia; 17 fathoms; Dec. 16, 1908; George W. Taylor; 1 male (39129).

Puget Sound; in animals of *Cardita borealis* Conrad; 2 specimens, one is a young female (17502).

Neah Bay, Washington; surface; *Albatross;* 1 female (18410).

Stewarts Point, California; in bivalve, *Kellia laperousii* Deshayes; July 11, 1911; W. F. Thompson; 1 male (Stanford Univ.).

Pacific Grove, California; from mantle cavity of *Mytilus edulis;* John C. Brown; 1 male (23929).

Off Santa Cruz Island, California; lat. 34° N.; long. 119° 29′ 30″ W.; 30 fathoms; P.; Feb. 6, 1889; station 2945, *Albatross;* 1 female y., from *Phallusia vermiformis* (45611). Identification ?.

False Bay, San Diego County, California; in mantle of *Mya californica;* June 4, 1882; C. R. Orcutt; 1 female holotype, 1 y. (17498).

San Diego Bay, California; Beacon No. 8, NW. by W. $\frac{1}{2}$ W. $\frac{8}{10}$ mile; 5 fathoms; fne. S. M. brk. Sh.; Mar. 19, 1894; station 3564; *Albatross;* 2 males, 1 female (25428).

San Diego, California; in *Donax levigatus;* C. R. Orcutt; 15 males, 13 females (50443).

PINNOTHERES PUBESCENS (Holmes).

Cryptophrys pubescens HOLMES, Proc. California Acad. Sci., ser. 2, vol. 4, 1894, p. 564, pl. 20, figs. 6 and 7 (type-locality, Muleje Bay, Gulf of California; type not extant).

Diagnosis.—Pubescent all over. Carapace subpentagonal, with two longitudinal furrows. Front with median notch. Legs subequal, dactyli curved.

Description of female.—Carapace subpentagonal, convex, median and cardiac regions tumid and separated from hepatic and branchial regions by a sulcus; a slight depression between gastric and cardiac regions and another behind the latter. Front slightly projecting, notched in middle. Antero-lateral margins not defined by a ridge; sides of carapace broadly rounded. Orbits nearly circular; eyestalks very short and stout. Antennæ shorter than width of front. Palp of outer maxilliped 2-jointed.[1]

Chelipeds moderate, exceeding first pair of legs; hand narrow, oblong, somewhat compressed, and concave on inner face; fingers about as long as palm, subcylindrical, hooked at tip, and not dentate on inner margin.

Legs subequal (last pair a little shorter than others), moderately slender, somewhat compressed, joints not unusually widened; dactyli rather slender, curved, and from one-half to two-thirds length of propodi.

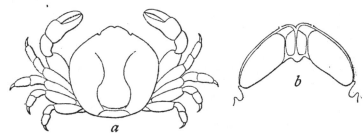

FIG. 43.—PINNOTHERES PUBESCENS, FEMALE HOLOTYPE. a, GENERAL OUTLINE, × 2⅕; b, BUCCAL AREA. (AFTER HOLMES.)

Abdomen rounded, slightly longer than broad, and covering entire sternal area. Body and legs covered with a uniform, short, dense pubescence. (After Holmes.)

Measurements.—Female holotype, length of carapace 9.75, width of same 10 mm. In Holmes's figure the carapace is shown longer than wide, perhaps due to flattening out the front in drawing.

Material recorded.—Muleje Bay, Gulf of California, Mexico; 1 female holotype (destroyed in San Francisco fire).

PINNOTHERES BARBATUS Desbonne.

Plate 19, figs. 8–11.

Pinnotheres barbata DESBONNE, in Desbonne and Schramm, Crust. Guadeloupe, pt. 1, 1867, p. 44 (type-locality, Guadeloupe; type probably not extant).

Diagnosis.—Heavily bearded below and on margins of legs. Orbits of female invisible from above. *D*actyli of legs falcate.

---

[1] It is likely that there is a small third segment which was overlooked, as was the case with *concharum*.

Description of female.—Carapace thin, yielding; transversely quadrilateral, with anterior margin arcuate and antero-lateral angles. rounded, postero-lateral corners truncated; glabrous; convex in both directions, anteriorly so deflexed that orbits and front are invisible in dorsal view. Orbits oval, larger than eyes. Extremity of front, as well as under surface of carapace, margin of abdomen and margins of legs except of dactyli, densely bearded; same bearding on upper margin of arm, inner margin of wrist, and along middle of inner-surface of palm.

First two articles of palp of outer maxilliped short and stout; last article about twice as long as wide, attached at inner distal angle of propodus.

Chelipeds short, stout; propodus convex above, sinuous below, fingers subconical, meeting when closed, a low tooth at middle of prehensile edge, tips hooked and crossing.

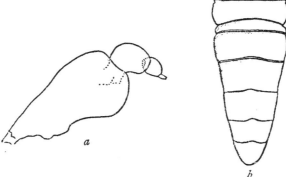

FIG. 44.—PINNOTHERES BARBATUS (23435). *a,* ENDOGNATH OF OUTER MAXILLIPED OF FEMALE, × 20; *b,* ABDOMEN OF MALE, × 7½.

Legs similar, broad, compressed, dactyli falcate; first leg very little shorter than second, third reaching to middle of propodus of second, fourth to middle of propodus of third.

Description of male.—Smaller and flatter than female and firmer though not hard, suborbicular, broader anteriorly, a large pit either side of middle, front convex and a little advanced as seen from above, eyes filling orbits and visible dorsally. Pubescence as in female, but covering chela also; propodus less sinuous below, thumb shorter. Last leg relatively smaller than in female. Second segment of abdomen narrower than first and third; sides of third convex; from first to seventh regularly tapering; extremity rounded.

Color.—Whitish or yellowish, spotted with black (Desbonne).

Measurements.—Female (23435), length of carapace 10.2, width of same 12.7 mm. Male (23435), length of carapace 7.6, width of same 8.4 mm.

Habitat.—Often found in pairs in stomach of *Turbo pica* (Desbonne).

Range.—St. Thomas; Guadeloupe.

Material examined.—St. Thomas, West Indies; Krebs, collector; received from Copenhagen Mus.; 1 male, 1 female (23435).

PINNOTHERES STROMBI Rathbun.

Plate 20, figs. 1 and 2.

Pinnotheres strombi RATHBUN, Proc. Acad. Nat. Sci. Philadelphia, 1905, p. 371, 2 text-figs. (type-locality, Clearwater Harbor, Florida, in *Strombus pugilis;* type in Mus. Phila. Acad. Sci.).

Diagnosis.—Carapace of female very wide, sides truncate. Eyes visible from above. Dactyli of legs falcate. Last leg much narrower than first three.

Description of female.—Carapace nearly one and a half times as broad as long, oblong, sides parallel, corners rounded; dorsal surface curving down toward margins, smooth and naked; integument very thin and easily wrinkled. Front less than one-fourth width of carapace, strongly deflexed, forming an obtuse angle at middle. Orbits suborbicular, eyes partly visible in dorsal view.

Outer maxillipeds with carpal and propodal joints short and stout, last joint small and almost terminal.

FIG. 45.—PINNOTHERES STROMBI, ENDOGNATH OF OUTER MAXILLIPED OF FEMALE HOLOTYPE. MUCH ENLARGED.

Chelipeds small, rounded; upper margin of propodus convex, lower margin sinuous, fingers slightly deflexed, stout, fitting close together, tips sharp and crossing.

Second pair of legs a little the longest, first pair stoutest, fourth pair much shorter and narrower than the others; dactyli short, falcate, with slender horny tips. Edge of front, anterior (or upper) margin of first three legs, and lower margin of carpus of first leg fringed with short dense pubescence, as are also the lower surface of carapace and edge of maxillipeds and abdomen.

Measurements.—Female holotype, length of carapace 6.6, width of same 9.6 mm.

Habitat.—Commensal in *Strombus pugilis* Linnaeus.

Material examined.—Clearwater Harbor, west coast of Florida; in living *Strombus pugilis;* 1905; H. S. Colton, collector; 1 female holotype (1629, Mus. Acad. Nat. Sci. Phila.).

PINNOTHERES SILVESTRII Nobili.

Pinnoteres silvestrii NOBILI, Boll. Mus. Zool. Anat. Comp. R. Univ. Torino, vol. 16, No. 402, 1901, p. 11 (type-locality, San Vicente, Chile; holotype in Mus. Zool. Turin); Revista Chilena Hist. Nat., vol. 6, 1902, p. 235 (repetition of original description).

Pinnotheres silvestrii RATHBUN, Proc. U. S. Nat. Mus., vol. 38, 1910, p. 587.

Diagnosis.—Entire body and appendages thoroughly calcified and hard. Carapace without furrows. *D*actyli of legs very short.

Description of female (after Nobili).—Carapace a little wider than long, completely calcified and therefore very hard, flat except anteriorly and toward the lateral and posterior margins where it is bent down; bare and smooth with a luster like porcelain. Frontal margin, pterygostomian region, infero-posterior parts and margin of abdomen covered with a grayish tomentum. No furrows on the carapace, but two very slight and rather wide depressions near gastric region, and two impressions at sides of cardiac region. Fronto-orbital border rather wide; front vertically deflexed and produced at sides where it meets orbit in a very distinct acute lobe; between this lobe and the interantennular partition, the frontal border is deeply sinuous. Orbits invisible from above, subcircular; ocular peduncles short, stout, obconical. Antero-lateral margin rather long, separated from the anterior margin by a fairly well-marked angle, and from the postero-lateral margins by a more distinct angle; these last are directed obliquely backward.

Maxillipeds slightly oblique and rather pilose; merus subellipsoid, broadly truncate at end; dactylus claw-shaped, exceeding in length the preceding segment.

Chelipeds rather large, subequal, calcified like the carapace; outside bare and smooth, inside thickly pilose; fingers somewhat curved, excavated and incurved at tip, prehensile surface bearing long, stout hairs.

Legs short, stout, calcified, pilose along lower surface, along articulations and on outer face of merus; dactyls very short. Abdomen calcified.

Measurements.—Female holotype, length of carapace 12, width of same 14 mm.

Range.—Known only from the single type female taken at San Vicente, Chile.

PINNOTHERES MARGARITA Smith.

Pinnotheres margarita SMITH, in Verrill, Amer. Nat., vol. 3, 1869, p. 245, footnote (type-locality, in the Pearl Oyster, *Margaritophora fimbriata*, of the Bay of Panama; holotype female in Peabody Mus. Yale Univ.); Trans. Connecticut Acad. Arts and Sci., vol. 2, 1870, p. 166.

Diagnosis.—Legs of second pair unequal in female, propodus and dactylus of right leg much longer than those of left. Body and limbs short-pubescent. Carapace uneven. Chelae long, subcylindrical.

Description (after Smith).—A stout, thick species, with firm integument, everywhere covered, except dactylus of right, second leg of female and tips of the other dactyli in both sexes, with a short, close, clay-colored pubescence, like a uniform coat of mud.

Female.—Carapace very convex in all directions; dorsal surface, beneath pubescence, smooth and shining. Cardiac region protuberant, separated from gastric region by a conspicuous sulcus, and from branchial regions by very marked and deep depressions which extend along cervical suture to hepatic region; branchial regions protuberaut along their inner side; front not protuberant, strongly deflexed, a slight median depression.

Maxillipeds more longitudinal and of a firmer consistency than usual; merus short and broad, inner margin angulated at middle; second segment of palpus large, widest at middle where is attached the terminal segment, which is slightly spatulate and reaches almost to tip of propodus.

Chelipeds very stout; hands long, nearly cylindrical; fingers somewhat cylindrical, nearly straight almost to tips, which are hooked by one another; dactylus with a small tooth near base which fits a slight sinus in propodal finger.

Legs stout; all ischial segments, as well as posterior margins of propodus and dactylus of last leg, are clothed with a long, woolly pubescence; first 3 dactyli short, curved and pubescent nearly to tips except in second right leg, where propodus is considerably longer than left leg of same pair, and dactylus very long, almost straight and naked; fourth dactylus long, straight, slender and pubescent.

Anterior margin of sternum with a broad, rounded sinus for reception of tips of palpi of maxillipeds. Abdomen orbicular, covering sternum.

Male.—Smaller than female, less thickly pubescent, cardiac and branchial regions less protuberant, separated from gastric region by slight depression; front slightly projecting, less deflexed than in female. Chelipeds and legs as in female, except that the right legs are like the left. Abdomen widest at third segment, margins from third to sixth straight and converging, sixth abruptly contracted, last segment nearly square.

Sterile female.—More like male than adult female, but narrower and more depressed, front more prominent and scarcely deflexed; dorsal surface slightly areolated, flat and clothed, except cardiac region and a small space in middle of gastric region, with an almost black, velvety pubescence. Cheliped stouter than in male and adult female, a black pubescence on upper surface of carpus and a small space at base of hand. Legs less pubescent than in male, propodus and dactylus of right second leg longer than in left leg, but not as long as in mature female. Abdomen not wider than in male, but margins slightly convex, sixth segment not contracted, extremity rounded. This form may be a distinct species.

Measurements.—Female, Pearl Islands, length of carapace 11.8, width of same 13.4 mm. Male, Pearl Islands, length of carapace 5.5, width of same 6.1 mm.

Habitat.—Lives in pearl oyster (*Margaritophora fimbriata* Dunker).

Range.—La Paz, Lower California, Mexico (Smith); Muleje Bay, Gulf of California (Holmes); Pearl Islands, Bay of Panama (Smith).

PINNOTHERES RETICULATUS, new species.

Plate 21, figs. 1 and 2.

Type-locality.—Gulf of California: off San Josef Island, Lower California, Mexico; lat. 25° 02' 15" N.; long. 110° 43' 30" W.; 17 fathoms; S. Sh.; March 17, 1889; station 3002, *Albatross;* 1 female (18217).

Diagnosis.—Female suborbicular, longer than broad; hands reticulated. Second leg longest, fourth dactylus longest, second and third propodus swollen at base.

Description of female.—Carapace suborbicular, longer than broad, soft, papyraceous, slightly convex; gastric region bluntly elevated on median line, and separated by a broad depression from the branchial region; 4 pits form a quadrilateral in this depression; a lunate depression near hepatic margin. Front slightly convex in dorsal view, scarcely projecting beyond curve of antero-lateral margin. Eyes orbicular, large, visible dorsally, corneae of moderate size.

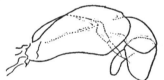

FIG. 46.—PINNOTHERES RETICULATUS, OUTER MAXILLIPED OF FEMALE HOLOTYPE, × 15½.

First two articles of palpus of outer maxilliped short and broad; dactylus curved and with subparallel margins and attached near proximal end of propodus which article it does not overreach.

Chelipeds and legs short-pubescent, chelipeds long-hairy inside, small; manus subcylindrical, increasing a little distally, lower margin straight, upper slightly convex; pubescence arranged in a reticulating pattern, the interstices of which are smooth and of a darker color (in alcohol); fingers stout, points hooked toward each other, prehensile edges each armed with two teeth, of which the one near base of dactyl is the largest.

Legs furnished with long hair on lower margin, especially on propodus of first leg; length expressed by 2. 4. 3. 1, the second leg being the longest; propodus of second and third very concave below, much thickened at proximal end; propodus of first and fourth straight below; dactyli 1, 2, and 3 similar, curved, and with slender tips, first shortest, second longest; dactylus 4 much longer than those preceding and longer than its propodus.

Abdomen longer and broader than carapace.

Color.—A triangular spot of lighter color embraces the front and orbits and is continued in a narrow stripe for a short way on the gastric region.

Measurements.—Female holotype, length of carapace 9.7, width of same 9.2 mm.

Range.—Known only from the type-locality.

PINNOTHERES MOSERI,[1] new species.

Plate 21, figs. 3 and 4.

Type-locality.—Port Royal, Jamaica; in black ascidian; 1893; R. P. Bigelow; female ovigerous (23440).

Diagnosis.—Carapace of female longer than broad. Dorsum subpentagonal, uneven. Legs slender, subequal.

Description of female.—Surface pubescent. Carapace thin, moderately firm but not hard, suborbicular with the front and intestinal region a little produced; surface uneven; gastric, cardiac, and intestinal regions elevated and well-defined; a low tubercle on either side of the anterior gastric slope; from a point on the postero-lateral margin a blunt slightly curved (concave outward) ridge runs forward and outward to the lateral angle or widest part of the carapace; below the ridge the surface descends steeply; the result being that the dorsum appears subpentagonal. Front narrow, arcuate, a median emargination.

FIG. 47.—PINNOTHERES MOSERI, ENDOGNATH OF OUTER MAXILLIPED OF FEMALE HOLOTYPE, × 17.

Palp of outer maxilliped two-jointed; outer edge very thick, inner edge thin; first joint short and broad, outer margin angled; second joint oblong, distally rounded. No trace of a dactylus could be found, so that this species presents an exception to the rule of a three-jointed palp in *Pinnotheres.*

Chelipeds elongate, rather slender; manus increasing distally, margins nearly straight; fingers nearly as long as top of palm, a little

---

[1] Named for Lieut. J. F. Moser, U. S. Navy, the first to collect this species.

deflexed, subcylindrical with tips turned inward almost at a right angle, a small gape at base formed by a sinus in each finger.

Legs very slender, of similar length, dactyli of first three legs with a straight base and long curved tip, of last leg with a long straight part and short curved tip; propodi with a thin fringe of hairs below and a fringe of hairs lying on the posterior surface which are attached near upper margin.

Abdomen longer than wide.

Measurements.—Female holotype, length of carapace 9, width of same 8.4 mm.

Habitat.—Commensal in black ascidian.

Range.—West coast of Florida; Jamaica.

Material examined.—

Off St. Martins Reef, West Florida; lat. 28° 43′ N.; long. 82° 56′ W.; 17 feet; rocky bottom covered with grass and thin layers of sand and mud; 1887; Lieut. J. F. Moser, U. S. Navy; 1 female ovig. (13056). "Very many sea squirts in the dredge."

Near Kingston, Jamaica; 1891; T. H. Morgan; 1 female (17238).

Port Royal, Jamaica; in black ascidian; 1893; R. P. Bigelow; 2 females ovig., 1 is holotype (23440).

PINNOTHERES SHOEMAKERI, new species.

Plate 22, figs. 1–4.

Type-locality.—St. Thomas; summer, 1915; Clarence R. Shoemaker, for whom the species is named; male holotype (49216).

Diagnosis.—Male and female, a deep median groove on front; legs similar, diminishing from 1 to 4; dactyli slender, curved. Male only, carapace very uneven; manus very swollen, strongly ridged.

Description of male.—Small species, pubescent all over; carapace subhexagonal, widest at middle; postero-lateral portion of branchial region forming a steeply inclined facet with smooth concave surface; dorsal surface cut by deep furrows into high regularly placed elevations; one furrow surrounds the circular cardiac region; one runs transversely across middle of carapace; another, transverse, on either side of median gastric areole; still another behind the front. Lateral angles broadly rounded and with antero-lateral margins bordered by a raised rim. Front thick, subtruncate, a minute median emargination, continued backward by an impressed line. Eyes spherical, projecting well out of orbits.

Palpus of outer maxilliped very small, compared to merus-ischium; shape akin to that of *P. moseri;* but the inner margin bears a very short and broad rudiment of a dactylus,

Chelipeds stout; manus inflated, a blunt crest above and two of the same on outer surface, the lower continued on thumb; fingers long, with slender, hooked tips; immovable finger horizontal; prehensile edges denticulate.

Legs short, similar, nearly of a length but diminishing slightly from first to fourth; lower margins of merus, carpus and propodus densely hairy; posterior surface of carpus-propodus of second and third legs with a thin fringe of hair attached near anterior margin; dactyli very slender throughout, curved, translucent, a few long hairs below, tips acuminate.

Abdomen narrow-triangular, second segment with a prominent, transverse, median tubercle; terminal segment subrectangular with rounded tip.

Description of female.—One damaged female showing a very thin and almost shapeless carapace, a broad abdomen and left cheliped

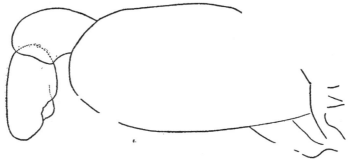

FIG. 48.—PINNOTHERES SHOEMAKERI, ENDOGNATH OF OUTER MAXILLIPED OF MALE HOLOTYPE, × 97.

and legs, is referred to this species. A little larger than male. Front and orbits as in male. Surface short-pubescent. Chela more elongate and manus less swollen, its ridges less evident than in male. Legs as in male.

Measurements.—Male holotype, length of carapace 3.7, width of same 3.3 mm. Female (49220), approximate length of carapace 4.7 mm.

Range.—West coast of Florida; St. Thomas, West Indies.

Material examined.—

Charlotte Harbor, Florida; Union College Collection (745, 746, 757, 763); 3 males, 1 female (deposit in U.S.N.M., Cat. Nos. 49217, 49218, 49219, 49220).

St. Thomas, West Indies; summer, 1915; Clarence R. Shoemaker; 1 male holotype (49216).

PINNOTHERES TAYLORI, new species.

Plate 21, figs. 5-8.

Type-locality.—Ucluelet, British Columbia; in transparent tunicate; Geological Survey of Canada; female holotype (40397).

Diagnosis.—Carapace in both sexes of subequal length and width. Chela inflated. Fourth leg shorter than the others which are nearly equal; dactyli slender, curved, similar; in male a fringe of long hair on legs 2 and 3. Male hairy all over.

Description of female.—Of small size. Carapace thin, but fairly firm, of equal length and width, very convex, subhexagonal, the side-margins parallel behind antero-lateral angles, the postlateral portion of the branchial region forming a steep facet strongly inclined to remainder of dorsal surface; cardiac region surmounted by two small tubercles side by side; cardiac and posterior gastric region defined by shallow grooves; posterior margin convex. In ventral view the posterior line of the antennular cavities, of the minute, basal segment of the antennae and the lower border of the orbit form a single transverse line.

Buccal cavity broadly hemispherical; penult segment of outer maxilliped broadly obliquely-truncate; last segment small, attached near distal end of the preceding and not overreaching it.

FIG. 49.— PINNOTHERES TAYLORI, ENDOGNATH OF OUTER MAXILLIPED OF MALE (40397). × 40.

Chelipeds hairy inside; manus much inflated, upper and lower margins convex; fingers equally stout, tips strongly hooked, a small tooth near base of dactyl, a larger tooth or lobe near middle of immovable finger. The chelipeds are shaped much as in *pugettensis* but are much smaller, specimens of subequal size compared.

Of the ambulatory legs, 1 and 4 are subequal, shorter than 2 and 3, which are subequal; propodi convex on anterior and straight on posterior margin; dactyli shorter than propodi, curved and with slender tips, that of first leg shorter then the others which are subequal to one another; propodi and dactyli hairy.

Description of male.—A little smaller than female and hairy all over; hair thick and matted on dorsal surface of carapace but not on oblique branchial facets or on ventral surface. Carapace widest at antero-lateral angles, the side margins converging slightly posteriorly; cardiac region defined by a furrow.

Chelipeds larger than in female and clothed with long hair; manus more inflated than in female. The legs of the first three pairs are nearly of a length, of the fourth pair shorter; besides the close, short hair with which they are covered, there is on the posterior surface of the last three segments of the second and third legs a thin fringe of long, fine hairs; it extends obliquely across carpus and near upper edge of propodus and dactylus; dactyli similar, curved, rather narrow with long, slender tips.

Segments of abdomen separate; lateral margins of third segment arcuate; terminal segment equilaterally triangular, blunt-pointed.

Measurements.—Female, holotype, length and width of carapace each 4.4 mm.; male (40397), length 4.3, width 4.2 mm.

Habitat.—In transparent tunicates.

Range.—British Columbia.

Material examined.—

Departure Bay, British Columbia; in a transparent tunicate; George W. Taylor, collector; 1 male, 1 young female (39127). "Common.'

Ucluelet, British Columbia; in transparent tunicates; Geological Survey of Canada; 1 male, 1 female holotype (40397).

Variations.—The immature female from *D*eparture Bay has an abdomen almost as narrow as the male, which in other respects it entirely resembles. In both the carapace has a more uneven surface than in the male from Ucluelet which is a little larger, and the lateral margin at widest part of carapace shows a short raised rim.

PINNOTHERES ORCUTTI, new species.

Plate 22, figs. 5 and 6.

Type-locality.—Manzanillo, Mexico; October, 1910; C. R. Orcutt, collector; 1 male holotype, dried (49215).

Diagnosis.—Longer than broad, hairy. Cheliped very stout. Second leg longest, fourth shortest; dactyli slender, curved, subequal. End joint of outer maxilliped minute, attached to distal, oblique margin of preceding segment.

Description of male.—Small. Carapace subhexagonal, longer than broad, broadest in posterior half, very uneven, postero-lateral portion of branchial region forming a steep triangular facet; dorsal surface densely hairy and bordered by a raised rim; cardiac region surrounded by a furrow except anteriorly, and surmounted by a median tubercle near its posterior end. Front truncate, corners rounded, a small but clear-cut, triangular, median notch; behind this runs a broad median furrow. Lateral margins long, convex; postero-lateral margins short, concave; posterior margin short and straight. Basal segment of antenna elongate and very obliquely placed.

Merus of outer maxilliped very wide and angled; propodus broadly subtriangular, having a minute linear dactylus inserted on the proximal portion of its oblique distal margin.

Cheliped (left only is present) stout, manus short, increasing greatly in width toward distal end where it is higher than superior length; lower margin of propodus straight except distal half of finger curved upward, well overlapping dactylus; edge of immovable finger with a few teeth of which the one at middle is largest; dactylus also wide at base, and strongly arched.

Legs narrow, length 2.1 and 3.4—that is, 2 longest, 1 and 3 next, and of about equal length; propodi with both margins convex except the hind margin of the second one which is straight; dactyli slender, curved, subequal, with spinelike tips.

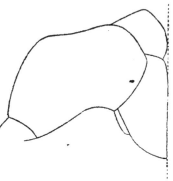

FIG. 50.—PINNOTHERES ORCUTTI, ENDOGNATH OF OUTER MAXILLIPED OF MALE HOLOTYPE, × 67.

Abdomen suboblong, terminal segment transversely oblong and longer than the sixth segment which has a shallow rightangled indentation on either side.

Color.—Dorsal surface of dried specimen a dark purple except across front, on posterior margin and posterior half of cardiac region.

Measurements.—Male holotype, length of carapace 3.6, width of same 3.1 mm.

Range.—Known only from the type-locality.

PINNOTHERES HEMPHILLI, new species.

Plate 23.

Type-locality.—Cedar Keys, Florida; between tides; Henry Hemphill; "seems rare"; 1 male (6420).

Diagnosis.—Male octagonal. Chelipeds stout. Second leg longest, as also its propodus and dactylus; propodus concave on posterior margin. Dactyli curved. Sternum sharply ridged.

Description of male.—Small and almost bare. Carapace suboctagonal, the posterior margin twice as long as anterior margin, midlateral margins parallel to median line, as long as postlateral margins, antero-lateral margins (to angles of front) longest. Carapace depressed, somewhat wrinkled.

Outer maxillipeds *in situ* short and wide; partly overlapped by sternum; the outer margin of ischium-merus broadly arcuate; palpus

100 BULLETIN 97, UNITED STATES NATIONAL MUSEUM.

with the first two joints short and broad, the dactylus small, inserted near middle of margin of propodus.

Chelipeds stout; a tuft of hair at distal, inner angle of merus; manus nearly as wide as superior length, and increasing distally in width a little, upper margin almost straight, lower convex, distal width greater than width across base of fingers; fingers broad at base, regularly tapering, denticulate on prehensile edges, a tooth at base of right dactyl.

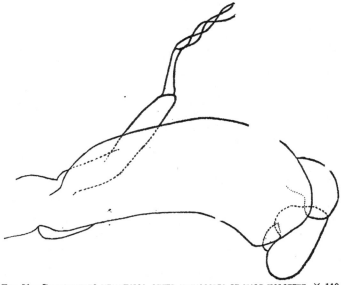

FIG. 51.—PINNOTHERES HEMPHILLI, OUTER MAXILLIPED OF MALE HOLOTYPE, × 110.

The length of the legs is expressed by 2.3.1.4; the second is distinctly longer than the others; and especially is its propodus, which differs also from the others in its concave posterior margin; posterior margin of propodus 1 convex, of 3 and 4 straight, anterior margin of all convex; dactyli curved, the second one longer and narrower than all the others; dactyli 1, 3, and 4 are very stout at base, subconical with short curved tip.

Sternum with a sharp obliquely transverse crest running inward from base of legs 1, 2, and 3.

Measurements.—Male holotype, length of carapace 2.3, width 3 mm.

Range.—Known only from the type-locality.

PINNOTHERES GUERINI Milne Edwards.

Pinnotheres guerini MILNE EDWARDS, Ann. Sci. Nat., ser. 3, Zool., vol. 20, 1853, p. 219 [185], pl. 11, fig. 9 (type-locality, Cuba; type in Paris Mus.).—GUNDLACH, Anales Hist. Nat., Madrid, vol. 16, 1887, p. 124.
Pinnateres guerini VON MARTENS, Arch. f. Natur., vol. 38, 1872, p. 105.

FIG. 52.—PINNOTHERES GUERINI, OUTER MAXILLIPED OF HOLOTYPE, ENLARGED. (AFTER MILNE EDWARDS.)

Diagnosis.—Female ?. Maxillipeds wide, palp very large, dactylus large, subspatulate. Hands bare, short and punctate (Milne Edwards). Carapace nearly one and one-half times broader than long (von Martens).

Habitat.—In oysters (Gundlach).

Range.—Cuba (Milne Edwards, von Martens); Porto Rico (Gundlach).

PINNOTHERES HIRTIMANUS Milne Edwards.

Pinnotheres hirtimanus MILNE EDWARDS, Ann. Sci. Nat., ser. 3, Zool., vol. 20, 1853, p. 219 [185] (type-locality, Cuba; type in Paris Mus.).

Diagnosis.—Female ?. Maxillipeds as in *P. guerini*. Hands elongate and strongly ciliated on lower margin. (Milne Edwards.)

Range.—Cuba (Milne Edwards).

Genus FABIA Dana.

Fabia DANA, Amer. Journ. Sci., ser. 2, vol. 12, 1851, p. 290; Proc. Acad. Nat. Sci. Philadelphia, vol. 5, 1851 (1852), p. 253; type, *F. subquadrata* Dana.

Raphonotus RATHBUN, Proc. Biol. Soc. Washington, vol. 11, 1897, p. 166; type, *R. subquadratus* Dana.

Female near *Pinnotheres*. Carapace marked by two longitudinal sulci which extend backward from the upper margin of the orbits, inclosing between them the median area. The anterior portion of the carapace is so deflexed that the orbital and antennal area is ventrally situated; fronto-orbital distance narrow; frontal margin flush with antennular surface, and not forming a projecting hood; orbits round, eyestalks spherical, corneas small. Terminal segment of palp of outer maxilliped articulating on inner margin of penult segment. Second ambulatory leg longest; dactyli short.

Male not known.

Restricted to North American waters.

Analogous species on opposite sides of the continent: *byssomiae* (Atlantic); *canfieldi* (Pacific).

KEY TO THE SPECIES OF THE GENUS FABIA.

A¹. Legs of second pair alike.
  B¹. A transverse pubescent sulcus across front. Palm distally widened.
    subquadrata, p. 102.
  B². No transverse sulcus or pubescence across front. Palm not distally widened _____*lowei*, p. 104.

A². Legs of second pair unlike, the right longer than the left.
  B². Second right propodite straight and tapering from proximal to distal end. slender. Chelae feeble, fingers horizontal_____*byssomiae*, p. 105.
  B³. Second right propodite straight and tapering from proximal to distal end. Legs not very slender, the first stout. Chelae not feeble, fingers deflexed_____*canfieldi*, p. 106.

FABIA SUBQUADRATA Dana.

Plate 24, figs. 1 and 3.

Fabia subquadrata DANA, Proc. Acad. Nat. Sci. Philadelphia, vol. 5, 1851, p. 253 (type-locality, Puget Sound; type not extant); Crust. U. S. Expl. Exped., vol. 13, Crust., pt. 1, 1852, p. 382; atlas, 1855, pl. 24, fig. 5 *a—e*.—HOLMES, Occas. Papers California Acad. Sci., vol. 7, 1900, p. 87 (part; not specimen from San Pedro, described).

Raphonotus subquadratus RATHBUN, Harriman Alaska Exped., vol. 10, 1904, p. 186 (part; not specimen from Monterey).

Diagnosis.—Front with transverse, pubescent sulcus. Last segment of maxilliped reaches end of penult segment. Palm widens

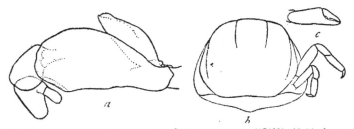

FIG. 53.—FABIA SUBQUADRATA, FEMALE. *a*, OUTER MAXILLIPED (17480), × 14; *b*, DORSAL VIEW OF BODY AND LAST TWO LEGS, × 1½; *c*, RIGHT CHELA, × 1½; *b* AND *c* FROM MONTEREY BAY, AFTER WEYMOUTH.

distally. Two rows of hair on lower surface of palm, the inner one continued to end of propodal finger.

Description of female.—Carapace smooth, glossy, membranaceous, subquadrate in outline with angles broadly rounded; space between longitudinal sulci longer than wide and slightly narrowed behind; antero-lateral margin rounded, and marked by a round cluster of pits. Front turned abruptly downward, the vertical part crossed by a shallow, pubescent, transverse sulcus, running between uppermost line of orbits. Eyestalks almost spherical, corneas small. Basal joint of antenna short and wide, situated mostly behind eye.

Penultimate joint of palp of outer maxilliped broad, flat, bearing the minute third joint near middle of inner margin.

Cheliped smooth, stouter than legs; hand elongate, increasing in width to distal end, margins almost straight, inner surface hairy; two rows of hair below, the innermost one continued to end of finger;

fingers much shorter than palm, a little gaping, tips crossing, immovable finger nearly horizontal, dactylus with a large tooth just behind middle which fits into a corresponding sinus in the propodal finger; this last has a small tooth at base.

Legs slender, glossy, both margins of merus and dactylus and posterior margins of carpus and propodus pubescent; relative length of legs 2. 3. 1. 4; propodites with convex anterior margins, first and fourth pairs with straight, and second and third with concave, posterior margin, dactyli curved, much shorter than preceding segment. Abdomen much larger than carapace.

Color.—Female (49627), the greater part of the carapace, especially the antero-lateral regions and a broad, longitudinal stripe on the abdomen, are orange chrome in a formalin-preserved specimen; remainder of specimen whitish.

Measurements.—Female (23928), length of carapace 10, width of same 12.5 mm.

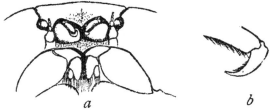

FIG. 54.—FABIA SUBQUADRATA. *a*, VENTRAL VIEW OF FRONT AND MOUTH, ENLARGED; *b*, END OF A LEG, ENLARGED. (AFTER DANA.)

Habitat.—Commensal in bivalve mollusks.

Range.—From off Akutan Pass, Alaska, to Laguna Beach, California. To a depth of 45 fathoms.

Material examined.—

Off S. entrance to Akutan Pass, Alaska; lat. 53° 56′ N.; long. 165° 56′ W.; 45 fathoms; brk. Sh. P.; temp. 43.5° F.; July 28, 1888; station 2843, *Albatross;* 1 female (17480).

Hammond Bay, Vancouver Island, British Columbia; in *Mya arenaria;* May 20, 1916; C. McLean Fraser; 1 female (49627).

Oyster Bay, Washington; from "butter" clam; Dec. 5, 1914; Biological Survey, U. S. Department of Agriculture; 1 female ovig. (coll. Biol. Surv.).

Pacific Grove, California; from mantle cavity of *Mytilus edulis;* John C. Brown; 2 females (23928).

Pacific Grove; 9 females (3 ovig.) and 1 y. female in *Mytilus* (Mus. Stanford Univ.).

Laguna Beach, California; William A. Hilton; sent to United States National Museum for identification.

FABIA LOWEI (Rathbun).

Plate 24, figs. 2 and 4.

Fabia subquadrata LOCKINGTON, Proc. California Acad. Sci., vol. 7, 1876 (1877), p. 155 [11], (part; specimen from San Diego).—HOLMES, Occas. Papers California Acad. Sci., vol. 7, 1900, p. 87 (part; specimen from San Pedro in *Tapes*, described).

Raphonotus lowei RATHBUN, Amer. Nat., vol. 34, 1900, p. 590 (type-locality, San Pedro Bay, California, in siphon of *Pholas pacifica;* holotype female, Cat. No. 23437, U.S.N.M.); Harriman Alaska Exped., vol. 10, 1904, p. 186, text-fig. 93.

Diagnosis.—Front smooth and naked, without sulcus. Last segment of maxilliped does not reach end of penult segment. Palm not widening distally. A single row of hair on lower surface of palm.

Description of female.—Carapace a little wider than in *subquadrata*, which it strongly resembles. Front without trace of shallow, transverse sulcus and without pubescence. Last segment of outer maxilliped does not reach end of penult segment. The palm does not widen distally, its margins being subparallel or a little convex; only one line of hair on lower surface, which is continued to end of finger. Fingers less arched and less gaping than in *subquadrata*. Posterior margin of propodus of first leg convex, of last leg slightly concave.

FIG. 55.—FABIA LOWEI, FEMALE (23437). *a*, ENDOGNATH OF OUTER MAXILLIPED, × 10; *b*, LEFT CHELA. × 4.

Color.—In life, whitish; carapace and abdomen largely covered with orange (Holmes).

Habitat.—Commensal in bivalve mollusks, namely, *Pholas, Pachydesma, Modiola,* and *Tapes=Paphia*.

Measurements.—Female (23437) length of carapace 10, width of same 12.5 mm.

Range.—California, from San Pedro to San Diego.

Material examined.—

San Pedro Bay; in siphon of *Pholas pacifica;* H. N. Lowe; 2 females ovig. (1 holotype) (23437).

San Pedro; in siphons of mussel, *Modiola modiolus;* H. N. Lowe; 29 females (14 ovig.) (29945).

San Pedro; H. N. Lowe; 1 female ovig. (32978).

Long Beach, California; H. N. Lowe; 15 females (13 ovig.) (49221).

Alamitos Bay; in mantle cavity of *Pholas californica;* Venice Mar. Biol. Sta.; 1 female (45583).

San Clemente Island; H. N. Lowe; 1 female (29944).

San Diego; from *Pachydesma crassatelloides;* H. Hemphill; 1 female ovig. (2272).

San Diego; H. Hemphill; 1 female (17299).

FABIA BYSSOMIAE (Say).

Plate 24, figs. 6 and 8.

Pinnotheres byssomiae Say, Proc. Acad. Nat. Sci. Philadelphia, vol. 1, 1818, p. 451 (type-locality, inhabits the *Byssomia distorta* [southern Atlantic coast of United States]; type in Mus. Phila. Acad. Sci.).

Diagnosis.—Legs of second pair unlike, right much longer than left. Chelipeds feeble, palm widest at distal end. Legs very slender.

Description of female.—Of small size. Longitudinal sulci limiting frontal region very deep, parallel for greater part of their length, convergent anteriorly and posteriorly where they terminate in 2 large pits. Posterior margin transverse.

First and second segments of palp of outer maxilliped of subequal size; dactyl attached behind middle of propodus and not reaching its extremity.

Chelipeds and legs nearly naked. Chelipeds feeble, but stouter than legs; palms widening a little distally, margins nearly straight, a row of long hair on inner side of lower surface, and continued to end of finger; fingers horizontal, shorter than palm, tips crossing, gape very slight, a tooth near base of dactyl.

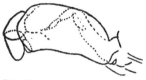

FIG. 56.—FABIA BYSSOMIAE, OUTER MAXILLIPED OF FEMALE (25648). × 21.

Legs slender; relative length, 2. 3. 1. 4; second leg on right side very much (half again) longer than on left side; propodites 3 and 4 and left 2 increase in width a little toward distal end, their anterior margins convex, posterior about straight; propodite right 2 half again as long as left 2, of uniform width and slightly curved; propodite 1 widest in middle, subchelate, hairy below distal half; other propodites and dactyls with a few scattered hairs below; dactyli falcate and similar except that of right 2 which is longer and less curved than its mate.

Abdomen covering whole lower surface of body.

Measurements.—Female (25648), length of carapace 5.4, width of same 6.8, length of right leg of second pair about 12, of left leg of same pair about 8 mm.

Habitat.—Say says of the *Byssomia* with which this crab is commensal: "It imbeds itself in the large *Alcyonium* of the southern coast, and between individuals of a species of aggregating *Ascidia.*"

Range.—" Southern coast "; west Florida; northwest Cuba.

Material examined.—

"Southern coast"; in *Saxicava arctica* (Linnaeus) = *Byssomya distorta;* 1 female holotype, dried and in very bad shape (Mus. Phila. Acad.).

Tampa Bay, Florida; 1901; *Fish Hawk;* 1 female (25648).

West end of Santa Lucia Bay, northwest Cuba; with mollusks; 2 to 5 fathoms; May 15, 1914; Henderson and Bartsch, *Tomas Barrera* Exped.; 1 female ovig. (48595).

FABIA CANFIELDI, new species.

Plate 24, figs. 5 and 7.

Raphonotus subquadratus RATHBUN, Harriman Alaska Exped., vol. 10, 1904, p. 186 (part: specimen from Monterey).

*Type-locality.—*Monterey, California; in folds of the keyhole limpet, *Lucapina crenulata;* Dr. C. A. Canfield; 1 holotype female, Cat. No. 3445.

*Diagnosis.—*Legs of second pair unlike, right much longer than left. First leg stoutest. Propodites of legs, except second right, thickening toward distal end. Fingers deflexed.

*Description of female.—*Carapace of unique specimen very soft and so crushed as to be for the most part beyond description. Front devoid of hair and with a short, longitudinal median depression.

FIG 57.—FABIA CANFIELDI, FEMALE HOLOTYPE, OUTER MAXILLIPED, × 21

Second segment of palp of outer maxilliped very small, shorter and narrower than first segment, and having the narrow terminal segment attached at about its middle.

As in *subquadrata* the palm increases in width to the distal end, the fingers are longer and inclined downward a little, the lower margin of the propodus being more markedly sinuous than in *subquadrata;* fingers not gaping, hairy along inner surface, a prehensile tooth at middle of dactyl and a smaller one at base of propodal finger.

Legs subcylindrical, naked, relative lengths 2. 3. 1. 4, the second leg on right side one-third longer than on left; first leg stouter than the others; propodites slightly curved, but, while having nearly parallel margins, they are a trifle stouter at distal end; an exception is that of second right leg which tapers to distal end; dactyli short, slender-conical, and little curved up to the hooked, horny tip, except the second right one, which is quite straight up to the tip and is also the longest.

Measurements.—Female holotype, carapace apparently about 7 mm. wide; approximate length of legs of second pair, left 8½, right 11 mm.

Range.—Known only from the type-locality.

Genus PARAPINNIXA Holmes.

Pseudopinnixa HOLMES, Proc. California Acad. Sci., ser. 2, vol. 4, 1894, p. 565; type, *P. nitida* (Lockington). Not *Pseudopinnixa* Ortmann, 1894.

Parapinnixa HOLMES, Proc. California Acad. Sci., ser. 2, vol. 4, 1894, p. 587. Name substituted for *Pseudopinnixa* Holmes, not Ortmann.

Carapace calcified, much broader than long, anterior margin nearly straight, frontal process deflexed. Orbits nearly round. Antennules obliquely or transversely plicate, fossettes communicating with each other beneath front. Buccal area small, very broadly subtriangular. External maxillipeds with ischium rudimentary, merus large, subtriangular; palp three-jointed, terminal segment joined to tip of preceding one. First leg largest, others successively diminishing in length, last pair very small. Abdomen of female small, not covering sternum.

Confined to America.

KEY TO THE SPECIES OF THE GENUS PARAPINNIXA.

A^1. Carapace more than twice as wide as long. Dactyli of ambulatory legs similar.
   B^1. Fingers gaping, prehensile edges without tooth. Propodus of second and third legs small, narrower than their respective merus joints.
   nitida. p. 107.
   B^2. Fingers meeting, movable finger with basal tooth. Propodus of second and third legs large, as wide as their respective merus joints.
   hendersoni, p. 109.
A^2. Carapace less than twice as wide as long.
   B^1. Eyes small. Dactyli of second and third legs longer and slenderer than those of first and fourth legs.
      C^1. Movable finger with tooth near middle_____*affinis*, p. 111.
      C^2. Movable finger with tooth near base_____*bouvieri*, p. 111.
   B^2. Eyes large. Dactyli of all legs similar_____*beaufortensis*, p. 112.

ANALOGOUS SPECIES ON OPPOSITE SIDES OF THE CONTINENT.

Atlantic.	Pacific.
hendersoni.	*nitida.*
bouvieri.	*affinis.*

PARAPINNIXA NITIDA (Lockington).

Pinnixa (?) *nitida* LOCKINGTON, Proc. California Acad. Sci. vol. 7, 1876 (1877), p. 155 [11], part: "male" (really female) only (type-locality, Angeles Bay, Gulf of California; type not extant).

Pseudopinnixa nitida HOLMES, Proc. California Acad. Sci., ser. 2, vol. 4, 1894, p. 566, pl. 20, figs. 8, 9.

Parapinnixa nitida HOLMES, Proc. California Acad. Sci., ser. 2, vol. 4, 1894, p. 587, explanation of figs. 8, 9.

Diagnosis.—Carapace more than twice as wide as long. Fingers without teeth and gaping. Dactyli of legs similar, short and hooked.

Propodus of second and third legs small, narrower than their respective merus joints.

Description of female holotype (after Holmes).—Carapace smooth, shining, a little over twice as wide as long, longitudinally convex, transversely plane, sides evenly rounded; anterior margin straight; front broadly triangular, short, much deflexed, not projecting, a groove behind margin. Orbits nearly as wide as long, inner hiatus wide and partly filled by base of minute antennae. Antennules plicated nearly transversely.

Epistome very short, curved. Buccal area broadly subtriangular, rounded in front, posterior portion covered by a projection of sternum. Maxillipeds subtriangular, ischium rudimentary; merus large, the portion nearer the mouth bent inward at a considerable angle to outer face; first joint of palp short and stout, second oblong, third very small.

Chelipeds rather short, moderately stout, smooth; merus not much longer than carpus; hand a little compressed, palm thickened, two

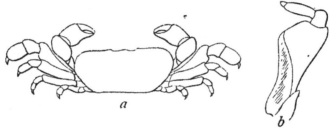

FIG. 58.—PARAPINNIXA NITIDA, FEMALE HOLOTYPE. *a*, GENERAL OUTLINE, × 2¾; *b*, OUTER MAXILLIPED, ENLARGED (AFTER HOLMES.)

longitudinal lines of short cilia on outer surface; fingers scarcely as long as palm, toothless, hooked at tip; on upper surface of movable finger a line of cilia roughened by minute projections.

Legs ciliated on margins; first leg stouter and a little longer than others, anterior surface of merus smooth and concave where it rubs against chelipeds, next two joints stout, dactylus short, subcylindrical, curved, and tapers rapidly to an acute, corneous tip; third leg a little shorter than second, both more compressed than first leg; dactyli similar in all three legs; fourth leg very short, not reaching distal end of merus of third, propodus relatively wider than in other legs, dactylus similar to others.

Abdomen small, not covering half of sternum; sides of posterior portion concave.

Color.—In spirits, bright orange, chelipeds and legs straw yellow.

Measurements.—Female holotype, length of carapace 5+mm., width of same 11 mm.

Locality.—Gulf of California at Angeles Bay, Mexico. Known only from type female, which was destroyed in San Francisco earthquake fire.

PARAPINNIXA HENDERSONI, new species.

Plate 26, figs. 1-5.

Type-locality.—Los Arroyas, Cuba; *Tomas Barrera* Exped.; holotype male, Cat. No. 48710, U.S.N.M.

Diagnosis.—Carapace more than twice as wide as long. Movable finger with basal tooth, immovable finger without teeth. Fingers not gaping. Dactyli of legs similar, short and hooked. Propodus of second and third legs large, as wide as their respective merus joints.

Description of male.—Carapace smooth, shining, a little over twice as wide as long, longitudinally very convex, transversely slightly

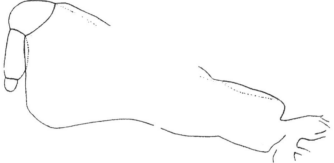

FIG. 59.—PARAPINNIXA HENDERSONI, FEMALE (48711), OUTER MAXILLIPED. × 61.

convex, sides arcuate, at the widest part margined by a thin rim which is pubescent; anterior margin nearly straight, a row of four distant pits behind margin, fronto-orbital width about one-third of carapace width; front broadly triangular, deflexed, point invisible in dorsal view, edge marginate and pubescent; further back a pubescent groove runs subparallel to frontal margin and terminates in orbital margin; below this groove are three pits in a triangle. Orbits circular, filled by eyes, corneae black, visible from above. Antennular cavities very large, not wholly separated from each other or from the orbits, bordering on the latter almost to their highest point and extending sideways farther than the minute antennæ.

Buccal cavity small, subtriangular; merognath transversely elongate, triangular; palp small, largely folding under merus, its outer edge straight and longitudinal, as is also the greater part of the distal border of merus; propodus very elongate, dactylus semioval.

Chelipeds short and stout, especially the merus, partly hairy inside, palm hairy outside in the distal part and in two bands leading toward carpus; palms very thick, protuberant inside at the middle; fingers stout, meeting when closed, tips hooking by each other; dactylus with a granulate, pubescent rim above, a shallow, basal tooth below which fits an excavation in the thumb.

Edges of legs more or less pubescent; on the second and third pairs a line of long hairs applied to the posterior surface of carpus and propodus, and attached near upper edge of the latter and obliquely on the surface of the former; first leg very large and thick, fourth very small, not exceeding merus of third, second and third more nearly of a size; all the dactyli are similar, falcate, and when flexed fit closely in a corresponding groove on the propodus, which in the first three pairs has a rounded end, but in the last pair is broadly truncate and as broad as long.

Abdomen broad at base, covering greater part of sternal width; third segment widest, ends rounded and longer than middle part; fourth segment partly invaginated in third; sides of abdomen concave from fourth to angle near base of seventh; latter very long, with rounded extremity.

Description of female.—Like male, except for the abdomen; this is small compared to sternum, widest at third segment, after that subtriangular with concave sides, last segment almost an equilateral triangle with blunt point; segments 1 to 6 of subequal length; margin fringed with long hair.

Color.—Specimens in alcohol show small, brownish pigment spots.

Measurements.—Male holotype, length of carapace 4, width of same 8.3 mm.; female (48711), length of carapace 3.5, width of same 8 mm.

Habitat.—Free-swimming, as all the specimens examined from Cuba were taken in plankton with a submarine light.

Range.—Cuba; Gulf of Mexico.

Material examined.—Taken on the northwest coast of Cuba, by Henderson and Bartsch while on the *Tomas Barrera* Expedition May, 1914; as follows:

Off Santa Lucia; May 12; 1 male, 3 females (48711).

Los Arroyas; May 19-20; station 8; 1 male holotype, 1 female (48710), 1 female (48709), 1 female (48712).

Ensenada de Cajon, off Cape San Antonio; May 22-23; station 11; 1 male (48713).

Gulf of Mexico: West coast of Florida; lat. 28° 45′ 00″ N.; long. 85° 02′ 00″ W.; 30 fathoms; gy. S. brk. Co.; Mar. 15, 1885; station 2405, *Albatross;* 1 female young (50166).

PARAPINNIXA AFFINIS Holmes.

Parapinnixa affinis HOLMES, Occas. Papers California Acad. Sci., vol. 7, 1900, p. 95 (type-locality, Dead Mans Island, San Pedro, California; type in Mus. Univ. California).

Diagnosis.—Carapace less than twice as wide as long. Movable finger with tooth near middle, immovable finger with two teeth near tip. Dactyls of first and fourth legs short and stout, of second and third legs longer and slenderer.

Description of female (after Holmes).—Near *P. nitida*, but carapace less than twice as wide as long; front triangular, with a short, median groove. Antennules oblique. Maxillipeds with perhaps a two-jointed palp, but this is uncertain.

Chelipeds stout, smooth; hand thickened, smooth, rounded above and below; dactyl hooked at tip, armed with a small tooth near middle of inner margin, pollex with two teeth at tip. First leg larger than others, dactyl short and stout; next two legs comparatively slender, dactyls longer; last leg small, reaching about to tip of merus of preceding leg, dactyl short and stout. Abdomen widest at third segment, behind which it is triangular, tip broadly rounded.

Size.—Not given.

Locality.—Dead Mans Island, San Pedro, California. Known only from type female.

PARAPINNIXA BOUVIERI,[1] new species.

Plate 25, figs. 4–10.

Type-locality.—Off Cape Catoche, Yucatan; 25 fathoms; station 2362, *Albatross;* holotype female, Cat. No. 23441, U.S.N.M.

Diagnosis.—Carapace less than twice or barely twice as wide as long. Movable finger with tooth near tip. Dactyls of first and fourth legs short and stout, of second and third legs longer and slenderer.

Description.—Minute. Carapace not more than twice as wide as long, otherwise much as in *hendersoni*.

The outer maxilliped when folded in place is triangular with the two free corners rounded and the longitudinal side about two-thirds as long as the posterior side, which is at right angles to it. The merus-ischium is obliquely truncate at the distal end, leaving the first joint of the palpus exposed; propodus elongate, distally tapering, dactylus very small, suboval; both these joints fold under the merus.

Chelae a little higher than in *hendersoni*, the propodus with convex, not sinuous, lower margin; immovable finger subtriangular, broad at base, a small tooth on prehensile edge near tip; dactyl with

---

[1] Named for Prof. E. L. Bouvier, Museum of Natural History, Paris, France.

a small basal tooth, remaining edge very finely and irregularly denticulate.

Legs similar to those of *hendersoni;* first leg shorter, especially merus and propodus, merus not reaching laterally beyond carpus of cheliped; propodus of last leg not so widely truncate at distal end as in *hendersoni;* the same applied hairs on second and third legs; dactyli of second and third legs longer and slenderer than of first and fourth.

Sides of male abdomen more gradually convergent than in *hendersoni;* seventh segment not more than one and one-half times as long as wide. Sides of triangular part of female abdomen straight, not concave.

FIG. 60.—PARAPINNIXA BOUVIERI, FEMALE HOLOTYPE, OUTER MAXILLIPED, × 100.

Measurements.—Female holotype, length of carapace, 1.6, width of same 3.1, fronto-orbital width 1.1 mm. Male (Charleston), length of carapace 2, width of same 3.5 mm.

Range.—Charleston, South Carolina, to Yucatan, Mexico.

Specimens examined.—

Charleston, South Carolina; off the bar; L. Agassiz; 1 male, with parasite in left branchial chamber (5744, M.C.Z.).

Off Cape Catoche, Yucatan; lat. 22° 08′ 30″ N.; long. 86° 53′ 30″ W.; 25 fathoms; Co. S.; Jan. 30, 1885; station 2362, *Albatross;* 1 female holotype (23441).

PARAPINNIXA BEAUFORTENSIS, new species.

Type-locality.—Off Beaufort, North Carolina; on fishing grounds, ¼ mile S. of Fishing Buoy (20 miles off Beaufort Inlet); 16 fathoms; hrd.; July 20, 1915; station 8293, steamer *Fish Hawk;* 1 male (Cat. No. 50170, U.S.N.M.).

Diagnosis.—Carapace about 1⅓ times as wide as long. Eyes large. Fingers of male not gaping. Dactyli of legs curved and having long, slender, horny tips.

Description of male.—Carapace rather regularly oval, convex, with the surface behind the anterior border depressed and plumose; regions indicated by a series of pits which are light brown in the specimen preserved in alcohol. Fronto-orbital width ⅔ as great as width of carapace; front about ⅖ width of carapace, a large emargination at its middle in dorsal view; edge of lobes sinuous. A tuft of hair on either side of dorsal surface near lateral margin, but not projecting sideways beyond that margin. A similar tuft, attached to the ventral surface, projects beyond the margin. Eyes large, of a bronze-brown color.

The outer maxillipeds are characteristic of the genus. Although it was not possible to get an exact drawing (see fig. 62), yet they seem to be near those of *P. nitida*.

Chelipeds stout; carpus squarish in dorsal

FIG. 61.—PARAPINNIXA BEAUFORTENSIS, MALE HOLOTYPE, DORSAL VIEW, WITH RIGHT CHELIPED DETACHED, × 20.

view, its outer distal angle prominent; palm inflated, margins convex, lower margin hairy, width greater at distal than proximal end, the upper distal angle higher than base of dactylus; dactylus about as long as upper edge of palm; both fingers much curved; when closed the tips cross and there is no gape, the thin and irregularly denticulate edges fitting together.

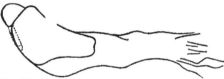

FIG. 62.—PARAPINNIXA BEAUFORTENSIS, OUTER MAXILLIPED OF MALE HOLOTYPE, × 176.

Legs fringed with long hair, especially on the propodites, where there is a border of hair on the lower margin, and in the second and third legs, a row of still longer hair attached near the upper margin on the posterior surface, the length of the hairs being twice as great as the width of the propodite. The legs diminish in stoutness from the first to the fourth. The first leg is a little longer than the second, the third about as long as the first. Propodites stout, upper margins con-

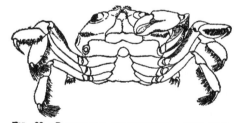

FIG. 63.—PARAPINNIXA BEAUFORTENSIS, MALE HOLOTYPE, VENTRAL VIEW, × 20.

vex. Dactylus of first three legs long, curved, the slender horny tips about half the entire length; dactylus of fourth leg similar in shape, but very much smaller.

Abdomen of male suboblong, distally tapering, at base not more than half the width of the sternum; sutures faint, except that marking the subtriangular terminal segment.

Measurements.—Male holotype, length of carapace 1, width of same 1.3 mm.

Locality.—Known only from the type-locality, given above.

Remarks.—This species is much narrower than any other *Parapinnixa*, and its eyes much larger. It is possible that the tiny specimen represents a postlarval stage of an unknown adult; that is, a stage having a crab form, but not the adult form; this idea seems to be indicated (1) by its small size; (2) by the eyes, large in proportion to the size of the body, as in most zoeae and megalops; (3) by the feathery ornamentation of the legs and carapace, which assist in maintaining a pelagic existence; and (4) by the thin prehensile edges of the fingers, which suggest that in another molt or two they might disappear and leave narrow, gaping fingers.

Genus DISSODACTYLUS Smith.

Dissodactylus Smith, Trans. Connecticut Acad. Arts and Sci., vol. 2, 1870, p. 172; type, *D. nitidus* Smith.

Echinophilus Rathbun, Amer. Nat., vol. 34, 1900, p. 590; type, *E. mellitae* Rathbun.

Carapace broader than long, pentagonal, broad behind, broadest at antero-lateral angles; surface not areolated; front narrow, horizontal, its margin continuous with arcuate antero-lateral margins. Eyes minute, superior margin of orbit slightly or not at all emarginate. Antennules transverse. Epistome usually very short, so that the labial border approaches very near the front, leaving only a narrow space, which is nearly filled by the antennulae. The labial border is not interrupted in the middle by any projection or emargination and is continuous with lateral margin of buccal area. Palate without longitudinal ridges.

Ischium and merus of maxillipeds coalescent; palpus composed of two or three segments, the dactylus, when present, small.

Chelipeds of moderate size; fingers longitudinal.

The ambulatory legs are small and differ little in length; dactyli of three anterior pairs bifurcate, those of posterior pair usually simple. In the male the sternum is flat and very broad, breadth between posterior legs much more than twice as great as breadth of basal segments of abdomen.

Male abdomen narrow and only three-jointed, the first and second segments anchylosed, and the third, fourth, fifth, and sixth also united in one piece; in the female abdomen the segments are usually all free.

Found only in America, where, so far as known, it lives on the outside of flat sea-urchins, its bifurcate dactyls enabling it to cling to the spines of the urchin.

Remarks.—The four new species are dedicated to those British carcinologists whose work has contributed most to a knowledge of the classification of decapod crustaceans.

The outer maxillipeds of the seven species fall into two distinct groups. In one group the distal portion of the merus is suboblong, its distal margin more or less truncate, while the palpus is of considerable size relatively, the carpus and propodus nearly equal in size, the latter quadrilateral, in two instances bearing a minute dactylus at the distal inner corner, in two instances lacking a dactylus, though in *nitidus* there is evidence of a rudiment. In the other group the distal portion of the merus is suboval, narrowing to the extremity and the palpus is very small, its three joints rather squarely end to end.

The form of the palpus is very difficult to determine in this genus as it is usually folded in a groove in the edge of the merus, and the segments, especially the dactylus, are easily broken off.

KEY TO THE SPECIES OF THE GENUS DISSODACTYLUS.

A^1. Dactyl of fourth ambulatory leg simple, not bifurcate. Carapace with no more than one dorsal ridge on each side. Antero-lateral margin entire, non-dentate.
  B^1. Carapace with a dorsal ridge on either side proceeding inward from the lateral angle. Legs not slender.
    C^1. Dorsal ridge oblique.
      D^1. Secondary spine of dactyls of legs 1, 2, and 3 of good size.
        E^1. Dactyls of legs 1, 2, and 3 bifurcate half way to their base.
          F^1. Edge of front transverse. Carapace flat for the most part and narrow, being only one-twelfth wider than long.
              nitidus, p. 116.
          F^2. Edge of front concave. Carapace very convex and wide, about one-fifth wider than long_____*mellitae*, p. 117.
        E^2. Dactyls of legs 1, 2, and 3 bifurcate less than halfway to their base_____*encopei*, p. 119.
      D^2. Secondary spine of dactyls of legs 1, 2, and 3 minute and remote from primary spine_____*borradailei*, p. 121.
    C^2. Dorsal ridge transverse_____*stebbingi*, p. 123.
  B^2. Carapace without dorsal ridge. Legs slender_____*alcocki*, p. 124.
A^2. Dactyl of fourth ambulatory leg bifurcate, as in the other pairs. Carapace covered with numerous transverse ridges. Antero-lateral margin dentate.
                            calmani, p. 125.

DISSODACTYLUS NITIDUS Smith.

Plate 26, figs. 6 and 7.

Dissodactylus nitidus SMITH, Trans. Connecticut Acad. Arts and Sci., vol. 2, 1870, p. 173 (type-locality, Panama; type in P.M.Y.U.).—RATHBUN, Proc. U. S. Nat. Mus., vol. 38, 1910, p. 545, pl. 48, fig. 6.

Diagnosis.—Carapace mostly flat. Dorsal ridge covering one-third of distance to median line. Dactyls of legs, 1, 2, and 3 bifurcate halfway to base. Terminal segment of male abdomen nearly an equilateral triangle. Palp of outer maxilliped 2-jointed.

Description of male.—Carapace broad posteriorly, breadth at posterior margin but little less than between lateral angles; postero-lateral margins about as long as antero-lateral. Dorsal surface naked and polished, slightly convex in front and along lateral margins, flat in middle and posteriorly. Antero-lateral border slightly arcuate,

FIG. 64.—DISSODACTYLUS NITIDUS, OUTER MAXILLIPED OF FEMALE (22113), × 50.

armed with an upturned margin which curves suddenly inward at lateral angle and extends one-third of way to middle of carapace; postero-lateral border nearly straight and armed with a slightly upturned margin.

Distal half of merus of outer maxillipeds suboblong, with straight and parallel sides, distal angles rounded; segments of palp long, the tip when flexed reaches anterior margin of sternum; second segment spatulate, distal end broad and squarely truncate, with what seems to be a rudiment of a dactylus at the flexor, distal angle.

Merus of chelipeds extends but little beyond margin of carapace; carpus short, smooth, unarmed; hands smooth, rounded, somewhat swollen; fingers slender, acutely pointed, slightly deflexed, prehensile edges minutely dentate; a small tuft of dense pubescence on lower edge of propodal finger near base.

Legs slightly hairy along edges, merus, carpus, and propodus joints somewhat compressed; dactyli 1, 2, and 3 smooth, naked, divided halfway to base; divisions cylindrical, acutely pointed, slightly curved, the anterior one of each leg somewhat longer than the posterior; dactylus 4 nearly straight, slightly compressed, sulcate above and below and naked.

First and second segments of abdomen narrower than third and fused; third to sixth fused, slightly expanded at base, considerably contracted at distal end; terminal segment small, nearly an equilateral triangle. Appendages of first segment reach almost to terminal segment, and are straight for basal two-thirds, terminal portion turned sharply outward at obtuse angle; basal portion hairy along outer edge, terminal portion on both edges. (Smith.)

Description of female.—The female resembles the male. The postero-lateral margin of carapace is a little concave. The abdomen covers the width of the sternum but does not quite reach the extremity; all segments free, second nearly twice as long as first, third, fourth, fifth, and sixth of subequal length and each twice as long as second; seventh triangular, one-third as wide and two-thirds as long as sixth, sides sinuous.

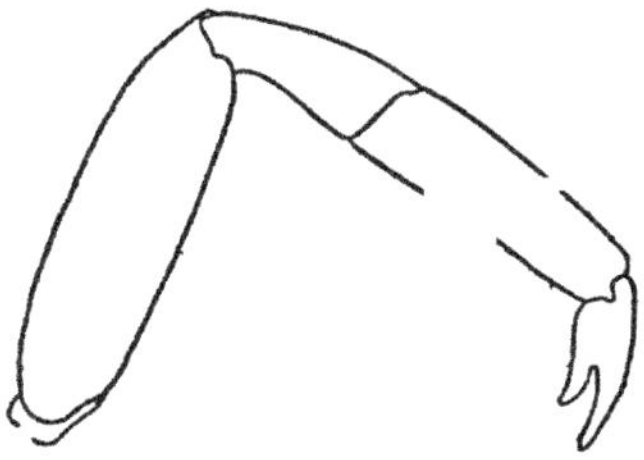

Color.—In alcohol, dirty white; carapace marked with irregular, transverse bands of purplish brown, and divisions of dactyli of first and third legs tipped with dark brown. (Smith.)

FIG. 65.—DISSODACTYLUS NITIDUS, THIRD LEG OF FEMALE (22113), × 11.

Measurements.—Male holotype, length of carapace 4.7, width of same 5.1 mm. Female (22113), length of carapace 5.5, width 6 mm.

Range.—From Santa Maria Bay, Lower California, Mexico, to Peru. To a depth of 5½ fathoms.

Material examined.—

Off Santa Maria Bay, Lower California; in boat dredge; Mar. 18, 1911; *Albatross;* 3 males, 1 female (49229).

Off Abreojos Point, Lower California; lat. 26° 42′ 30″ N.; long. 113° 34′ 15″ W.; 5½ fathoms; gn. M.; May 4, 1888; station 2835, *Albatross;* 2 females (22113).

Bay of Sechura, W. of Matacaballa, Peru; about 5 fathoms, in trawl; Apr. 8, 1907; R. E. Coker, collector; recd. from Peruvian Government; 1 female (40447).

DISSODACTYLUS MELLITAE (Rathbun).

Plate 28, figs. 7 and 8.

Echinophilus mellitae RATHBUN, Amer. Nat., vol. 34, 1900, p. 590 (type-locality, Pensacola, Florida, on *Mellita testudinata;* holotype, Cat. No. 23434, U.S.N.M.).

Dissodactylus mellitae RATHBUN, Bull. U. S. Fish Comm., vol. 20, for 1900 pt. 2 (1901), p. 21.

Diagnosis.—Edge of front concave. Dorsal ridge nearly as long as antero-lateral distance. Fingers of male gaping. Dactyli of legs 1–3 bifurcate for half their length. Male abdomen obtusely angled at sixth segment. Palp of outer maxilliped 2-jointed.

Description.—Carapace distinctly wider at lateral angles than posteriorly; antero-lateral and postero-lateral margins subequal. Dorsal surface nearly naked and smooth, very convex fore and aft, slightly so from side to side. Edge of front concave; antero-lateral borders arcuate and bearing a fine, raised, milled rim, which at the lateral angles curves inward on the carapace at an obtuse and rounded angle, and is continued on the dorsal surface for a distance nearly equal to antero-lateral distance; postero-lateral border with a straight rim; posterior margin sinuous.

Merus-ischium of outer maxilliped subspatulate; outer edge of carpus arcuate; propodus quadrate, widening distally; dactylus absent.

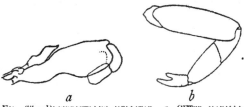

FIG. 66.—DISSODACTYLUS MELLITAE. *a*, OUTER MAXILLIPED OF FEMALE (40272), × 23; *b*, LEG OF PARATYPE (23434), × 15.

Chelipeds of male moderately stout; merus and carpus smooth, propodus roughened with short, oblique ridges, punctate and hairy on their superior-distal side; upper margin convex, lower sinuous; fingers a little deflexed, very narrowly gaping in proximal half, a depressed line of punctae on outer surface, edges entire, tips blunt. Chelipeds of female similar but feebler.

Legs hairy on edges, compressed; merus joints very wide; carpus-propodus with convex anterior and posterior margins, the propodus strongly tapering, especially in last leg; the dactylus of legs 1, 2, and 3 divided halfway to base, forks slender, curved, acute, anterior fork twice as long as posterior; dactylus 4 slender, straight, acuminate, furnished with long hair on posterior margin.

In male abdomen, segments 1 and 2 are partially fused, also 3 to 6 inclusive, which have convex margins with a very obtuse angle on sixth segment; seventh subtriangular with convex sides. Appendages of first segment overlap terminal segment; basal three-fifths stout and convergent; distal two-fifths slender and divergent. First segment of female abdomen linear, 2, 3, and 4 fused, 5 and 6 of subequal length, seventh broadly triangular, half as wide as sixth, sides sinuous. Eggs numerous, relatively large.

Color.—Light, with scanty dark mottlings which persist in alcohol and are then of a purplish color.

Measurements.—Male (40271), length of carapace 2.9, width of same 3.5 mm. Female ovig. (40272), length of carapace 3.3, width of same 4.5 mm.

Habitat.—Clings to the outside of key-hole urchin, *Mellita quinquesperforata=M. testudinata*, and the sand-dollar, *Echinarachnius parma.*

Range.—From Vineyard Sound, Massachusetts, to Pensacola, Florida.

Material examined.——

Off Menemsha Bight, Marthas Vineyard, Massachusetts; 1 specimen (U. S. Fisheries Lab., Woods Hole).

Western part of Vineyard Sound, Massachusetts; 11½ fathoms; 1908; F. B. Sumner; 1 specimen (U. S. Fisheries Lab., Woods Hole).

Narragansett Bay, Rhode Island; Beaver Tail Light S. E. by S., ⅜ mile; 8¼ fathoms; S. Sh.; Aug. 6, 1880; station 770, *Fish Hawk;* 2 males, 3 females, ovig. (40271).

Narragansett Bay; E. of Brentons Reef Lightship; 8½–10 fathoms; S. brk. Sh.; Aug. 23, 1880; stations 816–818, *Fish Hawk;* 7 males, 10 females (8 ovig.) (40272).

Beaufort, North Carolina: Union College Collection, No. 72c; 1 female ovig. (deposit, U.S.N.M., Cat. No. 42788). Union College Collection, No. 57c; 1 female (deposit, U.S.N.M., Cat. No. 42789).

Charleston, South Carolina; Feb. 8, 1852; L. Agassiz; 15 specimens (5729, M. C. Z.).

Pensacola, Florida; on *Mellita quinquesperforata;* J. E. Benedict; 20 specimens (1 male is holotype) (23434).

DISSODACTYLUS ENCOPEI Rathbun.

Plate 27, figs. 1–4.

Dissodactylus encopei RATHBUN, Bull. U. S. Fish Comm., vol. 20 for 1900, pt. 2 (1901), p. 22, text-fig. 5 *a–e* (type-locality, Stann Creek, 38 miles south of Belize, British Honduras; holotype, Cat. No. 23430, U.S.N.M.).
Dissodactylus crinitichelis MOREIRA, Arch. Mus. Nac. Rio de Janeiro, vol. 11, 1901, p. 37, pl. 3 (type-locality, State of Rio Grande do Sul, on *Encope*, sp., probably *E. emarginata;* type in Mus. Nac. Rio de Janeiro).

Diagnosis.—Dorsal ridge short, oblique. Legs stout; dactyls 1–3 bifurcate for less than half their length. Sides of segments 3–6 of male abdomen convex; last segment an equilateral triangle. Palp of outer maxilliped 3-jointed.

Description.—Carapace very wide, about one and four-tenths as wide as long, posterior width but little less than greatest width. Dorsal surface nearly naked and polished, strongly convex fore and aft, slightly so from side to side. Antero-lateral margins arcuate, marked by a very slightly raised line which after making a rounded turn at the lateral angle is continued obliquely on the dorsal surface for a short distance. Margin of front slightly convex, not advanced beyond the curve of the antero-lateral margins. Postero-lateral marginal raised line sinuous and oblique. Posterior margin sinuous.

Ventral surface and margins of carapace clothed with soft hair. Outer maxillipeds very small; inner distal angle of propodal segment fits against inner angle of merus, which is thickened and clothed with hairs; merus-ischium subspatulate, curved; propodus about as long as carpus, widening distally, truncate, and bearing on the extremity at inner angle, a short, stumpy dactylus.

Outer and upper surfaces of carpus and propodus of chelipeds crossed by oblique rugae fringed with hair; propodus elongate, sub-

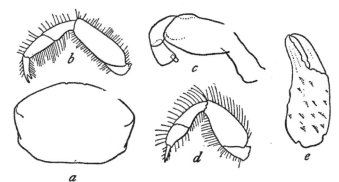

FIG. 67.—DISSODACTYLUS ENCOPEI, MALE (23430). *a*, OUTLINE OF CARAPACE, × 6; *b*, THIRD LEG, × 16; *c*, ENDOGNATH OF OUTER MAXILLIPED, × 46; *d*, FOURTH LEG, × 16; *e*, CHELA, × 16.

cylindrical; upper margin nearly straight, lower margin sinuous, fingers deflexed, grooved, meeting when closed, tips acute, crossing, a tooth at base of dactyl fitting in sinus of propodal finger. Chelipeds of female more slender than of male.

Legs sparingly fringed with long hair; dactyli 1–3 bifurcate for less than half their length, forks slender, curved, acuminate and largely horny; dactylus of last leg straight and styliform.

In male abdomen, segments 1 and 2 fused, slightly constricted at middle, 3–6 inclusive fused, sides slightly convex, last segment an equilateral triangle; appendages of first segment not reaching last one, straight and far apart until near tips where they curve strongly outward; tips acute and bearing a few hairs. All segments separate in female abdomen, first half as long as second, second two-thirds as long as third; third, fourth, fifth, and sixth nearly equal; last more than half as wide as sixth, very short-triangular.

Variations.—There is some variation in the width of carapace and legs and in the amount of bifurcation of dactyls.

Measurements.—Male (23430), length of carapace 4.6, width of same 6.6 mm. Female (23430), length of carapace 5, width of same 7 mm.

Habitat.—Clings to the outside of the thick keyhole urchin, *Encope emarginata*, and probably also of *E. michelini*, which replaces *emarginata* in the area between the Gulf of Mexico and Florida Keys and Yucatan.

Range.—From west coast of Florida to Rio Grande do Sul, Brazil. To a depth of 28 fathoms.

Material examined.—

West coast of Florida; lat. 26° 47′ 30″ N.; long. 83° 25′ 15″ W.; 28 fathoms; fne. wh. S. bk. Sp. brk. Sh.; Mar. 18, 1885; station 2410, *Albatross;* one female (23432).

West coast of Florida; lat. 26° 33′ 30″ N.; long. 83° 15′ 30″ W.; 27 fathoms; fne. wh. S. bk. Sp.; Mar. 18, 1885; station 2411, *Albatross;* one male (23433).

Jamaica; on underside of sand-dollars buried in sand; Sept. 1, 1910; E. A. Andrews; one male, three females (41751).

Off Vieques, Porto Rico; Cape San Juan Lighthouse NW. ¼ N. 14¾ miles; 11 fathoms; Co. S. Sh.; temp. 26.2° C.; Feb. 8, 1899; station 6084, *Fish Hawk;* one male y. (24524).

Stann Creek, 38 miles S. of Belize, British Honduras; attached to lower surface of *Encope emarginata;* W. A. Stanton; 14 specimens (23429); 20 specimens (1 male is holotype) (23430).

Sabanilla, Colombia; Mar. 16–22, 1884; *Albatross;* three females (23431).

DISSODACTYLUS BORRADAILEI,[1] new species.

Plate 27, figs. 5–8.

Type-locality.—Miami, Florida; 30 fathoms; John B. Henderson; 1 female holotype (Cat. No. 49230, U.S.N.M.)

Diagnosis.—Carapace very convex; postero-lateral surface concave. Dorsal ridges run from lateral angles halfway to median line. Legs short, stout; secondary spine of dactyls minute and remote from primary spine. Palp of outer maxilliped 3-jointed.

Description.—Near *encopei*. Carapace longer and narrower, and even more convex fore and aft. The antero-lateral margin turns at the lateral angles in almost a right angle to form a low ridge

---

[1] Named for Dr. L. A. Borradaile, Selwyn College, Cambridge, England.

which is continued obliquely inward and backward halfway to the median line; surface outside this ridge concave, so that the posterior dorsum appears very narrow; postero-lateral surface also hairy, concealing the postero-lateral margin, which is concave; carapace considerably narrower behind than across the middle; posterior margin sinuous; frontal margin convex, following the antero-lateral arch, but viewed from an oblique anterior direction the front is bilobed.

Buccal cavity anteriorly truncate. Outer maxillipeds large; merus subspatulate, broadly truncate at summit, inner corner rounded; carpus and propodus of good size, the latter widening a little toward distal end, outer two-thirds of distal margin truncate with rounded angles; inner third less produced and more oblique, bearing a short, broad dactylus which overreaches the propodus.

Carpus and propodus of chelipeds ornamented with oblique, rugose, hairy lines; propodus thick, distally diminishing; palmar portion

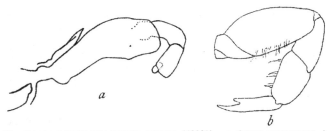

FIG 68.—DISSODACTYLUS BORRADAILEI, FEMALE (49231). a, OUTER MAXILLIPED, × 23; b, LEG, × 11.

very convex below in proximal two-thirds; fingers stout, furrowed, prehensile margins crenulate, meeting when closed, tips acute, turned inward and overlapping.

Legs short, stout, long-hairy; merus about twice as long as wide; dactyli nearly straight, tapering to a slender, slightly curved, horny tip; in the first three legs there is a slender, minute, subsidiary spine on the posterior margin, a little distal to the middle.

Abdomen of male with segments 1 and 2 fused, and also 3 to 6, inclusive; side margins of 2 to 6, inclusive, convex, 7 an equilateral triangle. Abdomen of female with segments separate, the second strikingly longer near middle than at the sides, seventh half as wide as sixth.

Measurements.—Male (23790), length of carapace 3.5, width of same 4.3 mm. Female (49231), length of carapace 6.2, width of same 8.1 mm.

Range.—Southern Florida; Jamaica. To a depth of 30 fathoms.

Material examined.—

Miami, Florida; 30 fathoms; John B. Henderson; 1 female holotype (49230).

West coast of Florida; lat. 26° 33′ 30″ N.; long. 83° 15′ 30″ W.; 27 fathoms; fne. wh. S. blk. Sp.; Mar. 18, 1885; station 2411, *Albatross;* 1 male (23790).

Off Montego Bay Point, Jamaica; dredged; June 28, 1910; E. A. Andrews; 1 female ovig. (49231).

Jamaica; received from Marine Biological Laboratory, Woods Hole; 1 female (47357).

DISSODACTYLUS STEBBINGI,[1] new species.

Plate 28, figs. 1 and 2.

*Type-locality.—*Sarasota Bay, Florida; Union College Collection, No. 629; 1 male holotype (deposit U.S.N.M., Cat. No. 49232).

*Diagnosis.—*Antero-lateral margin a thin rim. Dorsal ridges transverse, each covering one-third width of carapace. Sides of male abdomen from third to sixth segment, inclusive, nearly straight· terminal segment wider than long, margin arcuate. Palp of outer maxilliped 3-jointed.

FIG. 69.—DISSODACTYLUS STEBBINGI, OUTER MAXILLIPED OF MALE HOLOTYPE, × 63.

*Description of male.—*Carapace widest at the lateral angles; antero-lateral margin beginning at lower edge of orbit and forming a thin, sharp rim which at widest part of carapace turns in a rounded angle and is continued by the less prominent, concave, postero-lateral margin. A transverse, dorsal ridge on either side of carapace occupies a third of its width at the widest part and ends outwardly in the lateral margin; this ridge is anteriorly fringed with hair, while the upper surface of the marginal rim is pubescent and hairy. Front arcuate and continuous with upper margin of orbit.

Lower surface of carapace and maxillipeds pubescent. Merus-ischium of outer maxilliped spatulate, the blade of the spatula

---

[1] Named for the Rev. T. R. R. Stebbing, Tunbridge Wells, England.

broadly oval; palp small, articulated at summit of merus; propodus shorter than carpus, elongate-suboval, bearing a small, squarish dactyl at inner end of distal margin.

Chelipeds and legs not known.

Male abdomen and sternum punctate. First and second segments of abdomen completely fused, slightly constricted at middle; third to sixth, inclusive, completely fused, sides nearly straight, seventh wider than long, with arcuate margin.

Measurements.—Male holotype, length of carapace 3.3, width of same 4.3 mm.

Range.—Known only from Sarasota Bay, Florida.

DISSODACTYLUS ALCOCKI,[1] new species.

Plate 28, figs. 3 and 4.

Type-locality.—Gulf of Mexico, off Delta of Mississippi River; 35 fathoms; station 2388, *Albatross;* 1 female holotype (Cat. No. 23447, U.S.N.M.).

Diagnosis.—Carapace without dorsal ridge. Front bilobed. Legs slender and spinulous; secondary spine of dactyls 1 to 3 beneath primary spine. Palp of outer maxilliped 3-jointed.

Description.—Carapace convex fore and aft, nearly level from side to side; antero-lateral margin acute, with a milled rim and forming

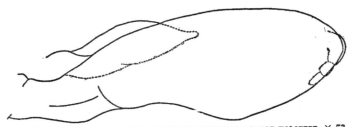

FIG. 70.—DISSODACTYLUS ALCOCKI, OUTER MAXILLIPED OF FEMALE HOLOTYPE, × 73.

a broad, rounded lateral angle and continued along the nearly straight and oblique postero-lateral margin; no dorsal ridge. Front bilobed, continuing the line of the antero-lateral arch, median sinus a broad, shallow, rounded sinus. Posterior margin deeply sinuous.

Lower surface of carapace and mouth parts and edge of female abdomen hairy. Outer maxillipeds compact, forming a triangle; merus-ischium spatulate, less dilated than in *stebbingi;* palp very small, folding under the thin edge of the merus and articulating near its extremity; propodus short, sides subparallel, end truncate; dactylus as long as propodus but narrower.

---

[1] Named for Dr. A. Alcock, London.

Cheliped roughened with a few faint rugae, a small tooth at inner angle of carpus, upper margin of propodus acute; propodus twice as long as wide, straight above, sinuous below, fingers slightly deflexed, stout, shutting close together, tips incurved, a minute tooth near base of dactyl.

Legs slender; merus joints dilated slightly; propodus tapering a little toward distal end; dactyls 1, 2, and 3 subcylindrical, little curved, ending in two short unequal spines, the smaller not in the same plane as the longer, but below it; posterior margin armed with two rows of spinules, a few of the same also on the corresponding margin of the propodus. Fourth dactylus straighter, simple, not bifurcate, both margins spinulous, as also margins of propodus and carpus.

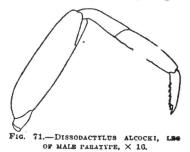

FIG. 71.—DISSODACTYLUS ALCOCKI, LEG OF MALE PARATYPE, × 10.

Male abdomen with segments 1 and 2 fused, 3 to 6 inclusive fused and with convex margins, 7 triangular with blunt tip. Female abdomen with segment 1 short, half as long as 2; 3, 4, 5, and 6 incompletely fused, 7 broadly triangular, half as wide as 6.

Measurements.—Female (23447), length of carapace 4.2, width 5.8 mm. Male smaller, too damaged for measurement.

Material examined.—Gulf of Mexico, off Delta of Mississippi River; lat. 29° 24′ 30″ N.; long. 88° 01′ 00″ W.; 35 fathoms; yl. S. bk. Sp.; Mar. 4, 1885; station 2388, *Albatross;* 1 male, 1 female holotype (23447).

DISSODACTYLUS CALMANI,[1] new species.

Plate 28, figs. 5 and 6.

Type-locality.—Grecian Shoals, Hawk Channel, Florida; 2 fathoms; station 7469, *Fish Hawk;* 1 female (Cat. No. 49233, U.S.N.M.).

Diagnosis.—Carapace sharply pentagonal, covered with rugae; antero-lateral margins dentate. Dactyl of legs 1 to 4 bifurcate, and with a marginal line of spinules. Palp of outer maxillipeds 3-jointed.

Description of female.—Appearance quite unlike those above described. Carapace pentagonal, deflexed in front of lateral angles, almost flat behind them, antero-lateral margins short, slightly convex, dentate, postero-lateral margins long, straight, and posteriorly converging, posterior margin sinuous; dorsal surface covered with short,

---

[1] Named for Dr. W. T. Calman, British Museum, London.

unequal, wavy, transverse rugae which at the margins tend to form teeth; a strong crest runs horizontally inward from the obtuse lateral angle. Antero-lateral margin cristiform, terminating at epistome and armed with three or four low teeth, the posterior of which is tooth-

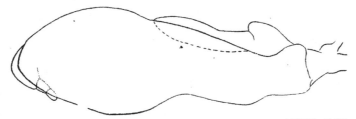

Fig. 72.—Dissodactylus Calmani, outer maxilliped of female (48570), × 50.

like, the others almost crenations. Front produced, thickened, edge nearly transverse, but faintly bidentate or even quadridentate. Upper margin of orbit narrow, cut deeply into carapace, outer angle blunt, thickened.

Lower surface of carapace hairy. Pterygostomian ridge sharp and prominent near epistome. Buccal cavity subtriangular, outer margins convex. Merus-ischium of the outer maxillipeds elongate-spatulate, extremity narrow; palpus relatively very small, folding in a groove in the edge of the merus; last 2 segments together shorter than carpus; dactylus nearly as large as propodus.

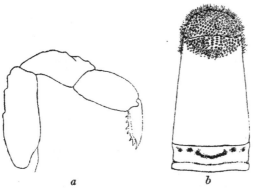

Fig. 73.—Dissodactylus Calmani. a, Last right leg of female holotype, × 14; b, abdomen of male (50168), × 12.

Chelipeds slender; merus, carpus and propodus rough with rugae, those of propodus oblique, the rugae arranged in 4 longitudinal ridges on outer and upper surface; palms narrow, fingers deflexed, not gaping, stout, grooved, tips acute, curved and overlapping.

Legs of nearly equal length and of moderate width, sharply margiued above, and punctate and hairy; merus joints a little dilated at middle; propodus slightly tapering distally; dactyls of all legs similar, curved, upper part flattened and ending in a slender, curved spine; below there is a laminar expansion which is in a plane oblique to the upper surface and terminates in a shorter spine and is bordered posteriorly by 7 slender spinules.

The abdomen covers the sternum, its segments are all separate, the seventh is half as wide as the sixth, beyond which it projects abruptly with an arcuate margin.

Male.—The tooth at the lateral angle and the ridge leading from it are more prominent in the male than in the female. The chelae are stouter in proportion to their length. The abdomen tapers regularly to the distal end; the terminal segment is semicircular.

Color.—A female (50168) has the legs banded with brown; in legs 1 to 3 one band covers the proximal half or two-thirds of the merus, another band occupies the major part of the propodus; on leg 4 the color occupies less than half the merus and the distal third of the dactylus. There are three transverse stripes of brown on the proximal third of the abdomen.

Measurements.—Female (holotype), length of carapace 4.8, width of same 6.4 mm. Female (50168), length of carapace 5.5, width of same 7.3 mm. Male (50168), length of carapace 4.6, width of same 5.7 mm.

Range.—East Florida; northwest Cuba. 2 to 4 fathoms.

Material examined.—

Grecian Shoals, Hawk Channel, Florida; ¼ mile S. W. by S. of Basin Hill; 2 fathoms; S. G.; Feb. 19, 1903; station 7469 *Fish Hawk;* 1 female holotype (49233).

Off Duck Key, Hawk Channel, Florida; 4 fathoms; mud bag; temp. about 69.5° F.; Dec. 20, 1912; station 4, haul 2, *Fish Hawk;* 1 male, 1 female ovig. (50168).

Northwest Cuba; Reef Laveros Italienas, opposite Cayo Laveros; 2 to 3 fathoms; Co. S. R.; June 2, 1914; Henderson and Bartsch, *Tomas Barrera* Exped.; 1 female (48570).

Subfamily PINNOTHERELIINAE Alcock.

Pinnotherelinae ALCOCK, Journ. Asiat. Soc. Bengal, vol. 69, 1900, p. 288.

Ischium of external maxillipeds usually distinct from merus, though smaller and sometimes imperfectly united with it. Merus longitudinal or little oblique. Palpus of good size, sometimes as large as merus-ischium. Carapace transverse, usually broadly so.

KEY TO THE AMERICAN GENERA OF THE SUBFAMILY PINNOTHERELIINAE.

A^1. Last joint of outer maxilliped articulated near proximal end of inner side of penultimate joint.
  B^1. Carapace much wider than long. Third leg longest.
    C^1. Third leg much longer than fifth. Lower or true antero-lateral margin forming an angle with postero-lateral margin_____*Pinnixa*, p. 128.
    C^2. Legs varying little in length. Lower or true antero-lateral margin curving gradually into postero-lateral margin_____*Scleroplax*, p. 170.
  B^2. Carapace very little wider than long. Second and third legs nearly equal.
    C^1. Ischium of outer maxilliped well defined_____*Opisthopus*, p. 172.
    C^2. Ischium of outer maxilliped either fused with merus, or only partly so _____*Pinnaxodes*, p. 174.
A^2. Last joint of outer maxilliped articulated at distal end of penultimate joint. Second leg longest.
  B^1. Proximal end of last joint of outer maxilliped narrow, distal end of penultimate joint wide. Carapace uneven. Spinules on chelipeds and legs _____*Tetrias*, p. 179.
  B^2. Proximal end of last joint of outer maxilliped same width as distal end of penultimate joint. Carapace smooth. No spinules on chelipeds or legs_____*Pinnotherelia*, p. 180.

Genus PINNIXA White.

Pinnixa WHITE, Ann. Mag. Nat. Hist., vol. 18, 1846, p. 177; type, *P. cylindrica* White.—RATHBUN, Bull. U. S. Fish Comm., vol. 20, for 1900,. pt. 2 (1901), p. 21.

Tubicola LOCKINGTON, Proc. California Acad. Sci., vol. 7, 1876 (1877), p. 55 [1]; type, *T. longipes* Lockington.

Carapace much wider than long; integument usually firm. Front narrow, nearly transverse, with a median groove. Orbit broadly ovate or nearly circular, with a wide inner hiatus, which is partly occupied by the basal antennal joint. Antennules transversely or obliquely plicated in wide fossettes which communicate with each other beneath the front. Eye-stalks very short. Epistome lineartransverse. Ischium of maxillipeds small, merus large, distal portion of outer margin convex; palp jointed to summit of merus; third joint articulated on inner side of the preceding one near base.

Chelipeds of moderate size; merus trigonous; carpus usually smooth; hand large, compressed. Second ambulatory leg larger than first; third largest of all; fourth much shorter than third and relatively stouter than first and second. Abdomen in both sexes usually 7-jointed and narrower at base than width of last sternal segment.

Habitat.—Live in bivalve mollusks, in tubes or holes of worms, in wormlike holothurians, and in mud.

The genus occurs in Japan and Australia as well as in America.

KEY TO THE AMERICAN SPECIES OF THE GENUS PINNIXA.

A^1. Carapace with a ridge entirely across hinder part of carapace, behind which the surface slopes steeply down.
  B^1. Second joint of palpus of maxilliped short and stout.
    C^1. Ridge on carapace sharp. Fingers not gaping. Last segment of male abdomen semicircular_____*transversalis*, p. 131.
    C^2. Ridge on carapace blunt. Fingers gaping. Last segment of male abdomen subtriangular_____*faxoni*, p. 133.
  B^2. Second joint of palpus of maxilliped long and narrow.
    C^1. Sharp ridge across front_____*patagoniensis*, p. 135.
    C^2. No sharp ridge across front_____*cristata*, p. 134.
A^2. Carapace without ridge on hinder part of carapace or with a ridge on cardiac region only.
  B^1. Pollex of cheliped distinctly developed.
    C^1. Dactylus of third leg strongly falcate, the corneous tip bent at an angle to the general outline of the segment. Females much larger and softer than males.
      D^1. Merus of third leg of male more than twice as long as wide. Fingers of female not gaping. Thumb of male horizontal. Carapace oblong_____*faba*, p. 142.
      D^2. Merus of third leg of male twice as long as wide. Fingers of female gaping. Thumb of male deflexed. Carapace pointed at sides.
        littoralis, p. 145.
    C^2. Dactylus of third leg straight or slightly curved, the corneous tip continuing the general line of the segment.
      D^1. Fourth leg when extended not reaching end of merus of third leg.
        E^1. Carapace very wide, nearly 3 times as wide as long. Third leg enormously large in proportion to body.
          F^1. Chela stout. Propodus of second leg wide_____*longipes*, p. 137.
          F^2. Chela feeble. Propodus of second leg narrow.
            tubicola, young, p. 165.
        E^2. Carapace twice or little more than twice as wide as long. Third leg not enormously large in proportion to body; merus less than twice as long as wide_____*floridana*, p. 138.
      D^2. Fourth leg when extended reaching end or beyond end of merus of third leg.
        E^1. Propodus of third leg slender, twice or more than twice as long as wide.
          F^1. Carapace twice as wide as long; a cardiac crest present.
            sayana, p. 156.
          F^2. Carapace less than twice as wide as long; no cardiac crest.
            minuta, p. 150.
        E^2. Propodus of third leg not slender, less than twice as long as wide.
          F^1. Propodus of third leg as wide as long or nearly so.
            G^1. A blunt ridge on cardiac region_____*cylindrica*, p. 159.
            G^2. No ridge on cardiac region.
              H^1. Carapace of female twice as wide as long. Propodus of last two legs squarish, much wider at distal end than proximal end of dactylus_____*tomentosa*, p. 141.

H². Carapace of female two and a half times as wide as long. Propodus of last two legs tapering at distal end almost to width of proximal end of dactylus_____*tubicola*, p. 165.
F². Propodus of third leg distinctly longer than wide.
G¹. Carapace very convex, sloping from the middle in all directions. Thumb strap-shaped. Manus much widened toward distal end. Fingers with wide, triangular gape__*barnharti*, p. 149.
G². Carapace more or less flattened, not sloping from the middle in all directions. Thumb not strap-shaped but diminishing from base.
H¹. Thumb deflexed.
J¹. Merus of third leg short and stout, about twice as long as wide _____*chaetopterana*, p. 151.
J². Merus of third leg elongate, much more than twice as long as wide.
K¹. Thumb subtriangular, the terminal spine forming one point of the triangle.
L¹. Two short, well separated cardiac crests. Thumb slightly deflexed_____*valdiviensis*, p. 154.
L². A single, bilobed cardiac crest. Thumb markedly deflexed_____*occidentalis*, p. 155.
K². Thumb composed of a short, high, quadrate base, followed by a short, slender spine at lower angle.
L¹. Palm strongly widened distally, lower margin nearly straight. Cardiac ridge blunt____*brevipollex*, p. 169.
L². Palm very convex below. Cardiac ridge acute.
affinis, p. 168.
H². Thumb not deflexed.
J¹. Thumb straight or nearly so. Lower margin of palm not convex.
K¹. A cardiac ridge present. No large spine on ischium of third leg in male_____*franciscana*, p. 161.
K². No cardiac ridge present. A large spine on ischium of third leg in male_____*retinens*, p. 139.
J². Thumb curved upward distally. Lower margin of palm convex.
K¹. Carapace more than twice as wide as long. Dactylus of third leg longer than propodus. Fingers of female gaping_____*hiatus*, p. 164.
K². Carapace less than twice as wide as long.
L¹. Carapace laterally tapering. Sides of male abdomen convex_____*schmitti*, p. 162.
L². Carapace laterally subtruncate. Sides of male abdomen not convex, first segment very wide at base.
weymouthi, p. 166.
B². Pollex of cheliped represented only by spiniform angle of palm.
monodactyla, p. 136.

Analogous species.—Besides several pairs of analogous species on opposite sides of the American continent, there are also in the genus *Pinnixa* close resemblances between certain species in the northern and southern hemispheres, as indicated below.

ANALOGOUS SPECIES OF PINNIXA ON OPPOSITE SIDES OF THE CONTINENT.

Atlantic.	Pacific.
brevipollex.	*affinis.*
faxoni.	*transversalis.*
cylindrica.	*franciscana.*
sayana.	*occidentalis.*
floridana.	*longipes.*

ANALOGOUS SPECIES OF PINNIXA IN NORTHERN AND SOUTHERN HEMISPHERES.

North Atlantic.	South Atlantic.
cristata. } *monodactyla.* }	*patagoniensis.*
North Pacific.	South Pacific.
occidentalis.	*valdiviensis.*

PINNIXA TRANSVERSALIS (Milne Edwards and Lucas).

Plate 29, figs. 1–3.

Pinnotheres transversalis MILNE EDWARDS and LUCAS, d'Orbigny's Voy. Amér. Mér., vol. 6, pt. 1, 1843, p. 23 (type-locality, Chili; type in Paris Mus.); vol. 9, atlas, 1847, pl. 10, figs. 3–3c.—NICOLET in Gay, Hist. de Chile, Zool., vol. 3, Crust., 1849, p. 156.

Pinnixa transversalis MILNE EDWARDS, Ann. Sci. Nat., ser. 3, Zool., vol. 20, 1853, p. 220 [186], pl. 11, fig. 5.—MIERS, Proc. Zool. Soc. London, 1881, p. 70, part; not specimens from Sandy Point.—ORTMANN, Zool. Jahrb., Syst., vol. 10, 1897, p. 329 (specimen described probably not a type).—LENZ, Zool. Jahrb., Suppl., vol. 5, 1902, p. 764.—RATHBUN, Proc. U. S. Nat. Mus., vol. 38, 1910, pp. 546 and 588, pl. 46, fig. 1 (after Milne Edwards and Lucas).

Pinnixa panamensis FAXON, Bull. Mus. Comp. Zoöl., vol. 24, 1893, p. 158 (type-locality, Panama; type in M.C.Z.); Mem. Mus. Comp. Zoöl., vol. 18, 1895, p. 30, pl. 5, figs. 1, 1a, 1b.

Diagnosis.—Sharp ridge entirely across hinder part of carapace. Ultimate segment of maxilliped much longer than penultimate. Chelae weak, fingers not gaping. Extremity of male abdomen much enlarged, last segment semicircular.

Description of male.—Carapace mostly smooth and punctate, granulated at sides; lateral angles forming a prominent shoulder, below which the carapace narrows at the sides; a groove behind the gastric region, otherwise carapace flat up to a prominent, sharp ridge across carapace at the cardiac region and ending at bases of last feet; behind ridge, carapace is inclined at angle of 45°. Front deflexed, transversely grooved, not produced. Orbits with a strong dorsal inclination. Antero-lateral margin a prominent granulate ridge in outer half only. Posterior margin short, concave.

Second segment of palpus of maxilliped very short, third very long, reaching proximal end of merus.

Chelipeds small; carpus squamose and setose anteriorly; chelae laterally compressed, a crest of tubercles on upper border, a longitudinal row of setiferous granules at lower third; above this the surface is granulate and setose, below it the surface is smooth and bare down to a row of tubercles on lower edge; fingers slender, closing together, outer margins granulate and setose, prehensile margins very finely denticulate.

First and second legs narrow, with straight dactyli and not very unequal; third and fourth legs of stouter build, dactyli short, very stout, short-pointed; merus of third with anterior border convex and pubescent, posterior border tuberculate; fourth leg very small, reaching to end of merus of third.

Between second and third segments of abdomen there is a row of hairs which lie flat on the second segment; some of third to sixth segments usually partly fused; end of abdomen from middle of sixth segment suddenly enlarged, end segment semicircular.

FIG. 74.—PINNIXA TRANSVERSALIS, OUTER MAXILLIPED OF FEMALE (40446), × 16½.

Sexual variation.—Female larger, carapace more swollen, frontal region more inclined, dorsal and antero-lateral ridges weaker, posterior margin wider, fingers longer, appendages more setose.

Color.—Violet; abdomen of male ash-colored, of female golden-yellow (Milne Edwards and Lucas).

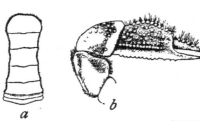

FIG. 75.—PINNIXA TRANSVERSALIS. *a*, ABDOMEN OF MALE, ENLARGED; *b*, RIGHT CHELIPED, ENLARGED. (AFTER FAXON.)

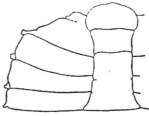

FIG. 76.—PINNIXA TRANSVERSALIS, ABDOMEN AND STERNUM OF MALE, ENLARGED. (AFTER MILNE EDWARDS AND LUCAS.)

Measurements.—Male (20625), length of carapace 4.7, width of same 9.7 mm. Female (40446), length of carapace 8, width of same 16.5 mm.

Habitat.—In tube of annelid resembling *Chaetopterus* (Coker).

Range.—Panama (Faxon) to Punta Arenas, Patagonia (Lenz).
Material examined.
Panama; Mar. 12, 1891; steamer *Albatross;* 1 male, 4 females ovig., cotypes (20625).
Peru; near northeast side of San Lorenzo Island; in piece of tube resembling that of *Chaetopterus;* 2½ fathoms; R. E. Coker; 1 female ovig. (40446).

PINNIXA FAXONI,[1] new species.

Plate 29, figs. 4-7.

Type-locality.—Trinidad; *Albatross;* male holotype; Cat. No. 7639, U.S.N.M.

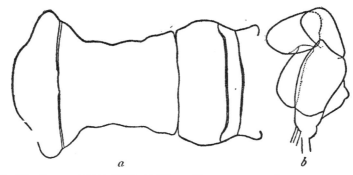

FIG. 77.—PINNIXA FAXONI (7639), × 13⅔. *a*, ABDOMEN OF MALE; *b*, OUTER MAXILLIPED.

Diagnosis.—Blunt ridge entirely across hinder part of carapace. Ultimate segment of maxilliped much longer than penultimate. Fingers gaping. Extremity of male abdomen much enlarged, last segment subtriangular.

Description of male.—Near *transversalis*, but much shaggier on all margins, especially on posterior margin of carapace and across distal end of second abdominal somite and thence in a straight line across sternum. Ridge across carapace at cardiac region not sharp, but bluntly rounded. Front more horizontal and advanced. Eyes and orbits shorter and wider.

Chelipeds similar, but palm more oblong, not narrowing distally; thumb rather short, tip upcurved; dactylus curved obliquely down, leaving a triangular gape when closed; a line of hair on lower edge of propodus, another a little above margin and continued on thumb, a third at middle of palm, upper portion densely hairy and granulate.

Legs similar, but a little shorter and wider, especially noticeable in merus of last two legs. Abdomen wider, fourth, fifth, and sixth

---

[1] For Dr. Walter Faxon, carcinologist of the Museum of Comparative Zoology, Cambridge, Mass.

segments fused, last segment subtriangular, widest, and laterally angled not far from sixth segment.

Sexual variation.—Female somewhat larger and more swollen than male, frontal region more inclined, posterior margin wider, and wider than in female *transversalis;* fingers elongate and very slightly gaping, less slender than in female *transversalis*.

Measurements.—Male holotype, length of carapace 5.6, width of same 11 mm. Female (7639), length of carapace 6.1, width of same 13.3 mm.

Range.—Trinidad (off Venezuela) and Monos Island, off NW coast of Trinidad.

Material examined.—Trinidad; shore; Jan. 30–Feb. 2, 1884; steamer *Albatross;* 7 males (1 is holotype), 7 females (2 ovig.) (7639). Monos Island, Trinidad; shore; Jan. 30–Feb. 2, 1884; steamer *Albatross;* 1 male (23436).

PINNIXA CRISTATA Rathbun.

Plate 29, figs. 8 and 9.

Pinnixa cristata RATHBUN, Amer. Nat., vol. 34, 1900, p. 589 (type-locality, Beaufort, North Carolina; holotype female, Cat. No. 42317, U.S.N.M.).

Diagnosis.—Sharp crest across posterior part of carapace. Fingers gaping, thumb short. Legs narrow; dactyls slightly curved except last pair.

Description of female.—Carapace very short, narrowed at either side; a high, sharp, transverse crest across carapace, ending above bases of last legs; surface punctate, wrinkled, and microscopically granulate; deep furrow behind gastric region; antero-lateral margin a raised crest, stopping short of hepatic region; posterior margin wide, concave. Front deflexed, not advanced; orbit no wider than half of front.

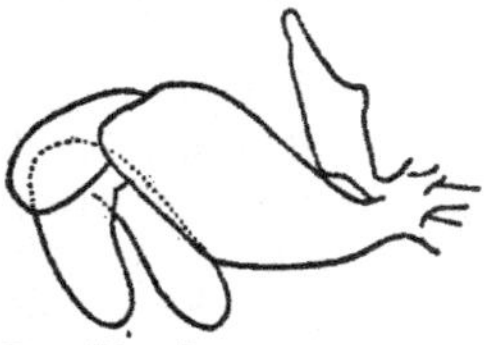

Chelipeds rather stout; palm suboblong, with upper and lower margins convex, surface covered with a reticulate pattern of fine granulations; thumb short, subtriangular, deflexed; a subbasal tooth on prehensile edge, also a small tooth near the tip, forming a truncate extremity against which the tip of the gaping dactylus fits; inner margin of dactylus bent in a right angle at middle, subentire.

FIG. 78.—PINNIXA CRISTATA, OUTER MAXILLIPED OF FEMALE HOLOTYPE, × 15.

Dactyli of legs slender, of fourth leg straight, of first, second, and third slightly curved; first leg slender, reaching to end of propodus of second; second intermediate in width between first and third, reaching one-third length of dactylus of third; merus of third leg $2\frac{1}{2}$

times as long as wide, tapering at both ends, marginate above, very hairy below; fourth leg overreaching by half its dactylus the merus of the third leg.

Measurements.—Female holotype, length of carapace 4, width of same 10.5 mm.

Specimen examined.—Beaufort, North Carolina; Union College collection (70c); 1 female ovig., holotype (42817).

PINNIXA PATAGONIENSIS, new species.

Plate 30, figs. 1-3.

Type-locality.—San Matias Bay, Patagonia; steamer *Hassler;* holotype male, Cat. No. 5741, M.C.Z.

Diagnosis.—Thumb reduced to spiniform angle of propodus; dactylus strongly curved to meet thumb. Sharp ridge across posterior part of carapace and across front.

Description.—Carapace narrowed to an acute angle at either side. A sharp, smooth crest entirely across hinder part of carapace at cardiac region; from it the carapace descends almost vertically to posterior margin and obliquely forward to gastro-cardiac furrow. A similar crest marks the antero-lateral margin of the dorsal surface, extending to outer angle of orbit whence a short crest continues obliquely downward

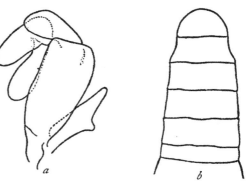

Fig. 79.—PINNIXA PATAGONIENSIS, MALE HOLOTYPE. *a*, OUTER MAXILLIPED, × 18¾; *b*, ABDOMEN, × 7½.

on the suborbital region; a transverse crest across the front is interrupted by the median furrow. This crest is connected with the antero-lateral crest by the upper part of the smooth orbital rim. Surface coarsely punctate, pubescent near outer angles. Front prominent, truncate, emarginate at middle. Orbit much larger than eye, externally angled. Antenna as long as width of front and one orbit.

Chela large; palm as high as superior length, upper and lower margins very convex, upper margin sharp-edged; thumb a short, much deflexed spine with obliquely truncate and bidenticulate tip; the truncate, distal end of manus with a broad, oblong tooth above digital spine; dactylus nearly straight and inclined obliquely downward at base, then abruptly recurved toward thumb leaving a high, narrow gape.

Dactyls of first three legs slightly curved, of last leg straight, with posterior or lower edge convex; all are longer than their propodi. First leg slender, reaching end of propodus of second; second with merus expanded below, reaches end of propodus of third; third much the stoutest, the lower edge of merus and of carpus-propodus much dilated; fourth leg slender, reaching end of merus of third; all margins sharp.

Variations.—In the small male, palm less high than in adult male, thumb of normal shape, triangular, as long as high, less deflexed, dactylus about as wide, moderately curved, when closed, making an insignificant gape.

Measurements.—Male, holotype, length of carapace 5.5, width of same 12.8 mm.

Specimens examined.—San Matias Bay, eastern Patagonia; *Hassler* Exped.; 1 male holotype, reproducing right cheliped (5741, M. C. Z.); 1 young, paratype (49248, U.S.N.M.).

Affinity.—Related to *P. cristata* in the sharp dorsal crest and to *P. monodactyla* in the spinelike thumb.

PINNIXA MONODACTYLA (Say).

Pinnotheres monodactylum Say, Journ. Acad. Nat. Sci. Philadelphia, vol. 1, 1818, p. 454 (type-locality, American; type probably not extant; said to be in "Richmond Museum").

Diagnosis.—Thumb reduced to spiniform angle of propodus; dactylus strongly curved to meet thumb. Carapace narrow, $1\frac{2}{3}$ times as wide as long.

Description of male.—Carapace subelliptical, narrowing each side to middle of lateral edge, which is rounded, a tubercle each side marking situation of antero-lateral angles. Surface punctate. Orbits suborbicular. Antennae subequal to width of front.

Chela oblong, somewhat quadrate; palm concave and ciliated at middle, a spiniform angle instead of a thumb, with a tooth at its base and another at base of dactylus larger. Dactylus abruptly incurved at base, rectilinear toward tip, with an angle at interior middle, tip acute, attaining tip of spiniform angle or thumb.

First, second, and fourth pairs of legs subequal, second rather the largest; third pair largest of all, and, as also the fourth pair, with somewhat dilated carpus-propodus.

Abdomen with a few larger punctae, terminal joint rounded at tip. (After Say.)

Measurements.—Male, length 0.3 inch (7.6 mm.), width 0.5 inch (12.7 mm.)

Range.—Probably Atlantic coast of the Southern States. Not seen since the type was taken.

PINNIXA LONGIPES (Lockington).

Tubicola longipes LOCKINGTON, Proc. California Acad. Sci., vol. 7, 1876 (1877), p. 55 [1] (type-locality, Tomales Bay, California, in tube of annelid; type not extant).

Pinnixa longipes LOCKINGTON, Proc. California Acad. Sci., vol. 7, 1876 (1877), p. 156 [12].—STREETS and KINGSLEY, Bull. Essex Inst., Vol. 9, 1877, p. 107.—HOLMES, Proc. California Acad. Sci., ser. 2, vol. 4, 1894, p. 573, pl. 20, figs. 19 and 20; Occas. Papers California Acad. Sci., vol. 7, 1900, p. 92.—RATHBUN, Harriman Alaska Exped., vol. 10, 1904, p. 188.—WEYMOUTH, Leland Stanford Jr. Univ. Publ., Univ. Ser. No. 3, 1910, p. 58, text-fig. 6 (part; not 2 young specimens from Monterey Bay).

Diagnosis.—Carapace nearly three times as wide as long. Third leg vastly larger than the others; fourth leg not reaching beyond merus of third.

FIG. 80.—PINNIXA LONGIPES, COTYPE, GENERAL OUTLINE, ENLARGED. (AFTER HOLMES.)

Description.—Carapace more than 2½ times as wide as long, somewhat flattened above, a depression behind the gastric region. A transverse groove just behind margin of front.

Chelipeds small, short, hairy; chelae ovate-oblong, distally tapering, upper margin convex, lower margin nearly straight, distally granulate; fingers subequal, subacute, fitting close together, a subbasal tooth on dactylus, a median tooth on thumb, and a subterminal notch into which the point of the dactylus fits.

First two legs slender and similar, with slender dactyli, the first leg the smaller. Third leg enormously developed, longer than width of carapace, merus thick, flat, and with a kind of flange on posterior margin; dactyl short, thick, curved, and shorter than the tapering propodus. Fourth leg shortest of all, not reaching beyond merus of third, broad, fringed with long hair, dactylus less curved than that of third leg and shorter than propodus.

FIG. 81.—PINNIXA LONGIPES, COTYPE, OUTER MAXILLIPED, ENLARGED. (AFTER HOLMES.)

Male abdomen tapers evenly from base to last segment, first two segments shortest, last segment broader than long, rounded at tip.

Measurements.—Ovigerous female (17862), length of carapace 2.2, width of same 5.7 mm.

Habitat.—Commensal in tubes of annelid worms (*Clymenella*).

Range.—California: Tomales Bay to San Pedro.

Specimen examined.—Tomales Bay, California, in tube of Maldanid; May, 1876; 1 female ovigerous (17862). Perhaps received from Lockington, but not so labeled.

PINNIXA FLORIDANA, new species.

Plate 30, figs. 4–7.

Pinnixa cylindrica KINGSLEY, Proc. Acad. Nat. Sci. Philadelphia, 1879, p. 402 (part: some specimens from Sarasota Bay).

Type-locality.—Marco, Florida; Henry Hemphill; female holotype, Cat. No. 6996, U. S. N. M.

Diagnosis.—Carapace twice or little more than twice as wide as long. Merus of third leg less than twice as long as wide. Dactyli of legs straight. Chelae weak, granulate.

Description of female.—Carapace appearing suboblong, the antero-lateral angle forming a sort of shoulder, the side walls steep

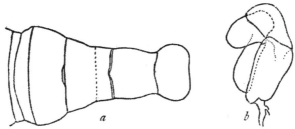

FIG. 82.—PINNIXA FLORIDANA. *a*, ABDOMEN OF MALE (49249) × 19; *b*, OUTER MAXILLIPED OF FEMALE PARATYPE (6996), × 19.

and tapering outwardly; surface smooth and punctate; gastro-cardiac groove shallow, cardiac region elevated but without ridge, antero-lateral margin, up to the cervical suture, a sharp, milled crest; posterior margin concave. Front truncate, not advanced. Orbits inclined forward outwardly in dorsal view, and downward in front view.

Chelipeds weak, hairy; chelae tapering distally; propodus with denticulate upper margin, a row of sharp granules near lower margin, continued feebly on finger, a row of granules at middle, and scattered granules above middle; fingers nearly horizontal, not gaping, tips hooked, two rows of sharp granules above dactylus, a few small teeth on prehensile edges.

Dactyls of ambulatory legs straight or nearly so, first two slender, last two stout. First leg narrower than second, reaching middle of dactylus of second; second leg to middle of propodus of third; third very wide, lower margins denticulate, merus and propodus marginate above, merus 1⅔ times as long as wide, propodus as wide as its anterior length, tapering rapidly toward dactylus; fourth leg not exceeding merus of third, wide in proportion.

A line of hair at distal end of second segment of abdomen is continned along sternum.

Sexual variation.—In male, carapace narrower, side wall relatively steeper, shoulder more prominent; palm wider proportionally, fixed finger a little shorter, fingers gape slightly. Abdomen wide at third segment whence it tapers unevenly to the last segment, which is wider than the sixth, truncate at extremity, strongly arched at sides, and fringed with long hair.

Measurements.—Male (49249), length of carapace 2.4, width of same 4.4 mm. Female (6996), length of carapace 3.5, width of same 8 mm.

Range.—Florida at Marco and Sarasota Bay.

Material examined.—

Marco, Florida; 1885; Henry Hemphill; 3 females ovig. (1 is holotype (6996).

Sarasota Bay, Florida: Union College collection (1015); 1 male, 1 female ovig., labeled *P. cylindrica* by Kingsley (49249). Union College collection (1039); 1 female ovig., labeled *P. cylindrica* by Kingsley (49250).

PINNIXA RETINENS,[1] new species.

Plate 41, figs. 1 and 2.

Type-locality.—Chesapeake Bay; off Poplar Island, Maryland; 20 fathoms; soft bottom; Apr. 25, 1916; station 8528; *Fish Hawk;* male holotype; Cat. No. 50167, U.S.N.M.

Diagnosis.—No cardiac ridge. Fingers slender, not deflexed. A strong, curved spine on ischium of third leg.

Description of male.—Carapace nearly twice as wide as long, nearly flat except toward margins, where it slopes gradually downward; regions indicated; lateral margin marked by a sharp, granulated ridge, up to the subhepatic sulcus; no cardiac ridge; a groove subparallel to, and a little in front of, the posterior margin.

Dactylus of outer maxilliped obliquely spatuliform and attached to the middle of the inner margin of the propodus.

---

[1] *Retinens,* holding fast, in allusion to the probable use of the ischial spine.

Chelipeds small, about as long as first leg; lower margin of propodus straight; manus suboblong, marginate below; fingers slender, subequal, not gaping, dactylus a little the longer, and having a tooth at the proximal third of its prehensile edge.

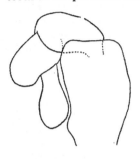

FIG. 83.—PINNIXA RETINENS, ENDOGNATH OF OUTER MAXILLIPED OF MALE HOLOTYPE. × 33½.

First and second legs similar, the second a little the longer, especially as to the last two segments; dactyli very slightly curved, long-pointed. Third leg stout, exceeding the second by the length of its dactyl and half its propodus; lower edge of ischium, merus and propodus armed with stout spinules; postero-distal end of ischium prolonged in a stout, curved spine, the point of which is directed upward and backward; dactylus more strongly curved than in preceding legs, and gradually tapering to a slender point. The small, posterior leg, if extended, would about reach the end of the merus of the third leg; its dactylus is stoutish, nearly straight, the tip curved slightly upward.

Abdomen constricted at base of second segment, widest between second and third segments, from which point it gradually narrows,

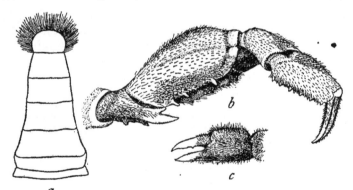

FIG. 84.—PINNIXA RETINENS, MALE HOLOTYPE, × 12. a, ABDOMEN; b, THIRD LEG, LOWER SIDE; c, LEFT CHELA.

the third, fourth, and fifth segments with sides partially concave; terminal segment much wider than long.

Anterior and posterior margins of sternal segments granulate; granules especially noticeable bordering posterior margin of carapace.

Sexual variation.—The only female at hand is immature and much smaller than the type male. The fingers are longer in proportion to the palm than in the male and the propodal finger is slightly deflexed; there is no spine at distal end of ischium of third leg; the fourth leg reaches slightly beyond the merus of the third leg.

Measurements.—Male holotype, length of carapace 4.2, width of same 7, length of third leg 10.5 (approx.) mm.

Range.—Chesapeake Bay; 20 fathoms.

Material examined.—

Chesapeake Bay; 20 fathoms; soft bottom; Apr. 25, 1916; station 8528, steamer *Fish Hawk;* 1 male holotype (50167).

Chesapeake Bay; 20 fathoms; sft. m.; Apr. 25, 1916; station 8522, steamer *Fish Hawk;* 1 female paratype (49641).

Affinity.—Allied to *P. floridana* in the character of the chelipeds and in lacking a carina on the carapace. Differs in the narrower legs and in the presence of a long spine on the ischium of the third leg.

PINNIXA TOMENTOSA Lockington.

Plate 30, fig. 8.

Pinnixa ? nitida LOCKINGTON, Proc. California Acad. Sci., vol. 7, 1876 (1877), p. 155 [11], part (female).

Pinnixa tomentosa LOCKINGTON, Proc. California Acad. Sci., vol. 7, 1876 (1877), p. 156 [12] (type-locality, Angeles Bay, Gulf of California; type not extant).—HOLMES, Proc. California Acad. Sci., ser. 2, vol. 4, 1894, p. 568, pl. 20, figs. 11-13.

Diagnosis.—Propodus of last two legs nearly square, much wider distally than proximal end of dactylus. Dactylus of first two legs longer, slenderer, and straighter than that of last two.

Description of female.—Carapace and legs covered with a short pubescence. Carapace nearly twice as wide as long, smooth, and rounding down to margins; a shallow depression behind gastric region; cardiac region swollen but not ridged. A transverse depression just behind margin of front; antero-lateral margin marked by a granulated line on the branchial region.

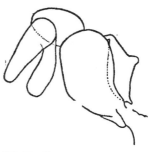

FIG. 85.—PINNIXA TOMENTOSA, OUTER MAXILLIPED OF FEMALE (29948), × 15½.

Hand oblong, compressed, margins convex, the lower one concave near the thumb. Fingers little over half length of palm, toothless; thumb wide, but abruptly narrowed near hooked tip; dactyl curved, subuncinate at apex, no longer than thumb. (Chela after Holmes.)

First leg much slenderer and shorter than second; second leg nearly as long as third and less stout, its propodus tapering distally: first and second dactyls slender, slightly curved, third and fourth dactyls stout, short, and more curved but not hooked; third leg very wide, merus a little more than one and a half times as long as wide, propodus nearly square; fourth leg similar in shape but much smaller.

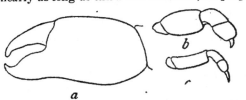

FIG. 86.—PINNIXA TOMENTOSA, FEMALE HOLOTYPE, ENLARGED. *a*, CHELA; *b*, THIRD LEG; *c*, FIRST LEG. (AFTER HOLMES.)

Measurements.—Female holotype, length of carapace 7.5, width of same 14 mm. (After Holmes.)

Range.—San Clemente Island, California, to Angeles Bay, Gulf of California.

Specimen examined.—San Clemente Island, California; H. N. Lowe, collector; 1 female without chelipeds (29948).

PINNIXA FABA (Dana).

Plate 31, figs. 1–4.

Pinnothera faba DANA, Proc. Acad. Nat. Sci. Philadelphia, vol. 5, 1851, p. 253 [7] (type-locality, Puget Sound; type not extant); U. S. Expl. Exped., vol. 13, Crust., pt. 1, 1852, p. 381; atlas, 1855, pl. 24, fig. 4 *a, b, c*.

Pinnixa faba STIMPSON, Journ. Boston Soc. Nat. Hist., vol. 6, 1857, p. 470 [30].—COOPER, Repts. Explor. & Survey, Miss. River to Pacific O., vol. 12, book 2, 1860, p. 387.—NEWCOMBE, Bull. Nat. Hist. Soc. British Columbia, 1893, p. 25.—HOLMES, Occas. Papers California Acad. Sci., vol. 7, 1900, p. 93.—RATHBUN, Harriman Alaska Exped., vol. 10, 1904, p. 188 (part; not from Sitka or San Pedro).—WEYMOUTH, Leland Stanford Jr. Univ. Publ., Univ. ser. No. 4, 1910, p. 59 (part), not specimen from Monterey Bay or text-fig. 7. Probably not *Pinnixa faba* Haswell, Cat. Australian Crust., 1882, p. 113, from Port Denison, Australia.

Pinnotheres faba BATE in Lord's Naturalist in Vancouver Id. and Brit. Col., vol. 2, 1866, p. 271.

Pinnixa littoralis WEYMOUTH, Leland Stanford Jr. Univ. Publ., Univ. ser. No. 4, 1910, p. 58 (part: text-fig. 5 and examples from Puget Sound).

Diagnosis.—Carapace oblong. Orbits oval. Dactyli of legs strongly curved. Thumb horizontal, a little shorter than movable finger. Fingers of female not gaping. Merus of third leg of male more than twice as long as wide.

Description of male.—Carapace firm, smooth, rounding steeply downward in all directions, truncate at the sides; gastric region high, not cristate. Antero-lateral margin marked by a low, blunt ridge, most prominent at antero-lateral angle, but disappearing near orbit. Front slightly advanced; orbits oval, filled by the eyes.

Chelipeds large; palm suboblong, widening distally, upper and lower margins slightly convex; thumb shorter than dactyl, slightly

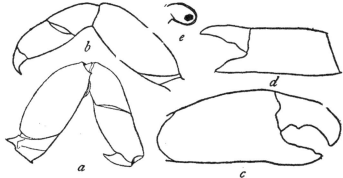

FIG. 87.—PINNIXA FABA (31599), × 6. *a*, THIRD LEG OF FEMALE; *b*, THIRD LEG OF MALE; *c*, RIGHT CHELA OF MALE; *d*, LEFT CHELA OF FEMALE; *e*, EYE IN ORBIT, FRONT VIEW.

deflexed with the end upcurved, edge finely crenulate and with a distal notch into which the tip of the dactylus fits; dactyl curved, a triangular tooth near its base, and forming when closed a moderate gape with the thumb; fingers hairy within and toward their base.

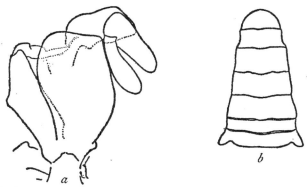

FIG. 88.—PINNIXA FABA (31599). *a*, OUTER MAXILLIPED OF FEMALE, × 16; *b*, ABDOMEN OF MALE, × 6½.

Legs similar in shape, except merus of first leg which is concave above instead of convex as in the other pairs. Propodi tapering distally, dactyli short and strongly curved. First leg narrow, reaching to end of propodus of second; second and third legs subequal in length, third a little wider, its merus over twice as wide as long; fourth leg shortest, reaching end of carpus of third.

Abdomen strongly narrowing distally for its first segment, then moderately tapering to the penultimate segment, which is constricted at middle; last segment rounded at tip.

Description of female.—Much larger than male. Carapace and abdomen soft and yielding; carapace more uneven, less deflexed posteriorly and without an antero-lateral marginal line. Median groove on gastric region very deep.

Chelipeds smaller than in male; palm narrower, its sides nearly parallel; fingers longer, less deflexed, not gaping, sharp-pointed, tips crossing; thumb without a terminal notch, but instead a shallow lobe at middle of prehensile margin; edge of dactylus feebly sinuous.

Legs more alike than in male; first leg reaching to middle of dactylus of second; fourth leg reaching beyond end of carpus of third; merus of third leg about twice as long as wide.

Variations.—In some males the antero-lateral angle is vertically compressed and correspondingly thin, forming a laterally projecting lobe.

Color.—In life, grayish-white (Cooper). Specimens in formalin from Doctor Fraser are as follows: General color of females orange-rufous with patches of scarlet on the gastric regions. Eggs orange-chrome. Male, orange-rufous, or dirty greenish-white with orange-rufous spots on carapace and a few of the same on chelipeds and legs. 1 female from Taylor Bay was entirely white in life.

Measurements.—Male (17468), length of carapace 7.6, width of same 13 mm. Female (17468), length of carapace 15.2, width of same 22.8 mm.

Habitat.—Commensal in clams and according to Holmes, in the cloaca of a large holothurian, *Liosoma arenata* [*arenicola*] Stimpson (now *Molpadia arenicola*). I think that Holmes's specimen may be *P. barnharti*.

Range.—From Prince of Wales Island, Alaska, to Humboldt Bay, California. San Pedro, California (Holmes).

Material examined.—

Cordova Bay, Prince of Wales Island, Alaska; taken in pairs from gill chambers of giant clams; June 16, 1897; steamer *Albatross;* 7 males, 5 females (4 ovig.) (21792).

Beaver Harbor, British Columbia; in shells of *Schizothaerus nuttallii;* July 12, 1888; steamer *Albatross;* 7 males, 9 females (7 ovig.) (17468).

Departure Bay, British Columbia; in *Schizothaerus nuttallii;* 1908; Geological Survey of Canada; 1 male, 1 female (40398).

Denman Island, British Columbia; from clams; May 12. 1914; steamer *Albatross;* 6 males, 8 females (6 ovig.) (48426).

Union Bay, Vancouver Island, British Columbia; in clams, *Saxidomus*, probably *giganteus;* Apr. 16, 1914; steamer *Albatross;* 1 male, boiled (48428).

British Columbia; 1916; C. McLean Fraser: Brandon Island, in *Schizothaerus;* May 18; 13 males, 10 females ovig., " of a dirty brown color" (49601); 9 males, 5 females ovig., " light-colored " (49602). Kanaka Bay; in *Schizothaerus;* May 29; 1 male ovig. (49603). Protection Island; May 29; 5 males, 4 females ovig., in *Schizothaerus* (49604); 2 y., in *Mya* (49605). Echo Bay, Newcastle Island; May 30; 16 males, 18 females (16 ovig.) in *Schizothaerus* (49606); 1 male in *Mya* (49607). Snake Island; in *Schizothaerus;* May 22; 1 male, 1 female ovig. (49608). Taylor Bay; in *Schizothaerus;* Apr. 5; 1 female ovig., "entirely white" (49609); both hands have been bitten off in the same place, a short distance from the carpus. South side of Mudge Island; in *Schizothaerus;* May 31; 19 males, 10 females (7 ovig.) (49610). False Narrows; in *Schizothaerus;* May 19; 33 males, 31 females ovig. (49611).

British Columbia; in *Schizothaerus nuttallii;* Geological Survey of Canada; 2 males, 1 female ovig. (40400).

Simeahmoo, Puget Sound; Dr. C. B. R. Kennerly, Northwestern Boundary Survey; received from Smithsonian Institution; 1 male, 1 female ovig. (1316, M. C. Z.).

Quarantine Dock, Washington; from clam; 1903; collected by sailors of steamer *Albatross;* 2 males, 4 females (3 ovig.) (31599).

Port Orchard, Puget Sound, Washington; from *Schizothaerus nuttallii;* July, 1889; O. B. Johnson; 4 males, 7 females (6 ovig.) (15107).

Quartermaster Harbor, Vashon Island, Puget Sound; from mantle of clam; Oct. 30, 1905; steamer *Albatross;* 1 female (48430).

Puget Sound, at Rosedale, Pierce County, Washington; in horse clam; Aug. 16, 1906; 2 males, 3 females (1 ovig.) (Stanford Univ.).

Yaquina, Oregon; Oregon State Agricultural College, Corvallis; 1 female (21619).

Humboldt Bay South, California, in *Tapes;* June 1, 1911; W. F. Thompson; 1 male, soft shell, 1 female (Stanford Univ.).

PINNIXA LITTORALIS Holmes.

Plate 31, figs. 5–8.

Pinnixa littoralis HOLMES, Proc. California Acad. Sci., ser. 2, vol. 4, 1894, p. 571, pl. 20, figs. 14–16 (type-locality, near Fort Bragg, Bodega Bay, California; cotype male, Cat. No. 20859, U.S.N.M.); Occas. Papers California Acad. Sci., vol. 7. 1900, p. 91.—RATHBUN, Harriman Alaska Exped., vol. 10, 1904, p. 188.—WEYMOUTH, Leland Stanford Jr. Univ. Publ., Univ. Ser. No. 4, 1910, p. 58 (part; not text-fig. 5 nor examples from Puget Sound).

Pinnixa faba Rathbun, Harriman Alaska Exped., vol. 10, 1904, p. 188 (part).

Diagnosis.—Carapace pointed at sides. Orbits pointed at outer end. Dactyli of legs strongly curved. Thumb of male very short, deflexed. Fingers of female gaping. Merus of third leg of male twice as long as wide.

Description.—Very near *faba*, with which it is associated. Differs as follows: Side walls of carapace less steep, and outline as seen from above not longitudinally truncate, but inclined obliquely backward and outward from the upper antero-lateral angle to a point on the true margin above the base of the second ambulatory. Eye similar, but orbit continued sideways beyond it and ending in an acute angle.

The male further differs from male of *faba* in having the thumb

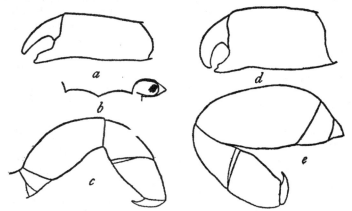

Fig. 89.—Pinnixa littoralis (31600), × 6. *a*, Left chela of female; *b*, front and eye in orbit, front view; *c*, third leg of female; *d*, left chela of male; *e*, third leg of male.

shorter, more deflexed, tip less upcurved, subterminal notch deeper; dactylus more strongly curved, prehensile edge entire. Legs more unequal than in *faba;* first leg reaching to distal third of propodus of second; second leg to dactyl of third; fourth overlapping carpus of third leg by length of dactylus; third leg noticeably stoutest, merus twice as long as wide.

The female differs from female of *faba* in having a little wider palm, thumb a little shorter, more deflexed, and with a very shallow subterminal notch, the dactylus in consequence more bent downward, leaving a narrow gape. Legs more unequal than in *faba*, but the difference is not so great as in the males of the two species.

Variations.—Immature females as well as immature males have chelae resembling those of adult males—that is, with the fingers widely gaping, thumb short, and dactyl strongly curved. The only exception is a female from San Francisco Bay, which has chelae similar to those of the adult female but with the thumb somewhat longer.

Color.—Can be told from *P. faba* by the constantly different color. In formalin, the carapace and appendages of female have a light greenish-yellow ground, with large blotches of coral red on anterior half. Eggs orange-vermilion, sometimes chestnut. First 3 ambulatory legs with a rufous band which embraces the propodus and part of the carpus. Male light dirty greenish-white, with bands on legs like those of female.

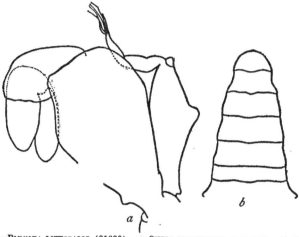

FIG. 90.—PINNIXA LITTORALIS (31600). *a*, OUTER MAXILLIPED OF FEMALE, × 14; *b*, ABDOMEN OF MALE, × 5⅔.

A pair from near the station at Departure Bay are said by Doctor Fraser to be yellow (in place of coral red), with a band of dull salmon color on the legs.

Measurements.—Male (31600), length of carapace 7.4, width of same 14 mm. Female (48427), length of carapace 16.3, width of same 26 mm.

Habitat.—Commensal in clams. Some were picked up on shore (Holmes). Of 500 specimens taken by Fraser, the adults with few exceptions inhabited the giant clam, *Schizothaerus*, while all the young of both sexes were taken from the small clam, *Mya*.

Range.—From Sitka, Alaska, to San Diego, California.

Material examined.—

Sitka, Alaska: W. H. Dall; 1 male, 1 female ovig. (17298). On clam; L. A. Beardslee; 1 female ovig. (17956).

Near Sitka: On *Saxidomus;* Sylvanus Bailey; 1 male, 1 female ovig. (17955). June 13, 1880; W. H. Dall; 1 male, 2 females (17954).

Cordova Bay, Prince of Wales Island, Alaska; in pairs in gill-chambers of giant clams; June 16, 1897; steamer *Albatross;* 8 males, 9 females (8 ovig.) (48431).

Beaver Harbor, British Columbia; in shells of *Schizothaerus nuttallii;* July 12, 1888; steamer *Albatross;* 6 males, 10 females (2 ovig.) (48432).

Victoria, British Columbia; C. F. Newcombe; 1 male, 1 female ovig. (15804).

Departure Bay, British Columbia; in *Mya arenaria;* George W. Taylor; 1 female y. (39128).

Ucluelet, British Columbia; in *Mya arenaria;* Geological Survey of Canada; 1 male, 1 female y. (40399).

Denman Island, British Columbia; in clams; May 12, 1914; steamer *Albatross;* 6 males, 12 females (5 ovig., 7 immat.) (48427).

Union Bay, Vancouver Island, British Columbia: In *Saxidomus,* probably *giganteus;* Apr. 16, 1914; steamer *Albatross;* 24 males, 22 females, all young (boiled) (48433). East of coal wharf; May 27, 1911; steamer *Albatross;* 1 female ovig. (49255).

British Columbia; 1916; C. McLean Fraser: Hammond Bay; in *Mya arenaria;* May 20; 68 males and females, all young (49612). Near Biological Station, Departure Bay; in *Schizothaerus;* May 16; 1 male, 1 female (49613). North shore, Departure Bay; in *Mya;* May 30; 16 males and females, all young (49614). Brandon Island; in *Schizothaerus;* May 18; 3 males, 2 females ovig., all light-colored (49615). Kanaka Bay; on *Mya;* May 29; 35 males and females, all young (49616). Protection Island; May 29; 2 males, 1 female, on *Schizothaerus* (49629); 124 males and females (all but 2 are immature), in *Mya* (49617). Echo Bay, Newcastle Island; May 30; 5 males, 4 females (3 ovig.), in *Schizothaerus* (49618); 30 males and females (only one is mature), in *Mya* (49619). Snake Island; in *Mya;* May 22; 32 males and females, all young (49620). Taylor Bay; Apr. 5; 3 males, 5 females (4 ovig.) in *Schizothaerus* (49621); 13 males and females, all young, in *Mya* (49622). Taylor Bay; May 17; 1 male, 1 female, in *Schizothaerus* (49630); 33 males and females, all young, in *Mya* (49623). South side of Mudge Island; in *Schizothaerus;* May 31; 45 males, 39 females (34 ovig.) (49624). S. E. end of Mudge Island; in *Mya;* May 31; 24 males and females, all young (49625). False Narrows; on *Schizothaerus;* May 19; 9 males, 7 females (4 ovig.) (49626).

Friday Harbor, Washington; 1916; Evelyn D. Way: July 6 and 11; found in shells of *Macoma nasuta, Mya arenaria* and cockles; 3 males (49950). July 15; in mantle of a big horse clam; 1 male, 1 female (49949).

Quarantine Dock, Washington; in clams; sailors of *Albatross;* 7 males, 5 females (2 ovig.) (31600).

Port Orchard, Puget Sound; from *Schizothaerus nuttallii;* O. B. Johnson; 1 male, 2 females (48429).

Humboldt Bay South, California; in *Tapes;* June 1, 1911; W. F. Thompson; 1 male (Stanford Univ.).

Near Fort Bragg, Bodega Bay, California; July, 1893; S. J. Holmes; 1 male cotype (20859).

Middle part of San Francisco Bay, California; from clams; 12½–10 fathoms); Feb. 6, 1912; station 5709, Str. *Albatross;* 1 immature female (48436).

Off Catalina Island, California; 50 fathoms; H. N. Lowe; 1 male 1 female, paper shell, with very small eggs (29946).

San Diego, California; H. Hemphill; 1 male (17501).

PINNIXA BARNHARTI,[1] new species.

Plate 32.

Pinnixa tumida STREETS, Bull. U. S. Nat. Mus., No. 7, 1877, p. 115; not *Pinnixa tumida* Stimpson, 1858.
Pinnixa faba RATHBUN, Harriman Alaska Exped., vol. 10, 1904, p. 188 (part: specimen from San Pedro).

Type-locality.—Under pier at Venice, California; from cloaca of a sea cucumber; female holotype, Cat. No. 45586, U.S.N.M.

Diagnosis.—Carapace convex, truncate at sides. Dactyli of legs nearly straight. Chelipeds large; fingers widely gaping; thumb strap-shaped.

Description of female.—Carapace hexagonal, very convex in both directions, sides truncate, antero-lateral margin a line of very fine granules not continned to hepatic region, side walls vertical, subhepatic region prominent, surface coarsely punctate toward the sides, furrow behind gastric region shallow, 3 deep pits on each side anteriorly, posterior margin very concave. Lobes of front prominent and arcuate, viewed from above. Orbits broadly oval, filled by the eyes. Antenna as long as width of front and one orbit.

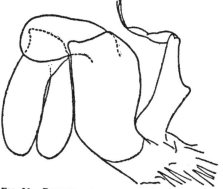

FIG. 91.—PINNIXA BARNHARTI, OUTER MAXILLIPED OF FEMALE (31510), × 15½.

---

[1] P. S. Barnhart, formerly naturalist at the Venice Marine Biological Station.

Chelipeds large; chelae rhomboidal, increasing greatly toward fingers, a sinus in lower margin near base of thumb; thumb subhorizontal, a little curved or convex beneath, of nearly equal width throughout, tip obliquely truncate, lower corner armed with a short, sharp tooth which is crossed by the sharp tip of the dactylus; dactylus oblique, making a large triangular gape with the thumb; a strong tooth at middle of dactylus, a fringe of hair above; a patch of long hair in the gape.

Merus of cheliped and legs hairy above; merus and propodus of last 2 legs hairy below, carpus and propodus of last leg hairy above. Legs thick, first nearly as long as second, but narrower, second reaching to dactylus of third, fourth to middle of carpus of third; merus of last 3 legs convex above, propodus of all tapering, dactylus short, nearly straight, broad at base, acuminate.

Variations.—Chela of male very like that of female except palm a little shorter; in small but mature female, palm still shorter.

Color.—The carapace in alcohol is largely a bluish-purple; a few patches of same color on chelipeds and first three legs.

Measurements.—Female holotype, length of carapace 10.7, width of same 16.2 mm. Male (M. C. Z.), length of carapace 10.2, width of same 15 mm.

Habitat.—Commensal with holothurians.

Range.—Venice, California, to Ballenas Bay, Lower California.

Material examined.—

Under pier at Venice, California; from cloaca of a sea-cucumber; Venice Marine Biological Station; 1 female ovig., holotype (45586).

San Pedro, California; 1901; T. D. A. Cockerell; 1 female (31510).

San Diego, California; from cloaca of *Liosoma arenata* [*arenicola*] Stimpson (now *Molpadia arenicola*); H. Hemphill; received March 15, 1874; 1 male, 1 female (5742, M. C. Z.).

PINNIXA MINUTA Rathbun.

Pinnixa minuta RATHBUN, Bull. U. S. Fish Comm., vol. 20, for 1900, pt. 2 (1901), p. 21, text-fig. 4 (type-locality, Mayaguez Harbor, Porto Rico; holotype, Cat. No. 23768, U.S.N.M.).

Diagnosis.—Carapace narrow, areolated. Dactyli of legs straight. Last 2 segments of outer maxilliped short and stout. Terminal segment of abdomen oblong.

Description of male.—Carapace less than twice as wide as long; an imaginary line connecting the lateral angles would run behind gastro-cardiac suture; middle portion of lateral margins subparallel. Regions separated by deep furrows; gastric, cardiac and branchial regions separately convex; cardiac region less than twice as wide as long, without crest; upper surface finely granulate, antero-lateral margins coarsely granulate. Front truncate, non-projecting. Antennae longer than width of front. Merus of maxilliped longitudinal.

Chelipeds granulate; palm stout, 1.5 times as long as wide, with longitudinal rows of pubescence; fingers horizontal, two-thirds as long as palm, slightly gaping, tips hooked.

Legs margined with broad bands of spinuliform granules and having straight dactyls; first two pairs narrow, dactyls slender; third pair similar but much larger; fourth pair narrowest, slightly overreaching merus of third, its dactylus shorter and relatively stouter.

Sternum granulate along margins of segments. A few granules on abdomen; sixth and seventh segments of abdomen narrowest; third, fourth, and fifth of subequal length. Surface of carapace, chelipeds and legs pubescent.

Measurements.—Male holotype, length of carapace 1.3, width of same 2.3 mm.

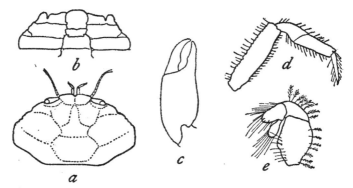

FIG. 92.—PINNIXA MINUTA, MALE HOLOTYPE. *a*, CARAPACE, ANTENNAE AND EYES, × 18½; *b*, LAST THREE ABDOMINAL SEGMENTS AND ADJACENT STERNUM, × 18; *c*, RIGHT CHELA, × 27; *d*, THIRD LEG, × 18; *e*, ENDOGNATH OF OUTER MAXILLIPED, MUCH ENLARGED.

Locality.—Porto Rico: Mayaguez Harbor, Point del Algarrobo E. 4⅝ miles; 16 to 17 fathoms; M. S.; temp. 23° C.; Jan. 20, 1899; station 6066, steamer *Fish Hawk;* 1 male holotype (23768).

PINNIXA CHAETOPTERANA Stimpson.

Plate 33, figs. 3–6.

Pinnixa cylindrica STIMPSON, Ann. Lyc. Nat. Hist. New York, vol. 7, 1859, p. 68. Not *P. cylindrica* (Say).—SMITH, Rept. U. S. Fish Commr., vol. 1, 1871–72 (1873), p. 546 [252] (part: ref. to South Carolina).

Pinnixa chaetopterana STIMPSON, Ann. Lyc. Nat. Hist. New York, vol. 7, 1860, p. 235 (type-locality, Charleston Harbor, South Carolina, on muddy or clayey shores, in tubes of *Chaetopterus pergamentaceus;* type not extant).—KINGSLEY, Proc. Acad. Nat. Sci. Philadelphia. 1878, p. 324 [9]; 1879 (1880), p. 402.—FAXON, Bull. Mus. Comp. Zoöl., vol. 5, 1879, p. 264, pl. 4, figs. 1–4 (zoea); pl. 5, figs. 8, 9 (chelipeds).—SMITH, Trans. Connecticut Acad. Arts and Sci., vol. 4, 1880, p. 250.—VERRILL, Amer. Journ. Sci., ser. 3, vol. 24, 1882, p. 371.

Diagnosis.—Carapace uneven. Two short ridges on cardiac region. Fingers of male widely gaping. Dactyli of first two legs slightly curved, of last two straight.

Description of male.—Surface densely pubescent and hairy, especially on chelipeds and margins of carapace and legs. Regions and their subdivisions well limited; subbranchial region advanced, forming a prominent shoulder with granulated edge; lateral angle a right angle; a submarginal groove parallel to posterior and postero-lateral margins; on each side of cardiac region a short sharp ridge arched upward; a few granules on antero-lateral margin; front little advanced, truncate.

FIG. 93.—PINNIXA CHAETOPTERANA, BUZZARDS BAY. *a,* CHELA OF ADULT FEMALE, × 6; *b,* CHELA OF ADULT MALE, × 4. (AFTER FAXON.)

Chelipeds strong, smooth; palm very wide, convex on margins except near base of fixed finger, which is very short, much deflexed, with a strong, triangular, prehensile tooth near its base and an obliquely truncate, notched tip; dactylus strongly curved, almost vertical and forming an oval gape when closed.

Dactyli of first two legs slender and slightly curved, of last two, stout and nearly straight; first and second legs slender and of nearly same size, second reaching to end of propodus of third; third stout, merus about twice as long as wide, nearly straight above, distally narrowed, posterior margin, as well as that of propodus, spinulous; fourth leg of similar build, reaching to middle of carpus of third.

Sixth segment of abdomen constricted at middle, seventh semicircular.

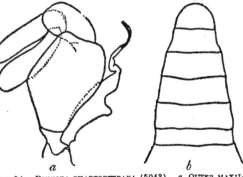

FIG. 94.—PINNIXA CHAETOPTERANA (5043). *a,* OUTER MAXILLIPED OF FEMALE, × 17; *b,* ABDOMEN OF MALE, × 6½.

Sexual variation.—Carapace more swollen in female, fingers moderately deflexed, not gaping, thumb longer and stouter, a small tooth at each end of truncate extremity, no subbasal tooth, a small tooth on dactylus behind middle.

Measurements.—Male (5043), length of carapace 6.3, width of same 14 mm. Female (18289), length of carapace 6.3, width of same 14 mm.

Habitat.—Commensal in tubes of annelids, *Amphitrite ornata* and *Chaetopterus pergamentaceus* " whose parchment-like sheath, expanded at the middle, is bent in the form of a semicircle, so that both extremities project to the surface of the mud" (Stimpson).

Range.—Wellfleet, Massachusetts, to Rio Grande do Sul, Brazil (Moreira).

Material examined.—

Wellfleet, Massachusetts; 1879; H. E. Webster; 3 males, 3 females (18808).

Woods Hole, Massachusetts; U. S. Fish Comm.: Mar., 1882; V. N. Edwards; 6 males, 6 females (5043). Aug. 31 1883; 2 males, 1 female ovig. (5812). In tubes of *Chaetopterus pergamentaceus;* Dec., 1882; V. N. Edwards; 24 males, 17 females (5824). Dredged; Nov. 17, 1884; V. N. Edwards; 1 male, 2 females (14455). 1875; 3 males, 3 females (19085). Mar. 28, 1888; V. N. Edwards; 1 male, 1 female (26105). Aug. 12, 1882; 1 female ovig. (31482). Aug., 1880; 1 male (36313).

Naushon Island, Massachusetts; U. S. Fish Comm.: In tube of *C. pergamentaceus;* 1882; 1 male, soft shell (34168). Shore; Aug. 31, 1882; 1 male, 1 female, with *Vesicularia* attached (38293). July 28, 1881; 1 male (40798). Aug. 10, 1882; 4 males, 1 female ovig. (40814).

Vineyard Sound, Massachusetts; U. S. Fish Comm.: In tubes of *C. pergamentaceus;* 1885; V. N. Edwards; 3 males, 4 females ovig. (10727). Aug. 1, 1886; 1 female ovig. (48434).

Quissett, Massachusetts; 3-5 fathoms; Sept. 3, 1882; U. S. Fish Comm.; 2 males (34188).

Buzzards Bay, Massachusetts; U. S. Fish Comm.: 8½ fathoms; br. M.; temp. 66° F.; Aug. 26, 1881; station 963; steamer *Fish Hawk;* 1 female ovig. (34057). 1875; 3 males, 2 females (35308). From West Island to north end of Woods Hole; 5-7 fathoms; bk. M. Sh. G.; temp. 71.1° F.; Aug. 15, 1887; stations 1211-1221; steamer *Fish Hawk;* 5 males, 5 females (12794).

Beaufort, North Carolina; in tubes of *Chaetopterus;* June 29, 1888; E. A. Andrews; 1 male, 1 female (18289).

Tangier Sound, South Carolina; 2⅓ fathoms; stky.; temp. 70° F.; June 4, 1891; station 1651; steamer *Fish Hawk;* 1 female (18212).

Punta Rassa, Florida; 1 fathom; Feb., 1884; H. Hemphill; 2 females young (23392).

Chandeleur Islands, Louisiana; L. R. Cary; 1 male, 1 female (33116).

Rio de Janeiro, Brazil; Krøyer; from Zool. Mus. Copenhagen; 1 female ovig. (31508).

Remarks.—The ovigerous female from Rio de Janeiro is only 7 mm. wide, the posterior margin (4 mm.) is wider than in typical specimens, the cardiac crest is a little weak; nevertheless I think this is a *chaetopterana.*

PINNIXA VALDIVIENSIS Rathbun.

'Plate 33, figs. 1 and 2; plate 34, figs. 5 and 6.

Pinnotheres transversalis CUNNINGHAM, Trans. Linn. Soc. London, Zool., vol. 27, 1871, p. 492; not *P. transversalis* Milne Edwards and Lucas, 1843.

Pinnixa transversalis MIERS, Proc. Zool. Soc. London, 1881, p. 70 (part: specimens from Sandy Point).

Pinnixa valdiviensis RATHBUN, Revista Chilena Hist. Nat., vol. 11, 1907, p. 45, text-fig. 1, pl. 3, figs. 2 and 3 (type-locality, Corral, Valdivia, Chile; cotypes male and female in Museum at Valparaiso, Chile; cotypes male, Cat. No. 32260, U.S.N.M.).

Diagnosis.—Two short cardiac ridges in male. Transverse hepatic ridge. Dactyli of legs 1 and 2 curved, of 3 and 4 straight. Chelae smooth. Thumb of male short, deflexed.

Description of male.—Regions faintly indicated, covered with very fine granules and scattered punctae, cardiac region with a short, blunt, transverse ridge either side of middle; a similar ridge across hepatic region. Front not advanced, widening at end. Orbits in front view inclined obliquely downward and outward.

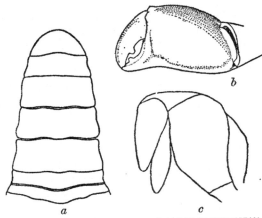

FIG. 95.—PINNIXA VALDIVIENSIS. *a*, ABDOMEN OF MALE (5740), × 5½; *b*, LEFT CHELA OF MALE (5740), × 5; *c*, OUTER MAXILLIPED OF COTYPE, MUCH ENLARGED.

Chelipeds stout, pubescent on inner side, palm widening distally, upper and lower margins convex; pollex short, triangular, deflexed, a ridge above lower margin, two teeth on prehensile edge near end; dactylus strongly curved down, edge with a tooth at middle; a slight gape when fingers are closed.

First two pairs of legs narrow, dactyli long, slender, and curved, but not strongly so; third leg a little longer than second, but half again as wide; fourth pair about as wide as second and a little over half as long; dactyli of third and fourth stouter than of first and second and nearly straight.

Abdomen nearly one-third as wide as sternum, sixth segment very short, sides concave.

Variations.—Female lacks cardiac ridges; digits longer than in male, longitudinal, and without prehensile teeth. In young male, thumb less deflexed, gape narrower.

Measurements.—Male cotype, length of carapace 3.7, width of same 6.8 mm. Male (Eden Harbor), length of carapace 8.2, width of same 16.5 mm.

Range.—Corral, Province Valdivia, Chile, to Sandy Point, Straits of Magellan (Cunningham).

Material examined.—

Chile: Corral, Province Valdivia; C. E. Porter; 3 males, 1 female, cotypes in Mus. Hist. Nat. Valparaiso; 2 male cotypes in U. S. Nat. Mus. (32260).

Eden Harbor, Smith Channel, Straits of Magellan; *Hassler* Exped., 1 male (5740, M. C. Z.).

PINNIXA OCCIDENTALIS Rathbun.

Plate 34, fig. 1.

Pinnixa occidentalis RATHBUN, Proc. U. S. Nat. Mus., vol. 16, 1893, p. 248 (type-locality, S. of Unimak Island, Alaska, 61 fathoms; type, Cat. No. 17474, U.S.N.M.), except specimen from San Diego; Harriman Alaska Exped., vol. 10, 1904, p. 187, pl. 7, fig. 4; pl. 9, figs. 6, 6a (except specimens from Cape Fox).—HOLMES, Occas. Papers California Acad. Sci., vol. 7, 1900, p. 89.

Pinnixa californiensis RATHBUN, Proc. U. S. Nat. Mus., vol. 16, 1893, p. 249 (type-locality, Monterey Bay, California, 37 fathoms; type, Cat. No. 17478, U.S.N.M.); vol. 21, 1898, p. 605; Harriman Alaska Exped., vol. 10, 1904, p. 187, pl. 7, fig. 3.—HOLMES, Occas. Papers California Acad. Sci., vol. 7, 1900, p. 90.—WEYMOUTH, Leland Stanford Jr. Univ. Publ., Univ. ser. No. 4, 1910, p. 56 (part; not young female from Pacific Grove).

Diagnosis.—Cardiac ridge biarcuate. Thumb deflexed. Dactyli of legs styliform. Merus of third leg narrow, from 2½ to 3½ times as long as wide.

Description of male.—Setose. Carapace laterally narrowed, surface uneven; hepatic and gastric regions well marked; on the cardiac region an acute, transverse crest which curves downward and backward in the middle. Median groove of front deep. Antero-lateral margin defined by an acute granulated ridge.

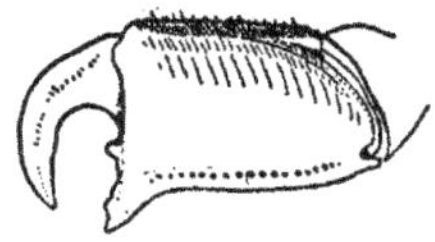

Chelipeds stout; manus widening a little distally, smooth outside, granulated above and below, lower margin nearly straight; thumb short, deflexed, a stout tooth at middle, a smaller one near tip; dactyl curved, with often, but not always, a tooth near middle.

FIG. 96.—PINNIXA OCCIDENTALIS, LEFT CHELA OF MALE (17470), × 3½.

Legs thick, granulate; dactyli styliform; third leg reaching by length of dactyl beyond second leg; merus with subparallel sides and three or more times as long as wide.

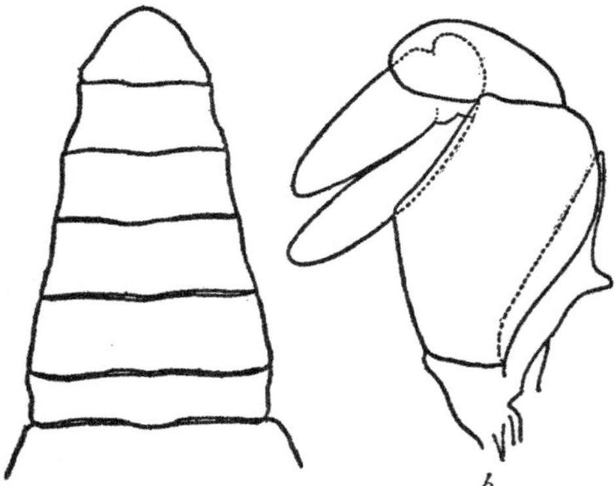

Abdomen of male narrowed a little at first suture, then tapering to the sixth segment which is constricted at middle.

Variations.—Cardiac ridge blunter in female than in male, immovable finger longer, third leg shorter and wider, about $2\frac{1}{2}$ times as long as wide. Specimens vary in the proportion of length to width of carapace, in the prominence of cardiac ridge, in the direction of the margins of the second abdominal segment of male, in the length of the immovable finger of the male.

FIG. 97.—PINNIXA OCCIDENTALIS, MALE HOLOTYPE. *a*, ABDOMEN, × 5½; *b*, OUTER MAXILLIPED, × 13⅜.

Measurements.—Male holotype, length of carapace 9.5, width 19.5, length of third leg 27 mm. Female paratype (17474), length of carapace 10.5, width 20.5, length of third leg, about 24 mm.

Habitat.—Commensal in burrows of the Gephyrean, *Echiurus* (Kincaid).

Range.—From Iliuliuk Harbor, Unalaska, to Magdalena Bay Lower California; 10 to 238 fathoms.

Material examined.—See page 157.

PINNIXA SAYANA Stimpson.

Plate 34, figs. 2–4.

Pinnixa sayana STIMPSON, Ann. Lyc. Nat. Hist. New York, vol. 7, 1860, p. 236 [108] (type-locality, off mouth of Beaufort Harbor, N. C., 6 fathoms, sandy mud; type not extant).—KINGSLEY, Proc. Acad. Nat. Sci. Philadelphia, 1878, p. 323 [8].—SMITH, Trans. Connecticut Acad. Arts and Sci., vol. 4, 1880, p. 252.—FOWLER, Ann. Rept. New Jersey State Mus., 1911 (1912), p. 596 (part; not refs. to *P. cylindrica* except Verrill and Smith, part).

Pinnixa cylindrica VERRILL, Rept. U. S. Commr. of Fish and Fisheries for 1871–1872 (1873), p. 367 [73] (part, not ref. to Stimpson).—SMITH, Rept. U. S. Commr. of Fish and Fisheries for 1871–1872 (1873), p. 546 [252], pl. 1, fig. 1 (female); not *P. cylindrica* (Say, 1818), nor ref. to South Carolina.

Pinnixa, species, FAXON, Bull. Mus. Comp. Zoöl., vol. 5, 1879, p. 263, pl. 4, figs. 5–15; pl. 5, figs. 1–7 (early stages).

THE GRAPSOID CRABS OF AMERICA.

Material examined of Pinnixa tubicola di.

Locality.	Lat. N.	Long. W.	Fath.	Bottom.	Temp.	Date.	Station.	Collector.	Specimens.	cat. No.
Iliuliuk Harbor, Unalaska	53 53 35	166 30 15	19	gn. M	°F. 43.9	Aug. 22, 1890	3333	Albatross	1 ♂	17477.
Do	53 59 36	166 29 43	85	gn. M	41.0	Aug. 15, 1890	3311	do	9 ♂ 2 ♀	17475.
Do	54 01 51	166 27 38	68	fne. bk. S	42.7	do	3313	W. H. Dall	1 ♂	17476.
Do, off Alaska			70–80	M. St				W. H. Dall	1 ♂ y	17514.
Do			20–30	M. Sh				do	1 ♂ y	17613.
Rider, Unalaska								Albatross	1 ♂ 1 ♀	19500.
South of Unimak Id	54 20 30	163 37 00	61	bk. S. M		May 21, 1890	3216	do	1 ♀	17474.
Between Unga and Shell Is Is, Sitka margins; in 1 path of Lurodes types	55 10 00	100 18 00	110	gn. M	41	July 31, 1888	2818		1 ♀	19824.
Chajala Co, Kadiak, Alaska								W. H. Dall	1 ♂	17512.
Port Iles, Alognak Bay, Alognak Island,			12–14	M. S				Albatross	1 ♂ 1 ♀	do
Juneau, Alaska			12–17	stky. M		Aug. 3, 1903	4272	W. H. Dall	2 ♂	17511.
Hot Springs,			12–18					Harriman Exp		
Sitka, Alaska			20					do		
Sitka, Alaska			13	G. M				W. H. Hall	2 ♂ 2 ♀	17510.
Queen Charlotte Sound, British Col	50 49 00	127 36 30	238	gy. S. P		Sept. 1, 1888	2862	T. Kincaid	2	25836.
Departure Bay, British Columbia			10–20		44.7			G. W. Taylor	16	870.
Quarantine Dock, Washington								Albatross	1 ♂	39132.
Off Johnson	47 52 00	121 44 00	31	gy. S	46.9	Sept. 21, 1888	2868	do	1 ♂ 3 ♀	17471.
Off Destruction Island, Washington	47 38 00	121 39 00	32	bk. S	48.4	do	2869	do	3 ♂ ♀	17472.
Off Grays Harbor, Oregon	40 44 00	121 32 00	58	rky	46.5	Sept. 23, 1888	2870	do	1 ♀ y	17473.
Off Cascade Head, Oregon			50	fne. rv. S		Sept. 3, 1914	H5732	do	1 ♂ y	50169.
So San Francisco Bay, California			68–80	gn. S		Oct. 15, 1912	D5788	do	1 ♂ y	49239.
Do			39–40	very fne. dk. gn. S		do	D5788	do	2 y	49240.
Point Ano Nuevo, California	37 08 00	122 28 10	46–33	very fne. gn. S		Oct. 21, 1912	D5789	do	1 ♂ y	49211.
Monterey Bay, California	36 56 00	122 06 00	47	br. M	51.3	Mar. 15, 1890	3148	do	1 ♂	17479.
Do	36 51 40	121 51 20	24	fn. gy. S. M	63	do	3141	do	1 ♂	21023.
Do	36 47 50	121 49 00	13	fa. S. M	54.5	Mar. 14, 1890	3134	do	2 ♂	20861.
Do	Point Pinos Light-house 5 6° E., 4.6 mi.		37	br. M	62.3	do	3133	do	21	17478.
Do			50–57	ft M. R	57	June 7, 1904	4550	do	1 ♀	48435.
Magdalena Bay, Lower California	24 38 00	112 17 30	51	gn. M		May 2, 1888	2833	Stanford Univ Albatross	4 ♂ 5 ♀ ovig	S. U. 22111.
									1 ♂	61.

[1] Type of occidentalis.   [2] A male st po of californiensis.

157

Diagnosis.—A sharp carina runs on the dorsum from anterolateral margin across cervical suture. A single, bilobed, cardiac carina. Dactyli of legs slender; posterior margin of last pair convex.

Description of male.—Allied to *chaetopterana*. Much less hairy. A sharp denticulated carina extends from base of third leg across branchial region and across cervical suture to hepatic region. A sharp bilobed carina on cardiac region. Fixed finger of chela more deflexed than in *chaetopterana*, its basal tooth weaker; dactylus more strongly curved. Legs long and slender; first leg reaches to end of propodus of second, second and third same length, fourth reaching beyond middle of carpus of third; merus of third leg about $3\frac{1}{2}$ times as long as wide, its lower margin as well as that of the propodus minutely denticulate; dactyli of first two legs slightly curved, of last two straight; anterior edge of dactylus of last leg straight, posterior edge convex. Abdomen wider, sides chiefly convex.

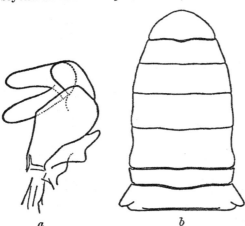

FIG. 98.—PINNIXA SAYANA, MALE (18211). *a*, OUTER MAXILLIPED, × 20; *b*, ABDOMEN, × 11$\frac{1}{2}$.

Sexual variation.—Cardiac carina less raised and less sharp in female than in male. Fingers not gaping; thumb longer, less deflexed, acutely pointed, widened at middle; dactylus less vertical than in male.

Color.—A rather fresh specimen is covered with very fine brownish dots, denser above than below.

Measurements.—Male (18211), length of carapace 4.8, width of same 9.7 mm. Female (18806), length of carapace 3.5, width of same 7 mm.

Development.—From the zoea direct to the crab form.

Habitat.—Either found free or dug out of the mud. In stomach of ocellated flounder (Verrill).

Range.—Vineyard Sound, Massachusetts, to Sarasota Bay, Florida. Shore to 26 fathoms.

Material examined.—

Vineyard Sound, Massachusetts; 1871; Verrill, Smith and Todd, U. S. Fish Comm.; 1 female (18806).

Woods Hole, Massachusetts; U. S. Fish Comm.: Sept. 18, 1882; 25 y. (40813). Surface; Oct. 2, 1882, 8 p. m.; 1 male y. (49223).

Woods Hole, Massachusetts, or vicinity; 1875; U. S. Fish Comm.: 1 male, 2 females (18807).

Buzzards Bay, Massachusetts; U. S. Fish Comm.: 1875; 1 female (35307). From West Island to N. end of Woods Hole; 6–8½ fathoms; bk. M. Sh. G.; temp. 71.1° F.; Aug. 15, 1887; stations 1211–1221; steamer *Fish Hawk;* 1 female (12847).

Narragansett Bay, Rhode Island; U. S. Fish Comm.: 8 fathoms; S. Sh.; temp. 62.5° F.; Aug. 6, 1880; station 772; steamer *Fish Hawk;* 3 males, 2 females ovig. (36323). 10¼ fathoms.; S. M. Sh.; temp. 69° F.; Aug. 6, 1880; station 774; steamer *Fish Hawk;* 1 female ovig. (34017). 20 fathoms; fne. sdy. M.; temp. 60° F.; Aug. 16, 1880; station 803; steamer *Fish Hawk;* 3 males, 2 females ovig. (36724). 14.5 fathoms; sdy. M. brk. Sh.; temp. 63° F.; Aug. 31, 1880; station 846; steamer *Fish Hawk;* 1 male, 1 female ovig. (36320). Shallows; 1880; 2 males (34003).

Long Island Sound: Branford Beacon, Connecticut, N.NW., 4¾ miles; 13.3 fathoms; sft.; temp. 70° F.; Sept. 17, 1892; station 1704; steamer *Fish Hawk;* 2 males (18211).

Chesapeake Bay: off Barren Island, Maryland; 26 fathoms; sft.; temp. 45° F.; Apr. 25, 1916; station 8523, *Fish Hawk;* 2 males (49642).

Chesapeake Bay: off Cove Point Light, Maryland; 4 fathoms; hrd.; temp. 51.9° F.; Apr. 25, 1916; station 8524, *Fish Hawk;* 1 female (49277).

Sarasota Bay, Florida; Union College collection; 1 male (48438).

PINNIXA CYLINDRICA (Say).

Plate 35, figs. 5 and 8.

Pinnotheres cylindricum SAY, Journ. Acad. Nat. Sci. Philadelphia, vol. 1, 1818, p. 452 (type-locality, Jeykill Island, Georgia; cotype in Brit. Mus.).—DE KAY, New York Fauna, Crust., vol. 6, 1844, p. 13.

Pinnixa cylindrica WHITE, List Crust. Brit. Mus., 1847, p. 33; Ann. Mag. Nat. Hist., vol. 18, 1846, p. 177.—STIMPSON, Ann. Lyc. Nat. Hist. New York, vol. 7, 1860, p. 235.—VERRILL, Rept. U. S. Fish Commr., vol. 1, 1871–72 (1873), p. 367 (part: ref. to Stimpson).—KINGSLEY, Proc. Acad. Nat. Sci. Philadelphia, 1878, p. 324 [9], (part. not ref. to Smith nor Long Island Sound); 1879 (1880), p. 402 (part; not refs. to Smith).—RATHBUN, Amer. Nat., vol. 34, 1900, p. 589.

Pinnixa laevigata STIMPSON, Ann. Lyc. Nat. Hist. New York, vol. 7, 1859, p. 68 (type-locality, near Fort Johnson, harbor of Charleston, South Carolina, with the lobworm, *Arenicola cristata;* type not extant).

Diagnosis.—A long, transverse crest across cardiac region. Fingers similar, hooked, unidentate. A sharp crest on thumb continued part way on palm. Merus of third leg less than twice as long as wide.

Description of male.—Carapace smooth, punctate, punctae small and scarce in middle third, large and more numerous elsewhere. A deep groove separates anterior cardiac region from gastric and branchial regions; a sharp, granulate antero-lateral crest which does not reach cervical suture; a transverse ridge crosses middle of cardiac region between bases of feet of last pair; outer angles hairy; posterior border short, a little concave. Front not prominent, truncate, with a submarginal groove, produced backward at middle.

Chelipeds smooth, punctate; chelae suboval; fingers horizontal, of subequal length, tips strongly hooked and overlapping when closed, leaving a gape; a sharp, finely milled crest runs from tip of pollex backward and upward, about two-fifths length of palm; a tooth on dactylus near middle of prehensile edge, one on pollex distal to the middle.

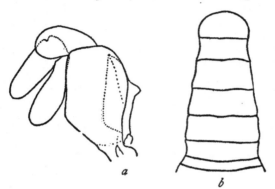

Fig. 99.—Pinnixa cylindrica, male (17952). *a*, Outer maxilliped, × 19; *b*, abdomen, × 7½.

Dactyli of legs straight in the main, the fourth one convex posteriorly and slightly concave anteriorly. First leg slender, reaching to end of propodus of second; second stouter, to middle of dactylus of third; third very stout, merus 1⅔ times as long as wide, distally narrowed, upper and lower margins finely granulate; fourth leg very short, reaching beyond merus of third by half length of dactylus.

Color.—Yellowish grey (Stimpson).

Measurements.—Male (17952), length of carapace 5.8, width of same 11.8 mm. Female (49126), length of carapace 7.8, width of same 15.3 mm.

Habitat.—"Lives with the lobworm (*Arenicola cristata*) in its hole in the sand, which is not lined by any tube. The young occur in the early spring on slimy shores at low-water mark" (Stimpson).

Range.—North Falmouth, Massachusetts; Chesapeake Bay to Sarasota Bay, Florida.

Material examined.

North Falmouth, Massachusetts; Aug. 5, 1914; F. M. Root; 1 female (49126).

Chesapeake Bay: off Point Lookout Light, Maryland; 20 fathoms; sft.; temp. 47.5° F.; Apr. 24, 1916; station 8520, *Fish Hawk;* 1 male (49643).

Sarasota Bay, Florida; Feb., 1884; Henry Hemphill; 1 male (17952).

? Florida; Union College collection (977); 1 male, identified by Kingsley (42790).

PINNIXA FRANCISCANA, new species.

Plate 35, figs. 1–4.

Type-locality.—Middle part of San Francisco Bay, California· 12½–10 fathoms; station 5709, *Albatross;* female holotype, Cat. No. 48450, U.S.N.M.

Diagnosis.—Granulate ridge near lower edge of propodus of cheliped; thumb horizontal. Merus of third leg wide. Orbital margin granulate.

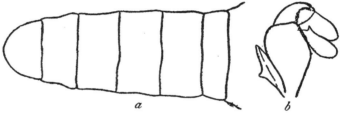

FIG. 100.—PINNIXA FRANCISCANA. *a,* ABDOMEN OF MALE (48445), × 13⅔; *b,* OUTER MAXILLIPED OF FEMALE HOLOTYPE, × 13⅔.

Description of female holotype.—Near *occidentalis,* but carapace smoother, cardiac ridge blunt and straight. Propodus of cheliped small, upper edge convex, densely granulate, lower margin straight from near the wrist to end of finger; outer surface with a granulate ridge just above lower edge, continued to end of finger and fringed above with hair, also a line of granules through the middle, and 2 lines of punctae and hairs on upper half. Fingers wide, not gaping, tips crossing, a wide triangular tooth at middle of dactylus, a similar tooth on distal half of fixed finger. First leg reaches nearly to middle of dactylus of second, second nearly to end of third, fourth to middle of carpus of third; merus of third leg 1⅔ times as long as wide, distally narrow.

Variations.—Male differs from female in sharpness of cardiac ridge. Cardiac ridge in female may be somewhat sharp, as in female, station 5743. The merus of second leg is very slender in the young female and increases proportionally more rapidly in width than the body in size. Cheliped of male not known.

Material examined of Pinnixa franciscana.

Locality.	Fathoms.	Bottom.	Temp. C.	Station.	Collector.	Specimens.	Cat. No.
Upper part San Francisco Bay	9¼–13½	fne. gy. dk. S.	13.29	5715	Albatross	1 juv.	48447.
Middle part San Francisco Bay	12¼–10	muddy S.	13.05	5709	do.	1 holotype.	48450.
Do.	10 –15½	fne. muddy S.	12.26	5743	do.	1	48448.
Do.	7¼–4½	sft. M.	12.01	5772A	do.	1	48446.
Do.	7 – 8¾	shelly, muddy S.	13.05	5824B	do.	3 ovig., 3 ♀	49242.
Lower part San Francisco Bay	9¼–12	very muddy S. bk. M.	12.69	5825	do.	6 ♀	48445.
			12.85	5723	do.	1 juv.	48449.

Measurements.—Female, holotype, length of carapace 5.7, width of same 11, length of third leg about 12 mm.

Range.—San Francisco Bay, California, 1¼–15½ fathoms.

Material examined.—See table.

PINNIXA SCHMITTI,[1] new species.

Plate 35, figs. 6, 7, and 9.

Pinnixa occidentalis RATHBUN, Harriman Alaska Exped., vol. 10, 1904, p. 187 (part: specimens from Cape Fox).

Type-locality.—Lower San Francisco Bay, California; 9½ to 11 fathoms; station 5723; *Albatross;* female holotype, Cat. No. 48441, U.S.N.M.

Diagnosis.—Palm stout, with convex margins. Thumb horizontal, long in female, short in typical male. Dactyli of legs nearly straight. Upper margin of leg joints moderately curved.

Description of female holotype.—Like *franciscana*, but carapace more oblong, as sides are less pointed and antero-lateral marginate crest is more prominent. Cardiac ridge obsolescent, broad and smoothly rounded; from a point behind each extremity, a smooth ridge curves forward to the anterior branchial angle, widening distally. Palm swollen, upper and lower margins convex, the latter becoming slightly concave under base of finger, which last inclines upward distally; outer and upper surfaces granulate, granules thinnest in middle, forming a line near lower edge, especially of finger, but very different from sharp, raised line in *franciscana;* prehensile edges of fingers meeting, sinuous. Relative lengths of legs much as in *franciscana*, but merus of third leg less dilated, twice as long as wide, margins granulate, as also of the propodus.

---
[1] Named for Mr. Waldo L. Schmitt, scientific assistant on the *Albatross*, 1911–14.

Variations.—In the fully developed male the chela is much higher in proportion to its length than in the female, its margins straighter, surface nongranulate, the thumb very short, the dactylus bent in an obliquely vertical direction, prehensile edges each obscurely bidentate. Abdomen suboblong, its sides slightly convex.

There are males in which the chelae are similar to those of the female, that is, swollen, granulate, thumb only a little shorter than dactylus; they differ from those of the female in being more swollen and dactylus shorter.

There is some variation (1) in the width of the legs, but as they are of the same general shape, I am disposed to think that the specimens listed below are all of one species; (2) in the length and prominence of the granulate, antero-lateral ridge; (3) in the inequalities of the dorsal surface of the carapace.

Measurements.—Female holotype, length of carapace 5, width of same 8.5, length of third leg about 9.7 mm. Male (25850), length

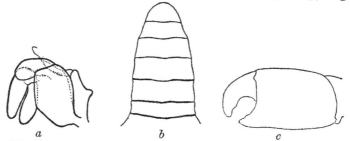

FIG. 101.—PINNIXA SCHMITTI, MALE (25850). *a*, OUTER MAXILLIPED, × 15½; *b*, ABDOMEN, × 7½; *c*, LEFT OR LARGER CHELA, × 7.

of carapace 5, width of same 9.2, length of third leg about 10 mm.

Range.—Port Levasheff, Alaska, to San Francisco Bay, California. 7 to 80 fathoms.

Material examined.—Port Levasheff, Unalaska, Alaska; 70 to 80 fathoms; M. St.; W. H. Dall; 1 male (48440). Left chela with medium thumb and slightly gaping fingers, right chela with long thumb and nongaping fingers, both palms granulate.

Cape Fox, Alaska; W. R. Coe, Harriman Exped.; 3 males, 1 female (25850). 2 males have short thumb, gaping fingers and smooth palm; smallest male has medium thumb, nongaping fingers and granulate chelae; female without chelae.

Puget Sound; 1908; 1 male, 1 female (Stanford Univ.). Small; male with medium thumbs, gaping fingers and granulate palm; female with long thumbs and granulate palm.

Upper San Francisco Bay, Cal., 9½ to 13½ fathoms; fne. dk. gy. S. sft. gy. M.; temp. 13.29° C.; Feb. 28, 1912; station 5715 *Albatross;* 1 male (48442). With longish thumbs, nongaping fingers and granulate palms.

Middle San Francisco Bay: 9 to 7 fathoms; fne., semiliquid brownish-gy. ʻM.; temp. 12.69° C.; Feb. 6, 1912; station 5706, *Albatross;* 1 male (48439); with shortish thumbs, fingers gaping, palms almost smooth. 7½ to 9 fathoms; sft. gy. M.; temp. 12.72° C.; Feb. 28, 1912; station 5718, *Albatross;* 1 female (48444); with long thumbs, nongaping fingers and granulate palms. 8½ fathoms; very muddy S.; temp. 12.69° C.; Dec. 18, 1912; station 5825 A, *Albatross;* 1 female (48443); with unequal chelae, long thumbs, nongaping fingers and granulate palms.

Lower San Francisco Bay; 9½–11 fathoms; bk. stky. M. Sh. Clinkers; temp. 12.85° C.; Mar. 6, 1912; station 5723, *Albatross;* 1 female holotype (48441). With long thumbs, non-gaping fingers and granulate palms.

PINNIXA HIATUS, new species.

Plate 36, figs. 1–4.

Type-locality.—Off Catalina Island, California; 50 fathoms; H. N. Lowe; 1 female holotype (Cat. No. 29949, U.S.N.M.).

Diagnosis.—Dactyli of legs styliform. Fingers gaping; propodal finger shorter, subtruncate at tip; a ridge near outer margin of each finger. Posterior margin of carapace long.

Description of female.—Carapace very wide, narrowed at sides, antero-lateral margin arcuate without a definite angle and marked by a raised and finely granulate edge up to hepatic region; posterior margin very long and straight; front advanced, widely emarginate in dorsal view. Surface smooth, sparingly punctate, a groove behind gastric region. Subhepatic region prominent. Orbits oval, filled by eyes.

FIG. 102.—PINNIXA HIATUS, OUTER MAXILLIPED OF FEMALE HOLOTYPE. × 19⅜.

Palm suboblong, upper margin convex, lower margin of palm and thumb sinuous, a granulated ridge on distal half just above lower edge. A corresponding ridge on dactylus. Thumb definitely shorter than dactylus and curving up toward it, extremity obliquely notched by means of a tuberculiform tooth. Dactylus curved, a tooth just behind middle; a wide gape when fingers are closed.

Merus of first leg slender, of other legs narrowed at distal end; carpus of all legs longer than propodus; propodus tapering distally, but wider at end than adjoining end of dactyl; dactyls styliform, more slender in first and second pairs; in third pair, the anterior edge of merus and posterior edge of propodus is finely saw-toothed, posterior edge of merus more coarsely granulate, anterior edge of dactylus spinulous. Posterior margin of merus of second leg finely granulate. Posterior margin of third leg and both margins of last leg fringed with hair.

Measurements.—Female holotype, length of carapace 3.6, width of same 7.7 mm.

Range.—Known only from the type-locality.

PINNIXA TUBICOLA Holmes.

Plate 36, figs. 5–8.

Pinnixa tubicola HOLMES, Proc. California Acad. Sci., ser. 2, vol. 4, 1894, p. 569, pl. 20, figs. 17 and 18 (type-localities, Trinidad, Cape Mendocino and Bodega Bay, California; types not extant); Occas. Papers California Acad. Sci., vol. 7, 1900, p. 91.—RATHBUN, Harriman Alaska Exped., vol. 10, 1904, p. 187.—WEYMOUTH, Leland Stanford Jr. Univ. Publ., Univ. ser. No. 4, 1910, pl. 57, text-fig. 4.

Pinnixa longipes WEYMOUTH, Leland Stanford Jr. Univ. Publ., Univ. ser. No. 4, 1910, p. 58 (part: two young specimens).

Diagnosis.—Carapace $2\frac{1}{2}$ times as broad as long in female, twice as broad as long in male. Propodus of third leg wider at distal end than base of dactylus. Chelae smooth, fingers hooked.

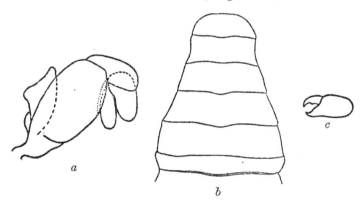

FIG. 103.—PINNIXA TUBICOLA. *a*, OUTER MAXILLIPED OF FEMALE (20860), × 22⅔; *b*, ABDOMEN OF MALE, TRINIDAD, × 15½; *c*, LEFT CHELA OF MALE, × 6, AFTER WEYMOUTH.

Description of female.—Carapace very wide, subcylindrical, sloping rather steeply on all sides toward the margins. Gastro-cardiac depression shallow. No cardiac ridge, only a smoothly rounded transverse elevation. Posterior margin very wide, slightly concave. Front not projecting beyond general contour. Outer portion of antero-lateral margin ridged.

Chelae smooth, narrowing distally, upper and lower margins of palm convex, a deep sinus below union of palm and immovable finger, fingers meeting or almost meeting, tips strongly hooked and crossing each other, each armed with a small tooth (which may be absent) the propodal tooth distal to the dactylar.

Of the ambulatory legs the first is feeble, reaching scarcely to middle 'of propodus of second leg; second leg much larger but of similar shape, reaching middle of dactylus of third leg; third leg very stout, with shorter dactylus than two preceding, the propodal joint much wider at distal end than base of dactylus; fourth leg shortest, not exceeding merus of third, its last 3 segments similar to those of third.

Description of male.—*D*iffers from female in smaller size, shorter posterior margin, slightly projecting front, with deep median groove, and longer antero-lateral marginal ridge. Abdomen wider at the first 3 segments than the sternum on either side, and tapering rapidly from base of third to middle of sixth segment, last segment wider than long, evenly rounded.

Measurements.—Female (20860), length of carapace 4, width of same 10, length of third leg about 10 mm. Male (Trinidad), length of carapace 3.2, width of same 6.8 mm.

Habitat.—Usually found commensal in leathery tubes of annelids (Holmes).

Range.—Puget Sound to San Diego, California.

Material examined.—

Puget Sound; 1907; 1 male (Stanford Univ.).

Friday Harbor, Washington; July 1, 1916; Evelyn D. Way; in tubes of *Amphitrite;* 4 females (49952).

Trinidad, California; June 2, 1896; S. J. Holmes, collector; 1 female, ovigerous, from calcareous tube of worm, cotype (20860).

Trinidad, California; June 27, 1911; W. F. Thompson; 1 male, 4 females (3 ovig.) (Stanford Univ.).

Mendocino, California; A. Agassiz; 1 female ovig. (1048, M. C. Z.).

Pacific Grove, California; 2 females y. (Stanford Univ.).

Off Point Conception, California; lat. 34° 25′ 25″ N.; long. 120° 20′ 00″ W.; 31 fathoms; gy. S. brk. Sh.; Jan. 8, 1889; station 2908, steamer *Albatross;* 1 female (24752).

PINNIXA WEYMOUTHI, new species.

Plate 36, figs. 9 and 10.

Pinnixa californiensis WEYMOUTH, Leland Stanford Jr. Univ. Publ., Univ. ser. No. 4, 1910, p. 56 (part: young female from Pacific Grove).

Pinnixa faba WEYMOUTH, Leland Stanford Jr. Univ. Publ., Univ. ser. No. 4, 1910, p. 59, text-fig. 7 (part; not synonymy).

Type-locality.—Monterey Bay, California; 5 fathoms; male holotype in Stanford University.

Diagnosis.—Carapace narrow, sides subtruncate. Chelae pubescent, thumb horizontal. *D*actyli of legs slightly curved.

Description of male.—Carapace smooth; cardiac region a transverse rounded elevation; sides subtruncate, antero-lateral angles rather prominent; front deflexed, advanced; antero-lateral margin sharp, from hepatic region backward; posterior margin little wider than base of abdomen.

Chelae pubescent; upper and lower margins of manus convex; fingers subconical, horizontal, curving toward each other near tips, edges sinuous, meeting, tips crossing.

Legs pubescent especially last 2 segments, propodi convex on both margins; first leg narrow, propodus no longer than wide, dactylus long and nearly straight; second leg wider, but similar, reaching end of propodus of third leg; third leg wide, dactylus slightly curved; fourth leg similar but smaller, reaching middle of carpus of third.

Abdomen with first segment very wide at base, its sides very oblique; second to sixth segment gradually diminishing, sixth constricted at middle; seventh short, margin broadly rounded.

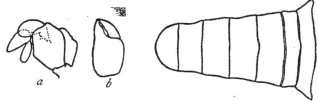

FIG. 104.—PINNIXA WEYMOUTHI, MALE HOLOTYPE. *a*, OUTER MAXILLIPED, × 20; *b*, LEFT CHELA, × 7½, AFTER WEYMOUTH; *c*, ABDOMEN, × 16.

Description of female.—The female referred here is about the same size as the male but has a thin shell, so that its shape is not well defined. Front less advanced and antero-lateral margin less sharply marked than in male. Chelae not pubescent outside, similar in shape to those of male, except that the thumb is a little shorter. Legs as in male.

Measurements.—Male holotype, length of carapace 3.3, width of same 5.3 mm.

Range.—Monterey Bay, California. Beach to 5 fathoms.

Material examined.—Monterey Bay, California, off Monterey wharf; 5 fathoms; June 28, 1907; 1 male holotype (Stanford Univ.). Third Beach, Pacific Grove, California; free on the beach; June 26, 1907; 1 female ovig. (Stanford Univ.).

Affinities.—Related to *P. tubicola*, *schmitti*, and *hiatus*, all of which have smooth carapaces, swollen palms, and nearly straight dactyli of the legs; *weymouthi* is narrower, its sides truncated, and antero-lateral angle more prominent. The male abdomen of *tubicola* is more triangular, that of *schmitti* has more convex margins than in *weymouthi*.

PINNIXA AFFINIS Rathbun.

Pinnixa affinis RATHBUN, Proc. U. S. Nat. Mus., vol. 21, 1898, p. 606, pl. 43, figs. 7–9 (type-locality, Panama Bay, 26 fathoms; female holotype, Cat. No. 21594, U.S.N.M.).

Diagnosis.—Lower margin of palm of female mostly convex; pollex short; posterior margin of merus of third ambulatory leg armed with spinules or small spines.

Description.—Allied to *P. occidentalis*. Carapace of female broader than in *occidentalis*, regions distinctly indicated, cardiac re-

FIG. 105.—PINNIXA AFFINIS, FEMALE HOLOTYPE. *a*, DORSAL VIEW, × 3; *b*, ENDOGNATH OF OUTER MAXILLIPED, × 22⅔.

gion crossed by a blunt, transverse, bilobed crest; surface punctate, punctae largest on branchial regions. Antero-lateral border of branchial region a granulate line. Front not advanced beyond line of subhepatic region. Third joint of palp of outer maxilliped articulated near proximal end of inner side of second.

Chelipeds smooth, pubescent; lower margin of palm convex except near base of pollex, which is short, very broad, deflexed, prehensile edge irregularly dentate, terminating in a short, acute spine; dactylus with a large tooth at basal third; fingers not gaping when closed.

First two legs slender, margins of propodal joints subparallel; first reaches to end of propodus of second; second to end of propodus of third. Third leg broadest; merus very hairy along margins; posterior margin armed with spinules, those near middle larger and spinelike; anterior margin granulous. Fourth leg reaches about to end of carpus of third; propodus narrow, as in first and second.

FIG. 106.—PINNIXA AFFINIS, RIGHT CHELA OF FEMALE HOLOTYPE, × 13.

Measurements.—Female holotype, length of carapace 3.4, width of same 7.3 mm.

Specimen examined.—Panama Bay, lat. 8° 27′ 00′′ N.; long. 79° 35′ 00′′ W.; 26 fathoms; gn. M.; Mar. 30, 1888; station 2803, *Albatross*; 1 female with eggs, holotype, Cat. No. 21594, U.S.N.M.

PINNIXA BREVIPOLLEX Rathbun.

Pinnixa brevipollex RATHBUN, Proc. U. S. Nat. Mus., vol. 21, 1898, p. 605, pl. 43, fig. 6 (type-locality, off Gulf of San Matias, Argentina, 43 fathoms, station 2768, *Albatross;* female holotype, Cat. No. 21593, U.S.N.M.).

Diagnosis.—Lower margin of palm nearly straight; pollex short; dactyl transverse. A transverse, blunt, cardiac crest.

Description.—Entire surface covered with a dense pubescent coat. Carapace uneven, punctate; gastric and hepatic regions bounded by very deep furrows; cardiac region high and crossed by a transverse crest, surmounted in male by two triangular tubercles compressed from before backward, subacute; in female, crest lower, blunt, divided in middle by a very shallow sinus. Frontal and hepatic re-

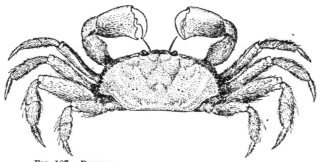

FIG. 107.—PINNIXA BREVIPOLLEX, FEMALE HOLOTYPE, × 2⅔.

gions granulated. Subhepatic region with a small depressed area or tubercle surrounded by a deep groove. Antero-lateral margin of branchial region armed with from 4 to 6 distant, blunt spinules beginning at the lateral angle and followed near hepatic region by smaller tubercles or granules. Inferior margin of carapace granulated. Margins of frontal lobes extend obliquely backward from middle. Antennae exceed width of front.

Last two joints of palpus of maxilliped oblong; terminal joint articulated near base of second joint, and overreaching it considerably, overlapping the sternum and in the male touching the tip of the abdomen.

Chelipeds of male unknown. Propodus of female very broad, flat, increasing in width distally, its greatest width equaling the superior length; upper margin slightly convex; lower also for proximal two-thirds; deflexed for its distal third; a very short high pollex with a small terminal spine, its distal margin transverse, armed with two tubercles, one near insertion of dactylus and the other at the middle.

*D*actylus transverse, a tubercle at middle of prehensile margin which fits against margin of propodus.

First and second legs narrow, second longer, larger, and reaching extremity of propodus of third pair. Merus of third pair very little dilated at middle. Fourth pair reaches middle of carpus of third pair.

Sides of male abdomen from third to fifth segments, inclusive, converge regularly; those of sixth segment still more convergent, segment very short; terminal segment narrowest, broader than long, rounded.

Measurements.—Female holotype, length of carapace 6.6, width of same 12.5 mm. Male (21593), length of carapace 5.5, width of same 11 mm.

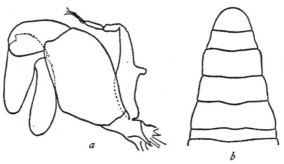

FIG. 108.—PINNIXA BREVIPOLLEX. *a,* OUTER MAXILLIPED OF FEMALE HOLOTYPE, × 18⅔; *b,* ABDOMEN OF MALE (21593), × 7½.

Material examined.—Off Gulf of San Matias, Patagonia; lat. 42° 24′ 00″ S.; long. 61° 38′ 30″ W.; 43 fathoms; dk. S. bk. Sp.; Jan. 14, 1888; station 2768, *Albatross;* 1 male, 2 females (1 female holotype), Cat. No. 21593, U.S.N.M.

Genus SCLEROPLAX Rathbun.

Scleroplax RATHBUN, Proc. U. S. Nat. Mus., vol. 16, 1893, p. 250; type, *S. granulata* Rathbun.

Carapace transverse, subpentagonal, hard, very convex, regions scarcely indicated, lower or true antero-lateral margin curving gradually into postero-lateral margin, not forming an angle with it as in *Pinnixa.* Ambulatory legs similar, third longest but not unusually long, fourth not noticeably reduced. Ischium of outer maxillipeds rudimentary, merus oblique, palpus three-jointed, the last joint articulating near proximal end of preceding joint.

Only a single species known.

SCLEROPLAX GRANULATA Rathbun.

Plate 37, figs. 1-3.

Scleroplax granulatus RATHBUN, Proc. U. S. Nat. Mus., vol. 16, 1893, p. 251 (type-locality, Ensenada, Lower California; female holotype, Cat. No. 17497, U.S.N.M.); Harriman Alaska Exped., vol. 10, 1904, p. 188, pl. 7, fig. 5.—WEYMOUTH, Leland Stanford Jr. Univ. Publ., Univ. Ser. No. 4, 1910, p. 59, text-fig. 8.

Pinnixa (Scleroplax) granulata HOLMES, Occas. Papers California Acad. Sci., vol. 7, 1900, p. 94.

Diagnosis.—Carapace high, not areolated, less than twice as wide as long. Chela of male very wide, of female very feeble. Legs similar.

Description of female.—Carapace smooth and punctate in center, finely granulate elsewhere; antero-lateral margin of the dorsum marked by an arcuate, granulate ridge, which stops short of the

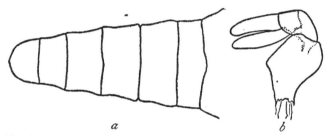

FIG. 109.—SCLEROPLAX GRANULATA. *a*, ABDOMEN OF MALE (49247), × 15½; *b*, ENDOGNATH OF OUTER MAXILLIPED OF FEMALE PARATYPE, × 15½.

postero-lateral margin, and at cervical suture; a separate, blunt prominence on hepatic margin. Front straight as seen from above, depressed at middle and outer corners; orbits ovate, less than half as wide as front. Merus of maxillipeds granulate, an alate expansion on either side; last two joints of palpus elongate, medially grooved, last joint reaching slightly beyond preceding. Chelipeds weak, granulate, especially on dorsal surface; thumb horizontal, with outer ridge, tip acute; fingers not gaping. Legs slender, third pair longest, dactyli slightly curved, last one nearly straight.

Sexual variation.—In male, chela much larger, propodus nearly smooth, with convex margins, thumb shorter than wide, with a large tooth occupying greater part of prehensile edge and largely filling gape caused by the curved dactylus. Abdomen constricted between last two segments, margin of last segment rounded. Carapace less extensively granulated. In the single male from San Francisco Bay the right chela and right legs, and probably the last left leg, are in process of regeneration.

Measurements.—Female (paratype), length of carapace 5.8, width of same 8.2 mm. Male (49247), length of carapace 4.6, width of same 6.2 mm.

Range.—Departure Bay, British Columbia, to Ensenada, Lower California.

Specimens examined.—

Departure Bay, British Columbia; George W. Taylor; 1 female (39130).

Friday Harbor, Washington; in *Mya arenaria;* July 1, 1916; Evelyn D. Way; 1 male, 1 female (49951).

Puget Sound; Aug., 1908; 3 males, 3 females, ovig. (Stanford Univ.).

Puget Sound, near Tacoma, Washington; 1907; 2 males, 2 females (1 ovig.), (Stanford Univ.).

Oyster Bay, Washington, from stomach of Bufflehead (*Charitonetta albeola*); Dec. 4, 1914; 1 female (Biol. Survey Coll., U. S. Dept. of Agr.).

Middle San Francisco Bay; 7–4 fathoms; large and small angular rock fragments; temp. 12.01° C.; Apr. 16, 1912; station 5775a · steamer *Albatross;* 1 male (49247).

Ensenada, Lower California; C. R. Orcutt; 3 females (holotype and paratypes) and fragment of male (17497).

Genus OPISTHOPUS Rathbun.

Opisthopus RATHBUN, Proc. U. S. Nat. Mus., vol. 16, 1893, p. 251; type, *O. transversus* Rathbun.

Carapace subquadrilateral, a little wider than long, firm, harder in male than in female, smooth; regions not defined; antero-lateral margins arcuate; front rounding down toward the epistome. Outer maxillipeds with ischium well developed and well defined, but partly fused with the merus; merus wide, palpus 3-jointed, the last joint articulated near the proximal end of the inner side of the penultimate joint. Ambulatory legs similar and differing little in size, second longest.

Perhaps should be united with *Pinnaxodes*.

Known only from the typical species.

OPISTHOPUS TRANSVERSUS Rathbun.

Plate 37, figs. 4 and 5.

Opisthopus transversus RATHBUN, Proc. U. S. Nat. Mus., vol. 16, 1893, p. 252 (type-locality, Monterey, California; female holotype, Cat. No. 3446, U. S. N. M.); Harriman Alaska Exped., vol. 10, 1904, p. 188, pl. 9, fig. 2, text-fig. 95.—HOLMES, Occas. Papers California Acad. Sci., vol. 7, 1900, p. 97.—WEYMOUTH, Leland Stanford Jr. Univ. Publ., Univ. ser. No. 4, 1910, p. 61, text-fig. 9.

Diagnosis.—Carapace little transverse, suboblong, convex, even, firm. Legs similar, second longest, but not notably so; dactyli curved.

Description of female.—Carapace highest in middle, whence it slopes downward in all directions. Some shallow depressions partially inclose the various regions. A border of pubescence along part of antero-lateral margin. True lateral margin of carapace not

FIG. 110.—OPISTHOPUS TRANSVERSUS, FEMALE (23927). *a*, DORSAL VIEW, × 1⅜; *b*, ENDOGNATH OF OUTER MAXILLIPED, × 8.

visible in dorsal view. Front not visible from above; in anterior view it is triangular, being longer in middle than at outer angles. Orbits a broad oval, filled by eyes.

Chelipeds of good size, pubescent, and with margins of merus and inner surface of carpus and propodus hairy; palm increasing slightly in width to distal end, lower margin convex; fingers elongate, nearly horizontal, not gaping, one small tooth at base of dactylus and two or three at base of pollex, tips hooked and crossing each other.

Legs thick, pubescent, margins hairy, second leg longest, third similar, reaching middle of dactylus of second, first pair a little shorter than third, fourth pair shortest, reaching middle of propodus. Margins of merus joints subparallel. Dactyli strongly curved.

Sexual variation.—Male much smaller and less convex than female, front less deflexed and pubescent. Chelae shorter and higher, thumb shorter. Abdomen tapers regularly from third to last segment. In the young female the hand approaches in shape that of male. The female may reach maturity while its abdomen is still as narrow as that of male (Weymouth).

Color.—Carapace covered with irregular and interrupted reticulating lines of red on a light ground. Spotted with vermilion (W. A. Hilton).

Measurements.—Female holotype, length of carapace 14, width of same 18 mm.; male (23051), length of carapace 9.8, width of same 11.8 mm.

Habitat.—Found in siphon of *Pholas*, the rock-boring mollusk; in mantle cavity of clam, *Mytilus edulis;* in the large key-hole limpet, *Lucapina crenulata;* and in the common holothurian, *Stichopus californicus.*

Range.—Monterey to San Diego, California.

Material examined.—

Monterey Bay, California: In mantle cavity of *Lucapina crenulata;* Harold Heath; received Nov. 25, 1898; 1 male (22880). In *Lucapina crenulata;* 4 males, 6 females (2 ovig., 1 of these has a narrow, subtriangular abdomen) (Stanford Univ.).

Monterey; C. A. Canfield; 1 male, 3 females (1 ovig., 1 is holotype) (3446).

Pacific Grove: John C. Brown; 6 males, 1 female y. (23926). In mantle cavity of *Mytilus edulis;* John C. Brown; 1 female (23927).

Two miles S. by E. of Point del Rey; Aug. 8, 1913; Venice Marine Biological Station; 1 male y. (49301).

San Pedro; H. N. Lowe; 1 female (49211).

Anaheim Landing; from siphon of a piddock (*Pholas*); H. N. Lowe; 1 male, 1 female (23051).

Off Catalina Island; 50 fathoms; H. N. Lowe; 1 male (29950).

Laguna Beach; W. A. Hilton, Pomona College; 1 male (50104).

Point Loma; Jan. 28, 1889; steamer *Albatross;* 1 female ovig. (17481).

San Diego; Henry Hemphill; received Apr. 15, 1874; 1 female (5743, M. C. Z.).

Genus PINNAXODES Heller.

Pinnaxodes HELLER, Reise Novara, vol. 2, Abth. 3, 1865, Crust., p. 67; type, *P. hirtipes* Heller=*P. chilensis* (Milne Edwards).

Carapace thin and either soft and yielding or firm and parchmentlike. Outer maxillipeds similar in form to those of *Pinnixa;* merus either fused or partly fused with ischium and nearly longitudinal in position; palpus of good size, terminal segment articulated on inner margin of preceding segment. Chelipeds much stouter than legs. Legs similar and not markedly unequal.

South America; Japan (Ortmann).

KEY TO THE AMERICAN SPECIES OF THE GENUS PINNAXODES.

A¹. Carapace subquadrate with corners rounded or obovate. Last segment of outer maxilliped overreaching penult segment. Legs not markedly slender, propodites not more than 2.5 times longer than wide; dactyl of last leg curved.
  B¹. Carapace soft in female, firm in male. Palms of female elongate. Legs narrow. Sixth segment of male abdomen constricted__*chilensis*, p. 175.
  B². Carapace firm in both sexes. Palms of female not much longer than broad. Legs short and broad. Sixth segment of male abdomen not constricted _____*meinerti*, p. 177.
A². Carapace of female circular. Last segment of outer maxilliped not overreaching penult segment. Legs very slender, propodites several times longer than wide; dactyl of last leg straight_____*tomentosus*, p. 178.

PINNAXODES CHILENSIS (Milne Edwards).

Plate 38.

Pinnotheres chilensis MILNE EDWARDS, Hist. Nat. Crust., vol. 2, 1837, p. 33 (type-locality, Valparaiso; type in Paris Mus.).—MILNE EDWARDS and LUCAS, d'Orbigny's Voy. l'Amér. Mérid., vol. 6, pt. 1, 1843, p. 23; atlas, vol. 9, 1847, Crust., pl. 10, figs. 2 and 2*a*.
Fabia chilensis DANA, U. S. Expl. Exped., vol. 13, Crust., pt. 1, 1852, p. 383.
Pinnaxodes hirtipes HELLER, Reise *Novara*, vol. 2, Abth. 3, Crust., 1865, p. 68, pl. 6, fig. 2 (type-locality, Ecuador, in an *Echinus*; type in Vienna Mus.).
Pinnaxodes chilensis SMITH, in Verrill, Amer. Nat., vol. 3, 1869, p. 245; Trans. Connecticut Acad. Arts and Sci., vol. 2, 1870, p. 170.—ORTMANN, Zool. Jahrb., Syst., vol. 7, 1894, p. 696, pl. 23, fig. 8.—RATHBUN, Proc. U. S. Nat. Mus., vol. 38, 1910, p. 587.
Pinnaxodes hirtipes ? RATHBUN, Proc. U. S. Nat. Mus., vol. 21, 1898, p. 607, pl. 43, figs. 10 and 11.

Diagnosis.—Carapace soft in female, firm in male. Palms of female elongate. Legs narrow. Sixth segment of male abdomen constricted.

Description of female.—Body very soft and yielding; carapace a little wider than long, subquadrate with rounded corners; posterior margin concave; anterior straight, interrupted by a median sulcus on the frontal region; lateral sulcus very short, not deep and continued to orbit, as in *Fabia*. Eyes, but not margin of front, partially visible in dorsal view. Ventral surface of body, including margin of front, also margins of chelipeds and legs, clothed with long, soft hair.

Chelipeds stout, elongate; palms increasing in width to distal end, upper margin straight, lower margin sinuous; fingers stout, a little deflexed, closing tight together, a subbasal tooth on dactyl.

Legs similar, of nearly equal length, 2 and 3 a little longer than 1 and 4; propodites tapering, sides nearly straight; dactyli moderately curved, last pair longest.

Abdomen covering sternum.

Description of male.—Firmer than female and of smaller size; fronto-orbital width nearly half of carapace width; chelae shorter, margins of palm convex; fingers more horizontal. Legs more unequal, the fourth definitely the smallest, the first stoutest; propodites faintly curved, dactyli longer than in female, third longest. Abdomen tapering regularly from third to sixth segment, the latter constricted at middle; terminal segment with end rounded.

Measurements.—Female (49238), length of carapace 13.2, width of same 15 mm. Male (22112), length of carapace 7, width of same 7.6 mm.

Habitat.—Parasitic in sea-urchins (*Caenocentrotus gibbosus* and *Loxechinus albus*). Nearly all the specimens of *C. gibbosus* in the

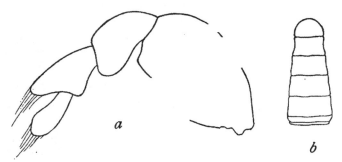

FIG. 111.—PINNAXODES CHILENSIS, MALE (22112). *a*, ENDOGNATH OF OUTER MAXILLIPED, MUCH ENLARGED; *b*, ABDOMEN, × 4½.

United States National Museum show evidence of parasitization. Verrill says[1] of a collection of 90 specimens of this urchin:

> An examination of the interior showed that in each specimen a crab (*Fabia chilensis*) * * * had effected a lodgment in the upper part of the intestine, which had thereby been greatly distended in the form of a membranous cyst, attached to one side of the shell and extending around to the lower surface near the mouth. The shell is usually swollen on the side over the cyst, and the anal area is depressed and distorted, with a large open orifice passing obliquely into the cyst, out of which the crab may thrust its legs at pleasure, but is apparently unable, when full grown, to come entirely out. All the specimens examined in the cyst were females, carrying eggs, but a very small crab found clinging among the spines appears to be the male. The crab probably effects an entrance into the intestine through the anus while quite young, and, by its presence and growth in that position, causes the gradual distortion of the shell and formation of the cyst.[2]

Range.—Ecuador to Port Otway, Patagonia.

---

[1] Amer. Jour. Sci., ser. 2, vol. 44, 1867, p. 126.
[2] For further notes on the habits of this species, see Verrill, in Trans. Connecticut Acad. Arts and Sci., vol. 1, 1866, p. 306, and Amer. Nat., vol. 3, 1869, p. 246.

Material examined.—

Pacasmayo, Peru; from *Caenocentrotus gibbosus* (L. Agassiz); beach, sandy at low water, cobblestones 5 to 6 feet deep at edge of high water; Oct. 7, 1884; Dr. W. H. Jones, U. S. N., U. S. S. *Wachusett;* fragments of 1 female (48250).

Bay of Tocopila, Chile; from *Loxechinus albus* (Molina); A. Hrdlička; 1 female (49237).

San Pedro, Chiloe Island, Chile; *Hassler* Exped.; 1 male (5737, M.C.Z.).

Port Otway, Patagonia; Feb. 9, 1888; *Albatross;* 1 male (22112). From *Loxechinus albus* (Molina); Feb. 9, 1888; *Albatross;* 1 female (49238). From *Loxechinus albus* (Molina); Feb., 1888; *Albatross;* 4 females (49235).

*Remarks.—*For discrepancies between the figures of Milne Edwards and Lucas and those of Heller, see Smith.[1] This species, as previously figured, represents the ischium and merus of maxilliped completely fused, but specimens examined by the writer show a faint or incomplete suture line.

The male from Chiloe shows the pits on the carapace more distinctly than the male from Port Otway; there are 4 in a square at the 4 corners of the cardiac region; in front of the anterior pair there is another pair, each of which forms a corner of a rhomb on each gastric region.

PINNAXODES MEINERTI Rathbun.

Plate 25, figs. 1–3.

Pinnaxodes meinerti RATHBUN, Proc. Biol. Soc. Washington, vol. 17, 1904, p. 162 (type-locality, Valparaiso, Chile; holotype male in Copenhagen Mus.).

*Diagnosis.—*Carapace firm in both sexes, wider than in *chilensis*. Palms of female short and broad. Legs short and broad. Sixth segment of male abdomen not constricted.

*Description of male holotype.—*Lower surface of body, inner surface of chelipeds and posterior margin of ambulatory legs covered with a thick felt-like coat; anterior margin of basal portion of legs clothed with long hairs. Carapace obovate, a little broader than long, thin but firm, and almost smooth. Fronto-orbital width about two-fifths width of carapace; front deflexed, its margin scarcely visible in dorsal view, and most produced at middle; orbits as wide as front, eyes showing from above. Merus and ischium of maxilliped fused, although the line of division is faintly indicated.

Chelipeds very stout; dactylus as long as upper margin of palm; lower margin of propodus concave at middle; fingers meeting when

---
[1] Trans. Connecticut Acad. Arts and Sci., vol. 2, 1870, p. 171.

closed, tips crossing, dactylus with a basal tooth; propodal finger with margin in part denticulate, a larger tooth near its middle.

Legs stout; propodites strongly tapering distally, that of third leg 0.6 as wide as its length on anterior margin; dactyli broad at base and considerably shorter than propodites.

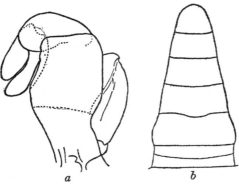

FIG. 112.—PINNAXODES MEINERTI. *a*, OUTER MAXILLIPED OF FEMALE (5760), × 15½; *b*, ABDOMEN OF MALE HOLOTYPE, × 8.

The abdomen diminishes regularly from third to seventh segment, the latter two-thirds as long as its basal width.

Description of female.—Larger than male, but general aspect similar. Palm longer and fingers shorter than in male, the dactyl shorter than upper edge of palm. Fingers gaping when closed. Dactyli of legs shorter and wider than in male. Abdomen longer but no wider than sternum.

Measurements.—Male holotype, length of carapace 6.8, width of same 7.9 mm. Female (M.C.Z.), length of carapace 11.7, width of same 13.8 mm.

Range.—Chile: Valparaiso to Talcahuano.

Material examined.—

Valparaiso; Mr. Krøyer; 1 male holotype (Copenhagen Mus.).
Talcahuano; *Hassler* Exped.; 1 female (5760, M.C.Z.).

PINNAXODES TOMENTOSUS Ortmann.

Pinnaxodes tomentosus ORTMANN, Zool. Jahrb., Syst., vol. 7, 1894, p. 697, pl. 23, figs. 9, 9*i* (type-locality, *Brasilien*; cotypes in Strassburg Mus.).

FIG. 113.—PINNAXODES TOMENTOSUS, FEMALE, COTYPE. *a*, OUTER MAXILLIPED, × ABOUT 10; *b*, GENERAL OUTLINE, NAT. SIZE. (AFTER ORTMANN.)

Diagnosis.—Carapace of female circular. Last segment of outer maxilliped not overreaching penult segment. Legs very slender, propodites several times longer than wide; dactyl of last leg longest and straight.

Description of female.—Entire body and feet covered with short, soft felt. Chelipeds of medium size, chela rather long, palm over twice as long as wide, toward the base somewhat smaller, margins rounded. Fingers shorter than palm, shutting together. Legs slender. propodus several times longer than wide; dactyli tolerably straight, except those of the third pair which are shorter and somewhat curved, and those of fourth pair which are longer and entirely straight. Abdomen very wide, covering the whole sternum.

Locality.—Brazil (2 females in Strassburg Mus.).

Remarks.—I think that this species is very likely a *Pinnotheres*.

Genus TETRIAS Rathbun.

Tetrias RATHBUN, Proc. U. S. Nat. Mus., vol. 21, 1898, p. 607; type, *T. seabripes* Rathbum.—ALCOCK, Journ. Asiat. Soc. Bengal, vol. 69, 1900, p. 335.

Carapace transversely oblong, hard, regions indicated, posterior two-thirds flattened, anterior third deflexed; antero-lateral angles rounded, sides steep. Ischium well developed, distinct from merus, palpus very large, joints end to end, the last two joints widening distally, last joint attached near inner end of distal margin of preceding. Chelipeds stout. Legs diminishing, from second to fourth, which is very small; first and third subequal. Abdomen of female suborbicular, at base only half width of sternum, tip overlapping buccal cavity.

Inhabits the Indo-Pacific and Gulf of California.

TETRIAS SCABRIPES Rathbun.

Plate 39, figs. 4 and 5.

Tetrias scabripes RATHBUN, Proc. U. S. Nat. Mus., vol. 21, 1898, p. 608, pl. 43, figs. 12–14 (type-locality, Gulf of California, 9½ fathoms; holotype female. Cat. No. 21595, U.S.N.M.).

Diagnosis.—Carapace transverse, oblong, uneven, hard. Palpus of maxilliped very large, segments end to end. Chelipeds and bases of legs spinulous. Second leg longest, fourth very small; dactyli curved.

Description of female.—Carapace covered with a short, dense coating of coarse, dark setae, beneath which the surface is punctate; regions indicated by impressed lines and pits, the deepest between cardiac and gastric regions; a tubercle on outer margin of hepatic region. Front projecting very slightly beyond anterior margin of carapace and bent down to form roof of antennular cavities, as in *Pinnixa*.

Maxillipeds bearded with long silken hairs, most noticeable on margins of last two palpar joints and in a transverse line on merus.

Outer surface of merus of chelipeds triangular, as wide as long; upper and lower margins rough with small spines or spinules; inner angles of carpus rectangular, each armed with two or three short spines. Palms wide, margins convex and subacute, outer surface covered with sharp tubercles in longitudinal lines, which extend to tips of fingers. Fingers finely dentate on prehensile edges and fitting close together.

Merus joints of legs narrow, margins subparallel; first leg reaches end of propodus of second, merus with inferior margin denticulate, superior margin with three spinules at proximal end, carpus and propodus short and wide, dactylus stout, half as long as propodus; second or longest leg a little longer than width of carapace, merus with spinule at proximal end of upper margin, lower margin somewhat roughened, carpus and propodus proportionally narrower than in first leg; third leg reaches to about middle of propodus of second, the segments resemble those of second pair but are unarmed; last leg much reduced, not reaching end of merus of second leg, segments proportionally rather wide, lower margin of ischium and merus

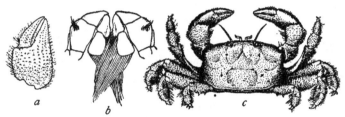

FIG. 114.—TETRIAS SCABRIPES, FEMALE HOLOTYPE. *a*, RIGHT CHELA, × 5; *b*, OUTER MAXILLIPEDS, × 5½; *c*, DORSAL ASPECT, × 2$\frac{1}{10}$.

armed with spines and spinules, dactylus very small. Legs covered with setae like those on carapace and fringed with hair; dactyli curved.

Abdomen of female fringed with long hair.

Measurements.—Female holotype, length of carapace 6. width of same 10, width of dorsal surface at middle 9.2 mm.

Material examined.—Southern part of Gulf of California; lat. 24° 12′ 00″ N.; long. 109° 55′ 00″ W.; 9½ fathoms; Sh.; Apr. 30, 1888; station 2826, steamer *Albatross;* 1 female ovig., holotype (21595).

Genus PINNOTHERELIA Milne Edwards and Lucas.

Pinnotherelia MILNE EDWARDS and LUCAS. in d'Orbigny's Voy. l'Amér. Mérid., vol. 6, pt. 1, 1843, p. 24; type, *P. laevigata* Milne Edwards and Lucas.

Carapace a little wider than long, antero-lateral margins rounded and rather prominent. Front between the eyes wide, deflexed, median projection fused with epistome. Orbits oval; eyes of medium size, somewhat elongate and tapering. Hiatus at inner angle of orbit filled by antennae; antennules transverse. Buccal area very convex. Outer maxillipeds large, longitudinally placed, ischium of good size and distinct from the elongate merus, joints of palpus large, end to end. Chelipeds stout, smooth. Legs similar, second longest. Sternum flat, forming an acute angle with the plane of the carapace. Abdomen of male triangular, 7-segmented, although fifth and sixth segments are partially fused; abdomen of female suborbicular, seventh segment invaginated in the sixth.

Contains only one species.

PINNOTHERELIA LAEVIGATA Milne Edwards and Lucas.

Plate 39, figs. 1–3; plate 40, figs. 1 and 2.

Cyclograpsus (?) *gnatherion* KINAHAN, Journ. Roy. Dublin Soc., vol. 1, 1857, p. 343 (type-localities, Chinchas Islands and Callao; types in Mus. Roy. Dublin Soc.).

Pinnotherelia laevigata MILNE EDWARDS and LUCAS, in d'Orbigny's Voy. l'Amér. Mérid., vol. 6, pt. 1, 1843, p. 25 (type-locality, Chile; type in Paris Mus.).

Pinnotherelia loevigata MILNE EDWARDS and LUCAS, in d'Orbigny's Voy. l'Amér. Mérid., vol. 9 (atlas), 1847, pl. 11, figs. 1–1e.

Diagnosis.—Carapace transversely oblong. Eyes far apart. Palp of outer maxilliped with all three joints large, end to end. Sternum flat, forming an acute angle with dorsal plane. Chelipeds heavy. Second leg longest.

Description.—Carapace subquadrilateral with the corners rounded off, a little wider at anterior than at posterior angles; hard and smooth, a longitudinal groove each side of the cardiac region, a shallow groove behind the gastric region; flat except in its anterior portion, which is strongly bent down; front two-fifths as wide as carapace, slightly arcuate in dorsal view, true edge seen only in front view, and transverse except at middle; posterior margin bent down, concave; lateral margin acute. Chelipeds and legs smooth without teeth or roughness of any kind. Outer surface of wrist rounded, palm inflated, upper and lower margins convex, distal margin very oblique; fingers similar, narrowly gaping, tips pointed, thumb shorter, horizontal. Legs thick, flattened, margins of merus subparallel, merus and propodus thickly hairy below, dactyli moderately curved; second leg longest, first next, then third and fourth.

Measurements.—Male (2435) length of carapace 10.7, width of same 12.7 mm.

Habitat.—Cast shells were found at Callao on rocks of inner side of natural dyke of rubble separating lagoon at mouth of river from ocean.

Range.—Peru and Chile; Porto Arenas, Patagonian Canal (Cano); Marquesas Island, south mid-Pacific Ocean.

Material examined.—
Callao, Peru; R. E. Coker; received from Peruvian Government; 17 males (40445).

San Lorenzo Island, Peru; U. S. Exploring Expedition; 4 males, 2 females (2435). While there is some doubt as to the origin of these specimens, the label is probably correct. Not recorded by Dana.

Tawhoe, Marquesas Islands; Dr. W. H. Jones, U. S. N.; 1 male (17951).

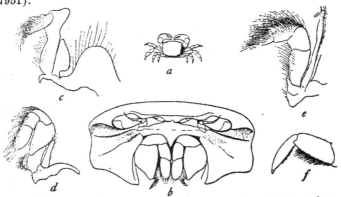

FIG. 115.—PINNOTHERELIA LAEVIGATA, MALE HOLOTYPE. *a*, DORSAL VIEW, × ⅜; *b*, ANTENNAL AND BUCCAL REGIONS; *c*, FIRST MAXILLIPED; *d*, OUTER MAXILLIPED; *e*, SECOND MAXILLIPED; *f*, EXTREMITY OF A LEG; *b–f* ARE ENLARGED. (AFTER MILNE EDWARDS AND LUCAS.)

Family CYMOPOLIIDAE Faxon.

Cymopoliidae FAXON, Mem. Mus. Comp. Zoöl., vol. 18, 1895, p. 38.
Palicae BOUVIER, Bull. Soc. Philom. Paris, ser. 8, vol. 9, 1898, pp. 56 [3] and 58 [5].
Palicidae ALCOCK, Journ. Asiat. Soc. Bengal, vol. 68, 1899, pp. 285 and 450.—RATHBUN, Bull. U. S. Fish Comm., vol. 20, for 1900, pt. 2 (1901), p. 12.—BORRADAILE, Ann. Mag. Nat. Hist., ser. 7, vol. 19, 1907, p. 482.

Carapace broadly transverse, subquadrilateral; antero-lateral margins dentate. Fronto-orbital width great, front dentate. Orbits and eyes large. Buccal cavity quadrate; outer maxillipeds not covering it; ischium strongly produced forward on the inner side; merus small, subtriangular, with a notch on the inner distal side for the articulation of the palpus. Afferent channels to branchiae opening at bases of chelipeds; efferent channels at antero-external angles of buccal cavity.

Chelipeds of moderate size, often unequal in male; usually tuberculate or granulate. Next three pairs of feet long, slender, and rough. Last pair either very short and slender, subdorsal, smooth, or similar in position and ornament to the other feet, and near the size of the first foot. Abdomen of male much narrower than sternum.

This family contains three genera, only one of which is American. Faxon and Bouvier place this family near the Dorippids in the Oxystomata, while Alcock and Borradaile, whom I follow, range it among the Brachygnatha.

Genus CYMOPOLIA Roux.

Cymopolia Roux, Crust. Médit., 1828, p. 77; type, *C. caronii* Roux. Name not invalidated by *Cymopolia* Lamouroux, Hist. Pol. Coral. Flex., 1816, p. 292, for a genus of algae.—ALCOCK, Journ. Asiat. Soc. Bengal, vol. 69, 1900, p. 450.—RATHBUN, Proc. Biol. Soc. Washington, vol. 28, 1915, p. 180.

Palicus PHILIPPI, Zweiter Jahresber. d. Vereins f. Naturk. in Cassel, 1838, p. 11; type, *P. granulatus* Philippi=*C. caronii* Roux.—RATHBUN, Proc. Biol. Soc. Washington, vol. 11, 1897, p. 165.—A. MILNE EDWARDS and BOUVIER, Mem. Mus. Comp. Zoöl., vol. 27, 1902, p. 40.

Carapace more or less depressed, broader than long, subquadrate, covered with granules and with symmetrical tubercles or rugosities that have a tendency to fall into transverse series.

Front broadly triangular, horizontal, usually lobed or toothed. Antero-lateral borders of carapace straight or little curved; and either lobed or toothed.

Orbits deep, the upper border cut into several teeth by two or three deep clefts; the lower border usually has two clefts. Eyes large, constricted in the middle and bearing two or more lobiform protuberances.

Antennules nearly transverse, interantennular septum a narrow plate. Basal antennal joint enlarged, standing in orbital hiatus; flagellum well developed.

Epistome sunken, not defined. Buccal cavity square; at its anterior angles there is a lobe formed by a prolongation of the pterygostomian region, and overlapping to a variable extent the inner lobe of the orbit; the buccal or pterygostomian lobe is sometimes horizontal, sometimes bent down toward a vertical plane, and in some species is deflexed in the young and horizontal in the old. The external maxillipeds do not close the buccal cavity anteriorly and do not always meet in the middle; merus much smaller and narrower than ischium; antero-internal angle of ischium and antero-external angle of merus much produced; the palp articulates near middle of concave summit of obliquely placed merus.

Chelipeds short, usually slender in female; in the adult male, the right one may be enlarged, sometimes to a great extent.

The two middle pairs of legs are much the largest, second usually longer than third; first pair shorter and slenderer, but otherwise similar to the middle pairs, fourth pair weak, sometimes filiform, and elevated above third pair.

Abdomen in both sexes consisting of 7 separate segments. In the female the genital openings are on the second segment of the sternum close to the suture between it and the first.

Habitat.—Taken on a variety of bottom—mud, sand, coral, rocks, broken shells, etc. The special use of the small, delicate, hind feet is not known. The carapaces are usually free from foreign substances; occasionally one bears an encrusting tube of a Serpulid. A very few instances of parasitism have been noted in the attachment of a Rhizocephalid to the abdomen.

Distribution.—Atlantic coast of America from off the southern coast of New England to Cape Frio, Brazil; eastern Atlantic and Mediterranean; Indo-Pacific from the western Indian Ocean to the Hawaiian Islands; Pacific coast of America from Gulf of California, Mexico, to Ecuador. 4 to 298 fathoms.

Characteristics of the American species.—The genus is divided into several groups ranging from the typical one, with rather flat, squarish carapace, with large and usually thin antero-lateral teeth which seem to form an integral part of it, and with legs of moderate length (not over twice as long as width of carapace), to the group with very convex, swollen carapace, small, thick, tuberculiform or spiniform, distant, antero-lateral teeth which project outside the periphery of the carapace, and with long slender legs (ranging from twice to three and one-half times width of carapace). To the first or typical group (1) belong *cristatipes, alternata, zonata, lucasii, faxoni,* and *affinis.* To the second or convex, long-legged group (2) belong *cursor, gracilis, floridana, gracilipes,* and *acutifrons.*

Between these extremes are various modifications as follows:

3. The group of *rathbuni, bahamensis,* and *isthmia,* which have small antero-lateral teeth, distant from each other, or from the orbit, or both; combined with a typical carapace and short legs; the three species agree in having the two frontal lobes very feebly subdivided.

4. Another small group allied to the typical one but with broad, projecting plates of abdomen and sternum which are conspicuous in dorsal view; this group includes *sica, angusta,* and *depressa.*

5. A group with convex carapace, prominent tubercles, approaching in aspect the most atypical group, but with short legs: *dentata, obesa, tuberculata.*

6. One long-legged form (*fragilis*) has a flatter carapace and larger, thinner, antero-lateral teeth than in the *cursor* group.

ANALOGOUS SPECIES ON OPPOSITE SIDES OF THE CONTINENT.

Atlantic.	Pacific.
alternata.	*zonata.*
faxoni.	*lucasii.*
obesa.	*tuberculata.*

KEY TO THE AMERICAN SPECIES OF THE GENUS CYMOPOLIA.

A^1. Length of second ambulatory leg not more than twice width of carapace.
B^1. Last sternal segment does not form a laminiform crest which is conspicuous in dorsal view.
C^1. The merus of second and third ambulatory legs has at its superodistal angle an obtuse lobe, more or less prominent, sometimes atrophied.
D^1. Carapace with four lateral teeth on each side (not counting extraorbital tooth), diminishing in size from front to back. Ambulatory legs with 3 or 4 large teeth on anterior margin__*cristatipes*, p. 186.
D^2. Carapace with two lateral teeth on each side, sometimes with rudiments of a third farther back. Ambulatory legs without large teeth on anterior margin except the distal tooth.
E^1. Pterygostomian lobe reaching well beyond inner suborbital lobe_____*rathbuni*, p. 198.
E^2. Pterygostomian lobe not reaching beyond suborbital lobe.
F^1. Antero-lateral teeth near together, wider than intermediate sinus.
G^1. Antero-lateral teeth blunt_____*alternata*, p. 188.
G^2. Antero-lateral teeth acute.
H^1. Outer margins of exorbital teeth converging anteriorly. Posterior margins of propodus and dactylus of first ambulatory bare in mature male_____*zonata*, p. 190.
H^2. Outer margins of exorbital teeth subparallel. Posterior margin of propodus and dactylus of first ambulatory clothed with shaggy hair in mature male_____*affinis*, p. 196.
F^2. Antero-lateral teeth distant, narrower than intermediate sinus_____*bahamensis*, p. 200.
C^2. The merus of second or second and third ambulatory legs has at its supero-distal angle a prominent lobe which ends in a sharp point.
D^1. Outer suborbital lobe strongly convex on anterior margin. Antero-lateral teeth blunt.
E^1. Last leg reaches end of merus of third leg. Two lateral teeth on carapace_____*obesa*, p. 205.
E^2. Last leg falls short of end of merus of third leg. Three lateral teeth on carapace_____*tuberculata*, p. 207.
D^2. Outer suborbital lobe truncate and nearly straight on anterior margin. Antero-lateral teeth acute.
E^1. Front quadridentate, teeth well separated. Antero-lateral teeth adjacent, sinus narrower than either tooth.
F^1. Outer orbital tooth pointing straight ahead. Tubercles of carapace very distinct from the prominences which bear them_____*dentata*, p. 202.
F^2. Outer orbital tooth with tips turned inward.
G^1. Lobe at supero-distal angle of merus of third leg acute_____*faxoni*, p. 194.
G^2. Lobe at supero-distal angle of merus of third leg obtuse. Posterior margin of propodus of first leg clothed with shaggy hair in mature male_____*lucasii*, p. 193.
E^2. Front bidentate, each tooth obscurely emarginate. Antero-lateral teeth distant, sinus as wide as tooth_____*isthmia*, p. 201.

B^2. Last sternal segment forms a thin, laminiform crest conspicuous in dorsal view. Carapace with 3 lateral teeth, exclusive of orbital tooth.
   C^1. Ridge above posterior margin of carapace one unbroken curve. Outer suborbital lobe with a margin slightly concave forward.
                                                                       depressa, p. 212.
   C^2. Ridge above posterior margin sinuous. Outer suborbital lobe a low triangle.
      D^1. Carapace wide, width 1.25–1.39 times length_____*sica*, p. 208.
      D^2. Carapace narrow, width 1.14 times length_____*angusta*, p. 210.
A^2. Length of second ambulatory leg more than twice width of carapace.
  B^1. Outer suborbital lobe much less advanced than the ear-shaped prominence formed by the pterygostomian region at its anterior angle.
   C^1. Outer suborbital lobe with margin concave forward.
      D^1. Front quadridentate_____*fragilis*, p. 213.
      D^2. Front bidentate_____*gracilipes*, p. 221.
   C^2. Outer suborbital lobe in form of a triangle.
      D^1. Outer suborbital lobe a low, obtuse triangle_____*acutifrons*, p. 223.
      D^2. Outer suborbital lobe a produced, equilateral triangle.
         E^1. One lateral tooth and one tubercle. Second leg 3½ times as long as width of carapace_____*gracilis*, p. 218.
         E^2. Three lateral teeth. Second leg 3 times as long as width of carapace_____*floridana*, p. 220.
  B^2. Outer suborbital lobe visible from above and almost as advanced as the pterygostomian lobe. One larger lateral tooth between two smaller lobes or denticles_____*cursor*, p. 215.

CYMOPOLIA CRISTATIPES A. Milne Edwards.

Cymopolia cristatipes A. MILNE EDWARDS, Bull. Mus. Comp. Zoöl., vol. 8, 1880, p. 28 (type-locality, Grenada, 92 fathoms; holotype, Cat. No. 6494, M.C.Z.).

Palicus cristatipes RATHBUN, Proc. Biol. Soc. Washington, vol. 11, 1897, p. 93.—BOUVIER, Bull. Soc. Philom. Paris, ser. 8, vol. 9, 1898, p. 65 [12].—A. MILNE EDWARDS and BOUVIER, Mem. Mus. Comp. Zoöl., vol. 27, 1902, p. 42, pl. 7, figs. 1–5.

Diagnosis.—Merus of ambulatory legs with three or four large teeth on anterior border. Carapace with four lateral teeth besides orbital. Suborbital margin oblique.

Description.—Five lateral teeth, including orbital, diminishing in size from anterior to posterior tooth; outer margin of orbital tooth very convex, an interval between that and next tooth, which is longitudinally truncate on distal half of posterior margin; third and fourth teeth subtriangular and subacute, last tooth rudimentary. Front divided by a deep emargination into two lobes, each of which is divided in two by a shallow sinus. Middle supra-orbital lobe broadly triangular, obtusely rounded at tip; outer lobe narrower, subacute. Regions of carapace well marked, tuberculate, the tubercles granulated; 12 tubercles in transverse row across cardiac and branchial regions, of which two tubercles are cardiac. Six transverse tubercles on posterior margin.

Infero-orbital border oblique and nearly straight, cut into two lobes by a narrow emargination, outer lobe one and one-half times as wide as inner; a denticle at inner end.

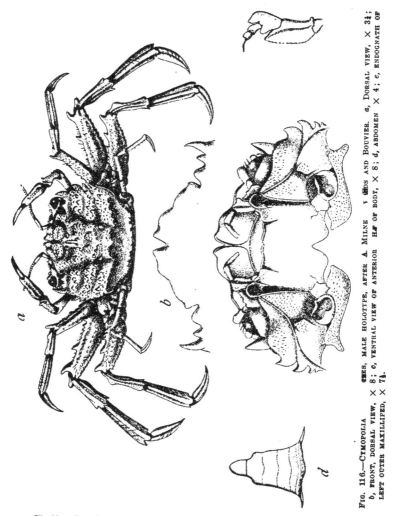

FIG. 116.—CYMOPOLIA CHES, MALE HOLOTYPE, AFTER A. MILNE EDWARDS AND BOUVIER. *a*, DORSAL VIEW, × 3¼; *b*, FRONT, DORSAL VIEW, × 8; *c*, VENTRAL VIEW OF ANTERIOR HALF OF BODY, × 8; *d*, ABDOMEN × 4; *e*, ENDOGNATH OF LEFT OUTER MAXILLIPED, × 7½.

Chelipeds of male narrow and rather short.

Ambulatory legs armed with three or four large teeth on anterior border of merus; the distal tooth is lamellate, obtuse, and truncate in the third leg, projecting forward in the other two; upper surface rough with denticles which tend to form longitudinal rows making two elevated regions separated from each other and from the margins by depressions; two lamellate lobes on anterior border of carpus, especially of second and third pairs. Last leg scarcely surpassing merus of preceding leg.

Measurements (after Bouvier).—Male holotype, greatest length of carapace 9, width 10.5, length of left cheliped 10.5, first ambulatory 11.5, second 18.5, third 16 mm.

Material examined.—Off Grenada, Windward Islands; lat. 11° 25′ 00″ N.; long. 62° 04′ 15″ W.; 96 (not 92) fathoms; Co. brk. Sh.; temp. 58.5° F.; Feb. 27, 1879; station 253, U. S. C. S. Str. *Blake;* one male holotype (Cat. No. 6494, M. C. Z.).

CYMOPOLIA ALTERNATA (Rathbun).

Plates 42 and 43.

Palicus alternatus RATHBUN, Proc. Biol. Soc. Washington, vol. 11, 1897, p. 95 (type-locality, lat 29° 11′ 30″ N.; long. 85° 29′ 00″ W., 26 fathoms; holotype, Cat. No. 19840, U. S. N. M.); Bull. U. S. Fish Comm., vol. 20, for 1900, pt. 2 (1901), p. 12 (part).

Palicus Blakei A. MILNE EDWARDS and BOUVIER, Bull. Mus. Hist. Nat. Paris, vol. 5, 1889, p. 123 (type-locality, Gulf of Mexico); Mem. Mus. Comp. Zoöl., vol. 27, 1902, p. 48, pl. 8, figs. 13–16 (type-locality, lat. 24° 43′ N.; long. 83° 86′ W., 37 fathoms; holotype at present in Paris Mus.).

Diagnosis.—Two blunt lateral teeth. Sides of carapace convergent anteriorly. Propodus of legs two and three slightly widening distally; anterior distal lobe of merus subrectangular. Posterior margin of propodus and proximal half of dactylus of leg one clothed with shaggy hair in adult male.

Description of typical specimens.—Elevations of carapace covered with small tubercles composed of a few granules and with single granules. Four frontal lobes well defined; median sinus an equilateral triangle, lateral sinuses small but well marked, lateral teeth much less advanced than median teeth and broadly rounded. Middle tooth of supraorbital margin broad, obliquely subtruncate, blunt, bounded on each side by a V-fissure very narrow toward the point; next tooth separated from the outer tooth of the orbit by a shallow sinus. Outer orbital tooth directed forward or a little outward, tip turned inward, outer margin mostly straight, sometimes a little concave. Next two antero-lateral teeth dentiform, with rounded tip, little projecting, second the smaller, sinuses small, V-form, after

FIG. 117.—CYMOPOLIA ALTERNATA, OUTLINE OF FRONT AND ORBITS OF MALE (19840), × 6.

second tooth a small tubercle or rudimentary tooth. Above posterior margin a thin, sinuous, elevated ridge, broken into a variable number of unequal transverse tubercles, with usually some granules interspersed. Suborbital margin oblique, its outer lobe nearly straight and separated from inner lobe by a small emargination; inner lobe bilobed, its inner angle produced in a small acute tooth beyond the triangular, pterygostomian lobe.

There are two forms of the male. In the first or stronger, the chelipeds are very unequal, the right large and heavy, the left slender and weak; both are tuberculate and pubescent, the carpus has an outer laminate crest, its outer surface covered with irregular laminiform lobes, manus surmounted by a double crest of same; right manus very thick, its width at distal end may equal one-half length of carapace, immovable finger short and wide, dactylus strongly bent down, overlapping the fixed finger and leaving a narrow gape; left manus little over one-third width of right, fingers long and narrow. In the second or weaker form of male, the right manus is about twice depth of left, its fingers long and slender. In the female the chelipeds are more nearly equal.

Of the ambulatory legs, the first reaches middle of propodus of second, third reaches middle of dactylus of second; merus joints rough, with squamose tubercles, a longitudinal groove on anterior edge, two of same on upper surface, an obtuse tooth at distal end, which in the first leg is subtriangular and produced a little distally beyond segment, in second and third legs is subrectangular, either straight or a little convex above and reaching in second leg just to, or not quite to end of segment, and in third leg not reaching end of segment; anterior proximal lobe of carpus rounded, somewhat triangular, anterior subdistal lobe low and rounded on first leg, triangular in second and third legs, posterior distal tooth inconspicuous; anterior margin of propodus convex, posterior straight; posterior margin of propodus and proximal half of dactylus of first leg clothed with shaggy hair in adult male.

In the first form of the male the appendages of the first abdominal segment are stout and twisted, tip bilobed, inner lobe thinner and longer than outer; in the second form the appendages are weaker and not twisted, tip less spreading.

Except for the chelipeds and abdominal appendages the two forms of the male agree. They may represent alternating generations, as in the crayfish of the genus *Cambarus*, or the weaker specimens may be simply immature.

Variety.—The species shows such great divergence from type that it seems almost possible to form a second species based on (*a*) the carapace being wider behind in proportion to its length, 1.15–1.24 : 1, instead of 1.09–1.21 : 1 in typical *alternata*, and the sides less parallel and more oblique; (b) lateral teeth of carapace subequal, second sometimes the larger and always pointing more outward than first. so that the sinus between the teeth is wider than in typical *alternata*. where the teeth are directed forward, the second always the smaller; (c) antero-distal tooth of merus of second and third legs more produced and slightly acutangular, though tip obtuse; (d) propodus of second leg longer and narrower than in typical, 1:3.9–4.75, while in typical, 1:3.27–4; (e) size larger.

Variation.—The characters c and d are not stable, and as to b, the teeth on opposite sides of the same specimen may be unlike. In some very small specimens of both sorts the pterygostomian lobe is less deflexed than in larger examples, and in consequence is more advanced than the inner suborbital lobe, as in the type of *P. blakei*. There is also considerable variation in other particulars, such as relative width of frontal teeth and sinuses and details of lower margin of orbit.

Measurements.—Male holotype, greatest length of carapace 6.6, width of same 7.6, length of second ambulatory 14; largest typical female (15281), greatest length of carapace 8, width of same 8.9; male, variety (49140), greatest length of carapace 11.9, width of same 14.3; largest female of variety (19854), greatest length of carapace 13.6, width of same 16.5 mm.

Range.—Off Cape Hatteras, North Carolina; Gulf of Mexico, along the west coast of Florida from Cape San Blas to Key West; 4 to 60 fathoms.

Material examined.—See page 191.

CYMOPOLIA ZONATA Rathbun.

Plate 44, fig. 3; plate 45, fig. 1.

Cymopolia zonata RATHBUN, Proc. U. S. Nat. Mus., vol. 16, 1893, p. 259 (type-locality, Gulf of California, 40 fathoms; holotype, Cat. No. 17484, U.S.N.M.).

Palicus zonatus RATHBUN, Proc. Biol. Soc. Washington, vol. 11, 1897, p. 94; Proc. U. S. Nat. Mus., vol. 21, 1898, p. 600.

Diagnosis.—Middle supra-orbital lobe obliquely truncate. A row of seven linear tubercles above posterior margin. Lateral teeth not projecting beyond line of carapace.

Description.—Carapace rather wide, sides distinctly convergent anteriorly; regions deeply separated, excepting the cardiac from the branchial; areolae granulate, depressions nearly smooth. Median sinus of front deep with end rounded, each lobe with a smaller emargination. Middle supra-orbital lobe quadrilateral, obliquely truncate, a narrow sinus either side; outer lobe of similar shape but rounded at inner angle and separated from the exorbital tooth by a

THE GRAPSOID CRABS OF AMERICA.

Material examined of Cynopolia alternata.

Lat. N.	Long. W.	Fath.	Bottom.	Temp.	Date.	Station.	Collector.	Specimens.	Cat. No.
° ′ ″	° ′ ″			°F.					
35 08 30	75 10 00	49	gy. S.		Oct. 17, 1885	2596	*Albatross*	1 ♀ ovig. (var.)	19842.
29 18 15	85 32 00	25	crs. gy. S. brk. Sh.		Feb. 7, 1885	2370	...do	5 ♂ 2 ♀ (typ.) 1 ♂ 1 ♀ (var.)	{19845. 19846.
29 15 30	85 29 30	27	G.		...do	2372	...do	4 ♂ 4♀ y(juv.) 1 ♂ (ae) 1	{19846. 19163.
29 14 00	85 29 15	25	Co.		...do	2373	...do	2 ♂ 2 ♀ (typ.) 1 ♂ (var.)	{19847. 19194.
29 11 30	85 29 00	26	S. G. brk. Sh.		Mar. 15, 1885	2374	..la	7 ♂ 8 ♀ (typ.)¹	19840.
28 45 00	85 02 00	30	gy. S. brk. Co.		...do	2405	...do	6 ♂ 7 ♀ (1 ovig.) typ.	19849.
28 46 00	84 49 00	26	crs. S. Co.		...do	2406	...do	4 ♂ 5 ♀ (1 ovig.) typ.¹	859.
28 47 30	84 37 00	24	Co. brk. Sh.		...do	2407	..la	3 ♂ (typ.)	19851.
28 33 30	84 15 30	27	fne. wh. S. bk. Sp.		Mar. 18, 1885	2411	...do	2 ♂ (var.)	19852.
26 18 30	84 08 45	28	fne. vh. S. bk. Sp. brk. Sh.		Mar. 19, 1885	2412	...do	3 ♀ (1 ovig.) typ.¹	19853.
26 33 30	84 10 00	36	sdy.	66	Apr. 2, 1901	7123	*Fish Hawk*	2 ♀ (var.)	5604.
26 19 00	84 33 00	30	S. Co.	69.5	Mar. 18, 1889	5106	*Grampus*	2 ♂ 1 ♀ (typ.)	15281.
26 06 00	84 11 00	30	S. Algæ	68	Mar. 21, 1889	5108	...do	2 ♀ (typ.) 2♂ 3 ♀ (var.)	15282.
25 44 32	82 48 15	32	hrd. S. Sh. M	70	Mar. 18, 1889	5101	...do	1 ♂ juv. (var.)	15280.
25 34 00	83 07 00	30	brk. Sh. fne S. Co.	66.5	Mar. 11, 1889	5085	...do	1 ♂ juv. (var.)	15279.
25 23 30	83 17 00	33.5	fn. S. blk. Sp.	68.5	Mar. 1, 1889	5078	...do	1 ♂ juv. (var.)	15278.
24 43 00	83 25 00	37	Sponge, Co., Crust.	63.5	Feb. 28, 1889	5072	...do	4 ♀	15277.
Off Dry Tortugas		16			1877-78	11	*Blake*	1 ♀ ovig. (typ.)	At present in Paris M.
Do.		16						1 ♂ 5 ♀ (2 ovig.) (var.)	49.
Do.		4					J. B. Henderson	1 ♀ 1♂ ovig. (var.)	49140.
Marquesas Key					1889		...do	1 ♀ (var.)	49139.
24 26 00	*81 48 15	37	Co.	74	jan. 15, 1885	2315	*Oryrus*	1 ♂ 1 ♀ (var.)	15283.
24 25 30	81 47 45	50	Co.	75	...do	2316	*Albatross*	1 ♀ (var.)	M. C. Z.
24 25 45	81 46 00	45	Co.		...do	2318	...d	3 ♀ ovig.	19844.
Off Key West, Sand Key Light, bearing WNW., Key West bearing N		60			June 19, 1893	24	State Univ. Iowa Exped	2 ♀	19854.

¹ One is infested with Rhizocephalid.   ² Type lot.   ³ Type of *blakei*.   S. U. I.

closed fissure. Edges of all the lateral teeth of carapace thin, straight, and in the same line· point of orbital tooth turned inward; three remaining teeth and rudiment of fourth, with interspaces small. Ridge above posterior margin parallel to it and broken into seven long tubercles, the end ones continuous, with postlateral margin. Outer suborbital lobe oblique, separated by a V-shaped sinus from inner lobe, which is more advanced and concave on margin, forming two lobes, the inner one sharp-pointed. Pterygostomian lobe less advanced and bent in an oblique plane, its upper surface concave.

Chelipeds very unequal in male, right the larger, its manus granulate, an interrupted ridge on upper margin of outer surface, and

FIG. 118.—CYMOPOLIA ZONATA, OUTLINE OF FRONT AND ORBITS OF MALE (22071), × 6.

below it a short ridge leading back from condyle of dactylus; fingers wide, gape small.

Ambulatory legs rather short; merus joints thick, with serrated ridges; first merus with a prominent, narrow, rounded, distal tooth; second and third merus with a thin, almost right-angled tooth, not projecting distally; proximal lobe of carpus low and rounded; distal lobe of second and third legs subrectangular; anterior margin of propodus slightly curved.

First four segments of male abdomen and penult segment of sternum (between legs of third pair) moderately cristate.

Color.—Legs banded with dark and light.

Measurements.—Male (22071), g r e a t e s t length of carapace 12.6, width of same 15.4, width of second ambulatory 25.5 mm.

Range.—Gulf of California, 8 to 40 fathoms.

Material examined.—See table.

Material examined of Cymopolia **zn.**

Bay	Lat. N.	Long. W.	Fath.	Bottom	Temp.	Date	Station	Collector	Specimens	Cat. No.
Off San Josef Island, Lower California	24 51 00	110 39 00	40	S. brk. Sh.	°F. 64	Mar. 16, 1888	2098	*Albatross*	1 ♂ holotype	81
Southern part Gulf of California	24 22 30	110 19 30	8	brk. Sh.		Apr. 30, 1888	2824	do	1 ♂ 1 ♀	22068
Do	24 11 45	109 55 00	10	Sh.		do	2827	do	1 ♂ 2 ♀	22069
Do	24 11 30	109 55 00	10	Sh.		do	2828	do	1 ♂	22070
Do	24 22 52		31	ky.	74.1	May 1, 1888	2829	do	1 ♂ 2 ♀	22071

CYMOPOLIA LUCASII (Rathbun).

Plate 44, figs. 1 and 2.

Palicus lucasii Rathbun, Proc. U. S. Nat. Mus., vol. 21, 1898, p. 600, pl. 43, fig. 2 (type-locality, off Cape St. Lucas, 31 fathoms, *Albatross* station 2829; holotype, Cat. No. 21590, U.S.N.M.).

Diagnosis.—Two acute lateral teeth. Posterior margin of propodus of first ambulatory clothed with shaggy hair in mature male. Dactyli of legs 2 and 3 with posterior margin sinuous. Lobe on merus of legs 1 and 2 acute, of leg 3 rounded.

Description.—Shape of carapace similar to *faxoni;* granulated tubercles distinct, those on cardiac and intestinal region have a transverse crest; fine granulation of depressions rather regular.

Fig. 119.—Cymopolia lucasii, male holotype, dorsal view, nearly 2.

Four frontal lobes well marked, median pair extremely narrow, separated by a deep and narrow triangular fissure which is rounded at the base; lateral fissures short, also rounded. The broad preorbital lobe which covers the eyestalk has a closed fissure, the outer piece of the lobe overlapping the inner and forming a slight emargination on the anterior edge. Middle and outer lobes of supraorbital margin subtriangular and blunt; middle lobe with inner edge slightly concave, outer slightly convex; outer lobe with nearly straight margins, the outer one about half as long as inner; exorbital tooth broad, trending inward, outer margin more or less convex. Antero-lateral lobes 2, dentiform, subequal, acute, outer margins convex, inner straight or a little concave, sinuses forming a small U; a third tooth is very small but dentiform. Lower margin of orbit oblique, the inner lobe slightly in advance of the line of the outer, which is a little convex; outer lobe with rounded outer corner and a short inner spine. Pterygostomian lobe in an oblique plane, broadly triangular, acute, concealing little of orbital lobe.

Right cheliped very heavy, palm nearly as wide as superior length, granules of outer surface chiefly in two longitudinal bands, one through the center and one near lower margin.

Ambulatory legs very wide; merus joints strongly cristate, one crest through middle of upper surface; distal lobe of legs 1 and 2 triangular, sharp-pointed, of third leg shorter, end rounded; lobes of carpus well developed, the distal lobe blunt in first leg, acute in first and second; propodus of leg 1 with a shaggy coat of hair along posterior border in male only; propodus of 2 and 3 very wide and widening distally; dactylus wide and with sinuous posterior margin.

Measurements.—Male holotype, greatest length of carapace 13.5, width of same 15.7, length of second leg 25.7 mm.

Range.—Lower California, Mexico: off Cape St. Lucas.

Material examined.—Off Cape St. Lucas, Lower California; latitude 22° 52′ 00″ N.; longitude 109° 55′ 00″ W.; 31 fathoms; rky.; temp. 74.1°; May 1, 1888; station 2829, *Albatross;* 3 males (1 is holotype), 4 females (21590).

CYMOPOLIA FAXONI (Rathbun).

Plate 45, figs. 2 and 3.

Palicus faxoni RATHBUN, Proc. Biol. Soc. Washington, vol. 11, 1897, p. 90 (type-locality, off Cape Hatteras, *Albatross* station 2596; holotype, Cat. No. 19841, U.S.N.M.).

Diagnosis.—Two acute lateral teeth. Merus of legs 1–3, with a large, flat, acute, distal spine. Tip of pterygostomian lobe sharp.

Description.—Carapace shaped as in the varietal form of *alternata*—that is, with the sides distinctly converging anteriorly; carapace of adult female very convex, surface hairy and with numerous tubercles and granules; the granules of depressions are smaller and more abundant than in *alternata.* Four lobes of front well marked, outer pair not much wider than inner, median emargination deeper than wide, its end rounded, lateral emarginations shallow. Middle and outer teeth of supraorbital margin triangular, subacute, middle one equilateral, outer one narrow and separated from outer tooth of orbit by a shallower sinus; this last

FIG. 120.—CYMOPOLIA FAXONI, OUTLINE OF FRONT AND ORBITS OF FEMALE HOLOTYPE. × 6.

tooth is directed forward, with the tip bent inward, its outer margin is mostly straight. Lateral teeth of carapace 2, similar, second a little smaller, both sharp-pointed, and with convex outer and concave inner margin. Lower margin of orbit oblique, line of inner lobe in advance of that of outer lobe, outer lobe slightly convex, inner lobe bilobed, outer portion rounded, inner portion a small acute tooth; inner lobe little obscured by the pterygostomian lobe, which is somewhat bent down and is triangular with a sharp tip, reaching just as far forward as inner angle of orbit.

Chelipeds in both sexes little unequal, right manus about twice as wide as left.

Ambulatory legs short and broad, a similar, triangular, sharp anterior tooth on the merus joints, the first one narrowest, the third lower than the second; posterior distal tooth of merus sharp; lobes of carpus prominent, the distal one of second and third legs acute; propodus of those legs widening greatly from proximal to distal end; dactyli wide, posterior margin sinuous.

Affinity.—On the whole, this species approaches nearest *Cymopolia affinis*, which has similar teeth on the carapace, but has not the large flat, triangular teeth of the merus of the legs.

Measurements.—Male (station xxx, *Blake*), greatest length of carapace 9.5, width of same 10 mm.; female holotype, greatest length of carapace 9.5, width of same 10.7, length of second leg 18 mm.; female (Yucatan Bank), greatest length of carapace 15.4, width of same 17.3 mm.

Range.—Off Cape Hatteras, North Carolina; off Yucatan, Mexico; off Cape Frio, Brazil (?).

Depth, 35 (?); 49 to 51 fathoms.

Material examined—See table.

Material examined of Cymopolia fazoni.

Locality.	Lat. N.	Long. W.	Fath.	Bottom.	Temp. °F.	Date.	Sta.	Collector.	Specimens.	Cat. No.
Off Cape Hatteras, N. C.	35 08	75	49	gy. S	69	Oct. 17, 1885	2596	*Albatross*	1 ♀ holotype	1941.
East of Cape Catoche, Yucatan	21 28	88	51	Co. S		1880	xxx	*Blake*	1 ♂ 1 ♀	4901, M. C. Z.
Yucatan Bank	10	50			1880		...do...	1 large ♀	4902, M. C. Z.
Off Cape Frio, Brazil [1]	23 40	35					*Hassler*	1 ♂ 1 ♀	3359, M. C. Z.

[1] Identification doubtful.

CYMOPOLIA AFFINIS (A. Milne Edwards and Bouvier).

Plate 46; plate 47, fig. 3.

Cymopolia dentata A. MILNE EDWARDS, Bull. Mus. Comp. Zoöl, vol. 8, 1880, p. 28 (part: specimen from Santa Cruz).

Palicus affinis A. MILNE EDWARDS and BOUVIER, Bull. Mus. Hist. Nat., Paris, vol. 5, 1899, p. 122, "*la mer des Antilles*"; Mem. Mus. Comp. Zoöl., vol. 27, 1902, p. 46, pl. 7, figs. 6–11; pl. 8, figs. 1–2 (type-locality, Santa Cruz, 115 fathoms; holotype, Cat. No. G493, M.C.Z.).

Palicus agassizi A. MILNE EDWARDS and BOUVIER, Bull. Mus. Hist. Nat., Paris, vol. 5, 1899, p. 124, "*la mer des Antilles*"; Mem. Mus. Comp. Zoöl., vol. 27, 1902, p. 47, expl. of pl. 8, figs. 5–12 (type-locality, Barbados, 69 fathoms; holotype, Cat. No. G507, M. C. Z.).

Palicus agassizii A. MILNE EDWARDS and BOUVIER, Mem. Mus. Comp. Zoöl., vol. 27, 1902, pl. 8, figs. 5–12.

Palicus alternatus RATHBUN, Bull. U. S. Fish Comm., vol. 20, for 1900, pt. 2 (1901), p. 12 (part: male specimen from St. Thomas).

Diagnosis.—Two acute lateral teeth. Posterior margin of propodus and dactylus of first ambulatory clothed with shaggy hair in mature male. Dactyli of legs 2 and 3 with posterior margin sinuous. Lobe on merus of legs 2 and 3 obtuse.

Description.—Near *C. alternata.* Carapace covered with tubercles and granules and short hairs. Median sinus of front a deep, narrow V. Middle and outer supra-orbital lobes subtriangular, blunt, the middle one larger, trending inward; outer orbital tooth inclined outward and forward,

FIG. 121.—CYMOPOLIA AFFINIS, OUTLINE OF FRONT AND ORBITS OF MALE (24515), × 6.

outer margin more or less convex, tip pointed. Two long, outstanding, triangular, lateral teeth, hooked forward and sharp, and followed by a rudimentary tooth. Ridge above posterior margin convex, irregular, composed of transverse, irregular tubercles. Suborbital border oblique, denticulate, inner lobe bilobed, the inner angle sharp and farther advanced than the pterygostomian lobe; this is inclined downward and forward and terminates in a subtriangular lobe with a tooth on inner slope.

Outer surface of larger manus covered with very fine, equal granules. Ambulatory legs with broad but not prominent lobes on anterior margin of carpus, a larger postero-distal tooth than in *alternata*, margins of propodus nearly parallel, posterior margins of dactylus sinuous; distal lobe of first merus long, narrow, either acute or obtuse, of second and third merus, broad, triangular and obtuse; posterior margin of propodus and dactylus of first leg of mature males, clothed with a shaggy coat of coarse, soft hair.

Length of sixth segment of male abdomen more than half its width.

THE GRAPSOID CRABS OF AMERICA.

Variations.—Large male (24515) more hairy than other specimens. Middle supraorbital lobe varying from subtruncate in large specimens through obtuse subtriangular in medium specimens to acute subtriangular in the smallest specimen. While the outer margin of outer orbital lobe is always convex, it may sometimes be partially straight or even a little concave. The tooth on merus of legs 2 and 3, although usually obtuse, is acute on right side only of a single specimen. The anterior distal lobe of carpus of the same leg varies in acuteness.

Color.—From a newly preserved specimen. Ambulatory legs banded with reddish orange-brown on a yellowish ground; third leg faintly banded, mostly yellowish. Carapace dull light brown mottled with orange brown. Chelipeds same uniform color as leg bands.

Measurements.—Male (24515), greatest length of carapace 12.7, width of same 14.8, length of second leg 24.4 mm.; female (49142), greatest length of carapace 9.9, width of same 11 mm.

Range.—Charlotte Harbor to Miami, Florida; St. Thomas; St. Croix; Barbados. 20–117 fathoms.

Material examined.—See table.

Remarks.—I think that the type of *agassizi* is a young specimen of *affinis;* compare, for example, Edwards and Bouvier's plate 8, figure 10, with the same leg on figure 1; in their plate 7, figure 6, the projection at inner angle of orbit appears exaggerated, while in plate 8, figure 6, the edge of the pterygostomian lobe is truncated because bent down away from the horizontal.

Material examined of Cymopolia affinis.

Locality.	Position.	Fath.	Bottom.	Temp.	Date.	Sta.	Collector.	Specimens.	Cat. No.
Off Charlotte Harbor, Fla.	26° 33' 00" N., 83° 10' 00" W.	23	sdy	68° F.	Apr. 2, 1901	7123	Fish Hawk	1 ♀, 3 ♂	49141, 94.
Off Miami, Florida		30					John B. Henderson.		49142.
Do.		60				1910	...do...	1 ♀, 1 ♂	315.
Western ... By Key West, Florida.		25		21.8 C.			...do...	1 ♂	
Off St. ..., West	Sail Rock W. x N. ½ N., 6 m.	20–23	Co.	65 F.	Feb. 6, 1899	6079	Fish Hawk	1 ♂ 1	61, M.C.Z.
Off Frederickstadt, St. ...	17° 37' 55" N., 64° 51' 20" W.	117	R.brk.Sh.	68	Jan. 5, 1879	132	Blake	1 ♀ ?	6307, M.C.Z.
Off Barbados	13° 04' 50" N., 59° 37' 40" W.	69	Co. Sh.		Mar. 6, 1879	273	...do...		

[a] Holotype of *P. agassizi* (not *P.*).

[b] Not 266.

[c] Holotype of *P.*.

CYMOPOLIA RATHBUNI (A. Milne Edwards and Bouvier).

Plate 48, figs. 1 and 2.

Palicus rathbuni A. MILNE EDWARDS and BOUVIER, Bull. Mus. Hist. Nat., Paris, vol. 5, 1899, p. 125, "*la mer des Antilles*"; Mem. Mus. Comp. Zoöl., vol. 27, 1902, p. 50, pl. 9, figs. 1-7 (type-locality, Barbados, 71 fathoms; Cat. No. 6508, M.C.Z.).

Palicus alternatus RATHBUN, Bull. U. S. Fish Comm., vol. 20, for 1900, pt. 2 (1901), p. 12 (part: female specimen from St. Thomas).

Diagnosis.—Ambulatory legs of second pair twice as long as width of carapace. Lobes on anterior margin of carpus of ambulatory legs obsolescent. Lateral emarginations of front minute. Outer suborbital lobe subtruncate, far back and exceeded by pterygostomian lobe and by lobe of exognath.

Description.—Carapace with areolations not sharply defined, ornamented with large granules and granulated tubercles, and devoid of hairs. The largest tubercles are four across cardiac region. Median emargination of front much wider than deep; lateral emarginations feeble, V-shaped. Middle lobe of supra-orbital margin subtriangular, obtuse, outer margin convex; next lobe narrower, inner margin convex, outer margin partially overlapping outer orbital tooth. This last is directed obliquely outward and forward with the tip obtuse and turned inward. Remaining teeth of lateral margin two in number, well separated, blunt, the first the larger, its anterior margin nearly straight and transverse, posterior margin oblique and convex; anterior margin of second tooth concave and transverse, posterior margin longitudinally oblique and straight; margin behind teeth crenulate. Posterior margin ornamented with a convex and slightly sinuous row of tubercles, two long ones, separated by two short ones and followed on either side by two of medium length. Outer lobe of suborbital margin transverse, truncate, slightly convex, far back, as is also the oblique inner lobe which is almost completely hidden by the pterygostomian lobe, which has a broadly rounded extremity, and by the lobe of the exognath. First movable segment of antenna strongly widened distally.

Chelae of female slender, feeble.

Ambulatory legs long and slender, lobes of carpus almost obsolete, margins of propodus nearly parallel, the segment widening very little in distal half; merus of first leg with a large triangular distal tooth, merus of second and third legs with an inconspicuous, blunt, distal tooth or lobe, surface sparingly granulate; dactyli long and narrow, almost wholly concave on posterior margin.

Abdomen of mature female rather narrow, exposing a considerable part of the sternum; first four segments with transverse prominences. Fifth segment of sternum laminate and slightly visible in dorsal view.

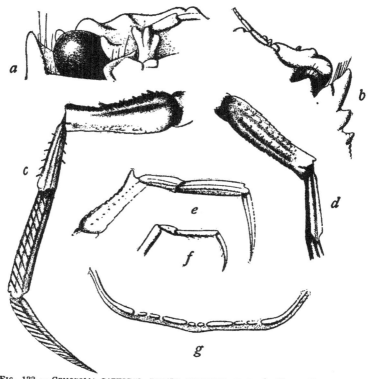

FIG. 122.— CYMOPOLIA RATHBUNI, FEMALE HOLOTYPE, AFTER A. MILNE EDWARDS AND BOUVIER. *a*, RIGHT OCULO-ANTENNAL REGION, VENTRAL VIEW, × 20; *b*, RIGHT ANTERO-LATERAL PORTION OF CARAPACE, DORSAL VIEW, × 14; *c*, SECOND LEFT LEG, UPPER FACE, × 11½; *d*, MERUS AND CARPUS OF THIRD RIGHT LEG, UPPER FACE, × 11½; *e*, FIRST LEFT LEG, LOWER FACE, × 11½; *f*, RIGHT LAST LEG, × 11½; *g*, POSTERIOR BORDER OF CARAPACE, × 14.

Measurements (after Bouvier).—Female holotype, greatest length of carapace 5; width of same, 6.2; length of second ambulatory leg, 12.4 mm.

Range.—Off St. Thomas and Barbados, 7 to 23 fathoms.

Material examined of Cymopolia rathbuni.

Locality.	Position.	Fathoms.	Bottom.	Temp.	Date.	Station.	Collector.	Specimens.	Cat. No.
Off St. Thomas, West Indies	Sail Rock W. by N. ¼ N., 1′ 11″, 25″ N.; 59° 38′ W.	20-23	Co	°C. 25.8	Feb. 6,1899	6079	Fish Hawk	1 ♀ immature.	......
Off Barbados		7 (not 71)	Co. S. brk. Sh.		Mar. 8,1879	287	Blake	1 ♀ holotype.	6508, M. C. Z.

Material examined.—See table.

Remarks.—The resemblance of the St. Thomas specimens to the figures of Edwards and Bouvier cited leads me to believe that it is the same species, granting, however, that in their plate 9, figure 2 (my fig. 122a), the most prominent lobe on the inner portion of the orbit is the pterygostomian lobe, and the one overlying it, the lobe of the exognath; the pterygostomian lobe conceals the lobe of the orbit, but in the St. Thomas specimen the outer corner of it is visible in ventral view.

CYMOPOLIA BAHAMENSIS (Rathbun).

Plate 47, figs. 1 and 2

Palicus bahamensis RATHBUN, Proc. Biol. Soc. Washington, vol. 11, 1897, p. 98 (type-locality, east of Andros Island, Bahamas, 97 fathoms; Cat. No. 11394, U.S.N.M.); Bull. Lab. Nat. Hist. State Univ. Iowa, vol. 4, 1898, p. 280 (except specimen from Barbados, which is *C. rathbuni*), pl. 9, fig. 2.

Diagnosis.—Ambulatory legs of second pair twice as long as width of carapace, of first pair in mature male clothed with shaggy hair on posterior margin of propodus and proximal half of dactylus. Lobes on anterior margin of carpus of ambulatory legs well marked. Pterygostomian lobe advanced, far exceeding inner suborbital lobe. Front bilobed.

Description.—Carapace very rough with coarse granulation. Front divided by a wide and deep V-shaped notch into two lobes, each of which is faintly emarginate. Two supra-orbital lobes triangular, obtuse. Outer orbital tooth long and narrow, with nearly straight margins and directed obliquely forward and outward. Lateral teeth two, distant, of subequal length, blunt (acute in the young), directed outward and only slightly forward. Crest above posterior margin sinuous, broken into seven irregular scallops. Outer suborbital lobe transverse and slightly convex. Lobe at angle of buccal cavity rounded and produced far beyond inner lobe of suborbital margin; the latter lobe is oblique, concave, and sharply pointed at inner angle.

Chelipeds of male unequal, right one large, manus bicristate above, outer surface granulate, lower margin convex; immovable finger slightly deflexed, its length less than width of manus.

Merus of first ambulatory terminating in a large, blunt tooth, of second and third armed on anterior margin with four or five curved, more or less spiniform denticles; terminal tooth large, subacute; posterior margin spinulous; upper surface covered with squamose granules; lobes on carpus small, but distinct; dactyli slender, posterior margin concave. Posterior margin of propodus and proximal half of dactylus clothed with shaggy hair in mature male.

Crests on first two abdominal segments of male trilobate and subparallel in dorsal view, the first the wider; third segment with lobe each side of middle, partially visible in dorsal view. Crest on fifth sternal segment, or that between bases of third legs, about half as wide as second abdominal segment, its posterior margin sinuous.

Measurements.—Male (22301), greatest length of carapace 7.4, width of same 9 mm.

Range.—Bahamas; dredged, to a depth of 97 fathoms.

Material examined.—

Southeast of Andros Island, Bahamas, in tongue of ocean; lat. 24° 02′ 00″ N.; long. 77° 12′ 45″ W.; 97 fathoms; wh. Oz.; temp. 73.4° F.; Apr. 13, 1886; station 2651, *Albatross;* 1 male holotype (11394).

Bahama Banks; May 18, 1893; State Univ. Iowa Exped.; 4 males, 4 females (22301); 12 males, 18 females (Mus. S. U. I.).

Off Green Cay, Bahamas; in oyster dredge; June 30, 1903; B. A. Bean, Exped. Geogr. Soc. Baltimore; 1 male (31060).

CYMOPOLIA ISTHMIA (Rathbun).

Plate 48, figs. 3 and 4.

Palicus isthmius RATHBUN, Proc. Biol. Soc. Washington, vol. 11, 1897, p. 97 (type-locality, near Aspinwall; lat. 9° 27′ N.; long. 79° 54′ W.; 25 fathoms; *Albatross*, station 2145; holotype, Cat. No. 7753, U.S.N.M.).

Diagnosis.—Lateral lobes of front obsolescent. Antero-lateral teeth two, thick, far apart. Distal lobe of merus of ambulatory legs acute; anterior lobes of carpus low.

Description.—Carapace covered with single granules and small tubercles, each composed of a few granules; granulation moderate. Front little advanced, median sinus shallow, much wider than long, median teeth very small, tuberculiform; lateral lobes broad and very shallow. Middle teeth of supraorbital margin triangular, obtuse, inner margin straight, outer margin slightly convex; outer tooth narrower, subtriangular, separated by a shallow sinus from the outer tooth of the orbit. This last is narrow, sharp-pointed, is directed forward, its outer margin nearly straight except near tip. Two antero-lateral teeth, small, thick, blunt, the first pointing forward, a very small sinus between it and orbital tooth, outer margin convex; the second tooth pointing outward, its anterior margin straight and almost transverse, posterior margin convex. Ridge above posterior margin composed of small, unequal tubercles. Outer lobe of suborbital margin transverse and slightly convex; inner lobe oblique, largely covered by pterygostomian lobe in ventral view, its outer portion a little convex, inner angle a small, sharp spine; pterygostomian lobe more produced than orbital lobe, arcuate, margin a little oblique, forming an obtuse angle at inner extremity.

FIG. 123.—CYMOPOLIA ISTHMIA, OUTLINE OF FRONT AND ORBITS OF FEMALE HOLOTYPE, × 10.

Chelipeds of male not known, of female very slender, subequal.

Merus joints of ambulatory legs rather slender, very rough; one or two of the more proximal spinules of the anterior margin are outstanding; the distal lobe is triangular and acute, and in the first leg has a long, sharp tip; anterior lobes of carpus inconspicuous; dactyli long and narrow, their posterior margins proximally straight.

Measurements.—Immature female, holotype, greatest length of carapace 5.8, width of same 7, length of second leg 13.7 mm.

Material examined.—Off Colon, Panama; lat. 9° 27' 00" N.; long. 79° 54' 00" W.; 25 fathoms; gn. M. brk. Sh.; Apr. 2, 1884; station 2145, *Albatross;* 1 female immature, holotype (7753).

CYMOPOLIA DENTATA A. Milne Edwards.

Cymopolia dentata A. MILNE EDWARDS, Bull. Mus. Comp. Zoöl., vol. 8, 1880, p. 28 (part; not specimen from Santa Cruz) (type-localities, Charlotte Harbor, 50 fathoms, and Barbados, 69 and 76 fathoms).

Palicus dentatus A. MILNE EDWARDS and BOUVIER, Mem. Mus. Comp. Zoöl., vol. 27, 1902, p. 53. pl. 9, figs. 15–17; pl. 10, figs. 1–6; pl. 11, figs. 1–3 (type-locality, Charlotte Harbor, 50 fathoms; holotype, Cat. No. 6492, M.C.Z.). In the synonymy, instead of "*obesus*," read "*dentatus*."

Diagnosis.—First antero-lateral tooth distant from orbital tooth. Supraorbital teeth triangular. Tubercles above posterior margin linear. Outer suborbital lobe subtruncate. Distal lobe of merus of ambulatories sharp; lobes of carpus well developed.

Description.—Carapace covered with definite tubercles which are ornamented with much finer granules than those elsewhere on the carapace. Front with four well-marked rounded lobes, median sinus deeper and wider than lateral. Two supraorbital teeth triangular, subacute, outermost sinus broader and a little shallower than the others; outer orbital tooth long, triangular, acute, outer margin

THE GRAPSOID CRABS OF AMERICA. 203

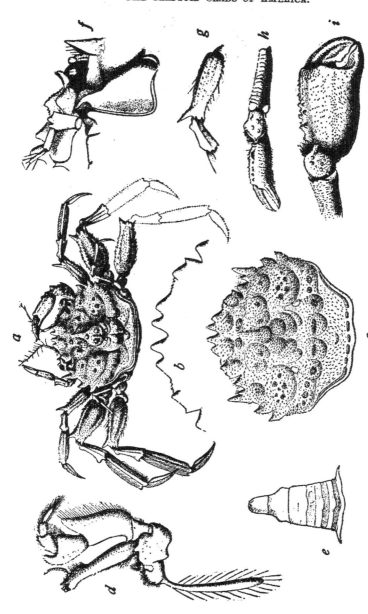

FIG. 124.—CYMOPOLIA DENTATA, AFTER A. MILNE EDWARDS AND BOUVIER. a, DORSAL VIEW OF FEMALE HOLOTYPE, × 3 ; b, FRONTAL BORDER OF THE SAME, × 7½; c, CARAPACE OF A ♂, DORSAL VIEW, × 3; d, 1ST OUTER MAXILLIPED, × 7½; e, ABDOMEN OF MALE (SAME AS c), × 7½; f, LEFT OCULO-ANTENNAL REGION OF SAME MALE, × 7½; g, ISCH AND CARPUS OF FIRST LEFT LEG OF FEMALE HOLOTYPE, USED IN ♀♂, × 5; h, LEFT CHELIPED OF ♂, OUTER FACE, × 4; i, ♂ CHELIPED OF ♂, OUTER FACE, × 4.

Material contained of *Cynopolta dentata*.

Locality.	Latitude.	Longitude.	Fathoms.	Bottom.	Date.	Station.	Collector.	Specimens.	Cat. No.
Off Charlotte Harbor, Fla. Off Sand Key, Fla.	Sand Key light bearing W. about 15 m.	" "	50 15		June 24, 1893		Stimpson, Bache S. U. I. Bahama Exped.	1 ♂ holotype. 1 ♂ immature	6192, M. C. Z. Mus. S. U. I.
Off Barbados	13 04 N.	59 37 40 W.	69	Co. Sh	Mar. 6, 1879	278	Blake	1 ♂, 1 sm. ♀	Paris do. 2771, M. C. Z.
Do	13 04 N.	59 36 45 W.	76	Co. brk. Sh	Mar. 5, 1879	272	do	1 ♂ 1 im. ♀	6,500, M. C. Z.

nearly straight or directed slightly outward. Antero-lateral margins very oblique, with two large acute teeth, the first one remote from orbital tooth; both have straight or concave anterior and convex posterior margins; a rudimentary third tooth. Ridge above posterior margin made of linear tubercles sometimes confluent. Outer suborbital lobe transverse, subtruncate, or a little arcuate; inner lobe more advanced, subtruncate, outer corner rounded, inner lobe with a small vertical tooth; pterygostomian lobe about half as wide and subtriangular.

Right cheliped of male very deep, between two-fifths and one-third as deep as carapace is long.

Merus of ambulatory legs armed with a large, acute, projecting tooth at anterior distal angle; granules of upper surface small, depressed; lobes of carpus prominent; propodus broad, not increasing much distally; posterior margin of dactylus strongly sinuous.

Color.—Yellowish-rose in alcohol.

Measurements.—Male type, greatest length of carapace 13, width of same 14 mm.

Range.—From Charlotte Harbor, Florida, to Barbados. 15 to 76 fathoms.

Material examined.—See table.

Affinity.—Near *C. obesa*, but carapace less convex, its tubercles very distinct from the elevations which bear them; the two divisions of each frontal lobe more developed and more deeply separated; outer angle of preorbital lobe dentiform; lateral teeth larger and more hooked forward; ridge above posterior margin formed of linear instead of swollen tubercles; merus and carpus of ambulatory legs shorter and broader, lobes of carpus better developed.

CYMOPOLIA OBESA A. Milne Edwards.

Plate 40.

Cymopolia obesa A. Milne Edwards, Bull. Mus. Comp. Zoöl., vol. 8, 1880, p. 27 (type-locality, lat. 23° 13′ N.; long. 89° 16′ W., 84 fathoms; holotype, immature female, not male, Cat. No. 6491, M.C.Z.).

Palicus obesus A. Milne Edwards and Bouvier, Mem. Mus. Comp. Zoöl., vol. 27, 1902, p. 51, pl. 9, figs. 8–14.

Diagnosis.—Carapace very convex, branchial regions laterally expanded. Granulation of carapace coarser between than upon

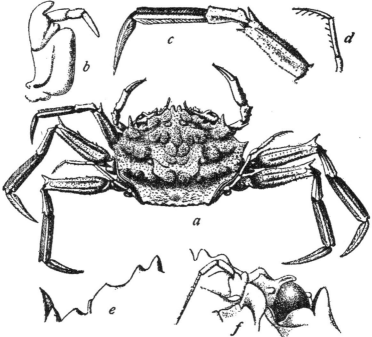

Fig. 125.—CYMOPOLIA OBESA, IMMATURE FEMALE (NOT MALE) HOLOTYPE, AFTER A. MILNE EDWARDS AND BOUVIER. *a*, DORSAL VIEW, × 3 ABOUT; *b*, ENDOGNATH OF RIGHT OUTER MAXILLIPED, × 6½; *c*, FIRST LEFT LEG, UPPER SIDE, × 5; *d*, LAST LEFT LEG, × 5; *e*, LEFT FRONTAL AND ORBITAL BORDER, FROM ABOVE, × 7½; *f*, LEFT OCULO-ANTENNAL REGION, VENTRAL VIEW, × 7½.

tubercles. Tubercles above posterior margin convex, protuberant. Outer suborbital lobe very convex. Lobes of carpus of ambulatory legs little developed.

Description.—Carapace very convex, antero-lateral margins very oblique, surface covered with very definite tubercles having a granulation much finer than that on remainder of carapace. Median sinus of front deep and narrow behind, lateral orbital hood without tooth at outer angle. Sinuses shallow, broadly -shaped. Pre-Teeth of supraorbital margin triangular, blunt, outer sinus somewhat shallower than the others. Outer orbital tooth very narrow, straight or a little convex, blunt, pointing straight forward. Antero-lateral teeth two, distant, thick, blunt, posterior margin convex, anterior straight or a little concave. Posterior margin with six or eight tubercles, of which four are large and prominent. Outer lobe of suborbital margin very convex, inner lobe very oblique, slightly convex, inner angle an acute tooth; this lobe is a little more than half concealed in ventral view by the large blunt-pointed pterygostomian tooth, which is slightly deflexed and is much more advanced than the lower margin of the orbit.

Chelipeds of male weak, the left manus two-thirds as high as the right, dactylus as long as upper margin of palm.

Merus of ambulatory legs narrow, angular, with three anterior carinae, upper surface scantily granulous, distal spine narrow, sharp, produced beyond the segment; anterior lobes of carpus feebly developed; propodus and dactylus of second and third legs wide, posterior margin of dactylus sinuous.

A transverse ridge on each of segments 1–6 of the male abdomen, that on sixth segment arched forward.

Measurements.—Male (17894), greatest length of carapace (approx.) 12.5, width of same 15.1, length of second leg 27 mm.; female, thin shell (*Blake*), greatest length of carapace 17.8, width of same 20.6 mm.

Range.—Gulf of Mexico, 13 to 88 fathoms.

Material examined.—See table.

Material examined of Cymopolia

	Lat. N.	Long. W.	Fath.	Bottom.	°F	Date.	Stn.	Collector.	Specimens.	Cat. No.
Off Mississippi R.r.	29 14 30	88 09 30	68	gy. M.		Feb. 11,1885	2378	*Albatross*	1 ♂	17894, M.C.Z.
South of Cape San Blas, Florida	28 42 30	88 35 29 00	88	gy. M.		Mar. 15,1885	2403	...do...	1 ♂ immature	17895, M.C.Z.
West coast of Florida			13						1 ♂ immature	008, M.C.Z.
North of Yucatan Bank	23 13 00	89 16 00	84		60	1877–78	36	*Blake*	(1 ♀ immature, 1 ♂ ♀ ?)	04, M.C.Z. Paris Mus.

¹ Type figured by Milne Edwards and ² Thin shell.

CYMOPOLIA TUBERCULATA Faxon.

Cymopolia tuberculata FAXON, Bull. Mus. Comp. Zoöl., vol. 24, 1893, p. 161 (type-locality, station 3355, *Albatross*, 182 fathoms; holotype, Cat. No. 4504, M.C.Z.); Mem. Mus. Comp. Zoöl., vol. 18, 1895, p. 38, pl. 6, figs. 3, 3*a*.

Diagnosis.—Carapace very convex, branchial regions laterally expanded. Granulation of carapace finer between than upon tubercles. Tubercles above posterior margin convex, protuberant. Outer suborbital lobe very convex. Lobes of carpus of ambulatory legs little developed.

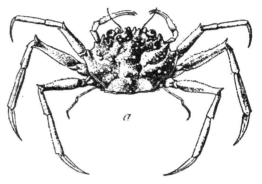

Description.—Allied to *C.* obesa. Carapace wider across branchial regions; granulation of surface finer between than upon tubercles while the reverse is true in *obesa;* median sinus of front more rounded at base; outer orbital lobes trending a little more inward; three antero-lateral teeth after the orbital, instead of two; in *obesa* a rudimentary tooth or tubercle may be present or absent after the second tooth; chelipeds equal; ambulatory legs slenderer than in *obesa*, especially as to propodal segments.

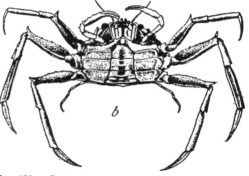

FIG. 126.—CYMOPOLIA TUBERCULATA, MALE, × 1½, AFTER FAXON. *a*, DORSAL VIEW; *b*, VENTRAL VIEW.

Color.—Ambulatories crossed by transverse bands of red, three of which cross the merus.

Measurements.—Male holotype, greatest length of carapace 13, width of same 18, length of second leg 34 mm.

Type-locality.—Bay of Panama; lat. 7° 12′ 20″ N.; long. 80° 55′ 00″ W.; 182 fath.; bk. G. Sh.; temp. 54.1° F.; Feb. 23, 1891; station 3355, A*lbatross;* 4 males, 1 female (4504, M. C. Z.).

CYMOPOLIA SICA A. Milne Edwards.

Plate 40, figs. 3 and 4.

Cymopolia sica A. MILNE EDWARDS, Bull. Mus. Comp. Zoöl., vol. 8, 1880, p. 29 (part), (type-localities, Sand Key, Yucatan Bank, Santa Cruz, Dominique, Grenade, Barbades; 56 to 138 fathoms).

Palicus sica RATHBUN, Proc. Biol. Soc. Washington, vol. 11, 1897, p. 97 (Dominique eliminated from type-localities); Bull. U. S. Fish Comm., vol. 20, for 1900, pt. 2 (1901), p. 13.—BOUVIER, Bull. Soc. Philom. Paris, ser. 8, vol. 9, 1898, p. 65 [12].

Palicus sicus A. MILNE EDWARDS and BOUVIER, Mem. Mus. Comp. Zoöl., vol. 27, 1902, p. 56, pl. 10, figs. 7–11; pl. 11, fig. 9 (type specified, male, Barbados, 82 fathoms, station 293 (by error, 243), *Blake;* holotype in Paris Mus.).

Diagnosis.—Abdomen and sternum showing prominent laminae behind carapace in dorsal view. Three subequal antero-lateral teeth. Inner suborbital lobe trilobed, inner sinus large. Carapace one-fourth to one-third wider than long; a sinuous posterior line of tubercles.

Description.—Carapace rather evenly but not strikingly convex, regions not deeply marked, granules of different sizes, sometimes clustered to form tubercles. Front very broad, not prominent, median teeth small, tuberculiform, median sinus a narrow V, lateral sinuses a little wider, shallow, rounded, lateral lobes shallow, rounded. Middle supra-orbital lobe broad, truncate, followed by a sinus rounded at base; outer lobe narrower, lobiform, separated by a shallow V-sinus from outer orbital tooth; this last is blunt and directed forward. Antero-lateral teeth three, small, distant, first tooth remote from orbit and from the second which is nearer the third. Above posterior margin a very sinuous line of tubercles. Lower margin of orbit very oblique, divided by a wide V into two lobes, the outer straight and rightangled, its inner margin concave, inner lobe with three marginal crenulations, inner angle subacute. Pterygostomian lobe very broad, bluntly obtusangled at tip, concealing in ventral view all but outer end of inner orbital lobe.

Chelipeds feeble and equal in both sexes, slenderer in female than in male.

First leg very short and slender, merus little rough and with a small, obtuse distal tooth; second and third legs with merus coarsely spinulous, especially along anterior margins, antero-distal angle almost rectangled, subacute; carpus very long and narrow, anterior lobes low; propodus sensibly widening distally; dactylus with posterior margin sinuous.

The most striking feature of this species is the arrangement of abdominal and sternal plates behind the carapace; first three segments of abdomen are carinated, the first carina has a shallow lobe

THE GRAPSOID CRABS OF AMERICA. 209

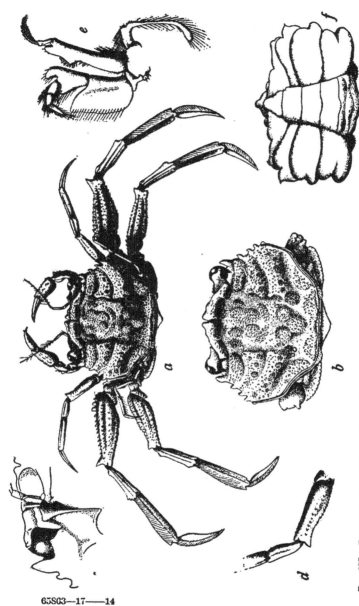

FIG. 127.—CYMOPOLIA SICA, AFTER A. MILNE EDWARDS AND ($\mathbf{M}$, a, ALE, DORSAL VIEW, × 3; b, CARAPACE OF SAME, DORSAL VIEW, × 4½; c, IBIT OCULO-ANTEN AL REGION OF SME, LOWER ME, × 7½; d, MERUS AND CARPUS OF FIRST RIGHT LEG OF SAME, UPPER ME, × 5; e, LEFT OUTER MAXILLIPED, × 8; f, STERNUM AND ABDOMEN OF LE, × 5½.

65803—17——14

behind each corner of the carapace, the second carina forms a prominent median lobe, while the third is invisible from above; the sternum has on either side, between the bases of the legs of the third pair, a wide plate which is prominent laterally and posteriorly. Abdomen of male wide, sides of third to fifth segments, inclusive, nearly straight.

Color.—Carapace light brown with a slightly bluish cast. Eyestalks reddish-brown. Merus joints of ambulatory legs with occasional tubercles of salmon color, sides of carpus and upper edge of propodus with faint dashes of brownish or salmon. (Notes on newly preserved specimens.)

Measurements.—Male (49149), greatest length of carapace 6, width of same 8.4, length of second leg 13.6 mm. Female (station 149, *Blake*), greatest length of carapace 9.8, width of same 13.5, length of second leg 23.4 mm.

Range.—From Gulf of Mexico to Florida Keys and Windward Islands; 15 to 125 fathoms.

Material examined.—See page 211.

CYMOPOLIA ANGUSTA (Rathbun).

Cymopolia sica A. MILNE EDWARDS, Bull. Mus. Comp. Zoöl., vol. 8, 1880, p. 29 (part: one specimen from Santa Cruz).

Palicus angustus RATHBUN, Proc. Biol. Soc. Washington, vol. 11, 1897, p. 97 (type-locality, off Santa Cruz, 117 fathoms, station 132 (not 32), *Blake;* holotype, Cat. No. 2930, M.C.Z.).

Diagnosis.—Carapace narrow, one-seventh wider than long; a sinuous posterior line of tubercles; three subequal antero-lateral teeth. Abdomen and sternum showing prominent laminae behind carapace in dorsal view. Inner suborbital lobe trilobed; inner sinus small.

Description.—Closely allied to *C. sica*, differs as follows: Carapace much narrower, ratio of length to width of carapace 1:1.14 as against 1:1.25 (smaller) to 1.39 (larger) in *sica;* sinuous line of tubercles above posterior margin less uneven; middle suborbital fissure narrower; ischium of maxilliped much wider; merus of second leg short and very broad in middle; median portion of second abdominal plate less prominent.

Measurements.—Holotype, greatest length of carapace 9, width of same 10.3 mm.

Specimens examined.—Off St. Croix, West Indies; lat. 17° 37′ 55″ N.; long. 64° 54′ 20″ W.; 117 fathoms; R. brk. Sh.; temp. 65° F.; Jan. 5, 1879; station 132, U. S. C. S. Str. *Blake;* 2 specimens, holotype and paratype (2930, M. C. Z.).

Locality not given; probably collected by the *Blake;* 1 male (3007, M. C. Z.).

Material examined of *Cymopolia sica*.

Locality.	Lat. N.	Long. W.	Fath.	Bottom.	Temp.	Date.	Station	Collector.	Specimens.	Cat. No.
Off Cape San Blas, Florida	28 42 30	85 29 00	88	gy. M.		Mar. 15, 1885	2403	*Albatross*	2 ♀	17892, M. C. Z.
Gulf of Mexico	28 31 00	85 53 00	119			1877-78	50	*Blake*	1	2928, M. C. Z.
West end of id.			42					*Bache*	2	90, M. C. Z.
Northern part of Yucatan Bank	23 32 00	88 05 00	95			1877-78	32	*Blake*	1 ♂ 1 ♀	29, M. C. Z.
Do.								...do...	1 ♂ 1 ♀	62, M. C. Z.
Off Key West, id.	23 13 00	80 16 00	84		60° F.	1877-78	36	J. B. Henderson.	6 ♂ 8 ♀ 3 ♀	Paris Mus. 61, M. C. Z. Paris id. 499.
Do...			71			1916		...do...	1 ♀ ovig.	49645,
Do...			100-123			1916		...do...	4 ♂ 16 ♀ (8 ovig.)	1 , id.
Off Wn Dry Rocks, Florida			65			1916		...do...	1 ♀ ovig.	1 , id.
South of Key West, Florida			80					...do...	1 ♂	497.
Southeast of Key West, id.			61					...do...	1 ♂	49148.
Gulf, off Key West	24 21 15	81 62 15	109	S.	54.5° F.	Feb. 19, 1902	7282	*Fish Hawk*	1 ♂ 1 ♀	49150.
Do.	24 21 55	81 58 25	93	S.	53° F.	Feb. 14, 1902	2279	...do...	1 ♀	49151.
Off Sand Key, Florida			80				9	Wm. Stimpson, *Bache*.	1	005, M. C. Z.
Do	Sand Key Light bearing W. about 8 miles.		15			June 24, 1893		State Univ. Iowa Bahama Exped.	1 ♂ 1 ♀	30, S. U. I.
Do			82			1913		J. B. Henderson.	1 ♂ 1 ♀	083.
Off Gart, Florida	25 11 30	80 10 00	60	G. S.	69.2° F.	Apr. 9, 1886	2641	*Albatross*	1 ♂.	1 892.
Mayaguez id, Porto id.	Pt. dol Algarrobo E. x N. 5¼ m.		97-120	Co.	24° C.	Jan. 20, 1899	6067	*Fish Hawk*	1 y. ♂	24316.
Off St. id, West	17 37 55	64 54 20	117	R. brk. Sh.	65° F.	aul. 5, 1879	132	*Blake*	1 ♂ 1 ♀	M. C. Z.
Off St. Christopher	17 16 40	62 43 48	61	fs. S. brk. Sh.	76° F.	Jan. 15, 1879	140	...do...		do.
Off Barbados	13 14 23	59 39 10	81	bk. Sp.	64.5° F.	Mar. 9, 1879	1 293	...do...	2 ♂ 1 ♀	Paris Mus.
Off id	11 25 00	62 04 15	96	Co. brk. Sh.	53.5° F.	Feb. 27, 1870	253	...do...	1 ♂ 4 ♀	do.

[1] 28 probably; not 28. [2] Not 43. [3] One is [4] Type designated by A. Milne Edwards and Bouvier, 902.

CYMOPOLIA DEPRESSA (Rathbun).

Cymopolia sica A. MILNE EDWARDS, Bull. Mus. Comp. Zoöl., vol. 8, 1880, p. 29 (part: specimens from *Blake* stations 192, 272, 292, and some from 132 and 293).

Palicus depressus RATHBUN, Proc. Biol. Soc. Washington, vol. 11, 1897, p. 98 (type-locality, off Dominica, 138 fathoms, station 192, *Blake;* Cat. No. 2701, M.C.Z.).—A. MILNE EDWARDS and BOUVIER, Mem. Mus. Comp. Zoöl., vol. 27, 1902, p. 58, pl. 11, figs. 4–8.

Diagnosis.—Carapace depressed; from one-fourth to one-third wider than long; a continuous, nontuberculous ridge above posterior

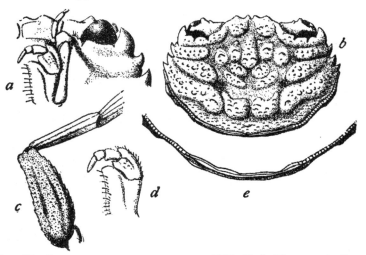

FIG. 128.—CYMOPOLIA DEPRESSA, YOUNG FEMALE (6505, M. C. Z.), AFTER A. MILNE EDWARDS AND BOUVIER. *a*, LEFT OCULO-ANTENNAL REGION, VENTRAL VIEW, × 15; *b*, CARAPACE,[1] DORSAL VIEW, × 8½; *c*, MERUS AND CARPUS OF SECOND RIGHT LEG, UPPER SIDE, × 11; *d*, ENDOGNATH OF LEFT OUTER MAXILLIPED, × 18; *e*, POSTERIOR BORDER OF CARAPACE, ENLARGED.

margin; three thin, subequal, antero-lateral teeth. Abdomen and sternum showing prominent laminae behind carapace in dorsal view. Inner suborbital lobe trilobed; inner sinus small. Tooth on merus of ambulatories atrophied.

Description.—Distinguished from *C. sica* and *C. angusta* as follows: Carapace about as wide as *sica*, more depressed than either; above the posterior margin a continuous ridge instead of a sinuous line of tubercles; inner suborbital fissure similar to that of *angusta*, the outer lobe appearing more transversely truncate, median portion of second abdominal plate less prominent than in the other species; ischium of maxilliped wide as in *angusta*, but widening more to ischium of maxilliped wide as in *angusta*, but widening more to-

---

[1] In the process of reproduction, *b* has become proportionally wider than the original.

ward the merus; merus of ambulatory legs with scarcely a trace of distal tooth; carpus with anterior lobes obsolescent. The anterolateral teeth are broader, thinner, and more acute than in *sica;* the areolations of the carapace are more sharply marked and the furrows between them almost smooth; intestinal region finely and evenly granulate, not irregularly, as in *sica*.

Measurements.—F e m a l e (station 293) greatest length of carapace 7, width of same 9.4, length of second leg 18.7 mm. Ratio of length to width varies from 1:1.26 to 1.34.

Range.—Lesser Antilles; 56 to 138 fathoms.
Material examined.—See table.

CYMOPOLIA FRAGILIS Rathbun.

Plate 51, figs. 2 and 3.

Cymopolia fragilis RATHBUN, Proc. U. S. Nat. Mus., vol. 16, 1893, p. 259 (type-localities, off Lower California, 58 and 71 fathoms, stations 2983 and 3011, *Albatross;* holotype, Cat. No. 17485, U.S.N.M.).—FAXON, Mem. Mus. Comp. Zoöl., vol. 18, 1895, p. 40, pl. 0, figs. 4, 4a.

Diagnosis.—Four subequal antero-lateral teeth. Carapace half again as wide as long. First ambulatory not reaching beyond merus of second.

Description.—Carapace very broad and not very convex, with four thin-edged antero-lateral teeth besides orbital tooth. Tubercles of carapace very well marked, high, trending forward, and definitely placed (see fig. 129a). Intervening space scantily filled with small inconspicuous granules and short hairs. Two median frontal teeth rounded and separated by a triangular sinus with rounded base; lateral sinuses shallower, equally wide, oblique; lateral teeth triangular, subacute. Sinuses of supraorbital margin triangular, the outer one rounded behind; middle tooth broad, obliquely truncate; next tooth narrower, rounded. Outer orbital teeth inclined forward and inward, outer margins convex.

Material examined of Cymopolia depressa.

Locality.	Lat. N.	Long. W.	Fath.	Bottom.	Temp.	Date.	Sta.	Collector.	Specimens.	Cat. No.
Off St. Croix	17 37 53	64 54 24	117	R. brk. Sh	°F. 65	Jan. 5, 1879	132	Blake	1 ♀	4902, M. C. Z.
Off Domínica	15 17 20	61 24 22	138	fne. S. M	63.75	Jan. 30, 1879	192	do	1 ♀ imm.	2701, M. C. Z.
Off Barbados	13 14 13	59 39 10	81	[a] S. brk. Sh	64.5	Mar. 9, 1879	203	do		Paris Mus.
Do	13 13 04	59 38 50	56	do	74.5	do	292	do	2 im.	2612, M. C. Z.
Do	13 13 12	59 38 45	76	Co. brk. Sb	64.75	Mar. 5, 1879	272	do	1 y. ♀ ♀.	6565, M. C. Z.

[1] Holotype.   [a] Figured by A. Milne Edwards and Bouvier.

Antero-lateral margin serrated, the teeth with straight anterior and curved posterior margins, teeth 1 to 3 trending forward, tooth 4 directed obliquely outward. Ridge above posterior margin broken into unequal, crenulated tubercles, formed in four scallops.

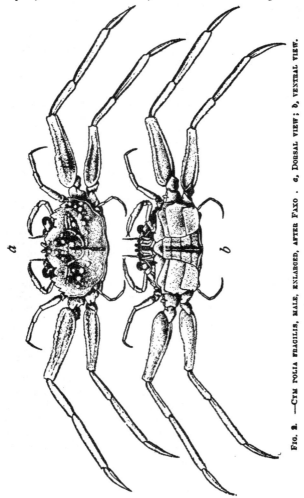

FIG. 2.—CYMOPOLIA FRAGILIS, MALE, ENLARGED, AFTER FAXO. *a*, DORSAL VIEW; *b*, VENTRAL VIEW.

Chelipeds very slender, equal. Sinuses of suborbital margin V-shaped; outer lobe oblique, bilobed; inner lobe more deeply bilobed, its outer angle more advanced than adjacent angle of outer lobe, and equally advanced with the broad pterygostomian lobe.

First leg unusually short and slender, not reaching beyond end of merus of second; merus with two anterior crests, dactylus very slender, strongly curved; second and third legs over twice as long as width of carapace, merus joints narrowing in distal half, four prominent, longitudinal crests, anterior distal end forming a right or slightly obtuse angle; anterior lobes on carpus obsolete, propodus enlarging little distally, dactylus with sinuous posterior margin.

Measurements.—Male holotype, greatest length of carapace 8.5, width of same 12.2, length of second leg 25.5 mm.

Range.—Lower California to Ecuador; 52 to 71 fathoms.

Material examined—See table.

Remarks.—There is no Atlantic prototype of this species known.

CYMOPOLIA CURSOR A. Milne Edwards.

Plate 52, figs. 1 and 2.

Cymopolia cursor A. MILNE EDWARDS, Bull. Mus. Comp. Zoöl., vol. 8, 1880, p. 29 (type-localities, Sand Key, Havane, St. Kitts, Dominique, Barbades, 138–245 fathoms).—SMITH, Proc. U. S. Nat. Mus., vol. 6, 1883, p. 21.

Cymopolia dilatata A. MILNE EDWARDS, Bull. Mus. Comp. Zoöl., vol. 8, 1880, p. 28 (type-locality, St. Kitts, 208 fathoms; Cat. No. 6496, M. C. Z.).

Palicus cursor RATHBUN, Proc. Biol. Soc. Washington, vol. 11, 1897, p. 95.—A. MILNE EDWARDS and BOUVIER, Mem. Mus. Comp. Zoöl., vol. 27, 1902, p. 64, pl. 12, figs. 6–13 (not 14),[1] (female, Barbados, 204 [209] fathoms, *Blake* station 274, selected as type; Cat. No. 6495, M.C.Z.).

Diagnosis.—Three lateral teeth. Second leg about three times as long as width of carapace. Suborbital lobes visible from above, inner one very narrow.

Description.—Carapace broad, very convex, ornamented with about 13 large round

---

[1] At the bottom of pl. 12, for 6–14, read 6–13; in explanation of pl. 12, for fig. 12, 13, 14, read fig. 11, 12, 13, respectively.

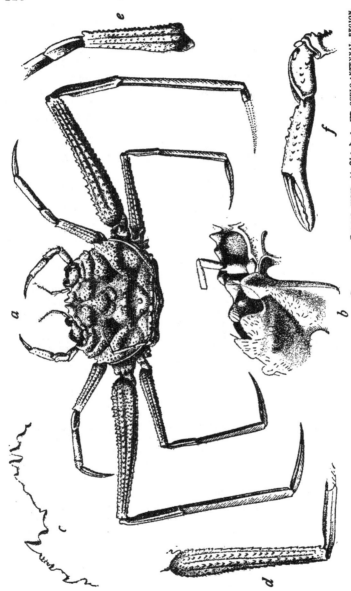

Fig. 130.—Cymopolia cursor, female, after A. Milne Edwards and Bouvier. a, Dorsal view, × 2½; b, left oculo-antennal region, ventral view, × 8; c, left half of frontal and lateral border, × 5; d, merus and carpus of third left leg, upper me, × 5; e, merus and carpus of first left leg, upper me, × 5; f, left cheliped, me me, × 8.

tubercles and several small ones, and some granules near the tubercles; depressions in the main finely and sparingly granulate. Frontal teeth tuberculiform, of subequal width, sinuses U-form and of subequal width, but the median sinus much deeper. Outer angle of preorbital lobe much depressed; next 2 lobes short, broadly rounded, the outer one most advanced; sinuses V-shaped, the middle one narrower than the others. Outer lobes of orbits strongly inclined toward each other, outer margins convex. Antero-lateral teeth three, small, thick, of which one is hepatic and two branchial, the latter near together and the anterior of the pair the larger. Ridge above posterior margin straight, formed of six or more round tubercles, with granules intervening. Lobes of suborbital margin very prominent, visible in dorsal view; outer lobe very large and rounded, inner lobe very narrow, with blunt extremity; sinuses broadly V-shaped,

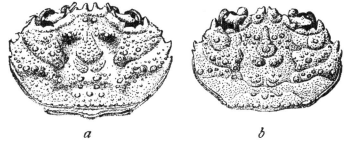

FIG. 131.—CYMOPOLIA CURSOR, AFTER A. MILNE EDWARDS AND BOUVIER. *a*, DORSAL VIEW OF BODY OF FEMALE SHOWN IN FIG. 130, × 3½; *b*, SAME VIEW OF FEMALE (6496), TYPE OF C. DILATATA, × 4.

the inner one rounded at base. Pterygostomian lobe large, rounded, a little constricted at base, covering all of the inner suborbital lobe except the tip, in ventral view.

Chelipeds slender in both sexes.

Of the ambulatory legs the second is very much longer and stronger than the third; the first reaches a little beyond end of merus of second, the third a little beyond the carpus of second, second about three times as long as width of carapace; merus joints long, slender, tapering to distal end, anterior margin concave, upper surface armed with sharp granules, most of which form four longitudinal rows, one on each margin and two above; carpus joints without anterior lobes; propodal joints with margins subparallel; posterior margin of dactylus slightly sinuous in second leg only.

Abdomen of female with a crest on somites 1–5.

Measurements.—Female (7818) greatest length of carapace 8.3, width of same 12.2, length of second leg 35 mm.

Variations.—Individuals differ in the prominence of the tubercles of the carapace, in the bluntness of the lateral teeth, and the distinctness of the longitudinal arrangement of spines on the merus of legs.

Range.—Off Cape Hatteras, North Carolina; from Gulf of Mexico to Florida Straits and Barbados; 107 to 290 fathoms.

Material examined.—See table.

CYMOPOLIA GRACILIS Smith.

Plate 50; plate 51, fig. 1.

Cymopolia gracilis SMITH, Proc. U. S. Nat Mus., vol. 6, 1883, p. 20 (type-locality, off Marthas Vineyard, Massachusetts, 142 fathoms, station 878, *Fish Hawk;* Cat. No. 18453, U.S.N.M.).

Palicus gracilis RATHBUN, Proc. Biol. Soc. Washington, vol. 11, 1897, p. 95.

Diagnosis.—One lateral tooth. Second leg three and one-half times as long as width of carapace. Inner suborbital lobe spiniform, not visible from above.

Description.—Near *C. cursor.* Carapace more convex and antero-lateral margins more oblique. Front narrower posteriorly, the upper margin of orbit, therefore, wider; frontal teeth of outer pair shorter, the sinuses between them and the median teeth very shallow. Upper sinuses of orbit U-shaped, the inner much the widest, the two teeth of subequal size, rounded. Only one antero-lateral tooth, and that on the branchial region, small and blunt; a tubercle on hepatic margin. Ridge above posterior margin more irregular and composed of fewer and smaller tubercles. The

Material examined of ¹*ympolia cursor.*

Locality.	Lat. N.	Long. W.	Fath.	Bottom.	Temp. °F.	Date.	Sta.	Collector.	Specimens.	Cat. No.
South of Cape Hatteras, North Carolina	34 39 15	75 33 30	107	gy. S. P.	...	d. 18, 1885	2601	*Albatross*	2 ♀ (1 ovig.)	18572
Off Cape San Blas, Florida	28 38 30	85 52 30	112	gn. M. brk. Sh.	...	M. 14, 1885	2401	...do	3 ♂ (1 ovig.)	17889
Off Sand Key, Florida			128		...		55	W. Simpson	1 ♂	3009, M. C. Z.
Off Havana, Cuba	23 09 00	82 21 00	242	wh. S. brk. Sh.	...	1877–78	2133	*Blake*	1 ♀ (ovig.)	3058, M. C. Z.
Southeast of Cuba	19 55 00	75 48 03	290	fne. gy. S. Oz.	52	Feb. 27, 1884	146	*Albatross*	...	7818
Off St. Christop et	17 22 36	62 54 12	215	fne. gy. S. blk. Sp.	55	Jan. 14, 1879	192	*Blake*	1 ♂	2611, M. C. Z.
Do	17 17 20	62 46 43	208	fs. S. M.	63	Jan. 30, 1879	148	...do	1 ♂ (ovig.)	6496, M. C. Z.
Off Dominica	15 12 00	61 24 00	138	cs. S.	49½	M. 9, 1879	291	...do	1 ♀	Paris Mus.
Off Barbados	13 12 00	59 41 00	210	...do	53½		291	...do	1 ♀	20, M. C. Z.
Do	13 00 00	59 36 00	209	fn. S.	53½	M. 5, 1879	274	...do	1 ♂	19l, M. C. Z.
Do	13 00 50	59 36 00	209	...do	53½	...do	274	...do	1 ♀	Paris Mus.
Do	13 00 50	59 36 00	209	...do	53½	...do	274	...do	1 ♂	6495, M. C. Z.
Do	12 54 48	59 36 30	288	brk. Sh.	46¾	M. 6, 1879	281	...do	1 ♀	2049, M. C. Z.
Do								...do	1 ♀	4896, M. C. Z.

¹ Type *C. dilatata.* ² Type of *C.* sex.

outer lobe of the suborbital margin is triangular and blunt, more advanced than outer angle of orbit but not conspicuously so; the inner lobe is small, spiniform, acute, separated by a very wide, deep sinus from the outer tooth, and quite invisible except in front view, being hidden by the large pterygostomian lobe, which is nearly as advanced as the basal joint of the antenna; outer sinus of lower orbit triangular.

The first leg reaches to end of merus of second, the third to end of carpus of second, second about three and one-half times as long as width of carapace; merus joints more cylindri-

FIG. 132.—CYMOPOLIA GRACILIS, OUTLINE OF FRONT AND ORBITS OF FEMALE (11411), × 6.

cal than in *cursor* and with posterior margin almost unarmed except at distal end; carpal joints cylindrical, not flattened and angled.

Crest of fifth abdominal segment of female obsolescent; first three segments of male abdomen cristate; fifth segment with side margins very obtusely angled.

Measurements. — Female (9735), length of carapace 10.6, width of same 16, length of second leg 57 mm.

Variations.—In smaller specimens the lateral sinuses of the front are shallower, the longitudinal grooves which run back from the inner orbital sinus and meet the transverse gastrocardiac groove are deeper, the few largest tubercles stand out more prominently than in larger specimens.

Range.—From off Marthas Vineyard, Massachusetts, to off Curaçao, via Gulf of Mexico; 100 to 280 fathoms.

Material examined.—See table on page 219.

CYMOPOLIA FLORIDANA, new species.

Plate 41, figs. 3 and 4.

Type-locality.—Off Sand Key, Florida; 120 fathoms; J. B. Henderson, collector; one female holotype (Cat. No. 50362, U.S.N.M.).

Diagnosis.—Three lateral teeth. Second leg three times as long as width of carapace. Inner suborbital tooth short, hidden by the lobe at the angle of the buccal cavity.

Description.—Near *C. cursor* and *C. gracilis*, but the carapace is more quadrate, the postero-lateral margins being more transverse and making less of an angle with the lateral margins. Tubercles and granules of carapace distinct from one another and acutely pointed; about 20 tubercles are large and are arranged as follows: 3 in a triangle on the mesogastric region; 2 cardiac, side by side; 1 intestinal; 4 in a row above the posterior margin; 5 on each branchial region, of which 4 form a rhomb near the center, and the fifth is near the cardiac region. Four acute teeth on frontal margin; those of middle pair slender, separated by a deep, V-shaped sinus. Outer orbital tooth narrow, pointing nearly straight forward; outer angle of preorbital tooth a right angle; the two superior teeth of the orbit are subtriangular, the outer one the larger; inner sinus U-shaped, other sinuses V-shaped. Three antero-lateral teeth, small, sharp, rather spiniform, the hepatic tooth a little further from the next tooth than the branchial teeth are from each other. Outer suborbital tooth subtriangular, set off by large sinuses; inner angle of orbit produced obliquely inward in a small, acute tooth which is hidden by the large, lobiform, subacute, pterygostomian tooth. First peduncular segment of antenna armed with a slender spine nearly as advanced as the middle spines of the front.

Legs slender. The propodus of the first leg reaches end of merus of second, the merus of second leg reaches a little past the middle of the propodus of the third. Second leg about three times as long as width of carapace. Merus joints of first three legs similar to those of *gracilis*. Last leg less slender than in *gracilis;* a small tooth on the coxa is visible in dorsal view in line with the row of tubercles above posterior margin of carapace (see fig. 3, pl. 41).

Measurements.—Female holotype, length of carapace 5.2, width of same 6.6 mm.

Range.—Known only from the type-specimen.

Affinities.—In *C. cursor* and *C. gracilis*, the teeth and tubercles of the carapace are all blunter and rounder and the inner suborbital tooth is more advanced; in *C. gracilis* there is only one lateral tooth and the legs are longer.

CYMOPOLIA GRACILIPES A. Milne Edwards.

Plate 52, figs. 3 and 4.

Cymopolia gracilipes A. MILNE EDWARDS, Bull. Mus. Comp. Zoöl., vol. 8, 1880, p. 29 (type-localities, lat. 23° 13' N.; long. 89° 16' W., 84 fathoms; Montserrat, 208 fathoms; Grenade, 92 fathoms).

Palicus gracilipes RATHBUN, Proc. Biol. Soc. Washington, vol. 11, 1897, p. 94.—A. MILNE EDWARDS and BOUVIER, Mem. Mus. Comp. Zoöl., vol. 27, 1902, p. 59, pl. 11, figs. 10–14 (type specified, female, lat. 23° 13' N.; long. 89° 16' W., 84 fathoms; Cat. No. 6497, M. C. Z.).

Diagnosis.—Lateral lobes of front not developed. One antero-lateral tooth. Second leg between two and one-half and three times width of carapace. Inner suborbital lobe prominent, bilobed. Two triangular prominences on third segment and one on fourth segment of male abdomen.

Description.—Carapace very wide behind, ornamented with few but prominent tubercles; the most prominent are the two large cardiac tubercles, side by side, separated by a deep furrow, and a cluster of six or eight smaller tubercles on the posterior branchial area; surface minutely granulated. Fronto-orbital margin less prominent and less deeply cut than in any other species; median sinus equilaterally triangular, median teeth tuberculiform, lateral lobes almost obsolete. Preorbital lobe little developed, concealing but a small part of eyestalk. Supra-orbital lobes very shallow, rounded, inner twice as wide as outer, inner and middle sinuses V-shaped, outer sinus a shallow bay. Outer tooth of orbit very short, not reaching middle of cornea. Only one antero-lateral tooth and that on the branchial region just in front of widest part of carapace; tooth thick, obtuse, prominent; behind it a groove runs parallel to postero-lateral margin. Outer lobe of suborbital margin shallow, triangular, blunt, separated from exorbital tooth by a broad, shallow sinus, and by a deeper sinus from the long inner lobe, which has a sinuous outer margin and at the extremity is divided into two subequal lobes, visible from above, the inner lobe more advanced and on a higher level. The rounded pterygostomian lobe does not conceal the extremity nor the outer portion of the inner orbital lobe.

Chelipeds equal and slender in both sexes, nearly smooth; fingers more filiform than usual, crossing far behind the tips; dactylus much curved, exceeding the fixed finger, each armed distally on prehensile edge with a few, small, distant teeth.

The first leg reaches beyond carpus of second, the third beyond propodus of second, the second is between two and one-half and three times width of carapace; legs without protuberances, merus joints strongly narrowed distally and moderately roughened, propodal joints widening very slightly toward distal end, dactyli sinuous on posterior margin.

Abdomen of male bears two compressed, triangular tubercles on third segment, which are visible in dorsal view, and a similar, median tubercle on fourth segment.

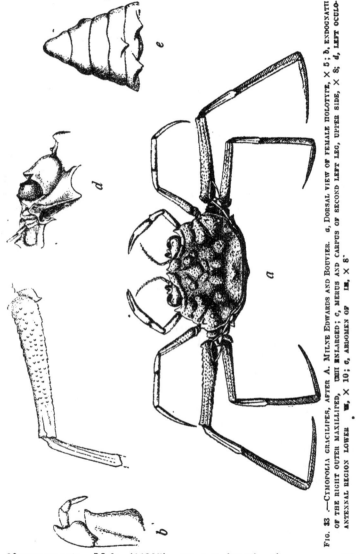

Fig. 33.—CYMOPOLIA GRACILIPES, AFTER A. MILNE EDWARDS AND BOUVIER. a, DORSAL VIEW OF FEMALE HOLOTYPE, × 5; b, ENDOGNATH OF THE RIGHT OUTER MAXILLIPED, UNH ENLARGED; c, MERUS AND CARPUS OF SECOND LEFT LEG, UPPER SIDE, × 8; d, LEFT OCULO-ANTENNAL REGION LOWER ᴍ, × 10; e, ABDOMEN OF ᴍ, × 8.

Measurements.—Male (11395), greatest length of carapace, 4.6; width of same, 7; length of second leg, 19 mm.

Material examined of Cymopolia gracilipes.

Locality.	Lat. N.	Long. W	Fath.	Bottom.	Temp.	Date.	Sta.	Collector.	Specimens.	M.
Northern part of Bk	23 13 08	80 16 00	84		°F. 60	1877–78.	36	Blake	1 ♂ 1 ♀ [1]	53, M. C. Z.
Do	23 13 08	80 16 00	84		60	1877–78.	36	....do...	1 ♀	Paris M.
Bahamas	17 16 40	62 43 48				1880.		Albatross	1 ♀	11395.
Off St. Christopher			61	fne. S. brk. Sh.	76	Jan. 15,1879	149	Blake	1 ♀	88, M. C. Z.
Off Montserrat	16 41 10	62 14 50	298	bk. Sp.	49.5	Jan. 16,1879	154	....do...	1	2632, M. C. Z.
Off Grenada	12 01 45	61 47 25	92	Lava S.	62	Mar. 1,1879	262	....do...	1 ♂ 1 ♀	6498, 609, M. C. Z., Paris M.
Do	12 01 45	61 47 25	92	fne. S.		....do...	262	....do...	1 ♂	

[1] Female is type figured by A. Milne Edwards and Bouvier, 1902.

THE GRAPSOID CRABS OF AMERICA. 223

Range.—Yucatan Bank and the Bahamas to Grenada; 61 to 298 fathoms.
Material examined.—See table.

CYMOPOLIA ACUTIFRONS A. Milne Edwards.

Cymopolia acutifrons A. MILNE EDWARDS, Bull. Mus. Comp. Zoöl., vol. 8, 1880, p. 30 (type-locality, lat. 11° 49′ S.; long. 37° 10′ W., 15 fathoms, *Hassler* Exped.; holotype, Cat. No. 6506, M. C. Z.).

Palicus acutifrons RATHBUN, Proc. Biol. Soc. Washington, vol. 11, 1897, p. 94.— A. MILNE EDWARDS and BOUVIER, Mem. Mus. Comp. Zoöl., vol. 27, 1902, p. 62, pl. 12, figs. 1–5.

Diagnosis.—Carapace covered with numerous fine granules. Middle lobes of front with narrow elongate tips. One antero-lateral tooth. Inner suborbital tooth not exceeding pterygostomian lobe. Chelipeds tuberculous.

Description of immature female.—Carapace everywhere covered with fine contiguous granules among which are some larger, more elevated granules. Like *C. gracilipes*, there is only one antero-lateral tooth, and in the same position but smaller. Median sinus of front deep, triangular, median lobes very narrow or bluntly spiniform at extremity; at their outer base, a tubercle. Supraorbital lobes small, triangular, separated by rounded sinuses. Outer lobe of orbit short, not reaching middle of cornea. Outer suborbital lobe low, broad, triangular, subacute, separated by a wide sinus from the exorbital lobe and by a wider, deeper sinus from the inner lobe, which was probably inconspicuous (specimen damaged). Pterygostomian lobe wide, very prominent and doubtless visible in dorsal view. Eye remarkably large, the corneal part wider than long, the peduncle dorsally flattened.

Chelipeds small, 4 or 5 small tubercles on carpus, a linear row of three or four on upper border of palm; fingers strongly dentate.

Second pair of ambulatory legs missing; merus of first and third pairs slender, cylindrical, spinulous on the anterior portion.

Measurements.—Immature female, greatest length of carapace 6, width of same 9 mm.

Material examined.—Off coast of Brazil, north of Bahia; lat. 11° 49' S.; long. 37° 10' W.; 15 fathoms; January 18, 1872; station 225, U.S.C.S. Str. *Hassler;* 1 immature female, holotype (6506, M. C. Z.).

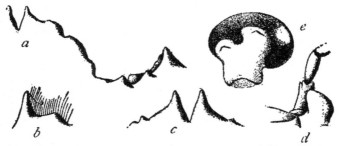

FIG. 134.—CYMOPOLIA ACUTIFRONS, FEMALE HOLOTYPE, AFTER A. MILNE EDWARDS AND BOUVIER. *a,* RIGHT FRONTAL BORDER, × 15 (EDGE BETWEEN MEDIAN TEETH AND ORBIT BROKEN); *b,* RIGHT OUTER ORBITAL ANGLE AND ADJOINING TOOTH OF INFRA-ORBITAL BORDER, × 15; *c,* MEDIAN PART OF FRONTAL BORDER, × 20; *d,* LEFT PTERYGOSTOMIAN LOBE AND CONTIGUOUS ANTENNAL PEDUNCLE, VENTRAL VIEW, × 15; *e,* RIGHT EYE, FROM ABOVE, × 15.

Family GRAPSIDAE Dana.

Grapsoidiens MILNE EDWARDS, Hist. Nat. Crust., vol. 2, 1837, p. 68.
Grapsidae DANA, Amer. Journ. Sci., ser. 2, vol. 12, 1851, p. 287; U. S. Expl. Exped., vol. 13, Crust., pt. 1, 1852, p. 329.—ALCOCK, Journ. Asiat. Soc. Bengal, vol. 69, 1900, pp. 283, 288, and 295, and synonymy.

The palp of the external maxillipeds articulates either at the antero-external angle or at the middle of the anterior border of the merus; exognath either very slender or very broad. Interantennular septum very broad; the division of the orbit into two fossae is accented.

Front of great breadth. Carapace usually quadrilateral, with the lateral borders either straight or slightly arched, and the orbits at or very near the antero-lateral angles. Buccal cavern square; there is generally a gap, often large and rhomboidal, between the external maxillipeds. Male openings sternal.

Littoral, among rocks; pelagic, in drift weed and timber; inhabiting estuaries and marshes, or rivers, rarely on land.

KEY TO THE SUBFAMILIES AND AMERICAN GENERA OF THE FAMILY GRAPSIDAE.

A^1. The antennules fold beneath the front in the ordinary way.
  B^1. No oblique hairy ridge on the exposed surface of the external maxillipeds.
    C^1. The lower border of the orbit runs downward toward the buccal cavern.
          Subfamily *Grapsinae*, p. 226.
      D^1. Front much less than half the greatest breadth of the carapace.
        E^1. Merus of external maxillipeds longer than broad.
          F^1. Fingers with broad, spooned tips_____*Grapsus*, p. 226.
          F^2. Fingers acute, not spooned_____*Geograpsus*, p. 231.
        E^2. Merus of external maxillipeds not longer than broad.
          Leptograpsus, p. 234.
      D^2. Front more than half, or about half, the greatest breadth of the carapace.
        E^1. Posterior surface of orbit concave.
          F^1. Antennae excluded from the orbit_____*Goniopsis*, p. 236.
          F^2. Antennae entering the orbit.
            G^1. Carapace depressed, distinctly striated___*Pachygrapsus*, p. 240.
            G^2. Carapace convex, almost smooth_____*Planes*, p. 253.
        E^2. Posterior surface of orbit bulging outward_____*Grapsodius*, p. 259.
    C^2. The lower border of the orbit does not run downward toward the buccal cavern, but is supplemented by a rather distant suborbital crest, which is in line with the anterior border of the epistome.
          Subfamily *Varuninae*, p. 260.
      D^1. First segment of male abdomen not entirely covering the sternum between the legs of the last pair.
        E^1. Antero-lateral margins arcuate. Merus of external maxillipeds as broad as long.
          F^1. Surface very uneven. Front strongly advanced. Suborbital crest not forming a stridulating ridge____*Cyrtograpsus*, p. 260.
          F^2. Surface little uneven. Front moderately advanced. Suborbital crest a stridulating ridge which scrapes against a short and usually horny ridge on the distal end of the arm.
            Hemigrapsus, p. 264.
        E^2. Lateral margins straight and parallel. Merus of external maxillipeds longer than broad_____*Tetragrapsus*, p. 273.
      D^2. First segment of male abdomen covering the sternum between the legs of the last pair.
        E^1. Propodus of large cheliped normal.
          F^1. Carapace subrotund_____*Glyptograpsus*, p. 275.
          F^2. Carapace squarish_____*Euchirograpsus*, p. 281.
        E^2. Propodus of large cheliped prolonged proximally far beyond its articulation with the carpus_____*Platychirograpsus*, p. 278.
  B^2. An oblique hairy ridge on the exposed surface of the external maxillipeds_____Subfamily *Sesarminae*, p. 283.
    C^1. Antennae lodged in the orbital hiatus.
      D^1. Epistome not projecting beyond the edge of the front.
        E^1. Carapace quadrate or subquadrate.
          F^1. Carapace convex_____*Sesarma*, p. 284.
          F^2. Carapace flat_____*Metopaulias*, p. 317.
        E^2. Anterior half of carapace with arcuate margin, posterior half rectangular.

F¹. Antero-lateral margins dentate.   Carapace convex.
                                            Sarmatium, p. 321.
F². Antero-lateral margins entire.   Carapace posteriorly flat.
                                            Cyclograpsus, p. 325.
 D². Epistome projecting beyond the edge of the front.
                                            Chasmagnathus, p. 329.
C². Antennae excluded from the orbit by the tooth at the lower inner angle of the orbit meeting or nearly meeting the front.
 D¹. Dactyli of legs of ordinary length. Abdomen of male subtriangular_____*Metasesarma*, p. 319.
 D². Dactyli of legs very short. Abdomen of male subcircular.
                                            Aratus, p. 322.
A². The antennules are visible in dorsal view in deep clefts in the front of the carapace_____Subfamily *Plagusiinae*, p. 331.
 B¹. Carapace broader than long_____*Plagusia*, p. 331.
 B². Carapace longer than broad_____ *Percnon*, p. 337.

Subfamily GRAPSINAE Dana (part).

Grapsacea MILNE EDWARDS, Ann. Sci. Nat., ser. 3, Zool., vol. 20, 1853, p. 163 [129] (part).

Grapsinae DANA, Amer. Journ. Sci., ser. 2, vol. 12, 1851, p. 287; U. S. Expl. Exped., vol. 13, Crust., pt. 1, 1852, p. 331 (part).—ALCOCK, Journ. Asiat. Soc. Bengal, vol. 69, 1900, pp. 288 and 295.

Front usually strongly deflexed. The lower border of the orbit runs downward toward the buccal cavern. Antennal flagellum very short. The external maxillipeds leave usually a wide rhomboidal gap between them, and are not traversed by an oblique hairy crest; the palp articulates at or near the antero-external angle of the merus, and the exognath is narrow and exposed throughout. The male abdomen fills all the space between the last pair of legs.

Genus GRAPSUS Lamarck.

Grapsus LAMARCK, Sys. Anim. sans Vert., 1801, p. 150; type, *G. pictus* Latreille=*G. grapsus* (Linnaeus).

Orthograpsus (part) KINGSLEY, Proc. Acad. Nat. Sci. Philadelphia, 1880, p. 194.

Goniopsis (part) DE HAAN, Fauna Japon., Crust., 1833, p. 5; 1835, p. 33.

Carapace little broader than long, much depressed, regions fairly well defined, branchial groove very clear, branchial regions with regular obliquely transverse ridges, gastric region with transverse squamiform sculpture. Lateral borders arched and armed with a tooth just behind the acute outer orbital angle.

Front about one-half the breadth of the anterior border of the carapace, strongly deflexed; along the line of flexion are 4 tubercles, the outer of which form the supraorbital angles.

Orbits of moderate size, deep, distinctly divided into two fossae; lower border notched near outer angle; the wide inner orbital hiatus is filled partly by the antennal peduncle and partly by a strong isolated tooth in the inner fossa.

Antennules nearly transverse in narrow fossae; interantennular septum very broad. Antennal flagellum short, lying in the orbital hiatus; excretory tubercle of basal joint very prominent.

Epistome deep, from front to back, well defined, its wings run up toward the orbital hiatus. Buccal cavity square with the anterolateral corners rounded off. External maxillipeds widely separated by a rhomboidal gap in which the mandibles are exposed; ischium and merus narrow, the latter slightly the shorter; and the palp, which is coarse, especially as to its carpus, articulates near the anteroexternal angle of the merus.

Chelipeds subequal; much shorter than the legs, but in the male much stouter; hands and fingers short and stout, tips of fingers broad and hollowed in a spoon.

Legs broad and compressed, especially the merus; dorsal surface of some of the joints has a striated or squamiform sculpture; dactvli thorny.

Abdomen with seven segments in both sexes; in the male its base is as broad as the sternum between the last pair of legs.

Found on rocks and reefs of all tropical and subtropical seas.

GRAPSUS GRAPSUS (Linnaeus).

ROCK CRAB; SALLY LIGHTFOOT.

Plates 53 and 54.

Pagurus maculatus CATESBY, Nat. Hist. Carolina, Florida, and the Bahama Islands, vol. 2, 1743, p. 36, pl. 36, fig. 1, "inhabit the rocks overhanging the sea."

Cancer grapsus LINNAEUS, Sys. Nat., ed. 10, vol. 1, 1758, p. 630 (type-localities, America and Ascension Island; types not extant).

Grapsus pictus LATREILLE, Hist. Nat. Crust., vol. 6, an XI [1802–1803], p. 69 (type-locality, *les îles de l'Amérique méridionale*; type in Paris Mus.).

Grapsus webbi MILNE EDWARDS, Ann. Sci. Nat., ser. 3, Zool., vol. 20, 1853, p. 16.

Grapsus (Goniopsis) pictus DE HAAN, Fauna Japon., Crust., 1835, p. 33.

Grapsus maculatus MILNE EDWARDS, Ann. Sci. Nat., ser. 3, Zool., vol. 20, 1853, p. 167 [133], pl. 6, figs. 1—1 *n* (type-locality, Antilles; type in Paris Mus.).

Grapsus webbi MILNE EDWARDS, Ann. Sci. Nat., ser. 3, Zoöl., vol. 20, 1853, p. 167 [133] (type-locality, Canary Islands; type in Paris Mus.).

Grapsus ornatus MILNE EDWARDS, Ann. Sci. Nat., ser. 3, Zool., vol. 20, 1853, p. 168 [134] (type-locality, Chili; type in Paris Mus.).

Grapsus altifrons STIMPSON, Ann. Lyc. Nat. Hist. New York, vol. 7, 1860, p. 230 (type-locality, Cape St. Lucas; cotypes in Mus. Comp. Zoöl. and Cat. No. 2022, U.S.N.M.).

Grapsus grapsus IVES, Proc. Acad. Nat. Sci. Philadelphia, 1891, p. 190.

Diagnosis.—Front vertical. Tooth on wrist with short, spiniform tip. Fingers spoon-shaped.

Description.—Carapace discoidal; transverse and oblique ridges prominent, intervening spaces reticulate. Front deep (depth variable), almost vertical, overhanging the epistome, partly concealing the antennules, its free edge crenate.

Length of epistome about one-third its greatest breadth. Tooth at inner angle of orbit subovate.

Chelipeds in male about as long as carapace, shorter in the female; inner border of ischium and arm strongly spinate; distal half of outer border of arm less deeply spined; wrist with scattered tubercles on its upper surface and with its inner angle produced in an ovate, falcate tooth with a short spiniform tip; palm nearly as high as long, its outer surface sculptured, its upper border culminating in a tooth; fingers with very broad rounded tips; length of dactylus in male nearly twice the length of the upper border of the palm.

First pair of legs shortest, fourth pair next, third pair longest, about twice as long as carapace. Only in the last pair of legs does the breadth of the merus approach half its length; the distal end of the upper border of each merus is a spine, while the same end of the lower border in the first three pairs is armed with two or three spines.

Fig. 135.—Grapsus grapsus, outer maxilliped, enlarged. (After Milne Edwards.)

Measurements.—Male (16032), length of carapace 77, width of same 87 mm.

Color.—Usually variegated with deep red and light greenish (in alcohol). Chelae an even, brilliant red. Sometimes the carapace and legs are entirely red without mottlings.

Habits.—According to Catesby, " these crabs inhabit the rocks overhanging the sea; they are the nimblest of all other crabs; they run with surprising agility along the upright side of a rock and even under rocks that hang horizontally over the sea; this they are often necessitated to do for escaping the assaults of rapacious birds which pursue them. These crabs, so far as I could observe, never go to land, but frequent mostly those parts of the promontories and islands of rocks in and near the sea, where by the continual and violent agitation of the waves against the rocks they are always wet, continually receiving the spray of the sea, which often washes them into it, but they instantly return to the rock again, not being able to live under water and yet requiring more of that element than any of the crustaceous kinds that are not fish."

Range.—Tropical and subtropical shores of America as well as of the eastern Atlantic. South Florida and Bahamas to Pernambuco, Brazil; Bermudas; Lower California (San Benito Island) to Chile.

A subspecies, *Grapsus grapsus tenuicrustatus*, inhabits the Indo-Pacific region.

Material examined.—

South Florida; 1879; S. Stearns; 1 male (3461).

Abaco, Bahamas; 1886; *Albatross;* 1 male (16314).

East side of Andros Island, Bahamas; in cave halfway between Smith's place and lighthouse south of South Bight; May 14, 1912 · Paul Bartsch; 2 females (45545).

Hog Island, opposite Nassau, Bahamas; June 19, 1903; B. A. Bean; 1 male, 2 females (31045).

San Salvador, Bahamas; 1886; *Albatross;* 4 males, 6 females (11359).

Rum Cay, Bahamas; 1886; *Albatross;* 3 males (11365).

Cabañas, Cuba; on coral reef, above water; May 31, 1900; William Palmer and J. H. Riley; 1 female (23806).

Cuba; 1914; Henderson and Bartsch, *Tomas Barrera* Exped. ·

On reef flat between Cayo Hutia and Little Cayo, northeast of light; May 20; 1 young male (48575).

Ensenada de Cajon, off Cape San Antonio; May 22; 2 females ovig. (48577).

Cape Cajon; 1 male (48601).

Jamaica; 1884; *Albatross;* 1 male (7334).

Montego Bay, Jamaica; rocky ledges along shore; called "Sally Lightfoot"; C. B. Wilson; 1 male (42880).

Jacmel, Haiti; C. F. Baker; 1 female (22559)

Porto Rico; *Fish Hawk:*

San Juan; Jan. 4, 1899; 1 female, shedding (24054). Jan. 11 and 12, 1899; 1 male, 1 female (24061).

Aguadilla; Jan. 18, 1899; 11 y. (24053).

Boqueron Bay; Jan. 28, 1899; 1 female (24052).

Guanica Bay; Jan. 28, 1899; 1 male, 1 female (24056).

Reefs at Ponce; Jan. 30 and 31, 1899; 11 males, 7 females (24060).

Playa de Ponce Reef; Feb. 1, 1899; 2 males (24059).

Playa de Ponce Lighthouse; Jan. 31, 1899; 3 males, 2 females (24062).

Arroyo; Feb. 4, 1899; 1 male, 2 females (24057).

Hucares; Feb. 14, 1899; 1 male (24055).

Caballo Blanco Reef, Vieques; Feb. 7, 1899; 4 males (24058).

St. Thomas, W. I.: Shore; Jan. 17–24, 1884; *Albatross;* 5 males, 4 females (7467); 1 y. female (18560). Shore near town; June 28, 1915; C. R. Shoemaker; 1 male (49808). Lagoons; July 7, 1915; C. R. Shoemaker; 1 male (49809).

Port Castries, St. Lucia; 1887; *Albatross;* 1 male (22090).

Swan Islands, Caribbean Sea; Feb., 1887; C. H. Townsend; 2 males, 3 females (14555).

Old Providence; Apr. 4–9, 1884; Albatross; 2 males, 1 female (7543)

Sabanilla, Colombia; Mar. 16–22, 1884; Albatross; 4 males, 1 female (7564).

Curaçao; Feb. 10–18, 1884; Albatross; 1 female (7840).

La Guayra, Venezuela; abundant on stones of breakwater; Lieut. Wirt Robinson, U. S. Army; July 26, 1895; 1 male (18816).

Trinidad; shore; Jan. 30 to Feb. 2, 1884; Albatross; 1 y. male (7635).

Fernando Noronha, Brazil; 1876–1877; R. Rathbun, Hartt Explorations; 3 males, 2 females, 4 y. (40587).

Hungry Bay, Bermudas; July–Sept.; F. G. Gosling; 1 male (25436).

Cape Verde Islands; U. S. Exploring Exped.; 1 female (2337).

Porto Grande, St. Vincent, Cape Verde Islands; Nov. 11, 1889; W. H. Brown, U. S. Eclipse Exped. to Africa; 1 y. (14866).

Azores; William Trelease; 3 males, 3 females (18625).

Horta, Fayal, Azores; Nov. 2, 1889; W. H. Brown, U. S. Eclipse Exped. to Africa; 1 female (14867).

Pim Bay, Fayal, Azores; beach at low tide in pools and under stones; Lewis Dexter; 1 female (17600).

Ascension Island; Mar. 20, 1890; W. H. Brown, U. S. Eclipse Exped. to Africa; 4 males, 4 females (14868).

San Benito Island, Lower California, Mexico; Mar. 28, 1897; A. W. Anthony; 1 male (20686).

South end Cedros Island, Lower California; Apr.. 1912; H. N. Lowe; 3 specimens (43851).

San Roque Island, Lower California; Apr. 17, 1897; A. W. Anthony; 1 male (20685).

Asuncion Island, Lower California; Apr. 16, 1897; A. W. Anthony; 2 males, 3 females (20687).

Magdalena Bay, Lower California; Apr. 8, 1889: Albatross; 2 males (15524).

Margarita Island, Lower California; May 2, 1888; Albatross; 1 male (22089).

Cape St. Lucas, Lower California; J. Xantus; 7 males, 2 females, 9 y., cotypes of *G. altifrons* Stimpson (2022); 1 male (5319).

La Paz, Lower California; L. Belding; 5 males, 1 female (5318).

La Paz Harbor, Lower California; Mar. 12, 1889; Albatross; 1 male (15525).

Socorro Island, W. of Mexico; Mar. 8, 1889; Albatross; 1 male, 1 female (15526).

Clarion Island, W. of Mexico; Mar. 4, 1889; *Albatross;* 2 males (16032).

Taboga Island, Panama: May 12–15, 1911; Meek and Hildebrand; 1 female, paper-shell (43990). June, 1914; J. Zetek; 1 male (48777).

Galapagos Islands:
Chatham Island; Aug. 16 and 17, 1884; Dr. W. H. Jones, U. S. Navy; 6 males, 1 female (14365). Apr. 4, 1888; *Albatross;* 8 males, 3 females (22083); 2 males, 3 females (20626). Jan. 4, 1905; *Albatross;* 1 male, 1 female (33171).

Hood Island; Apr. 7, 1888; *Albatross;* 5 males, 4 females (22086).

Albemarle Island; Apr. 10, 1888; *Albatross;* 3 males, 1 female, 1 y. (22088).

James Island; Apr. 11, 1888; *Albatross;* 6 males, 1 y. (22084).

Indefatigable Island; Apr. 12, 1888; *Albatross;* 2 males, 2 females (22085).

Duncan Island; Apr. 13, 1888; *Albatross;* 3 males, 3 females (22087).

Callao, Peru (?); U. S. Expl. Exped.; 1 female (2344).

N. end of Callao water front; from rocks in and above the surf; Dec. 27, 1906; R. E. Coker; received from Peruvian Government; 1 female, 2 y. (40436).

San Lorenzo Island, Peru; Jan., 1884; Dr. W. H. Jones, U. S. Navy; 1 male (13865).

Chincha Islands, Peru; July 13; R. E. Coker; received from Peruvian Government; 1 male (40437).

Mollendo, Peru; July 25, 1908; R. E. Coker; received from Peruvian Government; 1 female (40438).

Grapsus strigosus (Herbst) is recorded from Loretto, North America, by White (List Crust. Brit. Mus., 1847, p. 40), under the name, *Goniopsis strigosus*. This is probably *Grapsus grapsus*.

Genus GEOGRAPSUS Stimpson.

Geograpsus STIMPSON, Proc. Acad. Nat. Sci. Philadelphia, vol. 10, 1858, p. 101; type, *G. lividus* (Milne Edwards).

Orthograpsus KINGSLEY, Proc. Acad. Nat. Sci. Philadelphia, 1880. p. 194, part: *O. hillii* Kingsley=*G. lividus* (Milne Edwards).

Differs from *Grapsus* as follows:

Carapace more quadrate, the sides being very little arched, and also broader. Lobe at inner lower angle of orbit not so completely isolated. Antennal peduncle less massive. Epistome shorter fore and aft, and less well defined.

Chelipeds much more massive than the legs; fingers pointed. Thorns of the dactyli of the legs less crowded and less coarse than in

Grapsus. Between the coxae of the second and third pairs of legs is a narrow fossa fringed with hair leading to the branchial cavity. Inhabits tropical America as well as the Oriental region.

GEOGRAPSUS LIVIDUS (Milne Edwards).

Plate 55.

Grapsus lividus MILNE EDWARDS, Hist. Nat. Crust., vol. 2, 1837, p. 85 (type-locality, Antilles; type in Paris Mus.).—DANA, U. S. Expl. Exped., vol. 13, Crust., pt. 1, 1852, p. 340; atlas, 1855, pl. 21, figs. 5 *a–c.*

Grapsus brevipes MILNE EDWARDS, Ann. Sci. Nat., ser. 3, Zool., vol. 20, 1853, p. 170 [136], (*Patrie inconnue;* type in Paris Mus.).

Geograpsus lividus STIMPSON, Ann. Lyc. Nat. Hist. New York, vol. 7, 1860, p. 230.

Geograpsus occidentalis STIMPSON, Ann. Lyc. Nat. Hist. New York, vol. 7, 1860, p. 230 (type-locality, Cape St. Lucas; cotypes in Mus. Comp. Zoöl.).

Orthograpsus hillii KINGSLEY, Proc. Acad. Nat. Sci. Philadelphia, 1880, p. 194 (type-localities, West Indies; type in Mus. Phila. Acad. Sci., and Key West, Fla.).

Diagnosis.—Lateral margin well defined throughout its extent. Front little deflexed. Fingers pointed.

Description.—Carapace subquadrilateral, widening behind, a little convex, lateral borders well defined, and posteriorly continued on the dorsal surface nearly to its middle by a sinuous line near the posterior border. Transverse markings fine, obsolescent on the gastric region, absent on the cardiac region.

The four tubercles along the upper border of the front are rather prominent; edge of front in dorsal view nearly straight or slightly concave at the middle. Notch near outer end of lower border of orbit deep.

Chelipeds in both sexes a little unequal; about one and one-half times length of carapace, covered with transverse, more or less squamiform striae; upper surface of last three joints tuberculous; inner margin of arm expanded, proximally denticulate, distally with larger teeth; an acute tooth or spine at inner angle of wrist.

The greatest breadth of the merus joints of the legs is more than half their length. First pair of legs slightly shorter than the fourth, second pair longest, two and one-third times as long as carapace. Last three joints of all the legs conspicuous with long slender bristles.

Color.—Yellowish red with reticulating lines or patches of a darker red or purplish; sometimes wholly red.

Measurements.—Male (24041), length of carapace 24.8, width of same 30.2 mm.

Habits.—Lives among loose stones and rocks along shore above the water's edge. Probably nocturnal.

Range.—Florida Keys to São Paulo, Brazil; Bermudas; Cape Verde Islands, 10-30 meters (E. & B.); Lower California to Chile; Hawaiian Islands.

Material examined.—
Indian Key, Florida; 1885; H. Hemphill; 1 male (17601).

Ensenada de Cajon, off Cape San Antonio, Cuba; May 22-23, 1914; Henderson and Bartsch, *Tomas Barrera* Exped.; 1 male (48576).

Mariel, Cuba; 1900: Wm. Palmer and J. H. Riley; May 10; among stones above tide; 1 male (23807). June 10; 1 y. (23808).

Kingston Harbor, Jamaica: May to July, 1896; F. S. Conant: 1 male (19605). Dr. T. H. Morgan; 3 males, 1 female (17223).

Porto Rico; 1899; *Fish Hawk:*
Puerto Real; Jan. 26; 2 males (24042).
Reefs at Ponce; Jan. 30; 7 males, 3 females (24041).
Hucares; Feb. 13; 2 females (24044).
Ensenada Honda, Culebra; Feb. 9-10; 5 males, 2 females (24043).

Vieques Island, Porto Rico; Mar. 28, 1900; L. Stejneger; 1 male y. (23679).

St. Thomas, West Indies: Jan. 17-24, 1884; *Albatross;* 3 females (18561). Shore near town; June 28, 1915; C. R. Shoemaker; 1 male (49814). East shore of harbor; July 6, 1915; C. R. Shoemaker; 1 male, 1 female ovig. (49807). A. H. Riise; 1 female (2465)

Port Castries, St. Lucia, West Indies; Dec. 2, 1887; *Albatross;* 1 male (22092)

Monos Island, Trinidad; shore; Jan. 30-Feb. 2, 1884; *Albatross;* 1 male (7642).

Rifwater, Curaçao; one-half fathom; Mar. 6, 1905; J. Boeke; 1 male (42971).

Sabanilla, Colombia; 1884; *Albatross;* 9 males, 18 females (7344)

Old Providence, West Indies; Apr. 4-9, 1884; *Albatross;* 3 females (7549).

Iguape, São Paulo, Brazil; 1901: R. Krone; 1 male (47857).

La Paz, Lower California; L. Belding; 3 females (4623).

Pichilinque Bay, Lower California; Mar. 27, 1911; *Albatross;* 1 female ovig. (44581).

James Island, Galapagos Islands; Apr. 11, 1888; *Albatross;* 1 male, 2 females (22093).

San Lorenzo Island, off Callao, Peru: U. S. Expl. Exped.; 1 male (2338). Jan., 1884; Dr. W. H. Jones, U. S. N.; 1 female (15108).

Clipperton Island; lagoon; Nov. 23; Stanford University; 1 male, 1 female (25662).

Locality ?; U. S. Exploring Exped.; 1 male, 1 female (2381).

Genus LEPTOGRAPSUS Milne Edwards.

Leptograpsus (part), MILNE EDWARDS, Ann. Sci. Nat., ser. 3, Zool., vol. 20, 1853, p. 171 [137]; type, *L. variegatus* (Fabricius).

Differs from *Grapsus* as follows:

Region not strongly defined. Lateral borders armed with two teeth behind the orbital angle. Front moderately deflexed, its superior tubercles not prominent.

The tooth projecting from the lower wall of the inner fossa of the orbit is small, and the hiatus is largely filled by the greatly expanded first movable segment of the antenna. Epistome short fore and aft, laterally not nearly reaching the orbit. Merus of outer maxillipeds shorter and broader than the ischium, and as broad as it is long.

Chelipeds in the male much stouter than the legs. Tips of fingers hollowed out in shallow spoons.

Contains only one species.

LEPTOGRAPSUS VARIEGATUS (Fabricius).

Plate 56.

Cancer variegatus FABRICIUS, Entom. Syst., vol. 2, 1793, p. 450 (type-locality, *in Americae meridionalis Insulis;* type not extant).
Grapsus variegatus LATREILLE, Hist. Nat. Crust., vol. 6, an XI [1802–1803], p. 71.
Grapsus personatus LAMARCK, Hist. Nat. Anim. sans Vert., vol. 5, 1818, p. 249 (type-locality, New Holland; type in Paris Mus.).
Grapsus strigilatus WHITE, in Gray, Zool. Miscellany, June, 1842, p. 78 (type-locality, New Zealand).
Grapsus planifrons DANA, Proc. Acad. Nat. Sci. Philadelphia, vol. 5, 1851, p. 249 (type-locality, *ad oras juxta urbem* 'Valparaiso'; Cat. No. 2343, U.S.N.M.); U. S. Expl. Exped., vol. 13, Crust., pt. 1, 1852, p. 338; atlas, 1855, pl. 21, figs. 3a–3e.
Leptograpsus variegatus MILNE EDWARDS, Ann. Sci. Nat., ser. 3, Zool., vol. 20, 1853, p. 171 [137].
Leptograpsus verreauxi MILNE EDWARDS, Ann. Sci. Nat., ser. 3, Zool., vol. 20, 1853, p. 171 [137], (type-locality, *Australie;* type in Paris Mus.).
Leptograpsus ansoni MILNE EDWARDS, Ann. Sci. Nat., ser. 3, Zool., vol. 20, 1853, p. 171 [137], (type-locality, *Ile de Juan-Fernandez;* type in Paris Mus.).—DE MAN, Notes Leyden Mus., vol. 12, 1890, p. 84.
Leptograpsus gayi MILNE EDWARDS, Ann. Sci. Nat., ser. 3, Zool., vol. 30, 1853, p. 171 [137], (type-locality, Chili; type in Paris Mus.).

Diagnosis.—Subcircular. Two side teeth. Fingers spoon-shaped. Merus of maxillipeds as broad as long.

Description.—Carapace subcircular. Surface of front and anterior gastric region tuberculate. Surface between transverse and oblique ridges very finely reticulate. Edge of front transverse, crenate. The three teeth of the sides diminish in size from the orbital tooth backward.

Chelipeds in male one and one-half times as long as carapace, shorter in the female. Inner border of arm laminate, and dentate at the distal end. A tooth at extremity of outer border. Arm and wrist crossed by transverse striae; wrist tuberculate above, as are also the palm and the proximal half of the dactylus. Tooth at inner angle of wrist short, obtuse, and very deep. Outer face of hand almost smooth except for a longitudinal ridge running from the tip of the pollex nearly to the wrist; lower edge obscurely tuberculate; inner face finely striate, and with a few tubercles. Fingers widely gaping at base in the male, dactylus with a large, low, basal tooth; pollex with a similar large tooth just distal to that on the dactyl. Length of dactylus in male a little more than one and one-half times the length of upper border of palm.

First pair of legs shortest, third pair longest, about twice as long as carapace. Merus of last pair twice as long as broad, merus of other pairs longer. A spine at distal end of upper border of each merus, and two or three small spines or teeth at distal end of lower border of all save in the last pair.

Variation.—A variable species. Individuals differ in width of carapace and legs, in the curvature of the side margins and the size of their teeth, in the prominence of the frontal lobules and of the posterior cardiac region, and of the sculptural lines.

Color.—Variable. Red and yellow mixed, or dotted with violet-red, or sometimes whitish. " Bluish-gray, everywhere transversely lineated and blotched with black; feet often reddish " (Stimpson).

Measurements.—Male (2129), length of carapace 54.8, width of same 61.8 mm.

Range.—From Peru to Chile; Juan Fernandez. Pernambuco (Kingsley.) Also Easter Island; Australia; New Zealand; Norfolk Island; Shanghai (Heller).

Material examined.—

Callao or San Lorenzo, Peru; U. S. Expl. Exped.; 1 male (2328).
Chincha Islands, Peru; specimens in Copenhagen Mus.
Cobija, Chile; specimens in Copenhagen Mus.
Antofagasta, Chile; Nov., 1914; Dr. J. N. Rose; 1 male, 1 female ovig. (49057); 2 specimens (49061).
Valparaiso, Chile; U. S. Exploring Exped.; 1 male, 2 females, cotypes of *Grapsus planifrons* Dana (2343).
Chile: Gilliss, collector; 1 male (2414). C. E. Porter; 1 female (44520).
Chile or Peru; U. S. Exploring Exped.; 4 males, 1 female (2129).
Easter Island; *Albatross:* Dec. 16, 1904; 1 female (33196). Dec. 20, 1904; 2 males, 2 females (33195). La Perouse Bay; Dec. 17, 1904; 4 males, 3 females (33197).
Port Jackson, Australia; H. W. Parritt; 2 females (32439).

Genus GONIOPSIS de Haan.

Goniopsis DE HAAN, Fauna Japon., Crust., 1833, p. 5; 1835, p. 33; type, *G. cruentata* (Latreille).
Goniograpsus (part) DANA, Amer. Journ. Sci., ser. 2, vol. 12, 1851, p. 287; Proc. Acad. Nat. Sci. Philadelphia, vol. 5, 1851 (1852), pp. 247 and 249.

Carapace quadrate, much broader than long, convex, groove defining the branchial region deep. Outer two-thirds of dorsal surface traversed by sharp oblique ridges; anterior third crossed by transverse broken striae. Antero-lateral tooth acute; a similar tooth just behind it on the lateral border.

Front about half the width of the carapace, vertical, the superior lobes truncate and prominent.

Orbits of good width, at the corners of the carapace; lower border with two notches at outer end; the orbital hiatus is filled by a lobe belonging to the inner of the orbital fossae, and excluding the antenna from the orbit. First movable joint of antenna provided with a broad lateral expansion. Antennules folded transversely.

Epistome well defined, small, deeply concave. Buccal cavity square with the anterior corners rounded. The outer maxillipeds are narrow and separated by a very broad rhomboidal gap in which the mandibles are exposed. Merus and ischium subequal in length. The large palp articulates at the outer angle of the merus.

Chelipeds unequal, much more massive than the legs and about as long as the third pair; fingers slightly hollowed at tip.

Legs broad and compressed, especially the merus, which, like that of the cheliped, bears transverse markings; last three joints bristly, dactyli spinous.

As in *Geograpsus* there is, between the coxae of the second and third pairs of legs, a narrow fossa fringed with hair, which leads to the branchial cavity.

The abdomen in both sexes is composed of seven somites, and in the male covers the sternum between the last pair of legs.

Contains only two species, which are analogous species on opposite sides of the continent: *cruentata* (Atlantic); *pulchra* (Pacific).

KEY TO THE SPECIES OF THE GENUS GONIOPSIS.

A¹. Color yellow or red_____*cruentata*, p. 237.
A². Color purplish or brown. Carapace a little wider than in *cruentata*. Appendages of male abdomen straighter, tip more transverse.
pulchra, p. 239.

GONIOPSIS CRUENTATA (Latreille).

MANGROVE CRAB; TREE CRAB.

Plate 57.

Cancer ruricola DE GEER, Mém. pour servir à l'Hist. des Insectes, vol. 7, 1778, p. 417, pl. 25 (not *C. ruricola* Linnaeus).

Grapsus cruentatus LATREILLE, Hist. Nat. Crust., vol. 6, 1803, p. 70 (type-locality, *les îles de l'Amérique méridionale;* type in Paris Mus.).

Gecarcinus ruricola MACLEAY, Trans. Zool. Soc. London, vol. 1, 1835, p. 184, not *G. ruricola* (Linnaeus).

Grapsus (Goniopsis) cruentatus DE HAAN, Fauna Japon., Crust., 1835, p. 33.

Grapsus longipes RANDALL, Journ. Acad. Nat. Sci. Philadelphia, vol. 8, 1839 (1840), p. 125 (type-locality, Surinam; type in Mus. Phila. Acad. Sci.).

Goniopsis ruricola WHITE, List Crust. Brit. Mus., 1847, p. 40.—SAUSSURE, Mém. Soc. Phys. Hist. Nat. Genève, vol. 14, 1858, p. 30, pl. 2, figs. 18, 18a.

Grapsus pelii HERKLOTS, Addit. Faunam Afr. Occ., 1851, p. 8, pl. 1, figs. 6 and 7 (type-locality, *prope Boutry;* type in Leyden Mus.).

Goniograpsus cruentatus DANA, U. S. Expl. Exped., vol. 13, Crust., pt. 1, 1852, p. 342; atlas, 1855, pl. 21, fig. 7.

Goniopsis cruentata RATHBUN, Bull. U. S. Fish Comm., vol. 20, for 1900, pt. 2 (1901), p. 15, pl. 1 (colored).

Diagnosis.—Carapace quadrate, with one side tooth. Front vertical. Antenna excluded from orbit. Fossa fringed with hair between coxae of second and third legs. Color yellow or red.

Description.—Carapace widening a little behind, sides a little curved, branchial region swollen above the level of the cardiac and intestinal regions. Surface between ridges very finely wrinkled.

Median pair of frontal lobes a little wider than outer pair. Surface of front tuberculate, lower edge nearly straight, thin, projecting, crenulate.

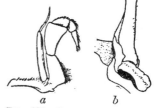

FIG. 136.—GONIOPSIS CRUENTATA. *a*, OUTER MAXILLIPED, ENLARGED, (AFTER MILNE EDWARDS); *b*, LEFT FIRST APPENDAGE OF ABDOMEN OF MALE (7677), LOWER SIDE, × 2.

Merus of chelipeds prominently ridged; inner margin expanded in a broad lamina with edge coarsely dentate or spinate; outer margin with smaller teeth. Carpus tuberculate, inner angle armed with a few denticles. Outer surface of chelae for the most part smooth and flat, with rows of spiniform tubercles above and below; inner face of palm sparingly tuberculate. Tips of fingers hollowed out in shallow, inconspicuous spoons.

The second or longest pair of legs is about twice as long as the carapace. The merus joints of all the legs are about twice as long as wide, and their anterior margins end in a sharp tooth, while the posterior margins of all but the last pair are feebly dentate at the distal end. Thorns of dactyli very strong.

Color.—A very showy crab. Carapace brownish-yellow or brick-red; legs red, with spots of a darker red, extremities yellow; chelipeds red, except the palms, which are almost white, and the fingers, which are yellow.

Measurements.—Male (7542), length of carapace 48.2, width of same 56, width of front 29 mm.

Habits.—Very abundant and very active in mangrove swamps on the roots and along the trunks of the trees and on wet, muddy shores of inlets of the sea.

Range.—Bahamas and Gulf of Mexico to Province of São Paulo Brazil; Bermudas; West Africa.

Material examined.—

Pensacola, Florida; in fish stomach; Silas Stearns; 1 male (13849).

Nassau, Bahamas: 1886; *Albatross;* 2 males, 1 female (16313). Dec. 25, 1898; *Fish Hawk;* 1 male, 1 y. (24064).

Spanish Wells, Bahamas; 1893; Biol. Exped. State Univ. Iowa; 1 female (Mus. S.U.I.).

Tampico, Mexico; "live in the soft mud of the river banks"; June 1, 1910; Edward Palmer; 1 male, 2 females (1 ovig.) (44503).

Near Belize, Honduras; W. A. Stanton; 1 male (21382).

Old Providence, Caribbean Sea; Apr. 4–9, 1884; *Albatross;* 5 males, 9 females (7542).

Curaçao; Feb. 10–18, 1884; *Albatross;* 4 males, 3 females (7586).

Mariel, Cuba; in swamp; "climbs trees"; May 10, 1900; William Palmer and J. H. Riley; 1 male (23805).

Cuba; 1914; Henderson and Bartsch, *Tomas Barrera* Exped.: Los Arroyas; May 20; 4 males, 4 females (2 ovig.) (48572). Ensenada de Cajon, off Cape San Antonio; May 22; 3 males, 1 female (48578).

Jamaica; Mar. 1–11, 1884; *Albatross;* 2 males, 1 female (7677).

Montego Bay, Jamaica: P. W. Jarvis; 1 y. (19055). Bogue Islands; C. B. Wilson; 1 female (42882). June 17, 1910; E. A. Andrews; 1 female ovig. (42874). Salt Pond; June 28, 1910; E. A. Andrews; 2 females, 2 y. (41749).

Kingston Harbor, Jamaica; T. H. Morgan; 1 male, 1 female (17222).

Porto Rico; 1899; *Fish Hawk:* Rio Bayamon, above Palo Seco; Jan. 16; 1 male (24063), 16 males, 13 females (24066). Hucares; Feb. 13 and 14; 4 males, 10 females (24065). Ensenada Honda, Culebra; Feb. 9; 1 y. (24034).

San Juan Bay, Porto Rico; in mangrove swamp; Jan. 18, 1899· Paul Beckwith; 3 males, 4 females (22793).

Brandons, Barbados; in mangrove swamp; H. M. Lefroy; 1 female (26410).

Brazil; Lieut. F. E. Sawyer, U. S. Navy; 5 males, 8 females (17596).

Pernambuco, Brazil; 1876–1877; R. Rathbun, Hartt Explorations; 2 males, 3 females, 2 y. (40608). On mangroves; Aug. 1, 1899; A. W. Greeley, Branner-Agassiz Exped.; 3 males (25703).

Plataforma, Bahia, Brazil; 1876–1877; R. Rathbun, Hartt Explorations; 1 male, 1 female (40607).

Rio de Janeiro, Brazil; U. S. Exploring Exped.; 1 male, 1 female (2345).

Terra de Masahe, Province of Rio de Janeiro, Brazil; Jan., 1912; H. von Ihering; 2 females (47835).

Iguape, Province of São Paulo, Brazil; 1912; R. Krone; 1 male (Mus. S. Paulo).

Bermuda; F. V. Hamlin; received from Wesleyan University; 1 male (4022).

Hungry Bay, Bermuda; July–Sept.; F. G. Gosling; 1 male (25437).

Dakar, West Africa; May 3, 1892; O. F. Cook; 1 male (21388).

Rock Spring, Monrovia; Apr., 1894; O. F. Cook and G. N. Collins; 5 males (20574).

Monrovia; March, 1895; O. F. Cook; 2 males (21389).

Mouth of Mesurado River, Monrovia; O. F. Cook; 1 female (20668).

Baya River, Elmina, Ashantee; Nov. 27, 1889; W. H. Brown, U. S. Eclipse Exped. to Africa; 4 males, 2 females (14881).

GONIOPSIS PULCHRA (Lockington).

Plate 58.

Goniograpsus pulcher LOCKINGTON, Proc. California Acad. Sci., vol. 7, 1876 (1877), p. 152 [8], (type-locality, Magdalena Bay, west coast of Lower California; type not extant).

Goniopsis cruentatus KINGSLEY, Proc. Acad. Nat. Sci. Philadelphia, 1880, p. 190 (part: specimen from west coast of Nicaragua).—CANO, Boll. Soc. Nat. Napoli, ser. 1, vol. 3, 1889, pp. 101 and 235 (not locality Amoy).

Goniopsis pulcher NOBILI, Boll. Mus. Zool. Anat. Comp. R. Univ. Torino, vol. 12, No. 280, 1897, p. 3.

Goniopsis pulchra RATHBUN, Proc. U. S. Nat. Mus., vol. 38, 1910, p. 547, pl. 47, fig. 3.

Diagnosis.—Like *cruentata*, but carapace wider; color purplish or brown; appendages of male abdomen straighter; tip more transverse.

Description.—Differs very little from *G. cruentata* of the Atlantic side, but the following characters have been observed

The width of the carapace is slightly greater in proportion to the length when specimens of equal size are compared.

The appendages of the first segment of the male abdomen are straighter, the tip a little more transverse.

The color is darker, purplish or dark brown predominating; the size of spots and blotches is variable, but they are of a citrine color, becoming yellowish on the legs.

Measurements.—Male (12467), length of carapace 33, width of same 40, width of front 22.5 mm.

Habits.—Like those of *G. cruentata.*

Range.—From Magdalena Bay, Lower California, to Peru.

Material examined.—

Magdalena Island, Lower California; Dec. 5, 1905; Nelson and Goldman, Biol. Survey, U. S. Dept. Agriculture; 1 male (33416).

Lower California (?); 1 male (12467).

Guaymas, Mexico; gulf side, beach, under stones, etc.; Feb. 23 and 27, 1891; P. L. Jouy; 1 male, 1 female (17294), 1 male y. (17306).

FIG. 137.—GONIOPSIS PULCHRA, ABDOMEN OF MALE (12467). × 1½.

Boca del Jesus Maria, Costa Rica; April, 1905; P. Biolley and J. F. Tristan; 1 male, 1 female (32284).

Santo Domingo, Gulf of Dolce, Costa Rica; April, 1896; H. Pittier; 1 male (19436).

Las Vacas, near Capon, Peru; on beach; Jan. 23, 1908; R. E. Coker; received from Peruvian Government; 3 females (40435).

"*Chanduya;* common on muddy beaches; noted especially about the mangrove swamps."

Genus PACHYGRAPSUS Randall.

Pachygrapsus RANDALL, Journ. Acad. Nat. Sci. Philadelphia, vol. 8, 1839 (1840), p. 127; type, *P. crassipes* Randall.

Goniograpsus (part) DANA, Proc. Acad. Nat. Sci. Philadelphia, vol. 5, 1851 (1852), pp. 247 and 249.

Carapace quadrate, broader than long, though sometimes slightly so, a little convex, regions not well defined. Fine oblique lines on the branchial regions, similar transverse lines on the anterior half. Antero-lateral angle acute; there is usually behind it on the lateral margin one or two teeth, but sometimes none.

Front broad, half or more than half the width of the carapace, deflexed; along the line of flexion are four lobes more or less prominent.

Lower border of orbit notched near its outer end. Orbital hiatus partly filled by a lobe belonging to the inner fossa, which does not, however, exclude the antenna from the orbit. Antennules almost transverse.

Epistome well defined, short fore and aft. Buccal cavity square with the anterior corners rounded off. Maxillipeds having a rhomboidal gap between them; merus shorter than ischium, and bearing the coarse palpus on its anterior border, nearer the outer than the inner angle.

Chelipeds either subequal or unequal, much more massive than the legs. Fingers slightly grooved inside, forming shallow spoons at the tip.

Legs broad and compressed, especially as to the merus, which, like the arm of the chelipeds, is transversely striated; the last three joints have bristly edges and the dactyli are thorny.

The abdomen in both sexes has seven segments, and in the male covers the sternum at its base.

Inhabits the Pacific and Atlantic coasts of America, the eastern Atlantic, thence eastward through the Mediterranean to the Indo-Pacific.

KEY TO THE AMERICAN SPECIES OF THE GENUS PACHYGRAPSUS.

A^1. Carapace with one or two side teeth behind the outer orbital tooth.
  B^1. Carapace with one side tooth.
    C^1. Distal end of posterior margin of merus of last pair of legs entire.
      D^1. Sides of carapace strongly sinuous_____*crassipes*, p. 241.
      D^2. Sides of carapace slightly sinuous_____*maurus*, p. 244.
    C^2. Distal end of posterior margin of merus of last pair of legs dentate.
      D^1. Edge of front sinuous_____*transversus*, p. 244.
      D^2. Edge of front simply convex_____*gracilis*, p. 249.
  B^2. Carapace with two side teeth.
    C^1. Striae of carapace naked_____*marmoratus*, p. 250.
    C^2. Striae of carapace hidden by short hairs_____*pubescens*, p. 252.
A^2. Carapace without side teeth_____*corrugatus*, p. 252.

ANALOGOUS SPECIES ON OPPOSITE SIDES OF THE CONTINENT.

Atlantic.	Pacific.
maurus.	*crassipes*.
marmoratus.	*pubescens*.

Species on both sides of the continent: *transversus*.

PACHYGRAPSUS CRASSIPES Randall.

Plate 59.

Pachygrapsus crassipes RANDALL, Journ. Acad. Nat. Sci. Philadelphia, vol. 8, 1839 (1840), p. 127 (type-locality, Sandwich Islands; this locality is probably erroneous; type in Mus. Phila. Acad. Nat. Sci.).—DE MAN, Notes Leyden Mus., vol. 12, 1890, p. 86, pl. 5, fig. 11.

Grapsus eydouxi MILNE EDWARDS, Ann. Sci. Nat., ser. 3, Zool., vol. 20, 1853, p. 170 [136], (type-locality, Chili; type in Paris Mus.).

Leptograpsus gonagrus MILNE EDWARDS, Ann. Sci. Nat., ser. 3, Zool., vol. 20, 1853, p. 173 [139], (*Patrie inconnue;* type in Paris Mus.).

Diagnosis.—One side tooth. Striae mostly long. Sides very convex in front. Merus of last leg entire behind.

Description.—A large species. Carapace not much broader than long, covered, except the cardiac and intestinal regions, with transverse and oblique striae; sides arched in their anterior half, strongly convergent posteriorly, armed with a strong tooth behind the larger orbital tooth.

Front half, or nearly half, as wide as the carapace, surface granulate; a small but pronounced lobe at the outer corners, the intervening margin convex or slightly sinuous and crenulate.

Orbits oblique, their major diameter less than one-third the width of the front, lower border denticulate.

Chelipeds normally subequal, much heavier than the legs, and about one and two-thirds times as long as the carapace. Arm and wrist finely striated, chela almost smooth. Inner margin of arm limbed, distal end of limb dentate. A subacute tooth at inner angle of wrist. Hand about one and one-half times as high as its superior length; upper border margined, adjacent inner surface obliquely striated, lower part of outer surface traversed by an obliquely longitudinal raised line which runs from the end of the thumb toward the wrist. Fingers narrowly and irregularly gaping. *D*actylus one and three-fourths times as long as the upper margin of the palm.

2. 3. 1. 4. represents the legs in order of length, the second pair the longest and one and two-thirds times as long as the carapace. Merus joints with a spine near the distal end of the anterior border, and a few teeth near the same end of the posterior border of all save the last pair.

Terminal segment of male abdomen triangular.

Color.—Dark crimson spotted with ecru.

Habits.—A very common species on rocky shores.

Measurements.—Male (17455), length of carapace 39, width of same 46 mm.

Range.—From Oregon to Gulf of California; Galapagos Islands· Chile; also Japan and Korea.

Material examined.—

Crescent City Bay, California; lat. 41° 42′ N.; 2 males (5328).

Monterey Bay, California: *Albatross;* 1 male, 1 female (15602). Under submerged rocks along shore; Harold Heath; 1 male (22879). Pacific Grove; Stanford Univ.; 2 males, 1 female (19076). Pacific Grove; June, 1905; J. E. Benedict; 56 specimens (32234).

San Simeon Bay, California; Apr. 4, 1890; *Albatross;* 1 male, 2 females (17456).

Santa Barbara, California; Yarrow, Henshaw and Shoemaker; 9 males, 7 females, 3 y. (2324).

Wilmington, California; 1880; D. S. Jordan; 1 male, soft shell (3080).

Venice, California; from the landward side of the breakwater at about high-tide level; Ray A. Carter; 1 male (43032).

Santa Catalina Island, California: Harbor, on beach; 1874; W. H. Dall; 2 males, 2 females (14825). 1880; D. S. Jordan; 1 male (3104). April, 1897; *Albatross;* 1 male (20161).

San Clemente Island, California: May 8, 1888; *Albatross;* 1 male, 4 females (22094). Thomas Casey; 4 y. (44528).

Lajolla, California; Mar. 6, 1898; *Albatross;* 1 female (21786).

Point Loma, California; *Albatross;* 1 male, 1 y. (17457), 32 specimens (18970).

Cardiff, near San Diego, California; Aug., 1914; Mrs. M. I. Sparks, collector; received from W. D. Webb; 1 female (49064).

San Diego, California; 1880; D. S. Jordan; 5 males and females (3564).

Southern California; W. H. Dall; 1 y. (17296).

San Quentin, Lower California, Mexico; L. Belding; 7 males, 7 females (5322).

Guadalupe Island, W. of Lower California; Feb. 28, 1889; *Albatross;* 2 males, 1 female (15567).

Ballenas Bay, Lower California; May 3, 1888; *Albatross;* 7 males, 7 females (17455).

Margarita Island, Lower California; Mar. 19, 1911; *Albatross*, 3 females (44579).

Turtle Bay, Lower California; July 30, 1896; A. W. Anthony; 4 males, 4 females (19509).

Tiburon Island, Gulf of California; Apr. 12, 1911; *Albatross;* 1 female (44578).

Tagus Cove, Albemarle Island, Galapagos Islands; 12 fathoms; Stanford Univ.; 1 male (25663).

"Sandwich Islands"; probably error for California coast; type-specimen in Mus. Phila. Acad. Nat. Sci.

Miura, Atami District, Japan; Mar., 1890; F. Sakamoto, collector; received from Garrett Droppers; 1 male, 1 female (18871).

Misaki, Sagami, Japan; 1906; *Albatross;* 1 female ovig. (44634).

Yenoshima, Japan; received from Univ. Tokyo; 1 female (45837).

Matsushima, Japan; shore; July 28, 1906; Bureau of Fisheries; 1 male (45854).

Fusan, Korea; P. L. Jouy; 2 males (15106).

PACHYGRAPSUS MAURUS (Lucas).

Plate 60, figs. 1 and 2.

Grapsus maurus LUCAS, Explor. Sci. Algérie, Zool., Insectes, vol. 1, 1846, p. 20, and vol. 4, (atlas), pl. 2, fig. 5 (type-locality, Oran; type in Paris Mus.).

Goniograpsus simplex DANA, Proc. Acad. Nat. Sci. Philadelphia, vol. 5, 1851 (1852), p. 249 (type-locality, Rio Janeiro ?; type not extant); U. S. Expl. Exped., vol. 13, Crust., pt. 1, 1852, p. 344; atlas, 1855, pl. 21, fig. 8.

Leptograpsus maurus MILNE EDWARDS, Ann. Sci. Nat., ser. 3, Zool., vol. 20, 1853, p. 173 [139].

Pachygrapsus simplex STIMPSON, Proc. Acad. Nat. Sci. Philadelphia, vol. 10, 1858, p. 102 [48].

Pachygrapsus maurus KINGSLEY, Proc. Acad. Nat. Sci. Philadelphia, 1880, p. 199.

Diagnosis.—One side tooth. Striae mostly short. Merus of last leg entire behind.

Description.—A small species. Carapace narrower than in *P. crassipes* or *P. transversus*, with which this species has been confused; sides less convergent behind, also less arched anteriorly than in *P crassipes;* striae fine and much broken up, but present except on the intestinal region.

Edge of front bilobed, its surface granulate; superior lobes little prominent.

Tooth of wrist acute.

Distal angle of posterior border of merus of last pair of legs entire; of the first three pairs, dentate or denticulate.

Color.—Dark brown, with isolated yellow spots; similar spots at the ends of the several joints of the legs.

Measurements.—Mature female (18627), length of carapace 6.4, width of same 7.3 mm. Male (Lucas), length of carapace 17, width of same 19.5 mm.

Range.—Rio Janeiro (Dana, Heller); also East Atlantic and Mediterranean; Madeira (Stimpson); Terceira, Azores (Rathbun); Algiers (Lucas); Oran (M. Edwards).

Material examined.—Terceira, Azores; William Trelease; 1 female (18627). Guia Island, Grand Canaries; September, 1894; 3 males, 1 female ovig., 4 y., with *Serpula* attached (44529).

PACHYGRAPSUS TRANSVERSUS (Gibbes).

Plate 61, figs. 2 and 3.

Grapsus transversus GIBBES, Proc. Amer. Assoc. Adv. Sci., vol. 3, 1850, p. 181 (type-locality, Key West; type not extant).

Pachygrapsus transversus GIBBES, Proc. Amer. Assoc. Adv. Sci., vol. 3, 1850, p. 182.

Goniograpsus innotatus DANA, Proc. Acad. Nat. Sci. Philadelphia, Vol. 5, 1851 (1852), p. 249 (type-locality, *ad oras Americae Australis !;* type, Cat. No. 2329, U.S.N.M.); U. S. Expl. Exped., vol. 13, Crust., pt. 1, 1852, p. 345; atlas, 1855, pl. 21, fig. 9 a–c.

Leptograpsus rugulosus MILNE EDWARDS, Ann. Sci. Nat., ser. 3, Zool., vol. 20, 1853, p. 172 [138], (type-locality, Brazil; type in Paris Mus.).
Pachygrapsus laevimanus STIMPSON, Proc. Acad. Nat. Sci. Philadelphia, vol. 10, 1858, p. 102 [48], (type-locality, *in portu Jacksoni vel Sydney Australiae;* type not extant).
Metopograpsus dubius SAUSSURE, Mém. Soc. Phys. Hist. Nat. Genève, vol. 14, 1858, p. 445 [29], pl. 2, fig. 16 (type-locality, St. Thomas; type in Geneva Mus.).
Metopograpsus miniatus SAUSSURE, Mém. Soc. Phys. Hist. Nat. Genève, vol. 14, 1858, p. 444 [28], pl. 2, fig. 17 (type-locality, St. Thomas; type in Geneva Mus.; its sides are distended by a parasite, one side more than the other).
Grapsus declivifrons HELLER, Verh. k. k. zool.-bot. Ges., Wien, vol. 12, 1862, p. 521 (type-locality, Rio Janeiro; type in Vienna Mus.).
Pachygrapsus intermedius HELLER, Crust. Reise Novara, 1865, p. 44 (type-locality, Rio Janeiro; type in Vienna Mus.).
Pachygrapsus socius STIMPSON, Ann. Lyc. Nat. Hist. New York, vol. 10, 1871, p. 114 (type-localities, Peru, Panama, San Salvador, Manzanillo, and Cape St. Lucas; cotypes from Cape St. Lucas and Panama in Mus. Comp. Zoöl.).
Pachygrapsus advena CATTA, Ann. Sci. Nat., ser. 6, Zool., vol. 3, 1876, p. 7, pl. 1, fig. 1 (type-locality, Marseille on vessel from Pondichery, by way of Cape of Good Hope; type probably not extant).

Diagnosis.—One side tooth. Sides strongly convergent. Front sinuous. Upper edge of movable finger smooth. Merus of last leg dentate at posterior distal end.

Description.—A small species. Carapace one-third broader than long, covered with oblique and transverse striae, the latter being short and faint on the cardiac and intestinal regions, and granulated on the anterior half of the carapace; intervening space granulate; sides slightly arched, strongly convergent posteriorly, armed with an acuminate tooth close behind the similar but larger outer orbital tooth.

Front slightly more than half as wide as carapace, edge sinuous and granulate, with three shallow sinuses, sides little oblique, surface smooth, except a transverse granulate line on each of the two marginal lobes. The superior frontal lobes of the middle pair are prominent, outer pair oblique, flattened.

Orbits oblique, about two-fifths the width of the front, lower border denticulate.

Chelipeds equal, much stouter than the legs, in the male one and two-thirds times as long as the carapace; arm and wrist, and also the merus joints of the legs, transversely striated. Inner margin of arm produced in a laminate expansion, denticulate on the edge, and distally laciniate. Carpal tooth blunt. Chela finely granulate; upper surface of palm with a marginal line, and oblique striae on the inner side; lower surface obliquely striated; an obliquely longitudinal line on the outer surface near the lower edge, runs the length of the propodus. Palm one and one-half times as high as its superior length; length of dactylus somewhat greater than this height. Fingers narrowly gaping except where interrupted by a large triangular tooth near the middle of the pollex. Inner surface of palm with fine markings.

Legs with a slender spine at the distal end of the anterior border, and two or three triangular spines or teeth at the same end of the posterior border. Second and third pairs subequal, one and three-fourths times as long as carapace; first and fourth pairs subequal, considerably shorter. Last three joints bristly and thorny.

Terminal segment of male abdomen broadly triangular.

Measurements.—Male (9370), length of carapace 11.4, width of same 14.9 mm.

Habits.—Found among stones, roots of mangroves, and on sandy shores.

Range.—From Bahamas and Florida Keys to Montevideo, Uruguay (Cano). From California (Kingsley) to Peru; Galapagos Islands. Also Bermudas; West Africa; and the Oriental Region.

Occasionally brought from further north; specimens taken from bottom of whaler at Provincetown, Cape Cod.

Material examined.—

On whaler, Cape Cod, Massachusetts; Sept. 3, 1879; U. S. Fish Comm.; 2 males (19023).

Charleston, South Carolina (M. C. Z.).

New Providence, Bahamas; 1886; *Albatross;* 1 female (17675).

Andros Island, Bahamas; May 14, 1912; Paul Bartsch: East side of island, near lighthouse, south of South Bight; 1 male (45550). Smiths landing, south side of east end of South Bight, Long Bay Key District; 1 male (45573).

Near Lake Kissimmee, Florida; A. M. Reese; 1 male (44508).

Cape Florida, Florida; Edward Palmer; 1 y. male (17898).

Broad Creek, Florida; ocean front; Dec. 17, 1906; Pine and Bean· 1 female (33147).

Salt Pond Key, Florida; 1884; Edward Palmer; 1 specimen (9371).

Harbor of Key West, Florida; 1884; Edward Palmer; 3 males, 1 female, 1 y. (9370).

Key West, Florida: Dec., 1883; D. S. Jordan, 1 male (6364). H. Hemphill; 4 males, 2 females, 1 y. (14441). 1884; *Albatross;* 2 males (17673). Feb. 3, 1901; B. A. Bean and W. H. King; 1 female, soft shell (44533).

Bird Key, Florida; Apr. 8, 1889; *Grampus;* 4 males, 2 females (15257).

Dry Tortugas, Florida; from outside of floating live-car; T. W. Vaughan, collector; 8 males, 2 females (49633).

Garden Key, Florida; coral rock and piling; Dec. 25, 1912; *Fish Hawk;* 1 male (49916).

Matagorda Island, Texas; J. D. Mitchell; 1 female (22818).
Tuxpan, Mexico; G. Lincecum; 3 females (2 ovig.) (44509).
Mariel, Cuba; May 10, 1900; William Palmer and J. H. Riley: Among rocks; 1 female (23816). On bushes and wharves; 5 males. 15 females (23817).
Point Vedado, Cuba; M. S. Roig; 1 male (46083).
Cuba; 1914; Henderson and Bartsch, *Tomas Barrera* Exped.: Esperanza; May 11; 1 male (48623).
Los Arroyas; May 19; 1 female (48622).
Ensenada de Cajon, off Cape San Antonio; 2 males, 1 female 1 y. (48620).
Reef at Cape San Antonio; 1 male, 1 female (48621).
Cabañas; on sand, shell, grass to mud bottom; June 8–9; 3 males, 3 females (48593); 1 male, 3 females (48591); 1 male (48619).
On reef flat between Cayo Hutia and Little Cayo, NE. of Light; 1 y. (49593).
Jamaica; Mar. 1–11, 1884; *Albatross*; 3 females (18567).
Montego Bay, Jamaica; rocks in front of sea view, laboratory; Aug. 30, 1910; E. A. Andrews; 4 males, 2 females (1 ovig.) (43055).
Bogue Islands, Montego Bay, Jamaica; 1910; C. B. Wilson: June 15; 1 female (42888). On the mangrove roots with sponges, ascidians, etc.; June 20; 1 male (42883). From sponge on mangrove roots; Aug. 10; 1 male, 2 females (1 ovig.) (42886).
Kingston Harbor, Jamaica: T. H. Morgan; 3 males, 2 females (17224). 1893; R. P. Bigelow; 1 male, 5 females, 2 y. (17981).
Port Royal, Jamaica; P. W. Jarvis; 2 males (19062).
San *D*omingo, W. I.; 1878; W. M. Gabb; 5 males, 7 females (3200).
Porto Rico; 1899; *Fish Hawk:* Beach, San Juan; Jan. 12; 5 males, 8 females (24011). Mayaguez; Jan. 19; 2 females, 2 y. (24029). On corals, Mayaguez; Jan. 21; 1 female (24027). On coral reef, Mayaguez; Jan. 23; 6 y. (24017). Porto Real; Jan. 27; 2 y. (24020). Boqueron Bay; Jan. 28; 1 female (24028). On coral reef, Boqueron Bay; Jan. 28; 1 male (24025). Guanica Bay; Jan. 28; 1 female (24023). Reefs at Guanica; Jan. 29; 1 male (24018). Playa de Ponce Reef; Feb. 1; 1 male, 2 females, 1 y. (24031). Reefs at Ponce; Jan. 30; 2 males, 3 females, 1 y. (24030). Ponce; Jan. 30 and 31; 9 males, 4 females (24015). Arroyo; Feb. 4; 1 female (24022). On Lighthouse Reef, Arroyo; Feb. 3; 2 y. (24019). Caballo Blanco Reef; Feb. 27; 1 male, 1 female (24026). Ensenada Honda, Culebra; Feb. 10, 1899; 1 female (24024). Culebra; Feb. 11; 1 female (24021). Fajardo; Feb. 17; 1 male (24032).

St. Thomas; 1915; C. R. Shoemaker: Shore near town; June 29; 1 male, 2 females, each with isopod parasite in branchial chamber (49819). Water Island; July 10; 1 male, 1 female ovig. (49815). Trinidad; shore; Jan. 30–Feb. 2, 1884; *Albatross;* 1 y. (17674). Monos Island, Trinidad; Jan. 30–Feb. 2, 1884; *Albatross;* 1 y. (18571). Curaçao; Feb. 10–18, 1884; *Albatross;* 3 males, 4 females (18555). Natal, Brazil; 1911; Fred Baker, Stanford Exped.; 1 female ovig. (44532).

Brazil; 1899; A. W. Greeley, Branner-Agassiz Exped: Mamanguape stone reef; June 23; 2 males (25704). Rio Parahyba do Norte, Cabedello; on mangroves; June 20; 1 male, 1 female (25705). Parahyba River; on mangroves; June 21; 1 male (25709). Pernambuco stone reef; July 7; 1 male (25707). Stone Reef at Boa Viagem, 5 miles south of Pernambuco; July 6; 1 male (25708). Rio Goyanna stone reef; June 18; 4 males (25706).

Bahia, Brazil; May, 1915; J. N. Rose; 1 male (48298).

Plataforma, Bahia, Brazil; 1876–1877; R. Rathbun, Hartt Explorations; 4 males, 1 female (40825).

Abrolhos Islands, Brazil; Dec. 27, 1887; *Albatross;* 5 males, 1 female (22095).

Nictheroy, Rio de Janeiro, Brazil; July 20, 1915; J. N. Rose; 4 males, 1 female (48301).

Rio de Janeiro (?), Brazil; U. S. Expl. Exped.; 2 males, cotypes of *Goniograpsus innotatus* Dana (2329).

Iguape, São Paulo, Brazil; 1902; R. Krone, collector; from H. von Ihering; 2 females (47852).

Bermudas: 1876–1877; G. Brown Goode; 20 specimens (42673). 1901; A. E. Verrill and party; 5 specimens (39247); specimens of *Leidya* taken from gill cavity.

Porto Grande, St. Vincent, Cape Verde Islands; Nov. 11, 1889; W. H. Brown; U. S. Eclipse Exped. to Africa; 4 males (14865).

Manly, near Sydney, Australia; Australian Museum; 2 males (17054); 4 specimens (43776).

Easter Island, South Pacific; shore; Dec. 16, 1904; *Albatross;* 1 female (33186).

Pichilinque Bay, Lower California, Gulf of California, Mexico; Apr. 29, 1888; *Albatross;* 1 male, 1 female (22096).

Agua Verde Bay, Mexico; Apr. 1, 1911; *Albatross;* 1 male (44576).

Punta Arenas (Pacific-Estero side), Costa Rica; Feb., 1905; J. F. Tristan; 1 female (32363).

Taboga Island, Panama; May 11–15, 1911; Meek and Hildebrand, Smithsonian Biol. Survey; 1 female (44172).

Reef north of Tagus Hill, Tagus Cove, Albemarle Island, Galapagos Islands; Mar. 16, 1899; Stanford Univ.; 1 male (25664).

Chatham Island, Galapagos Islands; Dr. W. H. Jones, U. S. Navy; 1 female (17682).

San Lorenzo Island, Peru; Dr. H. E. Ames, U. S. Navy.; 1 male (17683).

Oyster beds of Matapalo (near Capon), Peru; Jan. 23, 1908; R. E. Coker, collector; received from Peruvian Government; 1 female (40450).

PACHYGRAPSUS GRACILIS (Saussure).

WHARF CRAB.

Plate 60, fig. 3; plate 61, fig. 1.

Metopograpsus gracilis SAUSSURE, Mém. Soc. Phys. Hist. Nat. Genève, vol. 14, 1858, p. 443 [27], pl. 2, fig. 15 (type-locality, St. Thomas; type in Geneva Mus.).

Grapsus guadalupensis DESBONNE, in Desbonne and Schramm, Crust. Guadeloupe, p. 47, 1867 (type-locality, Guadeloupe; type probably not extant).

Pachygrapsus gracilis STIMPSON, Ann. Lyc. Nat. Hist. New York, vol. 10, 1871, p. 113.

Grapsus (Leptograpsus) rugulosus VON MARTENS, Arch. f. Naturg., vol. 38, pt. 1, 1872, p. 108; not *Leptograpsus rugulosus* M. Edw.

Diagnosis.—One side tooth. Sides strongly convergent. Front convex. Upper edge of movable finger tuberculate. Merus of last leg dentate at posterior distal end.

Description.—Same size and general appearance as *P. transversus;* but may be distinguished by the following characters:

Cardiac and intestinal regions smooth. Margin behind lateral tooth concave or nearly straight. Front nearly two-thirds as wide as carapace, edge convex, upper surface smooth, without granulated line; upper lobes obsolescent, outer pair considerably wider than inner pair. Orbit between one-third and one-fourth width of front. Inner projection of wrist a sharp spine. Upper margin of propodus and subinferior ridge stronger. *D*actyli spinulous above. Fingers irregularly toothed, nearly meeting when closed. Inner surface of palm rough with tubercles. Abdomen broader.

Color.—Pinkish.

Measurements.—Male (24014), length 12, width 15.9 mm.

Habits.—Found just above the water level on piles, wharves, etc.

Range.—From Bahamas and Florida to Rio Parahyba do Norte, Brazil. Bermudas.

Material examined.—

Green Turtle Cay, Bahamas; E. A. Andrews; 1 male, 2 females (20715).

Nassau, Bahamas; Dec. 25, 1898; *Fish Hawk;* 4 males (24014).

Mariel, Cuba; on bushes and wharves; May 10, 1900; William Palmer and J. H. Riley; 1 male (23818).

Montego Bay, Jamaica; from sponges and algae in brackish pond· July 2, 1910; C. B. Wilson; 1 female (42884).

Bogue Islands, Montego Bay, Jamaica; 1910: On the mangrove roots' with sponges, ascidians, etc.; June 20; C. B. Wilson; 1 male (42884). July 6; E. A. Andrews; 1 female (43054). From sponge on mangrove roots; Aug. 10; C. B. Wilson; 2 females (1 ovig.) (42887).

Kingston Harbor, Jamaica: T. H. Morgan; 1 female (17225). May–July, 1896; F. S. Conant; 2 males, 1 female (19601).

Port Royal, Jamaica; P. W. Jarvis; 1 female·(19061).

Boqueron Bay, Porto Rico; Jan. 27, 1899; *Fish Hawk;* 1 male, 1 female (24037).

Sabanilla, Colombia; Mar. 16–22, 1884; *Albatross;* 2 females, 1 y. (18447).

Rio Parahyba do Norte, Brazil; on mangroves; June 21, 1899; A. W. Greeley, Branner-Agassiz Exped.; 1 male (25710).

Plataforma, Bahia, Brazil; 1876–1877; R. Rathbun, Hartt Explorations; 2 males, 3 females ovig. (40603).

Bermudas; 1876–1877; G. Brown Goode; 2 specimens (49254).

PACHYGRAPSUS MARMORATUS (Fabricius).

Plate 62.

Cancer marmoratus FABRICIUS, Mant. Insect., vol. 1, 1787, p. 319 (type-locality unknown; type not extant).

Cancer marmoreus OLIVIER (after Fabricius), Encyc. Méth., Hist. Nat., Insectes, vol. 6, 1791, p. 161.

Cancer femoralis OLIVIER, Encyc. Méth., Hist. Nat., Insectes, vol. 6, 1791, p. 166 (type-locality, *sur les rivages de la mer méditerrannée;* type probably not extant).

Cancer marmoratus OLIVI, Zool. Adriat., 1792, p. 47, pl. 2, fig. 1 (type-locality, Adriatic; type probably not extant). Here described as a new species.

Grapsus varius LATREILLE, Hist. Nat. Crust., vol. 6, p. 67, an XI [1802–1803] (type-locality, *sur les côtes de la Méditerranée, près de Montpellier;* type in Paris Mus.).

Grapsus marmoratus DUMERIL, Dict. Sci. Nat., vol. 19, 1821, p. 322.

Goniograpsus varius ? DANA, U. S. Expl. Exped., vol. 13, Crust., pt. 1, 1852, p. 344.

Leptograpsus marmoratus MILNE EDWARDS, Ann. Sci. Nat., ser. 3, Zool., vol. 20, 1853, p. 171 [137], pl. 7, fig. 3.

? Leptograpsus berthcloti MILNE EDWARDS, Ann. Sci. Nat., ser. 3, Zool., vol. 20, 1853, p. 172 [138], (type-locality, *Iles Canaries;* type in Paris Mus.).

Pachygrapsus marmoratus STIMPSON, Proc. Acad. Nat. Sci. Philadelphia, vol. 10, 1858, p. 102 [48].

I have included this species here on account of its doubtful occurrence in Rio Janeiro recorded by *D*ana, and because the South

American fauna is only partially known. Dana's specimen is not extant; it may perhaps have come from Madeira or the Cape Verde Islands.

Diagnosis.—Quadrate. Two side teeth. Movable finger tuberculate above. Merus of last leg entire behind.

Description.—Carapace but little broader than long; sides a little arched in front, converging very little behind the posterior side tooth. Surface smooth in the posterior two-thirds of the middle portion. Two side teeth besides the orbital tooth, the three diminishing in size backwards.

Front as a whole convex, edge sinuous, faintly bilobed, and finely crenulate. Superior lobes prominent, outer pair oblique, narrower than inner pair.

The spines at the distal end of the arm and the inner angle of the wrist are sharp. Upper margin of palm and proximal half of finger tuberculate; a ridge on distal half of lower part of propodus, which is obsolete in old specimens. Edge of pollex prominent near its middle; fingers little gaping.

FIG. 138.—PACHYGRAPSUS MARMORATUS, OUTER MAXILLIPED, ENLARGED. (AFTER MILNE EDWARDS.)

Legs with the customary spine on the anterior border; first pair with one posterior spine; second and third pairs with three or four; last pair entire.

Color.—A mixture of shades of green, gray, brown, and white. One variety is covered with transverse bands of white; another is absolutely black. (After Risso.)

Measurements.—Male (14864), length of carapace 27.8, width of same 30.7 mm.

Habitat.—Littoral. Risso[1] says of these crabs that they are timid and cease their activity at the least sign of danger until assured that no one will molest them, when they resume their sports and combats; but if one makes the least movement to seize them they flee swiftly and seek fissures in the rocks to hide in, and threaten with their claws. They leave the water many times a day to walk in the sun, and at night they roam in search of dead animals left by the tide. The females produce each time 400 or 500 small eggs, and remain under the rocks until they are hatched.

Range.—Rio Janeiro? (*D*ana). Inhabits the Mediterranean; Black Sea; west coast of France; Azores; Madeira.

Material examined.—

Biarritz, France; Rev. A. M. Norman; 3 male and female (14497).

Terceira, Azores; June 29, 1894; William Trelease; 1 male, 2 females (18626).

---

[1] Crust. Nice, 1816, p. 22.

Horta, Fayal, Azores; Nov. 2, 1889; W. H. Brown, U. S. Eclipse Exped. to Africa; 2 males, 3 females (14864).

Pim Bay, Fayal, Azores; Lewis *D*exter; 1 male (17681).

Locality unknown; 2 females (2546).

PACHYGRAPSUS PUBESCENS Heller.

Plate 160, fig. 1.

Pachygrapsus pubescens HELLER, Crust. Reise *Novara*, 1865, p. 45, pl. 4, fig. 4 (type-locality, Chili; type in Vienna Mus.).

Diagnosis.—Quadrate. Two side teeth. Striae hairy. Movable finger smooth above. Merus of last leg entire behind.

Description.—Carapace almost quadrate, slightly broader than long,[1] flattened. Sides nearly straight, two teeth behind the orbital angle, which is large and very pointed, the other teeth gradually diminishing.

Front somewhat deflexed, convex, very finely crenulate. Superior lobes of front subequal, rather flat.

Surface transversely striate, ridges hidden by short hairs; posterior-medial area smooth.

Inner margin of arms crest-like, three or four sharp teeth at the extremity. Arm and wrist subsquamose outside; wrist with small isolated tubercles above, an acute tooth at inner angle; palm above obsoletely tuberculous, outside and below smooth; finger smooth.

Three last joints of legs hirsute, first to third pairs with merus dentate at extremity above and below; that of fourth pair likewise with one tooth above, but entire below.

Measurements.—Female type, length of carapace 20, width of same 23 mm. (After Heller.)

Range.—Chile (Heller).

PACHYGRAPSUS CORRUGATUS (von Martens).

Plate 160, fig. 4.

Grapsus (*Leptograpsus*) *corrugatus* VON MARTENS, Arch. f. Naturg., vol. 38, pt. 1, 1872, p. 107, pl. 4, figs. 8 and 8*b* (type-locality, Cuba; type, Cat. No. 3702, Berlin Mus.).

Pachygrapsus corrugatus KINGSLEY, Proc. Acad. Nat. Sci. Philadelphia, 1880, p. 200.

Diagnosis.—Quadrate. No side teeth. Striae hairy and strongly marked. Several ridges on palm.

Description.—Carapace little broader than long, subquadrate, side margins nearly straight, entire, and little converging posteriorly. Carapace, arm, chela, and merus joints of legs crossed by numerous sharp raised ridges, those of the carapace transverse at the middle,

---

[1] Heller's figure represents it longer than broad.

oblique at the sides; from each ridge arises a fringe of short hair, which lies flat.

Front almost straight-edged, with the corners obliquely cut off. Carpus armed with an inner spine. Three long and some short longitudinal ridges on outer surface of chelae, inner surface smooth; fingers gaping, strongly toothed. Upper margin of merus joints short-hairy, a moderate or feeble subterminal tooth, lower margin irregularly toothed at end, next two joints long-hairy; dactyli armed with stout thorns.

Measurements.—Type, length of carapace 12, width of same 14 mm.

Range.—Cuba (von Martens).

Genus PLANES Leach.

Planes BOWDICH, Excursions in Madeira and Porto Santo, 1825, pp. XI and 15, pl. 12, figs. 2a and 2b; type, *P. clypeatus* Bowdich=*minutus* (Linnaeus).

Nautilograpsus MILNE EDWARDS, Hist. Nat. Crust., vol. 2, 1837, p. 89; type, *N. minutus* (Linnaeus).

Carapace quadrate-oval, about as long as broad, convex, regions scarcely defined. Lines on the carapace faint. Sides convex. Antero-lateral angle acute; behind it a slight notch forming a small tooth.

Front about half width of carapace, gently deflexed, lobes at line of flexion almost obsolete.

Orbit with a slight notch below near the outer angle; inner hiatus wide. Palpus articulating at middle of anterior margin of merus of outer maxillipeds.

Legs much flattened, somewhat natatory. Teeth on merus joints more feeble than in *Pachygrapsus.*

Otherwise much as in that genus.

Contains two species which are pelagic, one of them of almost world-wide distribution. Species on both sides of the continent: *minutus.*

KEY TO THE SPECIES OF THE GENUS PLANES.

A^1. Carapace uniformly convex; postero-lateral margins arcuate_*minutus*, p. 253.
A^2. Carapace depressed about the middle; postero-lateral margins nearly straight_____*marinus*, p. 258.

PLANES MINUTUS (Linnaeus).

Plate 63.

GULF-WEED CRAB; TURTLE CRAB; COLUMBUS'S CRAB.

Cancellus marinus minimus quadratus SLOANE, Nat. Hist. Jamaica, vol. 2, 1725, p. 270, pl. 245, fig. 1.

Turtle Crab BROWNE, Hist. Jamaica, 1756, p. 421, pl. 42, fig. 1.

Cancer minutus LINNAEUS, Syst. Nat., ed. 10, vol. 1, 1758, p. 625 (type-locality, *in Pelagi Fuco natante;* type not extant).

Cancer pusillus FABRICIUS, Syst. Entom., 1775, p. 402 (type-locality, *in Oceano boreali;* type in Kiel Mus.).

Cancer glaberrimus HERBST, Naturg. Krabben u. Krebse, vol. 1, 1790, p. 262, pl. 20, fig. 115 (type-locality unknown; type not extant).

Pinnotheres minutus BOSC, Hist. Nat. Crust., vol. 1, an X [1801–1802], p. 244.

Pinnotheres pusillus BOSC, Hist. Nat. Crust., vol. 1, an X [1801–1802], p. 244.

Pinnotheres glaberimus BOSC, Hist. Nat. Crust., vol. 1, an X [1801–1802], p. 244.

Grapsus minutus LATREILLE, Hist. Nat. Crust., vol. 6, an XI [1802–1803], p. 68.

Grapsus cinereus SAY, Journ. Acad. Nat. Sci. Philadelphia, vol. 1, 1817, p. 99. Not G. cinereus BOSC, 1802.

Grapsus pelagicus SAY, Journ. Acad. Nat. Sci. Philadelphia, vol. 1, 1818, p. 442 (type-locality, Gulf Stream; type not extant).

Planes clypeatus BOWDICH, Excursions in Madeira and Porto Santo, 1825, p. 15, pl. 12, figs. 2a and 2b (type-locality, between Lisbon and Madeira, on logs; type not extant).

Grapsus testudinum ROUX, Crust. Médit., 1828, p. (52), pl. 6. figs. 1–6 (type-locality, *dans le roisinage des côtes de la Sardaigne;* type probably not extant).

Grapsus pelagicus ROUX, Crust. Médit., 1828, p. (55), pl. 6, figs. 7–9.

Grapsus (Grapsus) pusillus DE HAAN, Fauna Japon., Crust., 1835, p. 59, pl. 16, fig. 2 (type-locality, Japan; type no longer in Leyden Mus.).

Nautilograpsus minutus MILNE EDWARDS, Hist. Nat. Crust., 1837, vol. 2, p. 90.

Grapsus diris COSTA, Fauna Napoli, Crust., 1838, pl. 4, fig. 1 (and corresponding text) (type-locality, Gaeta; type probably not extant).

Nautilograpsus major MCLEAY, in Andrew Smith's Zool. South Africa, Annul., 1838, p. 66 (type-locality, South Africa; type probably not extant).

Nautilograpsus smithii MCLEAY, in Andrew Smith's Zool. South Africa, Annul., 1838, p. 67 (type-locality, South Africa; type probably not extant).

Planes minutus WHITE, List Crust. Brit. Mus., 1847, p. 42.

Planes linnaeana BELL, Brit. Stalk-eyed Crust., 1851, p. 135 (type-locality, Devon and Cornwall; type in Brit Mus.).

Planes cyaneus DANA, Proc. Acad. Nat. Sci. Philadelphia, vol. 5, 1851 (1852), p. 250 (type-locality, *in mari Pacifico, lat. bor. 28°, long. orient. 174°;* type not extant).

Nautilograpsus angustatus STIMPSON, Proc. Acad. Nat. Sci. Philadelphia, vol. 10, 1858, p. 103 [49], (type-locality, *in mari Pacifico, lat. bor. 34°,. long. occ. 155°;* type not extant).

Planes minutus VERRILL, Trans. Connecticut Acad. Arts and Sci., vol. 13, 1908, p. 325, text-fig. 7, pl. 13, figs. *a–j';* pl. 27, fig. 6.

Diagnosis.—Carapace uniformly convex; postero-lateral margins arcuate.

Description.—The carapace varies from a little longer than broad to a little broader than long; strongly convex in both directions; almost smooth; faint oblique lines on the outer part of the branchial region, and very short transverse lines on the anterior portion. The front may be a little more, or a little less than half the width of the carapace, edge convex, slightly bilobed, granulate. Orbits little oblique. Lateral tooth obtusely angled, sometimes obsolete. Lower margin of orbit granular. Tooth at inner angle equilateral, subacute.

Merus of outer maxillipeds much broader than long.

Chelipeds equal. Inner margin of arm dentate, distal end coarsely so. Inner angle of wrist subacute. Palms with several oblique lines above and below, a longitudinal line on the pollex. continued feebly and brokenly on the palm. Pollex with a prominence near the middle of its prehensile edge. Fingers narrowly or not at all gaping.

Second and third legs subequal, one and one-half times as long as carapace, fourth leg shortest. Merus joints with an inconspicuous anterior subterminal tooth, and a few posterior denticles, which diminish in strength from the first to the fourth pair, where they are minute. Last three segments thorny and with a dense fringe of hair on the anterior edge.

Abdomen of male regularly triangular from the middle of the third segment to the end.

Color.—Extremely variable. Usually irregularly mottled or blotched with light greenish-yellow or pale yellow on a darker olive-green ground color, the carapace thus imitating the olive-green colors of the gulfweed (*Sargassum*) and the whitish patches of Bryozoa (*Biflustra*) with which the *Sargassum* is commonly covered (Verrill).

For variation and color see page 671 and Plate VI of Murray and Hjort, "*Depths of the Ocean*," 1912.

Measurements.—Male (17712), length 18.9, width 18.7 mm.

Habits and distribution.—Pelagic, common in gulfweed, especially in the Sargasso Sea; occasionally on turtles, floating logs and sticks, or jellyfishes; and in living sponges. In all tropical and temperate seas.

Occasionally found on shore, as at Sakonnet Point, Rhode Island, an egg-bearing female, under rocks; and at Salina Cruz, Mexico, on sand beach.

Material examined.—See pages 256 and 257.

Material examined of Planes minutus.

Locality.	Lat. N.	Long. W.	Depth in meters.	Temp. surface.	Date.	Station.	Collector.	Spec.	Cat. No.
Wds Hole, Massachusetts				°F.	1899	Under rock	B. A. Bean	1♂	44672
Sakonnet Point, Rhode Island					Sept. 13, 99.	2532	W. N. e. 4.	1♀	31476
Southeast of Gorges				87	July 14, 83.	2243	do	1♂	11038
South of Marthas Vineyard	40 34 30	66 48 00		64	Sept. 26, 1884.	2200	do	2♂3♀	7285
Off Mas Vineyard	40 10 00	70 26 00		68	Sept. 20, 83.			2.	5422
Do	39 59 40	70 41 00	70–72			935,037	Fish Hawk	2♂1♀	4658
	39 45 00	69 44 45			Aug. 4, 88.				
South of Blk Island	19 00 25	69 00 00		68	Sept. 8, 88.	994	do	1♂4♀	40519
Off Marthas Vd	39 40 00	71 30 00		74	Aug. 19, 88.	2203	Albatross	6	8220
Off Nantucket Shoals	39 34 15	71 41 15		69¼	Sept. 30, 88.	2065	do	1	15052
D	39 29 00	70 58 40		72	July 30, 88.	2041	do	10	6616
Ede of Gulf Stream off Cape	39 22 50	68 25 20		81	July 28, 88.	2039	do	2♂	40520
Off North Carolina	38 19 26	68 68 30			June 6, 88.		B. A. Bn	3♀	31090
Do	36 20 24	74 46 30		69	June 4, 85.	2425	Albatross	5♂6♀	17896
Off east est of	32 43 00	71 51 00			1885.		do	14.	16053
Do	31 16 00	71 50 00			ek. 22, 88.		do	25.	22998
Bahamas, in	28 00 00	75 00 00			ek. 15, 88.		Bean and Riley	3♂2♀	31081
	About 120 mi. N. by E. of Abaco Id.				June 15, 1903		do	12♂13♀	31081
Do	ek. N. of Goat Abaco Island				June 16, 69.		B. A. Bn	5♀	31070
Southwest of Goat Abaco Island	27 57 30	77 27 30			May 2, 1886.	94	Albatross	9.	11399
Gen Turt e Cy, Bahamas					Jly 21, 1903		E. A. Andrews	1♂2♀	20714
Between Nassau and Elbow Key, Bahamas, in Sargassum	24 30 43	76 23 45			1880.		B. A. Bean	3 y	31052
Southwest of Abra Island, Bahamas					Apr. 15–17, 1884		do		11409
Key West, Florida					May–July, 1896.		F. S. Gent	1 ♀ v.	18448
Kingston Harb r, Jamaica							Dr Kershner	1♂1♀	19600
Northeast of Pto Rico	20 00 00	63 00 00			nd. 17–	234.	Albatross	12.	8714
St. Thomas, West Ides					187o–7		G. Bn Goode.	1 y	18563
Bermudas					Jan. 12, 13.		Dr Geo Hawes.	25.	43047
Bermudas, in fl ating seaweed, north shore					ad. 13, 83.		Dr F. V. Hamlin.	10.	5170
Bermudas							Albatross	14.	4027
Southwest of Bermudas	31 15 42	67 39 10				M Sta. 37.		100.	7744
From Bermudas to Bahamas				°C.		106.	Bache	2♂	49957
Do	32 33 00	72 14 00	Surface.	19 15	Jan. 30, 1914.	10178.	do	10♂	49944
Do	32 30 00	65 48 00	do	19.2	ek. 58, 19.	19.	d.	5♂1♀	49942
Do	32 29 00	71 29 00	do	18.95	ek. 1, 1914.	10171.	do	11♂4♀	49946
Do	32 27 00	71 55 00	do	18.9	Feb. 2, 19.	10182.	do	3♂4♀	49945
	30 27 00	66 06 00						(3 ov ig.)	
Do	29 27 00	66 05 00	1000–0	20 12	do 19.	1e.	do	1♂2♀	49948
Do	29 17 00	67 07 00	Surface.	20.07	ek. 20, 19.	19.	do	1♂2♀	49943
								(3 o vig.)	

Locality	Latitude	Longitude	Bottom	Temp. °F.	Date	Station	By whom collected	Sex	Cat. No.
Do	28 51 00	70 08 00	...do...	19.47	Feb. 24, 1914	10185	do	1♂1♀	4994?
Do	28 51 00	70 08 00	30-0	19.47	do	do	do	2♀2♀	49941
Do	28 51 00	75 13 00	50-0	21.55	Feb. 28, 1911	10194	do	1♀	49940
Atlantic (?) gulf w de	21 00 00	51 00 00					U. S. Expl. Exped.	1♂1♀	2325
Atlantic Ocean							Dr. C. C. Craft.	3	44471
Three ... al Harbor, Azores, from ... ing log	41 00 00	141 00 00			Ih, 98		Loris Del.	3♂3♀	17712
California					1880		D. S. Jordan.	1	3124
At sea, west of Humboldt Ba..., on ...					Ab. 7, 1891		W. H. all.	1♀v	1651
Off anta Catalina Is..., Ca...	32 47 30	118 10 00			Ab. 28, 90		Albatross	1♂1♀	4474
San Pedro, California	30 56 00	139 50 00					Il Albatross	1♂1♀	2975
Point Loma, California					Jan. 28, 1889		Albatross	4	17749
Off San ... (a...				59	Apr. 23, 1884	2929	do	1♂1♀	17450
At sea, ... of Lower California					Aug 1, 1886		Lieut. G. M. Stone	1♂1♀	17?17
San Benedicto Is and, Lower California					Apr. 20, 1888		A. W. Anthony.	3	29695
orte Bia, ... (a...					Feb. 10, 1891		do	1	19517
San Jose de Cabo, Lower California					Oct		do	1♀	29?94
Gulf of California	20 00 00	106 12 00					Albatross	1♀	22101
Off Acapulco, Me... o, from ... ift e				81–83	Ab. 25, 1898	4587	do	2♂3♀	29?24
... Is and, Mexico					Oct. 17, 98		A. W. Anthony.	18	33214
Salina Cruz, Mexico, sand each.	20 00 00	92 13 00		85	May 2–15, 1911	4605	Chas. C. am	1♂♀	21700
Off Guatemala	12 20 00	85 20 00					A b. !!, ... k and Hilde-brand.	1	323?2
... Is and, Panama						Surf. Sta. 18	Albatross	1♂	33215
Off Coombia	1 03 00	80 15 00			Apr. 1, 98		do	2♀o	44173
Off Galapagos Is ... from Green Turt e	5 17 S.	85 20 00		70–71	Nov. 10, 1904	4649	Pavt Herrendean.	2♀	22997
Galapagos ...					Aug 12, 1884		Dr. W. H. Jones, tr S.N.	1♂1♀	22100
Off Peru									5046
Payta, Peru									33213
Northeast of Hawaiian Is'ands, on ...	23 23 00	141 41 05		60	Mar 19, 1902	3800	Mr.	1♂	29?319
Northeast of Hawaiian Is ands ... net	23 23 00	141 41 05		61	Ab. 28, 1912	3900	io	1♂	29347
South east of Oahu Is ... Hawaiian Is ands.	Diamond Head Light, N. 76° E. 2.2'			74		3813	do	1♂	29346
South coast of Ahu Is and, ... on floating sticks	Lae o Ka Laau Light, N. 72° W. 15.6'			76	Apr. 2, 09	3833	do	1♂	23345
Nukuhiva Is and, Marquesas ... In ... of Cory-phaena.					Ab. 5, 899	13	do	1♀	44673
Japan?								1♀	13728
Indian Ocean	Equator, Lat. E. 90°				Feb. 1884		Dr. F. C. Da'e, U. S. S. Palos.	14	19314
Mauritius							H. A. Ward. Esp.	1♀	17697
Kerguelen land					Ab. 28, 1874		Dr. J. H. Kder, U. S. N.	1	15054

1 Large specimens.

PLANES MARINUS Rathbun.

Plate 64.

Planes marinus RATHBUN, Proc. U. S. Nat. Mus., vol. 47, 1914, p. 120, pl. 3 (type-locality, at sea, west of Lower California, in lat. 23° 49′ N.; long. 127° 50′ W.; holotype, Cat. No. 6065, M. C. Z.).

Diagnosis.—Carapace depressed about the middle; postero-lateral margins nearly straight.

Description.—Carapace convex antero-posteriorly and from side to side; surface covered with punctae and fine reticulations; coarser striae cross the anterior half transversely and nearly all the branchial region obliquely. Surface of front covered with short striae and minute granulation; free edge arcuate and faintly bilobed, each lobe appearing in front view slightly bilobed; edge a raised finely granulated rim; post-frontal lobes low. Antero-lateral margins convex, with one blunt tooth behind the tooth at the angle of the orbit; postero-lateral margins nearly straight, convergent.

Chelipeds equal, massive; upper and lower margins of arm transversely striated, inner expansion irregularly denticulated; outer surface of wrist finely striated, tooth at inner angle blunt; surface of palms nearly smooth, shining, punctate, upper surface rounded, covered with finely granulated longitudinal lines which become oblique proximally. Fingers stout, prehensile edges narrowly gaping, dentate, a larger tooth at middle of fixed finger.

Legs short and broad; third foot one and one-half times as long as carapace; merus of third pair three-fifths as broad as long; dactyli short and stumpy, armed with coarse spines.

Many species of *Planes* have been described in the past, all of which are referable to variations of *P. minutus;* but this form appears to be distinct. It has a great resemblance to *Pachygrapsus* also, and forms a link between the two genera.

From *Planes minutus* it differs in its broader carapace, somewhat depressed about the middle instead of uniformly convex; in the postero-lateral margins being nearly straight, as in *Pachygrapsus*, not arcuate as in *Planes minutus;* in the more extensive striation of the dorsal surface; in the broader basal joint of the antenna; the broader merus-joint of the outer maxilliped, both its inner and outer lobes being more strongly developed; in the feebler dentation of the distal end of the inner expansion of the arm.

Measurements.—Holotype male, length of carapace 17.6, width of same 19.3, width of front 10 mm.

Habitat.—Pelagic.

Material examined.—West of Lower California, Mexico; lat. 23° 49′ N.; long. 127° 50′ W.; surface; D. D. Raulet, collector; male holotype (6065, M. C. Z.); 1 male, 1 female, paratypes (22833, U.S.N.M.).

Genus GRAPSODIUS Holmes.

Grapsodius Holmes, Occas. Papers California Acad. Sci., vol. 7, 1900, p. 83; type, *G. eximius* Holmes.

Carapace striated above, with the sides converging behind, and armed with a single tooth behind the postorbital. Front broad, not deflexed, but with the median portion depressed. Eye-peduncles short. Orbits with the posterior surface bulging outwards instead of concave. Antennules and antennae unknown. Maxillipeds narrow, widely gaping, and devoid of an oblique piliferous ridge; merus subcordate, shorter than the ischium, the antero-internal angle produced; palp jointed near the middle of the distal margin of the merus. Dactyls spinulous. Abdomen of the male seven-jointed. (Holmes.)

Known only from the type-species.

GRAPSODIUS EXIMIUS Holmes.

Grapsodius eximius Holmes, Occas. Papers California Acad. Sci., vol. 7, 1900, p. 84, (type-locality, San Diego; type in Mus. Univ. Cal.).

Diagnosis.—Sides converging behind, unidentate. Posterior surface of orbit bulging outward. Maxillipeds slender.

Description.—Carapace undulated in front, flattened behind where it is more strongly striated; sides strongly converging posteriorly.

Front over half width of carapace, outer angles more or less projecting and rounded; anterior edge thin and minutely granulated, viewed from above it is nearly straight, being slightly convex on either side of the middle where it is a little concave; viewed from in front it sags downward in the center.

The orbits are remarkable in being swollen outward so that there is no hollow receptacle, as is usually the case, for the reception of the eyes; superior margin marked by a fine ridge; inferior margin by a line of granules from the postorbital tooth to the buccal area.

Maxillipeds slender and wide apart. Ischium much longer than merus, but not so wide; merus with outer margin convex and antero-external angle broadly rounded, inner margin straight and antero-internal angle produced in a prominent narrow lobe. First joint of the palp strongly convex near the middle of the inner margin. Exognath at the base about half as wide as ischium and tapering regularly to the tip which reaches slightly beyond the middle of the merus.

Chelipeds subequal. Merus short, trigonal, outer surface transversely striated, the inner margin produced into a laminate expansion which is distally truncated and dentate. Carpus with an inner spine. Hands smooth, inflated; upper margin of palm broadly rounded, but bears a fine ridge; a very fine ridge on the lower side of the outer surface extends upon the pollex. Fingers subcylindrical, not ridged nor grooved, and armed within with small teeth.

Merus of legs dilated and compressed, with the upper margins acute and ending in a tooth a little behind the supero-distal angle; infero-distal angle, in all but last pair, dentate. Carpal joints with a few small spines near the distal end of upper margin. Propodi with sides strongly convex, upper and lower margins spiny. Dactyls rather narrow, shorter than the propodi; strongly spinous above and below, and terminating in slender claws.

Abdomen of male widest at third segment, from which it tapers to the tip, near which it converges more rapidly; first segment much longer than the second; third segment about as long as the fourth, sides strongly convex; fifth segment scarcely longer than the fourth, and shorter than the sixth; last segment triangular, acute.

Measurements.—Male holotype, length of carapace 18.5, width of same 21, width of front 11.2 mm.

Range.—Known only from Doctor Holmes's description of the single male from San Diego, California. (After Holmes.)

Subfamily VARUNINAE Alcock.

Grapsinae DANA, Amer. Journ. Sci., ser. 2, vol. 12, 1851, p. 287 (part); U. S. Expl. Exped., vol. 13, Crust., pt. 1, 1852, p. 334 (part).
Varunacea MILNE EDWARDS, Ann. Sci. Nat., ser. 3, Zool., vol. 20, 1853, p. 175 [141].
Cyclograpsacea MILNE EDWARDS, Ann. Sci. Nat., ser. 3, Zool., vol. 20, 1853, p. 191 [157].
Grapsacea MILNE EDWARDS, Ann. Sci. Nat., ser. 3, Zool., vol. 20, 1853, p. 163 [129], (part: *Euchirograpsus*).

Front moderately or little deflexed, sometimes sublaminar. The declivous postero-lateral portion of the branchial region is set off from the rest of that region by a line more or less distinctly marked. The suborbital crest, which supplements the defective lower border of the orbit, is rather distant from the orbit and usually runs in a line with the anterior border of the epistome. Antennal flagellum usually of good length. External maxillipeds moderately or slightly gaping, without an oblique hairy crest; the palp articulates with the middle of the anterior border, or near the antero-external angle, of the merus, and the exognath while typically broad and exposed throughout, is in American genera rather narrow, and sometimes partly concealed. The male abdomen rarely covers all the space between the last pair of legs.

Genus CYRTOGRAPSUS Dana.

Cyrtograpsus DANA, Amer. Journ. Sci., ser. 2, vol. 12, 1851, p. 288; Proc. Acad. Nat. Sci. Philadelphia, vol. 5, 1851 (1852), pp. 247 and 250; type, *C. angulatus* Dana.

Carapace broader than long, narrowing anteriorly, fronto-orbital distance less than the length; convex; surface uneven, regions except the hepatic, well marked. Antero-lateral borders armed with three teeth behind the acute orbital angle, a raised line running inward on the carapace from the last tooth. Carapace widest at the last tooth.

Front occupying half the distance between the postorbital angles, rather prominent, slightly inclined, edge divided into two shallow lobes.

Orbits a good deal larger than the eyes; lower margin divided into two parts which are discontinuous. Near the inner angle a triangular lobe arises from the orbital fossa and with an adjacent lobe of the first movable joint of the antenna, half fills the inner hiatus of the orbit; the second movable joint occupies the remainder of the hiatus, permitting the flagellum to enter the orbit.

Antennular cavities broadly triangular, the antennules folding obliquely; interantennular septum narrow.

Epistome transverse, and shallow fore and aft. Buccal cavity with sides parallel, anterior margin arched beyond the outer maxillipeds, which have a moderate gape; merus and ischium of equal length, merus wider than ischium and as wide as long, subcordate, the thickened palp articulating at the middle of the anterior margin.

Chelipeds equal, in the male heavy, from one and two-thirds to one and three-fourths times as long as the carapace. Palms widening distally, fingers narrow, hollowed underneath.

Legs rather narrow, the second and third pairs much the longest. Dactyli unarmed.

Abdomen with seven segments in both sexes; and in the male leaves the sternum exposed at its base. First segment with a prominent transverse granulated ridge visible in a dorsal view.

Comprises two species inhabiting South America. Species on both sides of the continent: *angulatus*.

KEY TO THE SPECIES OF THE GENUS CYRTOGRAPSUS.

A^1. Antero-lateral teeth prominent, projecting beyond the general outline of the carapace_____ _____*angulatus*, p. 261.

A^2. Antero-lateral teeth not prominent, not projecting beyond the general outline of the carapace_____*altimanus*, p. 262.

CYRTOGRAPSUS ANGULATUS Dana.

Plate 65; plate 159, figs. 7 and 8.

Cyrtograpsus angulatus DANA, Proc. Acad. Nat. Sci. Philadelphia, vol. 5, 1851 (1852), p. 250 (type-locality, Rio Negro, Patagonia; holotype, Cat. No. 2346, U.S.N.M.); U. S. Expl. Exped., vol. 13, Crust., pt. 1, 1852, p. 352; atlas, 1855, pl. 22, figs. 6 *a–e.*

Cyrtograpsus cirripes SMITH, Trans. Connecticut Acad. Arts and Sci., vol. 2, 1869, p. 11, pl. 1, fig. 3 (type-locality, Rio de Janeiro; male holotype, Cat. No. 6125, in Mus. Comp. Zoöl.).

Diagnosis.—Antero-lateral teeth prominent. A postero-lateral tooth. Front strongly bilobed.

Description.—Carapace somewhat hexagonal; margins and surface granulate; three short transverse elevations on either side of the gastric region; mesogastric and cardiac regions prominent. Front with each lobe curved upward, sides (inner border of orbit) oblique. Orbit with a superior notch; outer tooth acute with convex outer margin. Next tooth on the side smaller, less curved; second tooth still smaller, and short-pointed; third tooth most prominent, acuminate, a single curved line of granules runs inward from it. On the postero-lateral margin is a low, blunt, obtuse-angled tooth.

Chelipeds granulate. Inner surface of carpus flattened, obtuse-angled. Propodus of adult male a little higher than its superior length, flattened above; a longitudinal ridge above the lower margin. Dactylus one and one-fourth times as long as upper margin of palm; pollex slightly deflexed. Fingers with a narrow gape; teeth small and low.

Third pair of legs the longest. Merus joints of all the feet with subparallel sides, and a subterminal tooth on the anterior border. Last three joints of last pair fringed with hair; propodus and base of dactylus of first and second pairs fringed posteriorly.

Measurements.—Male holotype, length of carapace 37, width of same 44.5 mm.

Range.—From Rio de Janeiro, Brazil, southward to Patagonia, thence northward on the Pacific coast to Peru.

Material examined.—

Rio de Janeiro, Brazil; Capt. Harrington; 1 male, holotype of *C. cirripes* (6125, M. C. Z.); received from Peabody Acad. Sci., Nov., 1885.

Montevideo, Uruguay; 1897; Bisege, collector; from Mus. S. Paulo; 1 female (48318).

Rio de la Plata; Capt. Page, collector; 2 males identified by W. Stimpson (2469).

Rio Negro, Patagonia; U. S. Exploring Exped.; 2 males, 2 females (1 male is holotype) (2346).

San Lorenzo Island, Peru; Dr. H. E. Ames, U. S. Navy; 5 males, 1 female (18624).

CYRTOGRAPSUS ALTIMANUS Rathbun.

Plate 66.

Cyrtograpsus altimanus RATHBUN, Proc. U. S. Nat. Mus., vol. 47, 1914, p. 121, pl. 4 (type-locality, San Matias Bay, Patagonia; holotype male, Cat. No. 6126, M. C. Z.).

Diagnosis.—Antero-lateral teeth not prominent. No postero-lateral tooth. Front faintly bilobed.

This species while closely related to *C. angulatus*, which inhabits the same region, is much smoother and less ornate so that there is no likelihood of their being confused.

The carapace is not strongly areolated though the regions are well defined; the gastric region lacks the beaded transverse ridge characteristic of the older species. The surface is densely covered with fine depressed granules and somewhat less numerous punctae; it appears almost smooth to the naked eye, while in *angulatus* the surface is obviously roughened with coarser granules. As to shape, the carapace has no sharp lateral angles, the antero-lateral margins are shorter than in *angulatus*, and the postero-lateral margins are longer and subparallel to each other. The antero-lateral margins have four teeth, including the orbital tooth, but they are small, especially the last two, and do not project beyond the marginal line; the intervals between the teeth diminish successively in length. There is no indication of a postero-lateral tooth.

The front is relatively wider than in *angulatus* and is feebly emarginate at the middle; the orbits are correspondingly smaller.

The outer maxillipeds have much the same shape in the two species, but in *altimanus* they are shorter and wider and the gape narrower.

The palms in the adult male are much higher in our species, especially at the distal end, and the movable finger is strongly deflexed; the immovable finger is nearly horizontal; there is a triangular space between the fingers for their proximal half only.

Legs narrower than in *angulatus*, second and third pairs subequal; propodal joint and proximal part of terminal joint of first three pairs fringed with hair on the posterior margin; last two joints and distal part of carpal joint of last pair fringed with hair on both margins.

The abdomen of the male is narrower and more oblong than in *angulatus*, and the appendages of the first segment slenderer.

Measurements.—Male holotype, length of carapace 16.8, width of same 18.4 mm.

Range.—Rio Grande do Sul, Brazil, and San Matias Bay, Patagonia.

Material examined.—

Rio Grande do Sul, Brazil; Capt. George Harrington; 1 male (6127, M. C. Z.); received from Essex Inst., in exchange.

San Matias Bay, Patagonia; *Hassler* Exped.; male holotype (6126, M. C. Z.); 2 males, paratypes (22835, U.S.N.M.).

Genus HEMIGRAPSUS Dana.

Hemigrapsus DANA, Amer. Journ. Sci., ser. 2, vol. 12, 1851, p. 288; Proc. Acad. Nat. Sci. Philadelphia, vol. 5, 1851 (1852), pp. 247 and 250; type, *H. crassimanus* Dana.

Lobograpsus A. MILNE EDWARDS, Ann. Soc. Entom. France, ser. 4, vol. 9, 1869, p. 173; type, *L. crenulatus* (Milne Edwards).

Carapace broader than long, quadrate with the antero-lateral margins rounded and dentate; depressed; an oblique ridge runs inward and backward from the postero-lateral margin. Front less than half width of carapace. Orbits of moderate size.

Antennules folding obliquely. Tooth at inner angle of orbit well developed. Antenna filling the orbital hiatus; flagellum entering the orbit. The suborbital crest forms a tuberculated or striated stridulating ridge which scrapes against the distal end of the arm. Epistome well developed.

Buccal cavity quadrate with the anterior corners rounded. Outer maxillipeds moderately gaping; merus broader but very little shorter than the ischium, and as broad as, or broader than long; its inner and outer margins convex, anterior margin excavate, with the palpus inserted toward its outer angle.

Chelipeds equal or subequal, stout. Palms often with a patch of hair inside. Fingers hollowed out beneath in a shallow groove.

Legs of moderate length, and almost unarmed.

The abdomen of the male does not cover the sternum at its base.

Distribution.—Temperate shores of the Pacific Ocean; Hawaiian Islands; South Atlantic coast of America.

KEY TO THE AMERICAN SPECIES OF THE GENUS HEMIGRAPSUS.

A^1. Three lateral teeth behind the orbital angle_____*affinis*, p. 264.
A^2. Two lateral teeth behind the orbital angle.
   B^1. Postero-lateral margins strongly convergent. Legs with thick fringes of hair_____*crenulatus*, p. 266.
   B^2. Postero-lateral margins not sensibly convergent.
      C^1. Legs devoid of hair. Two deep sinuses in the epistome___*nudus*, p. 267.
      C^2. Legs more or less hairy. Two shallow sinuses in the epistome.
                                                       oregonensis, p. 270.

HEMIGRAPSUS AFFINIS Dana.

Plate 67.

Hemigrapsus affinis DANA, Proc. Acad. Nat. Sci. Philadelphia, vol. 5, 1851 (1852), p. 250 (type-locality, *Porta* 'Rio Negro' *Patagoniae;* type not extant); U. S. Expl. Exped., vol. 13, Crust., pt. 1, 1852, p. 350; atlas, 1855, pl. 22, fig. 5.

Diagnosis.—Small, narrow; three side teeth; upper surface of palm flattened; fingers subacute; sixth segment of male abdomen shorter than fifth.

Description.—A very small species. Carapace very little broader than long, sides slightly convex and convergent to the acute anterolateral angles. Surface sloping gently downward to the four sides, uneven, partly granulated, the cardiac and posterior gastric regions well marked. Hepatic region depressed; a transverse depression opposite the last lateral tooth. Three notches in the lateral margin form three small teeth; distance from first tooth to orbital angle one and one-third times as great as distance from first to second tooth, which last is twice as great as distance from second to third tooth. The oblique raised line on the branchial region arises immediately behind the last tooth. Above the bases of the last two legs a granulated line runs parallel to the margin.

Front distinctly less than half width of carapace, edge convex, bilobed, outer angle obtuse.

Antennules very large. The basal and succeeding joint of the antenna together fill the orbital hiatus. The posterior margin of the epistome is regularly arcuate as seen from below. Merus of outer maxilliped much wider than ischium, its outer margin very arcuate, the palpus inserted not far from the middle.

The stridulating ridge is crenulate, the crenules becoming finer toward the outer end; it scrapes against a short portion of the distal border of the inner margin of the arm. Arms short, stout, unarmed, having a groove near the distal end. Wrist with inner margin rounded. Chelae heavy, palms strongly widened distally, wider than superior length, upper surface partly flat. Fingers rather narrow, edges crenulate, a triangular gap at base in the male, no gap in female, fingers subacute; several longitudinal rows of punctures, inclosing on the outer side of the pollex a fairly distinct ridge. In the female this ridge is more pronounced and is continued along the palm.

Legs slender, especially the dactyli; last pair much the shortest; unarmed, except for a feeble tooth near the end of the merus joints; furnished with a few hairs; dactyli grooved, with long slender horny tips.

Sixth segment of male abdomen shorter than fifth, seventh elongate, very little widened at base.

Measurements.—Male (22104), length of carapace 5.8, width 6, width of front 2.5 mm.; length 7 lines (*D*ana).

Range.—From Cape St. Roque, Brazil, to Gulf of San Matias, Patagonia: $10\frac{1}{2}$ to 52 fathoms.

Material examined.—See page 266.

HEMIGRAPSUS CRENULATUS (Milne Edwards).

Plate 68.

Cyclograpsus crenulatus MILNE EDWARDS, Hist. Nat. Crust., vol. 2, 1837, p. 80 ("*habite?*"; type in Paris Mus.)

Trichodactylus granarius NICOLET in Gay, Hist. Chile, Zool., vol. 3, 1849, p. 151; atlas, vol. 2, pl. 1, fig. 3 (type-locality, Chile; cotypes in Paris Mus.).

Hemigrapsus crenulatus DANA, U. S. Expl. Exped., vol. 13, Crust., pt. 1, 1852, p. 349; atlas, 1855, pl. 22, fig. 3.

Trichodactylus granulatus MILNE EDWARDS, Ann. Sci. Nat., ser. 3, Zool., vol. 20, 1853, p. 216 [182]. Probably a *lapsus pennae* for *T. granarius*.

Lobograpsus crenulatus A. MILNE EDWARDS, Ann. Soc. Entom., ser. 4, vol. 9, 1869, p. 173.

?*Heterograpsus barbimanus* CANO, Boll. Soc. Nat. Napoli, ser. 1, vol. 3, 1889, pp. 99 and 243, Chiloe; not *H. barbimanus* Heller.

Heterograpsus sanguineus LENZ (not Milne Edwards), Zool. Jahrb., Suppl., vol. 5, 1902, p. 765.

Diagnosis.—Two side teeth. Sides converging behind. Chela of male hairy inside. Legs with thick fringes of hair.

Description.—Of medium size; carapace sensibly broader than long; antero-lateral margins more strongly curved than in *H. affinis;* postero-lateral margins strongly convergent; flat behind, sloping down on front and sides; cardiac and posterior gastric regions well defined; surface covered with coarse, close-set granules, except posteriorly where they are replaced by punctae connected by fine impressed lines. Two notches in the lateral margin (the anterior the larger) form two acute teeth which are not prominent; interval equal to that between the first and the orbital angle, which is nearly a right angle. The oblique branchial ridge is considerably behind the last tooth and is more longitudinal in direction than in *H. affinis;* ridge above the last two legs short and sinuous.

Front two-fifths as wide as carapace, very slightly arched.

Basal joint of antenna very broad, with an outer lobe; not filling the orbital hiatus.

Posterior border of epistome sinuous in front view; with two reentering lobes in ventral view. Merus of maxilliped obcordate.

Stridulating ridge thick, its outer half divided in the male into three or four dentiform tubercles, inner portion finely crenulate; in the female the ridge is broken into about 15 tubercles which are longer in the outer half. It scrapes against a short curved horny ridge at the distal end of the inner margin of the arm.

Chelipeds granulate. Upper border of arm, inner border of wrist and inner surface (in the male) of chela, clothed with thick hair. Legs margined with fringes of hair. Inner margin of wrist bluntly angled. Palm inflated, very convex below, upper surface rounded, a faint longitudinal ridge runs obliquely on to the pollex. Fingers narrowly gaping, teeth irregular, tips horny.

Legs of medium width; dactylus of last pair proportionally broader than the others.

Sixth segment of male abdomen longer than fifth, but equalling the seventh.

Measurements.—Male (25032), length of carapace 27.5, width of same 32 mm.

Range.—Chile; west coast of Patagonia; also New Zealand.

Material examined.

Lota, Chile; Albatross; 1 y. (22103).

Cavancha, Chile; Plate, coll.; received from Lubeck Mus.; 1 female (33157).

Chiloe Island, Chile (M. C. Z.).

Port Otway, Patagonia; Feb. 9, 1888; *Albatross;* 47 males, 9 females (22102).

New Zealand; Otago Univ. Mus.; 1 male, 1 female (16230).

Bay of Islands, New Zealand; U. S. Exploring Exped.; 2 males (2330).

Port Chalmers, New Zealand; Charles Chilton; 2 males, 3 females (18165).

Dunedin, New Zealand; George M. Thomson; 1 male, 1 female (25032).

Locality unknown; U. S. Exploring Exped.; 2 males, 2 females (2432).

HEMIGRAPSUS NUDUS (Dana).

Plate 69.

Cyclograpsus marmoratus WHITE (not *Cancer marmoratus* Fabricius), List Crust. Brit. Mus., 1847, p. 41, *nomen nudum* (type-locality, Sitka; type in Brit. Mus.).

Pseudograpsus nudus DANA, Proc. Acad. Nat. Sci. Philadelphia, 1851, p. 249 (type-locality, *in Oregoniae freto* "Puget"; type, Cat. No. 2331, U. S. Nat. Mus.); U. S. Expl. Exped., vol. 13, Crust., pt. 1, 1852, p. 335; 1855, atlas, pl. 20, fig. 7a-e.

Hetcrograpsus marmoratus MILNE EDWARDS, Ann. Sci. Nat., ser. 3, Zool., vol. 20, 1853, p. 193 [159].
Heterograpsus nudus STIMPSON, Proc. Acad. Nat. Sci. Philadelphia, vol. 10, 1858, p. 104 [50].
Hetcrograpsus sanguineus KINGSLEY (not de Haan), Proc. Acad. Nat. Sci. Philadelphia, 1880, p. 208.
Brachynotus nudus HOLMES, Occas. Papers California Acad. Sci., vol. 7, 1900, p. 81.
Hemigrapsus nudus RATHBUN, Amer. Nat., vol. 34, 1900, p. 587.

Diagnosis.—Two side teeth. Sides subparallel behind. Epistome with two deep notches. Chela of male with hairy patch inside. Legs bare.

Description.—Carapace flat and punctate behind, deflexed and granulate in front. A curved line of light-colored pits arches forward from the last side tooth to the end of the H impression. Front convex, bilobed with a shallow emargination, edge thickened. Epigastric lobes not very prominent. Antero-lateral margin strongly curved, with two teeth behind the orbital tooth, intervals equal, last tooth very small.

Posterior margin of epistome with two broad, deep sinuses near the outer ends. Outer margin of merus of maxilliped moderately curved. Stridulating ridge composed of about 15 tubercles in the male, 20 in the female; in the male they are largest in the middle, most distant at the outer end; in the female they are nearly equal. The ridge scrapes against a smooth horny ridge on the edge of a lobe at the lower distal angle of the inner surface of the arm.

Chelipeds smooth. Inner angle of wrist bluntly angular. Propodus with a longitudinal ridge on the lower part of the outer surface, which is stronger in the female than in the male. Palms of male inflated, furnished with a patch of long fine hair at distal end of inner surface, and extending on the pollex. Fingers irregularly toothed, narrowly gaping in the male, tips horny.

Legs stout, rather short, smooth, and nude. *D*actyli stout, scabrous, those of the last pair especially short, their tips upturned.

Color.—Variable, generally of a mahogany red, but may be purplish, dark red, or red marbled with white, occasionally almost entirely white. Chelipeds mottled above with small, round, red spots. (Holmes.)

Measurements.—Male (32233), length of carapace 48, width of same, 56.2 mm.

Habitat.—Among rocks near the shore.

Range.—From Sitka, Alaska, to Gulf of California.

Material examined.—

Kodiak, Alaska; W. G. W. Harford; 5 male and female (3264). Doubt has been expressed as to this locality.

Sitka, Alaska: W. H. Dall; 9 males, 13 females (14804). Aug., 1892; *Albatross;* 7 males, 4 females (19335).

Departure Bay, Alaska; Feb., 1882; Dr. W. H. Jones, U. S. Navy, U. S. S. *Wachusett;* 4 males and females (5316).

Wrangel, Alaska; rocky beach, under stones, low water; May 25, 1882; Dr. W. H. Jones, U. S. Navy, U. S. S. *Wachusett;* 11 males and females (5331).

Steamer Bay, Etolin Island, Alaska; Bureau of Fisheries: June 18, 1914; 1 male, 1 female ovig. (49905). June 1, 1911; 2 females (49904).

Naha Bay, Revillagigedo Island, Alaska; May 30, 1904; Bureau of Fisheries; 2 females (1 ovig.) (44537).

Ward Cove, Revillagigedo Island, Alaska; Dr. T. H. Streets, U. S. Navy; 2 males, 13 females ovig. (14805).

Union Bay, Bayne Sound, British Columbia; May 14, 1906; *Albatross;* 14 males, 3 females (2 ovig.) (44500).

Denman Island, British Columbia; *Albatross:* Shore; June, 1903; 1 male (31588). North end, on shore, between tide marks; Apr. 18, 1914; 20 males, 19 females (9 ovig.), 7 y. (49911). North side, May 12, 1914; 2 males, 2 females (1 ovig.), 3 y. (49915).

Gabriola Island, Taylor Bay, British Columbia; June 20, 1903; *Albatross;* 12 males, 6 females (31590).

Nanaimo, British Columbia; July 10, 1888; *Albatross;* 24 males 4 females (15528). Beach; Aug. 2, 1897; Harlan I. Smith; 4 females (21367). Shore; June 19, 1903; *Albatross;* 6 males, 3 females (31589).

Victoria, British Columbia; Mr. Nichols; 3 males (1 paper shell) (44518).

Gordon Head, 7 miles from Victoria, British Columbia; May 1, 1905; J. E. Benedict; 1 male, 1 female (31505).

Safety Cove, British Columbia; Sept. 15, 1895; F. W. True; 1 male, 1 female (19496).

Straits of Fuca; 1880; D. S. Jordan; 4 males and females (3069).

Neah Bay, Washington; Apr. 27, 1914; *Albatross;* 11 males, 19 females (11 ovig.) (49912).

Puget Sound: U. S. Exploring Exped.; 3 specimens, all much broken; cotypes (2331). 1880; D. S. Jordan; 6 males and females (3098). *Albatross;* 1 female (2187).

Sucia Island, Washington; May 6, 1894; *Albatross;* 4 specimens (18962).

Sequin (Seguin), Washington; J. M. Grant; 10 specimens (49068).

Port Townsend, Washington; June 27, 1903; U. S. Fish Commission; 1 male (44519).

Port Orchard, Washington; July, 1889; O. B. Johnson; 9 males, 5 females (14969).

Angel Island, San Francisco Bay, California; Nov. 7, 1890; *Albatross;* 3 males, 3 females (20162).

Monterey, California: Trowbridge; 2 males (2056). H. Hemphill; 1 male (2287).

Monterey Bay, California; under submerged rocks along shore; Harold Heath; 1 male (22878).

Pacific Grove, California: Stanford Univ.; 1 male (19075). July, 1895; J. O. Snyder; 5 males, 2 females (19819). John C. Brown; 3 specimens (23922). June, 1905; J. E. Benedict; 28 specimens (32233).

Santa Barbara, California; 1880; D. S. Jordan; 1 female, ovig. (3049).

Catalina Harbor and Island, California; 1874; W. H. Dall; 1 male, 1 female (14802).

Pillar Point, San Mateo County, California; June 30, 1903; C. F. Baker; 1 male, 1 female (29311).

Point Loma, San Diego Bay, California; *Albatross;* 7 specimens (18961).

Turtle Bay, Lower California; Aug. 1, 1896; A. W. Anthony: 6 males, 10 females (19512).

HEMIGRAPSUS OREGONENSIS (Dana).

Plate 70.

Pseudograpsus oregonensis DANA, Proc. Acad. Nat. Sci. Philadelphia, vol. 5, 1851 (1852), p. 248 (type-locality, *in Oregoniae freto* " Puget "; type male, Cat. No. 2333, U.S.N.M.); U. S. Expl. Exped., vol. 13, Crust., pt. 1, 1852, p. 334; atlas, 1855, pl. 20, fig. 6a–b.

Brachynotus oregonensis HOLMES, Occas. Papers California Acad. Sci., vol. 7, 1900, p. 82.

Hemigrapsus oregonensis RATHBUN, Amer. Nat., vol. 34, 1900, p. 587.

Diagnosis.—Broad in front. Two side teeth. Surface uneven. Front with two prominent lobes. Palm of male with hairy patch inside. Legs hairy.

Description.—Differs from *H. nudus* as follows:

The distance across the front and orbits is greater, the anterolateral margins being less curved. Surface more uneven, especially anteriorly, sparsely punctate. A raised curved line of granules runs between the last side tooth and the end of the H impression. Edge of front margined and bearing two prominent thickened lobes. Epigastric lobes very prominent. The two lateral teeth are more prominent and their sinuses more deeply cut.

Posterior edge of epistome with two shallow V sinuses. The outer maxillipeds are nearer together, the merus is no broader than long.

Stridulating ridge of the male with three or four long smooth tubercles near the middle, and granules or small tubercles at each end; in the female the ridge is finely divided as in *H. nudus*. The lobe at the distal end of the arm which bears the complementary ridge is very prominent. Inner angle of wrist bluntly rounded. Palm of male with hairy patch inside. Legs with margins hairy.

Color.—Dull gray mottled with ferrugineous spots. Spots on legs small, but there may be blotches of considerable size on the carapace. Very young occasionally marked with large blotches of white. (Holmes.)

Measurements.—Male (2320), length of carapace 28.4, width of same 34.7 mm.

Range.—From Prince William Sound, Alaska, to Gulf of California, Mexico.

Material examined.—

Fox Island, Alaska; Harriman Alaska Exped.; 1 male (23842).

Sitka, Alaska: L. A. Beardslee; 1 male (14807). W. H. Dall; 32 males, 34 females, 1 y. (14810). *Albatross;* 1 male (19347).

Wrangel, Alaska; beach, low tide; July, 1882; Dr. W. H. Jones, U. S. Navy, U. S. S. *Wachusett;* 2 males (5330).

Alert Bay, Alaska; Feb. 22, 1882; Dr. W. H. Jones, U. S. Navy, U. S. S. *Wachusett;* 17 males and females (5317).

Klinquan, Prince of Wales Island, Alaska; June 17, 1897; *Albatross;* 1 male (21790).

Boca de Quadra, head of Mink Arm, Alaska; July 6, 1903; *Albatross;* 2 females (31594).

Fort Rupert, Beaver Bay, British Columbia: June 25, 1903; *Albatross;* 1 female (31672). Between high and low tide lines, among seaweed, kelp, etc.; Harlan I. Smith; 2 males, 1 female (22590).

Comox, British Columbia; 1893; *Albatross;* 80 specimens (18295).

Union Bay, Bayne Sound, British Columbia; May 14, 1906; *Albatross;* 62 males, 61 females (44499).

Denman Island, British Columbia; 1914; *Albatross:* North end, on shore, between tide marks; Apr. 18; 18 males, 8 females (2 ovig.) (49908). North side; May 12; 17 males, 8 females (3 ovig.) (49906)

Gabriola Island, Taylor Bay, British Columbia; June 20, 1903; *Albatross;* 8 males, 3 females (31591).

Nanaimo Bay, British Columbia; shore; June 19, 1903; *Albatross;* 2 males (31593).

Nanaimo, British Columbia: July 10, 1888; *Albatross;* 2 males (15529). Beach; Aug: 2, 1897; Harlan I. Smith; 3 males, 4 females (21366).

Victoria Harbor, British Columbia; 1 specimen (14466).

Gordon Head, 7 miles from Victoria, British Columbia; May 1, 1905; J. E. Benedict; 2 females (13504).
Puget Sound: *Albatross;* 8 males, 4 females, 1 y. (12788). U. S. Exploring Exped.; 1 male holotype (2333).
Sucia Island, Washington; May 6, 1894; *Albatross;* 10 specimens (18964).
Olga, Washington; Richard Willis; 10 specimens (31488).
Sequin (Seguin), Washington; J. M. Grant; 10 specimens (49069).
Port Townsend, Washington; June 27, 1903; *Albatross;* 1 male (44536).
Marrowstone Point, near Port Townsend, Washington, June 29, 1903; *Albatross;* 1 male, 7 y. (31592).
Port Ludlow, Washington; S. Bailey; 6 males and females (5324).
Port Orchard, Washington; July, 1889; O. B. Johnson; 13 males, 6 females (14968).
Oyster Bay, Washington; Dec., 1914; W. L. McAtee; 7 specimens (49130).
Crescent City Bay, California, lat. 41° 42′ N.; 6 males and females (5326).
Tomales Bay, Marin County, California; April 28, 1897; N. B. Schofield; 2 males, 1 female (49103).
Angel Island, San Francisco Bay, California; November 7, 1890; *Albatross;* 26 specimens (20163).
Oakland, California; H. Hemphill; 7 males (2288).
Off San Mateo, San Francisco Bay, California; 3 fathoms; *Albatross;* 10+ specimens (20165).
San Francisco Bay, near Palo Alto, California; July 4, 1903; C. F. Baker; 2 males (29315).
Monterey Bay, California; in sandy locations under rocks at mean and low tide marks; Harold Heath; 1 male, 4 females (22877).
Santa Barbara, California; June, 1875; Dr. H. C. Yarrow, Exploration west of 100th meridian; 25+ males and females (2320).
Playa del Rey, Los Angeles County, California; October, 1915; E. J. Brown; 3 males (48818).
Off Catalina Island, California; 50 fathoms; H. N. Lowe; 1 male (29961).
Laguna Beach, California; W. A. Hilton, Pomona College; 1 male (50103).
Zuninga Shoals, off Point Loma, California; shore; Mar. 3, 1904; *Albatross;* 1 female (44517).
San *D*iego, California; Mar. 9, 1898; *Albatross;* 1 male, 1 female (21789).
San *D*iego Bay, California; *Albatross:* Mar. 21, 1894; 3 fathoms; fne. S. brk. sh.; Sta. 3567; 1 female (19341). Mar. 21, 1894; 6

fathoms; fne. S. brk. Sh.; Sta. 3577; 2 specimens (19348). Mar. 24, 1894; 3 fathoms; R. oyster Sh.; Sta. 3589; 1 male, 2 females (19340). Mar. 31, 1896; 5 fathoms; M. Sh.; Sta. 3616; 2 females (20167). Mar. 31, 1896; 4 fathoms; M. Sh.; Sta. 3619; 2 males (20168). Lat. 32° 32' N.; long. 117° 07' W.; Monument 258, Mexican Boundary Survey; July 18, 1894; Dr. E. A. Mearns, U. S. Army; 4 males (18663).

San Bartolome Bay, Lower California; Apr. 11, 1889; *Albatross;* 1 male (17454).

Todos Santos Bay, Lower California; H. Hemphill; 2 males (19499).

Puerto Refugio, Angel Island, Gulf of California; Mar. 29, 1889; *Albatross;* 1 male (17453).

San Luis Gonzales Bay, Gulf of California; Mar. 27, 1889; *Albatross;* 50+ specimens (17452).

Guaymas Bay, Sonora, Mexico; shore; Feb. 20, 1904; Wm. Palmer; 2 females (31511).

Inner harbor, Guaymas, Mexico; Feb. 23, 1891; P. L. Jouy; 35+ specimens (17292, 17293).

Locality unknown; 1 y. (17685); and 1 male, U. S. Exploring Exped. (2444).

TETRAGRAPSUS, new genus.

Type.—T. jouyi (Rathbun).

Carapace quadrate, with the side margins parallel, dentate. Front more than half width of carapace, curved gradually downward, and not projecting beyond the antennules or epistome. Orbit outwardly very incomplete; eye slightly elongate. Suborbital ridge prominent, tuberculate. Maxillipeds gaping, merus longitudinally grooved, longer than wide, and as long as, but wider than, the ischium.

The maxillipeds resemble those of *Hemigrapsus* in form and surface but are not approximate. In the shape of carapace, chelipeds and legs and in the character of the suborbital ridge this genus approaches *Brachynotus* de Haan,[1] but in the latter the front is laminate, the maxillipeds smooth, their merus short and broad.

TETRAGRAPSUS JOUYI (Rathbun).

Plate 71.

Brachynotus (Heterograpsus) jouyi RATHBUN, Proc. U. S. Nat. Mus., vol. 16, 1893, p. 247 (type-locality, Guaymas, Mexico; type, Cat. No. 17496, U.S.N.M.).

Diagnosis.—Sides parallel, dentate. Front rounding down to antennular cavities. Maxillipeds gaping; merus elongate, grooved.

---
[1] Fauna Japon., Crust., 1833, p. 5; 1835, p. 34.

Description.—Carapace considerably broader than long, sides with two teeth behind the orbital tooth. Surface finely punctate and pubescent. Regions little marked; a gastro-cardiac sulcus; a depression at the inner angle of the hepatic region; a sulcus between the epigastric lobes.

Lower edge of front sinuous as seen from above, being arched over the antennules; convex in front view; outer angles rounded; sides oblique. Upper margin of orbit sloping backward and outward. The orbital and two side teeth are sharp pointed, deeply separated, and diminish in size from the first to the last; the first and second are similar, their outer margin convex; the last is slender, outer margin straight. Epistome very narrow, and in front view concave each side of the middle.

Inner half of postorbital stridulating ridge in male very finely striated; outer half formed of two elongated tubercles, the inner of which is twice as long as the outer, and is continuous with the striated portion; in the female there are four elongated dentiform tubercles and at the inner end a few denticles.

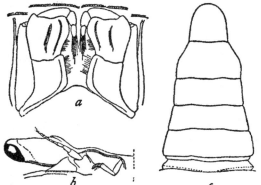

FIG. 139.—TETRAGRAPSUS JOUYI, MALE (17496). *a*, BUCCAL CAVITY, × 8½; *b*, ORBITAL AND ANTENNAL REGION, × 6½; *c*, ABDOMEN, × 6½.

Outer surface of arm crossed by short squamiform rows of fine granules; a very prominent lobe on the distal half of the inner margin bears a horny ridge. Inner angle of wrist produced in a very small tooth in the male, a spine in the female. Palms of male swollen, smooth, above proximally marginate; fingers stout, with fine prehensile teeth and a linear gape; a large patch of long hair occupies the center of the inner surface of the chela. The chela of the female is very small, upper margin of propodus complete, a ridge on lower part of outer surface; fingers faintly grooved, more broadly spooned than in male; inner surface of palm with a few hairs.

Legs slender, hairy, carpal and propodal joints, save in the last pair, densely pilose; merus joints crossed by short, granulated lines; dactyli slender, of last pair curved upward at tip.

Abdomen of male very narrow, last joint suboblong.

Measurements.—Male, holotype, length of carapace 10.4, width of same 13 mm.

Habitat.—Under stones. Scarce. (Jouy.)

Range.—Known only from the type-locality.

Material examined.—Guaymas, Mexico, Gulf side; Feb. 27, 1891; P. L. Jouy; 6 males, 5 females (1 male is holotype) (17496); paratype in M. C. Z.

Genus GLYPTOGRAPSUS Smith.

Glyptograpsus SMITH, Trans. Connecticut Acad. Arts and Sci., vol. 2, 1870, p. 153; type, *G. impressus* Smith.

Areograpsus BENEDICT, Johns Hopkins Univ. Circ., vol. 11, No. 97, 1892, p. 77; type, *A. jamaicensis* Benedict.

Carapace thick, much broader than long and broader behind than in front; dorsal surface distinctly areolated. Lateral margins strongly arcuate, quadridentate anteriorly. Front arched above the antennae and antennulae, but excavated and deflexed in the middle.

Epistome high and nearly perpendicular, crossed transversely by a sharp groove. Labial border straight as seen in a front view, broken by a distinct notch in the middle, as seen from below. At the sides of the epistome, in the antero-lateral angle of the buccal area, there is a deep and narrow notch, which serves as an efferent orifice. Palate without longitudinal ridges.

Basis of antenna movable, filling the whole space between the small, triangular, inner, suborbital lobe and the front; remainder of antenna within the orbit.

Outer maxillipeds with their inner margins almost meeting; ischium and merus of nearly equal length; both very broad, the merus broader than long, its antero-lateral angle not expanded; palp articulated near that angle.

Legs rather long; dactyli quadrangular, their angles armed with spines.

Abdomen of male with seven distinct segments.

Two analogous species only, inhabiting opposite sides of tropical America: *jamaicensis* (Atlantic); *impressus* (Pacific). Fluviatile in part, at least.

KEY TO THE SPECIES OF THE GENUS GLYPTOGRAPSUS.

A^1. Second and third side teeth slender. Legs nearly naked___*impressus*, p. 275.

A^2. Second and third side teeth not slender. Legs hairy____*jamaicensis*, p. 277.

GLYPTOGRAPSUS IMPRESSUS Smith.

Plate 72, figs. 1 and 2.

Glyptograpsus impressus SMITH, Trans. Connecticut Acad. Arts and Sci., vol. 2, 1870, p. 154 (type-locality, Acajutla; holotype in P. M. Y. U.).

Glyptograpsus spinipes CANO, Boll. Soc. Nat. Napoli, ser. 1, vol. 3, 1889, pp. 102 and 241 (type-locality, *Isola dello Perle;* type in Naples Mus.).

Diagnosis.—Small. Fronto-orbital width more than two-thirds of greatest width. Second and third side teeth slender, acute. Legs nearly naked.

Description.—Dorsal surface uneven, with numerous irregular, shallow punctures, and along the lateral borders with small tuberculous elevations. Cervical suture a distinct sulcus. Mesogastric region separated from protogastric lobes by deep sulci, which unite between these lobes and extend down the front as a broad and deep depression. Epigastric lobes very prominent, their anterior margins transverse and precipitous. Protogastric lobes well indicated, having an outer lobe separated as a small but very distinct tuberculiform elevation opposite the fissure of the orbit. Epibranchial lobes uneven and partly separated from the mesobranchial by well-marked, out short, depressions. Posterior portion of branchial region divided by a longitudinal ridge into a flat inner area, and a broad precipitous portion between the ridge and the lateral margin.

Front, as seen from before, very sinuous, and broken in the middle by a broad, deep, rounded sinus; its outer angles, as seen from above, obtusely rounded, the margin continuous to the inner angle of the orbit, where it passes abruptly downward beneath the ocular peduncle as a sharp ridge, leaving a distinct notch, above which the margin begins again and is continuous to the acutely triangular antero-lateral tooth, which is prominent and directed straight forward. Next tooth, or first tooth of the lateral margin, broad, obtusely rounded, and above the plane of the orbital tooth; second and third teeth slender, acute; last tooth on the postero-lateral margin small, acute, and somewhat below the level of the preceding ones. Inferior margin of orbit straight, finely dentate. Infero-lateral regions granulous, slightly hairy.

Chelipeds short, very unequal. Merus short, not reaching beyond margin of carapace, triquetral, angles denticulate. Carpus small, outer face granulous, inner edge slightly margined. In the larger hand the propodus is short and very stout, outer surface convex and finely granulous; pollex very short, its prehensile edge directed obliquely downward. *D*actylus straight, rather slender, granulous. Both fingers obtusely tubercular on the prehensile edges; tips horny, slightly excavate. Smaller hand slender, somewhat flattened, the angles armed with sharp spinules.

Abdomen of male broadest at its base, from which it tapers to the last segment, which is longer than broad and rectangular, except that the extremity is rounded. Appendages of first segment stout, suddenly constricted near the end on the under side, and curved outward and strongly downward to the tip.

Measurements.—Male holotype, length of carapace, including frontal lobes, 12.4, width of same 15, antero-lateral width 11.5, length of legs, first 19, second and third 25, fourth 21 mm. (After Smith.)

Distribution.—Acapulco, Mexico (M. C. Z.); Acajutla, Salvador (Smith); Panama (M. C. Z. and Cano).

Material examined.—Tobago Island, Panama; May 11-15, 1911; Meek and Hildebrand, Smithsonian Biological Survey of the Panama Canal; 2 males, 2 females (44174).

GLYPTOGRAPSUS JAMAICENSIS (Benedict).

Plate 72, fig. 3.

Areograpsus jamaicensis BENEDICT, Johns Hopkins Univ. Circ., vol. 11, No. 97, 1892, p. 77 (type-locality, Kingston Harbor; type, Cat. No. 17226, U.S.N.M.).

Glyptograpsus jamaicensis RATHBUN, Ann. Inst. Jamaica, vol. 1, 1897, p. 29.

Diagnosis.—Large. Fronto-orbital width less than two-thirds of greatest width. Second and third side teeth short, not slender. Legs hairy.

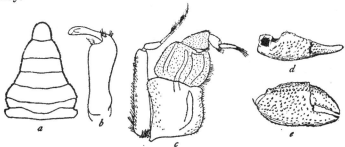

FIG. 140.—GLYPTOGRAPSUS JAMAICENSIS, MALE. *a*, ABDOMEN, HOLOTYPE, × 1½; *b*, FIRST RIGHT ABDOMINAL APPENDAGE (42881), INNER SIDE, × 2½; *c*, OUTER MAXILLIPED OF HOLOTYPE, × 3½; *d*, TOP VIEW OF RIGHT CHELA OF SAME, SLIGHTLY ENLARGED; *e*, OUTER VIEW OF SAME, SLIGHTLY ENLARGED.

Description.—Closely allied to *G. impressus*. Larger and narrower in front. The gastric region, as well as the lateral regions, is tuberculous. The longitudinal ridge which separates the flat from the precipitous portion of the branchial region is not well marked. Each of the two lobes of the front is bilobed. There is a break in upper margin of orbit at the entering angle, but no downward continuation of the inner part of the margin. The second lateral tooth is much shorter than the first, obtuse-angled, and is the most elevated of all; third tooth acute, with transverse anterior margin; last or postero-lateral tooth minute. Below the outer angle of the orbit there is a deep sinus which opens into a shallow gutter beneath the lateral teeth.

The arms reach a little beyond the carapace. The roughness of wrist and hands is caused by tubercles, the outer face of the palm is rather flat and separated by a ridge from the oblique upper face, inner face of large palm strongly protuberant on the upper half; its supero-distal part tuberculate. Fingers stout and rather stumpy, with a small gap between them, tips broadly excavate.

Legs densely hairy on the lower surface of the propodi and proximal end of the dactyli. Merus joints armed above with a tooth. Legs very granulous; the marginal spinules of the last three segments are very fine and numerous.

First segment of male abdomen transversely cristate; second segment not so wide as the third.

Measurements.—Male holotype, median length 32, width 37, anterolateral width 23, approximate length of legs, first 48, second 60, third 62, fourth 50.5 mm.

Range.—Known only from Jamaica.

Material examined.—

Montego Bay River, Jamaica: 2 feet, stony bottom; Aug 1, 1910; C. B. Wilson; 1 male (42881). Fresh water; Aug. 25, 1910; E. A. Andrews; 1 female (41524).

Kingston Harbor, Jamaica; T. H. Morgan; 1 male holotype (17226).

Genus PLATYCHIROGRAPSUS de Man.

Aspidograpsus KRØYER, MS., Copenhagen Museum; type, *A. typicus* Krøyer.
Platychirograpsus DE MAN, Zool. Anz., No. 500, 1896, p. 292; type, *P. spectabilis* de Man.

Very near *Glyptograpsus*. Differs as follows: The labial border is not straight in front view, but the two halves are arched upward. The outer maxillipeds have between them a narrow rhomboidal gape and are further from the epistome.

The propodus of the large cheliped is prolonged proximally far beyond its articulation with the carpus, and its outer surface is perfectly flat.

Fluviatile (in part, at least), inhabiting the Atlantic slope of Mexico, and also West Africa.

PLATYCHIROGRAPSUS TYPICUS Rathbun.

RIVER CRAB.

Plate 73

Aspidograpsus typicus KRØYER, MS., Copenhagen Museum; Bay of Mexico.
Platychirograpsus spectabilis RATHBUN (part). Proc. U. S. Nat. Mus., vol. 22, 1900, p. 279; not *P. spectabilis* de Man, 1896.
Platychirograpsus typicus RATHBUN, Proc. U. S. Nat. Mus., vol. 47, 1914, p. 122, text-fig. 3, pl. 5 (type-locality, Macuspana River, Montecristo, Tabasco, Mexico, 140 miles from the sea, altitude over 100 feet; holotype, Cat. No. 23761, U.S.N.M.).

Diagnosis.—Palm prolonged far beyond articulation with wrist. Dactylus of fourth leg very broad and flat.

THE GRAPSOID CRABS OF AMERICA. 279

Description.—Dorsal surface very uneven, covered with irregular punctures, and toward the lateral borders with tubercles, a few tubercles on the gastric region. Interregional grooves deep. Epi-

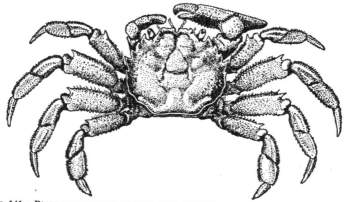

FIG. 141.—PLATYCHIROGRAPSUS TYPICUS, MALE (19863), DORSAL VIEW, SLIGHTLY REDUCED.

gastric lobes high, granulate; between them a deep furrow runs to the margin of the front. A very small protogastric lobule is situate near the upper fissure of the orbit. Outer portion of branchial region rather steeply inclined.

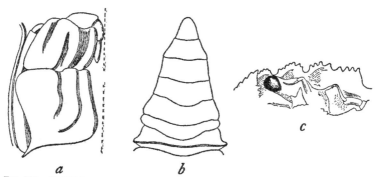

FIG. 142.—PLATYCHIROGRAPSUS TYPICUS, MALE HOLOTYPE. *a*, OUTER MAXILLIPED (DOTTED LINE SHOWING MEDIAN LINE), × 4; *b*, ABDOMEN, × 1½; *c*, RIGHT ORBITAL AND ANTENNAL REGION, VENTRAL VIEW, × 2½.

Front bilobed, a deep U-shaped median sinus; lobes irregular, highest near the outside, margined with horny-tipped spinules, as are also the upper margin of the orbit and the outer margin of the lateral teeth. Front separated from the inner border of the orbit by a

U sinus; outer orbital tooth strong, sharp and curved slightly inward. The first three teeth of lateral margin elevated, acute, the second the smallest, the third very prominent; last tooth very small, acute. Margins behind that tooth subparallel. Lower margin of orbit oblique, crenulate; outer sinus very broad and deep; inner tooth large, spinulous on the outer margin. Infero-lateral regions densely hairy.

Chelipeds strikingly unequal in length and thickness. The margins of the chelipeds and legs are spinulous, and their surfaces are more or less tuberculous. The inner margin of the upper surface of the wrist is rounded, but below the angle there are a few spinules on

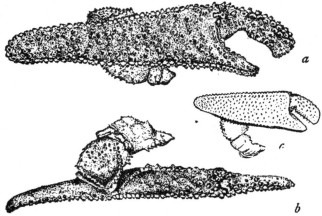

FIG. 143.—PLATYCHIROGRAPSUS TYPICUS, RIGHT OR MAJOR CHELIPED. a, ANTERIOR VIEW OF SPECIMEN IN HALIFAX MUSEUM, NATURAL SIZE; b, UPPER VIEW OF SAME, NATURAL SIZE; c, ANTERIOR VIEW OF 19863, × 1¼.

the inner surface. The propodus of the large chela is enormously developed, being expanded proximally in a projection which may equal in length the distance from the superior articulation with the carpus to the dactylus; outer face quite flat, three times as long as high, widening distally, covered with tubercles; upper surface flat, at right angles to the outer, narrowing a little distally. Fingers short and thick, pollex triangular, dactylus of nearly equal width throughout, slightly excavate at base of prehensile edge. Edges of both fingers set with low teeth and hollowed in shallow grooves. Smaller chela normal, palm widening distally, not much longer than the fingers measured horizontally.

The supero-distal spine on the merus of the legs is strong and acute, and most recurved on the first leg. Legs sparingly hairy; a short fur on the margins of the last three joints of the last pair. Dactylus of last pair ovate-lanceolate, much broader and flatter than in the other pairs.

The four American specimens differ from *P. spectabilis* from Gaboon in having the postero-lateral margins less convergent behind the third side tooth and that tooth a little farther forward, and also the dactylus of the fourth leg is flattened and broadened so that it does not resemble the other dactyli as much as in the figured type.

Measurements.—Male, holotype, median length of carapace 42.5, width of same 51 mm.

Habitat.—Lives in holes in clay banks, just above water line (Nelson and Goldman).

Range.—Mexico.

Material examined.

Macuspana River, Montecristo, Tabasco, Mexico; 140 miles from the sea, altitude over 100 feet; May 7, 1900; Nelson and Goldman, Biol. Survey, U. S. Dept. Agri.; 2 males (1 is holotype) (23761).

Mexico: Mexican Exhibit at World's Columbian Exposition, 1893; 1 male (19863).

"Gulf of Mexico"; 1 male (Copenhagen Mus.).

Locality unknown; 1 large claw (Halifax Mus.).

Genus EUCHIROGRAPSUS Milne Edwards.

Euchirograpsus MILNE EDWARDS, Ann. Sci. Nat., ser. 3, Zool., vol. 20, 1853, pp. 175 [141]; type, *E. liguricus* Milne Edwards.

Surface hairy. Carapace nearly square, slightly convex, regions indistinct. Antero-lateral angle acute; lateral margins tridentate. Front not more than half width of carapace, advanced, horizontal, bilobed. Orbit with an outer notch; the customary tooth near the inner angle is minute. The basal joint of the antenna closes the orbital hiatus, the antenna entering the orbit. Antennules nearly transverse. Epistome well defined, very short fore and aft, lower edge arched upward. Buccal cavity broader than long, widening anteriorly. Outer maxillipeds little separated. Ischium longer than merus, and longer than broad, widening anteriorly. Merus broader than long, outer angle rounded and prominent, inner side oblique, antero-internal angle notched for insertion of palp. The merus of the first pair of maxillipeds is not bilobed, being destitute of the inner lobe.

Chelipeds subequal, shorter and stouter than the legs. Hands angular, fingers pointed. Legs slender, flattened; dactyli spinous and acute.

The third to sixth segments of the abdomen are more or less fused.

Distribution.—Mediterranean, eastern Atlantic Ocean, West Indian region.

EUCHIROGRAPSUS AMERICANUS A. Milne Edwards.

Plate 74.

Euchirograpsus americanus A. MILNE EDWARDS, Bull. Mus. Comp. Zoöl., vol. 8, 1880, p. 18 (type-locality, Barbados, 69 fathoms, station 278, *Blake*; 1 male, 2 females, cotypes in M. C. Z.).—A. MILNE EDWARDS and BOUVIER, Résult. comp. Scient. de *l'Hirondelle*, fasc. 7, 1894, p. 46, pl. 4, figs. 10–14.

Diagnosis.—Square, hairy. Merus of first pair of maxillipeds not bilobed. First and second segments and base of third segment of male abdomen covering the sternum.

Description.—Carapace slightly broader than long; sides parallel; flattened posteriorly, sloping gently down toward the sides and front · surface covered with granules and short soft hair.

Front advanced, lamellar, with a narrow median notch; outer end of each lobe sloping backward to the small tooth at the inner angle of the orbit. Orbit large; upper margin oblique, sinuous, outer angle spiniform; behind it three smaller, subequal spines on the lateral margin, the last much reduced. Eye large, filling the orbit; lower margin of orbit little oblique, denticulate. Third joint of antenna hollowed out on the inner side.

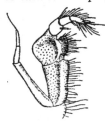

FIG. 144.—EUCHIROGRAPSUS AMERICANUS, OUTER MAXILLIPED, ENLARGED. (AFTER A. MILNE EDWARDS AND BOUVIER.)

Cheliped one and one-half times as long as carapace. Surface of arm, as well as the same joint of the legs, crossed by fine granulated lines; margins spinulous, a superior subdistal spine, four or five large spines on the distal half of the inner margin. Inner distal margin of wrist spinulous. Palm with three spinulous ridges above, a ridge near the lower edge, and another, less distinct, through the middle. Fingers rather slender, grooved, pointed, teeth low, irregular, fitting together, dactylus a little longer than upper margin of palm.

Legs slender, compressed, second pair over twice as long as carapace. Merus joints armed with a subdistal spine above and two distal spines below, one inner, the other outer; inner spine absent in last pair; first pair spinulous below, and armed with one or a few long spines on the distal half. Legs clothed with long hairs as well as short ones, dactyli armed with long thorns.

Color.—Yellowish-gray, arranged in marblings on the carapace and in transverse bands, alternately light and dark, on the feet. Eyes black. (Edwards and Bouvier.)

Dimensions.—Female (17671), length of carapace 14.1, width of same 15.5 mm.

Range.—From South Carolina to Caribbean Sea. Eastern Atlantic. 42 to 278 fathoms.

Material examined.—See table, page 283.

THE GRAPSOID CRABS OF AMERICA.

Subfamily SESARMINAE Dana.

Sesarmacea MILNE EDWARDS, Ann. Sci. Nat., ser. 3, Zool., vol. 20, 1853, p. 181 [147].
Cyclograpsacea MILNE EDWARDS, Ann. Sci. Nat., ser. 3, Zool., vol. 20, 1853, p. 191 [157], part.
Sesarminae DANA, U. S. Expl. Exped., vol. 13, Crust., pt. 1, 1852, p. 333.

Front strongly deflexed. The lower border of the orbit commonly runs downward toward the angle of the buccal cavern. Side walls of carapace finely reticulated with granules and hairs or hairs only. External maxillipeds separated by a wide rhomboidal gape; an oblique hairy crest traverses them from a point near the antero-external angle of the ischium to a point near the antero-internal angle of the merus; the palp articulates either at the summit, or near the antero-external angle, of the merus, and the exognath is slender and either partly or almost entirely concealed. The male abdomen either fills or does not quite fill all the space between the last pair of legs.

Fritz Müller[1] explains the function of the reticulation of the side walls and the hairy ridges on the maxillipeds in these crabs, which spend much of the time out of the water:

The small wart-like elevations and geniculated hairs form a fine net or hair-sieve extended over the lower surface of the carapace. Thus when a wave of water escapes from the branchial cavity (through the orifices at the anterior angles of the buccal frame) it immediately becomes diffused in this network of hairs and is then conveyed back to the branchial cavity by vigorous movements of the appendage of the outer maxilliped, which works in the fissure entering the cavity from above the last pair of legs. While the water glides in this way over the carapace in the form of a thin film, it

---
[1] Facts and Arguments for Darwin. Translation by W. S. Dallas, London, 1869, pp. 32–33.

Material examined of E. americanus

Locality	Lat. N.	Long. W.	Fath.	Bottom.	Temp.	Date.	Sta.	Collector.	Specimens.	Cat. No.
Off South Carolina	32 55 00	77 54 00	79	crs. S. bk. Sp.	59.1° F.	Jan. 5, 1885	2311	*Albatross*	1 ♀	17871
Off Cape ᴀ, Florida	2¼ m 3SE Fowey Rocks Lt.		45	rky	70°	Mar. 25, 1903	7511	*Fish Hawk*	4 ♂ 3 ♀	44675
Off Havana, Cuba	23 10	82 20	121	fn. gy. Co		Jan. 17, 1885	2330	*Albatross*	2 ♂	9501
Do	23 10	82 30	122	Co		Apr. 30, 1884	2162	do	2 ♂	7776
Off Arrowsmith aBk, Yucatan	23 59	88 45	130	Co		Jan. 22, 1885	2354	do	1 ♀	17672
Mar ? Mulillo, Colombia	9 30	76 20	42	gn. M. S.		Mar. 23, 1884	2142	do	1 ♂	7769
Off St. Lucia			278					*Blake*	2 ♀	6132, M. C. Z.

will again saturate itself with oxygen, and may then serve afresh for the purpose of respiration. In order to complete this arrangement the ridges on the right and left sides of the outer maxillipeds form together a triangle with the apex turned forward—a breakwater by which the water flowing from the branchial cavity is kept away from the mouth and reconducted to the branchial cavity. In very moist air the store of water contained in the branchial cavity may hold out for hours, and it is only when this is used up that the animal elevates its carapace in order to allow the air to have access to its branchiae from behind.

Genus SESARMA Say.

MARSH CRABS (F. Müller).

Sesarma SAY, Journ. Philadelphia Acad. Nat. Sci., vol. 1, 1817, p. 76; type, *S. reticulatum* (Say).
Pachysoma DE HAAN, Fauna Japon., Crust., 1833, p. 5; 1835, p. 33; type, *P. bidens* de Haan. Not *Pachysoma* MacLeay, 1821, a genus of Coleoptera.
Chiromantes GISTEL, Natur. Thierreichs, 1848, p. X; type, *C. bidens* (de Haan).
Holometopus MILNE EDWARDS, Ann. Sci. Nat., ser. 3, Zool., vol. 20, 1853, p. 187 [153]; type, *H. haematocheir* de Haan.
Holograpsus MILNE EDWARDS, Arch. Mus. Hist. Nat. Paris, vol. 7, 1854, p. 158. A slip of the pen for *Holometopus*.
Sesarma DE MAN (subgenus), Zool. Jahrb., Syst., vol. 9, 1895, p. 143; type, *H. haematocheir* de Haan.
Episesarma DE MAN (subgenus), Zool. Jahrb., Syst., vol. 9, 1895, p. 165; type, *S. tetragonum* (Fabricius).
Parasesarma DE MAN (subgenus), Zool. Jahrb., Syst., vol. 9, 1895, p. 181; type, *S. quadrata* (Fabricius)=*Cancer quadratus* Fabricius, 1798, not 1787=*S. plicatus* Latreille, 1802-1803.
Perisesarma DE MAN (subgenus), Zool. Jahrb., Syst., vol. 9, 1895, p. 208; type, *S. bidens* (de Haan).

Carapace squarish; sides usually straight and sometimes parallel, but sometimes convex; surface flattened; gastric region well delimited, divided into five subregions, of which the four antero-lateral subregions form four prominent postfrontal tubercles. Posterolateral regions usually crossed by oblique parallel ridges.

Side walls finely reticulate; due to a multitude of small uniform granules arranged in pairs in close-set parallel rows; between the pairs of granules is a little row of bristles, one of which in each row is long and points diagonally forward.

Front equals half or more than half of the anterior border and is obliquely or vertically deflexed. The deep oval orbits occupy the rest of the anterior border of the carapace; below their outer angle is a deepish gap leading into a system of grooves which open into a notch at the antero-lateral angle of the buccal cavern.

Antennules transverse, fossae narrow; septum transverse. Antero-external angle of second joint of antennal peduncle a good deal produced; flagellum slender and rather short, lying in the orbital hiatus.

Epistome well defined, prominent, rather short fore and aft. Buccal cavern nearly square. The external maxillipeds leave between them a large rhomboidal gap, which is a good deal filled up by a hairy fringe; they are obliquely traversed, from a point behind the antero-external angle of the ischium to the antero-internal angle of the merus, by a conspicuous line of hairs; the coarse palp is attached to the rounded summit of the obliquely directed merus.

Chelipeds massive in the male, subequal; palm high and short; fingers subacute and hollowed at tip.

Legs differing little in length, third pair longest; meropodites thin.

Abdomen in both sexes with seven separate segments, in the male occupying the whole breadth of the sternum between the bases of the last pair of legs. In the female the last segment is small and more or less impacted in the broad sixth segment.

Distribution.—Tropical and subtropical seas; often distant from the shore.

KEY TO THE AMERICAN SUBGENERA AND SPECIES OF THE GENUS SESARMA.

A^1. Carapace with a lateral tooth behind the outer orbital tooth.
  B^1. Manus with oblique, coarsely pectinated ridges on upper surface.
                        Subgenus *Chiromantes*, p. 287; *africanum*, p. 287.
  B^2. Manus without oblique, pectinated ridges on upper surface.
                        Subgenus *Sesarma*, p. 288.
    C^1. Sides converging anteriorly. Legs narrow and very long.
                      verleyi, p. 288.
    C^2. Manus with a definite marginal line above.
      D^1. Front more than half as wide as carapace.
        E^1. Carapace deeply grooved.
          F^1. Lateral margins nearly straight, curving in at the orbital tooth. About 11 strong blunt spinules on upper surface of movable finger_____ *sulcatum*, p. 289.
          F^2. Lateral margins sinuous. About seven to nine depressed spiniform tubercles on upper surface of movable finger.
                      reticulatum, p. 290.
        E^2. Carapace slightly grooved.
          F^1. Spines on upper surface of movable finger extending its entire length and about eight in number_____*aequatoriale*, p. 292.
          F^2. Spines on upper surface of movable finger restricted to basal half or three-fifths.
            G^1. Carapace four-fifths or less than four-fifths as long as wide.
              H^1. Width of male abdomen gradually diminishing from third to sixth segment. Granules on upper edge of movable finger continued a little on distal half_____*curacaoense*, p. 293.
              H^2. Male abdomen wider at third segment, abruptly contracted at fourth segment. Granules on upper edge of movable finger not continued on distal half_____*rhizophorae*, p. 294.
            G^2. Carapace more than four-fifths as long as wide.
                      crassipes, p. 294.

D². Front just half or less than half as wide as carapace.
  E¹. Front just half as wide as carapace_____*ophioderma*, p. 297.
  E². Front less than half as wide as carapace.
    F¹. Outer surface of palm not densely pilose.
      G¹. Upper margin of palm broken into several ridges.
        bidentatum, p. 295.
      G². Upper margin of palm a single sharp crest_____*jarvisi*, p. 296.
    F². Outer surface of palm densely pilose_____*barbimanum*, p. 298.
A². Carapace without a lateral tooth behind the outer orbital tooth.
  Subgenus *Holometopus*, p. 298.
  B¹. Movable finger not extraordinarily enlarged along its proximal half.
    C¹. Merus joints of legs less than twice as long as broad____*rectum*, p. 298.
    C². Merus joints of legs more than twice as long as broad.
      D¹. Front widening toward the lower margin.
        E¹. Merus of third leg less than three times as long as wide.
          F¹. Upper margin of hand not a sharp crest.
            G¹. Merus of third leg more than 2½ times as long as wide.
              cinereum, p. 300.
            G². Merus of third leg less than 2½ times as long as wide.
              H¹. Carapace a little wider posteriorly than anteriorly. Frontal lobes prominent _____*miersii*, p. 303.
              H². Carapace wider anteriorly than posteriorly. Frontal lobes not prominent _____*magdalenense*, p. 305.
          F². Upper margin of hand a sharp crest_____*occidentale*, p. 299.
        E². Merus of third leg three times, or more than three times, as long as wide.
          F¹. Third leg 2½ or nearly 2½ times as long as carapace.
            G¹. Deflexed front four times as wide as high. Fifth abdominal segment of male longer at middle than at sides.
              miersii iheringi, p. 304.
            G². Deflexed front higher, less than 4 times as wide as high. Fifth abdominal segment of male of even length throughout_____*biolleyi*, p. 306.
          F². Third leg twice as long as carapace_____*ricordi*, p. 308.
      D². Front not widening toward the lower margin, its sides subparallel.
        E¹. Carapace nearly square, its length nearly or quite equal to its width.
          F¹. Carapace a little wider than long.
            G¹. Manus without a definite marginal line above except in proximal part.
              H¹. Margins of merus joints of legs converging from the middle toward the carpus joints_____*angustipes*, p 311.
              H². Margins of merus joints of legs subparallel for their distal half_____*roberti*, p. 312.
            G². Manus margined above by a denticulated crest___*festae*, p. 313.
          F². Carapace a little longer than wide_____*angustum*, p. 314.
        E². Carapace distinctly transverse.
          F¹. Carapace wider anteriorly than posteriorly. Merus of third leg about twice as long as wide_____*hanseni*, p. 315.
          F². Carapace with parallel sides. Merus of third leg nearly three times as long as wide_____*tampicense*, p. 307.
  B². Movable finger extraordinarily enlarged along its proximal half.
    benedicti, p. 316.

ANALOGOUS SPECIES ON OPPOSITE SIDES OF THE CONTINENT.

Atlantic.	Pacific.
reticulatum.	aequatoriale.
curacaoense.	rhizophorae.
cinereum.	occidentale.
miersii iheringi.	biolleyi.
roberti.	angustum.

Species on both sides of the continent: *angustipes*.

Species indeterminable: *subintegra* White (List Crust. Brit. Mus., 1847, p. 38), Brazil, *nomen nudum*.

Subgenus CHIROMANTES Gistel (=PERISESARMA de Man).

Carapace with a lateral tooth behind the outer orbital tooth. Manus with oblique, coarsely pectinated ridges on upper surface.

SESARMA (CHIROMANTES) AFRICANUM Milne Edwards.

Plate 75.

Sesarma africana MILNE EDWARDS, Hist. Nat. Crust., vol. 2, 1837, p. 73 (type-locality, Senegal; type in Paris Mus.); Ann. Sci. Nat., ser. 3, Zool., vol. 20, 1853, p. 185 [151].

Sesarma (Perisesarma) africanum RATHBUN, Proc. U. S. Nat. Mus., vol. 22, 1900, p. 280.

Diagnosis.—Large, hairy. Transverse ridge inside hand. Merus joints of legs about two and one-half times as long as wide.

Description.—Carapace five-sixths as long as wide, a little narrowed at the anterior angles; the four post-frontal lobes prominent, the middle pair wider than the outer pair; surface uneven, crossed by short thick ridges of hair. Lateral tooth strong; behind it, a faint indication of another tooth.

Front vertical, concave, its greatest height nearly one-fourth its greatest width, lower edge very sinuous, sides subparallel, corners prominent and rounded.

Chelipeds in the male much more massive and in the female slightly more massive than the legs. Outer surface of arm and wrist crossed by granulated striae, outer surface of hand coarsely tuberculate. Arm with a sharp subterminal tooth above, and a row of stout teeth on the lower margins. Inner border of arm with a few inconspicuous denticles. Upper surface of palm with about four oblique, subparallel granulated ridges; inner face tuberculate, the upper half with a very prominent ridge, coarsely tuberculate. Palm of male much enlarged, wider than its length at middle. Dactylus less than twice the length of the upper border of the palm; its dorsal surface is crossed by about fifteen oblique, blunt ridges each of which is crossed by fine impressed lines. Fingers broad and flat, moderately gaping, each with two enlarged teeth. In the female the palm is much less enlarged, devoid of prominent ridge within, upper surface of dactylus simply granulate, fingers almost meeting.

Legs very broad; third pair twice as long as carapace, its merus two and one-half times as long as wide and its dactylus four-fifths as long as the preceding joint. Propodites of first and second pairs densely furry on anterior or lower surface, especially in male; carpopodites and propodites of all the legs with hairy margins, a few of the hairs long.

Terminal segment of male abdomen longer than wide.

Measurements.—Male (14870), length of carapace 29, width of same 33.2 mm.

Range.—Barbados. West Africa, from Senegal to Benguella (Osorio).

Material examined.—

Barbados; 1 large female (Paris Mus.).

Senegal; M. Delambre; 1 large female cotype, Paris Mus.; 1 male cotype, Cat. No. 20281, U.S.N.M.

Rock Spring, Monrovia; Apr., 1894; O. F. Cook and G. N. Collins; 12 males, 2 females (20572).

Beyah River, Elmina, Ashantee; Nov. 27, 1889; W. H. Brown, U. S. Eclipse Exped. to Africa; 2 males, 1 female (14870).

Gabun; M. Duparquet; 1 small male, 1 large female (Paris Mus.).

Subgenus SESARMA (=EPISESARMA de Man).

Carapace with a lateral tooth behind the outer orbital tooth. Manus without oblique, coarsely pectinated ridges on upper surface.

SESARMA (SESARMA) VERLEYI Rathbun.

Plate 76.

Sesarma (Sesarma) verleyi RATHBUN, Proc. U. S. Nat. Mus., vol. 47, 1914, p. 123, pl. 6 (type-locality, Mulgrave, Jamaica; holotype female, Cat. No. 24940, U.S.N.M.).

Diagnosis.—Carapace elongate, narrowed anteriorly. Merus of third leg four times as long as wide.

Description.—Carapace nearly nine-tenths as long as wide, strongly narrowed anteriorly, convex fore and aft, regions and supra-frontal lobes fairly well marked; of the latter the outer pair are narrower than the inner pair and their anterior margin more strongly marked. Anterior part of carapace granulous, postero-lateral area finely striated.

The front occupies a little more than half the fronto-orbital width, lower margin forming two prominent lobes in dorsal view, sides oblique, angles rounded off.

Upper margin of orbit directed outward and forward, outer angle broad and obtuse, the margin between it and the lateral tooth convex; this tooth is subrectangular with thickened tip.

Chelipeds of female narrow. Outer surface of arm and wrist crossed by short lines of granules, upper and outer margins rough with short, oblique and parallel lines of granules, inner margin irregularly spinulous. Palms longer than wide, sparingly covered with depressed granules, more numerous above and toward the carpus, where they are arranged somewhat in rows. Fingers as long as the middle length of the palm, prehensile edges irregularly toothed except the distal third which has a straight, horny edge, tips curved toward each other.

Legs unusually long; the third leg is three and one-fifth times as long as the carapace; its merus is four times as long as wide. The legs (as well as the carapace) are nearly naked, only the margins of the last two joints (proximal end of the propodus excepted) bordered with short hair with a few longer ones intermingled. The carpus has two prominent lines of granules on the upper surface, the posterior of which is continued somewhat obliquely on the next joint near its margin.

Measurements.—Female holotype, length of carapace 20, width of same 22.8, fronto-orbital width 16.5, width of front 9.1 mm.

Material examined.—Jamaica: Mulgrave (a small village in the Cockpit County near Ipswich, St. Elizabeth); 1 female, holotype, collected by Miss Verley, and received through Mr. P. W. Jarvis; Cat. No. 24940, U.S.N.M.

SESARMA (SESARMA) SULCATUM Smith.

Plate 78, figs. 3 and 4.

Sesarma sulcata SMITH, Trans. Connecticut Acad. Arts and Sci., vol. 2, 1870, p. 156 (type-locality, Corinto, west coast of Nicaragua; type in Mus. Comp. Zoöl.).

Diagnosis.—Carapace large, deeply grooved and hairy. Front wide. Side tooth acute. Eleven prominent blunt spinules on movable finger.

Description.—Carapace about four-fifths as long as wide, a little narrowed at anterior angles; middle pair of post-frontal lobes wider and more prominent than outer pair; regions well marked, surface uneven, covered with small tufts of coarse black hair. Lateral tooth strong, acute.

Front steep, concave, its greatest height one-fifth of its width, median sulcus very wide, lower edge very sinuous, with two strongly projecting, thickened lobes, sides parallel with the lower angles rounded and prominent.

Chelipeds of male very heavy. Outer surface of arm and wrist crossed by short rugose lines, outer surface of chela almost smooth. Upper margin of arm with a blunt subterminal tooth, lower surface bounded on three sides by short, thick spines. Palm nearly twice as high as its superior length, its upper margin acute, crenulate, distally projecting in a tooth, inner surface sparsely tuberculate. Pollex and adjacent portion of palm flattened. Fingers moderately gaping, each with two of its prehensile teeth enlarged. Upper surface of dactylus armed with about eleven blunt spinules. Chelipeds of female rather small.

Legs very broad; third pair two and one-third times length of carapace, its merus two and one-third times as long as wide. Merus joints each with a strong subterminal spine above. Legs, especially last three joints, with rows of hair thickest on margins of last two joints.

Terminal segment of male abdomen as long as broad.

Measurements.—Male (4631), length of carapace 33.3 mm., width of same 41.2 mm.

Range.—From La Paz, Lower California, Mexico, to Panama.

Material examined.—

La Paz, Lower California; L. Belding· 1 male (4631).

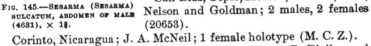

FIG. 145.—SESARMA (SESARMA) SULCATUM, ABDOMEN OF MALE (4631), × 1⅜.

San Blas, Tepic, Mexico; June 14, 1897; Nelson and Goldman; 2 males, 2 females (20653).

Corinto, Nicaragua; J. A. McNeil; 1 female holotype (M. C. Z.).

Boca del Rio Jesus Maria, Costa Rica; Apr., 1905; P. Biolley and J. F. Tristan; 1 male (32315).

La Capitana, Canal Zone; H. Pittier; 2 males (45532, 45569).

SESARMA (SESARMA) RETICULATUM (Say).

Plate 77.

Ocypode reticulatus SAY, Journ. Acad. Nat. Sci. Philadelphia, vol. 1, 1817, p. 73, pl. 4, fig. 6 (type-locality, "muddy salt marshes" [east coast of United States]; type not extant).

Ocypode (Sesarma) reticulatus SAY, Journ. Acad. Nat. Sci. Philadelphia, vol. 1, 1818, p. 442.

Sesarma cinerea WHITE (part), List Crust. Brit. Mus., 1847, p. 38, not *S. cinereum* (Bosc). Specimen presented by Thomas Say, perhaps cotype.

Sesarma reticulata GIBBES, Proc. Amer. Assoc. Adv. Sci., vol. 3, 1850, p. 180.

Diagnosis.—Very convex. Sides sinuous, tooth blunt. Front wide. Legs thickly tomentose.

Description.—Carapace about four-fifths as long as broad, convex fore and aft, sides concave behind the small, obtuse, lateral tooth, convergent at the orbital angles. Regions well marked, surface punctate.

Front obliquely inclined, superior lobes smooth, inner pair wider; lower edge sinuous, bilobed, sides parallel.

Chelipeds with outer surface of arm and wrist lightly rugose, palm almost smooth to naked eye, indistinctly granulate. Arm with an obtuse subterminal tooth above, both lower margins denticulate; inner angle of wrist rounded. Palm of male a little higher than its middle length, upper edge a single granulate line, inner surface with a short irregular ridge of tubercles near the distal end. Dactylus with seven to nine depressed spinules above on the basal two-thirds. Fingers with a widish gap, and an enlarged tooth near each end. In the female the palm is half again as high as its middle length, the dactylus has about four granules above near the base, the fingers gape narrowly.

Legs of the third pair about twice as long as carapace, their merus two and one-third times as long as wide. Subdistal spine on the merus sharp; last three joints densely tomentose.

Last segment of abdomen a trifle longer than wide.

Measurements.—Male (New Haven), length of carapace 23, width of same 28.3 mm.

Range.—From Woods Hole, Massachusetts, to Calhoun County, Texas.

Habits.—Burrows in muddy salt marshes.

Color.—Carapace, dark plum-colored or bluish-black speckles crowded on a grayish ground, the grayish color showing very little except on posterior part. Upper part of chelipeds, as shown when flexed, similarly colored, but brighter; upper part of legs like carapace. Greater part of palm yellowish. Under part of crab for the most part grayish.

Material examined.—

Woods Hole, Massachusetts; July 10, 1901; V. N. Edwards; 4 males, 2 females ovig. (45530).

Vicinity of Woods Hole, Massachusetts; M. J. Rathbun; 1 male (32482).

Wareham, Massachusetts; July 21, 1887; 4 males, 1 female (12782).

New Bedford, Massachusetts; September, 1882; Willard Nye, jr.; 1 female (40517).

Acushnet River, near New Bedford, Massachusetts; 1882; Willard Nye, jr.; 3 females (5784).

Newport, Rhode Island; shore; 1880; U. S. Fish Comm.; 1 male (36747).

Long Island Sound; Verrill and Smith (79, P. M. Y. U.).

Beesleys Point, New Jersey; Verrill and Smith (1201, P. M. Y. U.).

Smiths Island, Virginia; Paul Bartsch; 1 male (20062).

South or North Carolina; *Fish Hawk;* 1 specimen (17180).
Winyah Bay, South Carolina; January, 1891; *Fish Hawk;* 17 specimens (17490).
Myrtle Bush Creek, South Carolina; February 7, 1891; *Fish Hawk;* 2 specimens (17181).
Grand Isle, near New Orleans, Louisiana; G. Kohn; 2 specimens (2257)
Kellers Bay, Calhoun County, Texas; J. D. Mitchell; 1 male, returned to J. D. Mitchell.

SESARMA (SESARMA) AEQUATORIALE Ortmann.

Sesarma aequatorialis ORTMANN, Zool. Jahrb., Syst., vol. 7, 1894, p. 722, pl. 23, figs. 14, 14k, 14z (type-locality, Ecuador; type in Strassburg Mus.).

Diagnosis.—Carapace narrowed behind. Side tooth small. Spines on whole length of movable finger. Legs short.

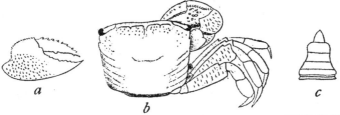

FIG. 146.—SESARMA (SESARMA) AEQUATORIALE, NATURAL SIZE. *a*, RIGHT CHELA OF MALE, OUTER VIEW; *b*, DORSAL VIEW; *c*, ABDOMEN OF MALE. (AFTER ORTMANN.)

Description (condensed from Ortmann).—Allied to *S. reticulatum*. Carapace narrowing a little posteriorly; side tooth small. Suprafrontal lobes well separated, nearly equal; lower margin of front slightly concave at middle.

Outer surface of arm and wrist with transverse rugae. Upper margin of arm granulate, distally with a small angle but no spine; anterior margin sharply granulate. Inner angle of wrist a right angle. Upper margin of palm cristate, finely granulate, with 30 to 40 granules, very fine near the wrist; outer surface smooth above and near the fingers, granulate below and in the middle; inner surface regularly granulate. In the female the granules of inner and outer surfaces are more feebly developed.

Legs short, the longest less than twice length of carapace; meropodites two and one-third times as long as wide; dactyl only a little shorter than propodi.

Measurements.—Male type, length of carapace 23, width of same 28 mm.

Locality.—Ecuador.

SESARMA (SESARMA) CURACAOENSE de Man.

MANGROVE CRAB.

Plate 78, figs. 1 and 2; plate 160, fig. 3.

Sesarma curacaoensis DE MAN, Notes Leyden Mus., vol. 14, 1892, p. 257, pl. 10, figs. 6, 6*a*, 6*b* (type-locality, Curaçao; type in Leyden Mus.).

Diagnosis.—Small and broad, almost smooth. Spines on movable finger confined to basal half or three-fifths. Legs tomentose.

Description.—A small species, resembling *S. reticulatum* in shape. Carapace four-fifths as long as wide, nearly smooth, regions slightly indicated, punctate, with scattered patches of tomentum. Suprafrontal lobes almost obsolete. Front oblique, lower margin bilobed in dorsal view, convex in front view. Side tooth well marked, obtuse.

Outer surface of arm and wrist rugose; upper margin of arm bluntly angled near the distal end, inner margin tuberculose, outer margin finely granulate. Palm higher than its middle length, outer surface coarsely punctate, upper margin a line of fine granules, inner surface with tubercles on distal end, scattered granules elsewhere. Upper surface of dactylus with 7 to 9 depressed spinules. Fingers subconical.

FIG. 147.—SESARMA (SESARMA) CURACAOENSE, LEFT APPENDAGE OF FIRST SEGMENT OF MALE ABDOMEN (17678), LOWER VIEW, × 4½.

Third leg a little more than twice as long as carapace, its merus two and one-fourth times as long as wide. Meral spine sharp; last three joints clothed with a thick tomentum, with longer hairs interspersed.

Measurements.—Male (17678), length of carapace 12.2, width of same 15.1 mm.

Habitat.—Under fallen leaves and other rubbish among mangrove roots. Very sluggish in its movements. (Jarvis.)

Range.—Cuba; Jamaica; Porto Rico; Curaçao; Bahia, Brazil.

Material examined.—

Cabañas, Cuba; June 2, 1900; Wm. Palmer and J. H. Riley; 1 male (23814).

Montego Bay, Jamaica; P. W. Jarvis; 1 male, 1 female (19422).

Old Harbor, Jamaica; P. W. Jarvis; 1 male (19423).

Jamaica; T. H. Morgan; 1 female (soft shell) (19419).

Bogue Islands, Jamaica; June 20, 1910; C. B. Wilson; 1 male (42890).

Porto Rico; 1899; *Fish Hawk;* 1 male (24033).

Curaçao; Feb. 10–18, 1884; *Albatross;* 4 males, 2 females, 3 y. (17678).

Mapelle, Bay of Bahia, Brazil; 1876–1877; R. Rathbun; 1 female ovig. (40823).

SESARMA (SESARMA) RHIZOPHORAE Rathbun.

MANGROVE CRAB.

Plate 79.

Sesarma (Sesarma) rhizophorae RATHBUN, Proc. Biol. Soc. Washington, vol. 19, 1906, p. 99 (type-locality, Boca del Jesus Maria, Costa Rica holotype male, Cat. No. 32491, U.S.N.M.).

Diagnosis.—Lateral tooth acute. Middle pair of supra-frontal lobes wider than outer pair. Legs slender.

Description.—The Pacific representative of *S. curacaoense*. Carapace and legs densely tomentose. The middle pair of post-frontal lobes are more prominent and much wider than the obscure outer pair. Lower margin of front less convex in front view. Lateral tooth acute. The granules on the upper surface of the dactylus, 9 or 10 in number, are confined to the basal half. The legs are longer and more slender, the third leg being two and one-fourth times as long as carapace, with its merus two and one-half times as long as wide; spine on the merus long and slender; dactyli long and slender. Abdomen of male narrower, the appendages of the first segment ending in long needle-like points.

Measurements.—Male type, length 10.9 mm., width 13.7 mm.

Habitat.—In the mud of mangrove swamps.

Locality.—Costa Rica; Boca del Jesus Maria.

Material examined.—Boca del Jesus Maria, Costa Rica: In mangrove swamps; Apr., 1905; J. F. Tristan; 1 male (32318). On mangroves in mud; Jan. 1906; J. F. Tristan and P. Biolley; 1 male (32491).

SESARMA (SESARMA) CRASSIPES Cano.

Sesarma crassipes CANO, Boll. Soc. Nat. Napoli, ser. 1, vol. 3, 1899, pp. 93 and 244 (type-locality, Pernambuco; type in Naples Mus.).

Diagnosis.—Rather narrow. Supra-frontal lobes equal. Side margins straight, tooth small.

Description (condensed from Cano).—Of medium size. Carapace rather narrow, its length 0.84 of its width. Surface rugulose and pubescent. Mesogastric area well defined, epigastric and protogastric lobes equally wide.

Front deeply sinuous. Lateral margins almost straight. Tooth at orbit acute and advanced, lateral tooth small and elevated above the first; behind it, a rudiment of another tooth.

Surface of arm rugulose, inner margin denticulate, lower margin armed with a tooth. Carpus granulous-squamose. Hand elongate, scarcely rugose, a crenulated crest above. Dactylus toward the base spinulous.

Legs rather robust; merus broad, with a superior spine. surface squamose and scarcely villose; three remaining joints villose, above hirsute.

Measurements.—Length of carapace 18.5, width of same 22, width of front 13 mm.

Locality.—Brazil; Pernambuco.

SESARMA (SESARMA) BIDENTATUM Benedict.

FRESH-WATER CRAB.

Plate 80.

Sesarma bidentata BENEDICT, Johns Hopkins Univ. Circ., vol. 11, 1892, p. 77 (type-locality, Kingston Harbor, Jamaica; holotype female, Cat. No. 17281, U.S.N.M.).

Diagnosis.—Carapace narrowed anteriorly. Side tooth acute. Palm tuberculate. Legs narrow.

Description.—Distinctly narrowed from posterior to anterior angles, regions distinct, surface coarsely punctate, and toward the front and sides granulate. Lateral tooth acute, behind it a trace of a second tooth.

Front just half or a little more than half the fronto-orbital distance, five times as wide as high, superior lobes well marked, their anterior margin zig-zag, the outer pair much narrower and more advanced than inner pair; lower margin sinuous, in dorsal view bilobed; sides parallel.

Outer face of arm and wrist crossed by short granulate lines; an obtuse angle near the distal end of the upper border of arm; lower margins marked by spiniform tubercles; inner angle of wrist rounded. Palm of male swollen and tuberculate except near the fingers; upper margin a more or less interrupted line of granules; inner surface in part coarsely tuberculate. Fingers narrow, irregularly gaping; prehensile teeth irregular, the largest one near the tip; upper surface of dactylus spinulous for three-fourths its length.

Legs narrow; third leg a little more than twice the length of carapace, its merus two and three-fourths times as long as wide. Tooth on meropodites spiniform; last two joints tomentose, especially on the margins.

Abdomen of male broadly triangular.

Measurements.—Male (32285), length of carapace 23.5, width of same 28, width of front 12.6 mm.

Habitat.—Lives in fresh-water rivers and mountain streams, under loose stones in the shallows.

Range.—Known only from Jamaica. To an altitude of 4,500 feet.

Material examined.—Jamaica:

Small streams near Troy, about 2,000 feet altitude, found under loose stones in the shallows; November, 1905; W. Harris; 4 males, 1 female (32285).

Accompong (St. Elizabeth); P. W. Jarvis; 1 male (19052).

Mandeville; Henderson and Simpson; 3 males (18573).
Near Kingston Harbor; T. H. Morgan; 1 female holotype (17281).
Mountain Spring, St. Andrew; P. W. Jarvis; 1 female (19053).
Yallahs River, St. Andrew; J. E. Duerden, collector; specimens returned to him.
Clyde Spring, Blue Mountains; altitude 4,500 feet; C. B. Wilson; 8 males, 10 females (42889).
Jamaica; altitude 4,000 feet; June 10, 1910; E. A. Andrews; 1 male, 1 female (41752).

SESARMA (SESARMA) JARVISI Rathbun.

Plate 81.

Sesarma (Sesarma) jarvisi RATHBUN, Proc. U. S. Nat. Mus., vol. 47, 1914, p. 124, pl. 7 (type-locality, Mount Diablo, St. Anns, Jamaica; holotype male, Cat. No. 24941, U.S.N.M.).

Diagnosis.—Narrowed anteriorly. Sides of front oblique. Merus of third leg about three times as long as wide.

Description.—Carapace about four-fifths as long as wide, narrowed anteriorly, rather flattened; regions and supra-frontal lobes well marked, outer pair of lobes very narrow. Surface irregularly punctate and sparingly covered with lumps each of which bears the stumps of tufts of hair. The oblique ridges usually found on the branchial regions are few and are broken into short irregular lines.

Front a little less than half as wide as carapace, diminishing in width below, lower margin convex in front view, bilobed in dorsal view. Lateral tooth well-marked, blunt.

The chelipeds in the small type male are not much enlarged. The outer surface of arm and wrist are finely rugulose, margins of arm finely granulate or denticulate and not prominent. The chelae are elongate (the specimen is perhaps not full grown); manus rough with a few scabrous granules outside, a sharp crenulated ridge above. Fingers irregularly toothed within, nearly meeting; upper surface of dactylus finely spinulous almost to the tip.

Legs very slender, spine of merus acuminate, surface scabrous, sparingly hairy. Third leg about two and one-half times as long as carapace, its merus a little over three times as long as wide.

Measurements.—Male holotype, length of carapace 10.7, width of same 12.7, anterior width 10.7, width of front 5.2 mm.

Range.—Jamaica, in the interior.

Material examined.—Mount Diablo, St. Anns, Jamaica; P. W. Jarvis; 1 male holotype (24941). Jamaica; received from Chicago Acad. Sci.; 1 female (32440).

SESARMA (SESARMA) OPHIODERMA Nobili.

Sesarma (Sesarma) ophioderma NOBILI, Boll. Mus. Zool. Anat. Comp. R. Univ. Torino, vol. 16, No. 415, 1901, p. 44 (type-locality, Esmeraldas, Ecuador; holotype, young female, in Turin Mus.).

Diagnosis.—Length and width of carapace equal. Front half as wide as carapace, its superior lobes forming a convex arch. Two lateral teeth.

Description.—Greatest width of carapace is near insertion of second pair of ambulatory legs and is equal to the length. Carapace flat; entire surface rough with rather large, robust granules; branchial regions bounded by a deep sulcus: mesogastric sulcus narrow, deep, reaching frontal lobes. Front one-fourth as high as wide, deflexed vertically and bordered on its lower margin by a very prominent crest, which in the middle is deeply emarginate; vertical surface concave and rather smooth. Frontal lobes disposed in a convex curve, not straight as in most of the Sesarmas; the lobes are subequal, those of the lateral pair slightly larger, but instead of being in a straight line, those of the median pair are convex and very prominent when seen from below; the margin of the lobes is without granulation other than that common to the rest of the surface. Outer orbital tooth short, inclined a little forward and outward; behind this tooth are two minute teeth or rudiments of teeth of which the second is the better developed.

Chelipeds very slender; merus rugose outside, its three margins denticulate, the inner one with a small dentiform projection toward the apex; carpus without inner projection, its surface coarsely granulous; manus somewhat convex outside, clothed with very distinct and rather acute granules, inner surface apparently smooth, upper margin little accented, more distinct in the male than in the female; digits longer than palm, teeth rather large, apices corneous and excavated.

Legs long and slender; merus externally rugulose, length two and one-half times width, upper margin bearing a strong acute spine, lower margin unarmed; length of propodus about one and one-half times that of dactylus, the latter slender and long-acuminate; both have a few setae at the end.

Color.—General color, olive-brown; legs olive-gray annulated with brown. Fingers of cheliped orange.

Measurements.—Female holotype, length of carapace 9, width of same 9, exorbital width 8.75, width of front 4.5, height of same 1 mm. (After Nobili.)

Range.—Known only from the type-specimen from Esmeraldas, Ecuador.

SESARMA (SESARMA) BARBIMANUM Cano.

Sesarma barbimana CANO, Boll. Soc. Nat. Napoli, ser. 1, vol. 3, 1889, pp. 93 and 245 (type-locality, Payta; type in Naples Mus.).

Diagnosis.—Palms pilose outside.

Description (after Cano).—Carapace subquadrate, surface rugulose and pubescent, a rudimentary tooth behind the lateral tooth. Front less deflexed than in *S. crassipes*.

Chelae subrotund, outside densely and coarsely pilose as far as the base of the fingers, the prehensile borders of which are armed with small acute teeth.

Merus joints of legs with a broad dilatation below, which ends in a rather strong tooth. Next two joints short-hirsute, dactyli incurved, armed below with spinules.

Measurements.—Female type, length of carapace 6, width of same 7, width of front 3.3 mm.

Range.—Payta, Peru.

Subgenus HOLOMETOPUS Milne Edwards (=SESARMA de Man, not Say).

Carapace without a lateral tooth behind the outer orbital tooth. Manus without oblique, coarsely pectinated ridges on upper surface.

SESARMA (HOLOMETOPUS) RECTUM Randall.

Plate 82.

Sesarma recta RANDALL, Journ. Acad. Nat. Sci. Philadelphia, vol. 8, 1839, (1840), p. 123 (type-locality, Surinam; male holotype in Mus. Phila. Acad. Sci.).—ORTMANN, Zool. Jahrb., Syst., vol. 10, 1897, p. 331, pl. 17, figs. 8 and 8a.

Sesarma mullerii A. MILNE EDWARDS, Nouv. Arch. Mus. Hist. Nat. Paris, vol. 5, 1869, p. 29 (type-locality, "Destero (Brésil);" type in Paris Mus.).

Diagnosis.—Large, coarse. Hands broad, a single line of granules above. Merus joints of legs less than twice as long as wide.

Description.—A large species. Carapace distinctly broader than long, diminishing a little in width from front to back, deeply grooved; surface rough with pits and transverse ridges, which on the anterior part are short and squamose, and which are in large part furnished with tufts of hair. A short oblique ridge, not far behind the antero-lateral angle, meets the lateral margin so as to form a slight swelling which is not a true tooth.

Supra-frontal lobes deeply separated, the inner pair not much wider than the outer pair. Front about six times as wide as high; sides parallel, lower margin cut into two lobes very prominent in dorsal view. Upper margin of orbit directed strongly backward and outward, outer angle prominent.

Chelipeds strong, upper and outer margins denticulate, with a deep subterminal sinus; inner margin expanded at the middle, dentate; outer surface of merus and carpus crossed by granulated rugae· inner angle of wrist acute. Hand nearly twice as high as its superior width; densely but obscurely granulate outside, a short longitudinal ridge near the middle, upper margin acute, granulate, the granules in pairs in a single line; inner surface irregularly dentate, pollex broad and flat, upper surface of dactylus with 14–16 spinules, which reach nearly to the tip.

Legs very broad and nearly naked; third leg a little over twice as long as the carapace, its merus less than twice as long as broad. Margins of last two joints tomentose.

Measurements.—Male (22839), length of carapace 26.4, width between orbital angles 30, width of front 18 mm.

Range.—From Trinidad, West Indies, to Desterro, Brazil.

Remarks.—*S. rectum* is nearest *S. sulcatum*, a species with a lateral tooth, and narrower and more hairy legs.

Material examined.—

Trinidad; Jan. 30–Feb. 2, 1884; *Albatross;* 1 male, 1 female (19476).

Surinam; Dr. Hering; 1 male holotype (Mus. Phila. Acad. Sci.).

Pernambuco, Brazil: 1876–1877; R. Rathbun; 1 male young (40817). Aug. 1, 1899; on mangroves; Branner-Agassiz Exped.; 1 male, 1 female (25711).

Mapelle, Bay of Bahia, Brazil; 1876–1877; R. Rathbun; 1 female (40819).

Plataforma, Bahia, Brazil; 1876–1877; R. Rathbun; 1 male (40820).

Salt Lagoa, Caravellas, Bahia, Brazil; 1876–1877; R. Rathbun; 1 female (40818).

Terra de Masahe, Rio de Janeiro, Brazil; Jan., 1912; E. Garbe; 2 males (47862).

Rio de Janeiro, Brazil; Thayer Exped.; 1 male (22839).

Ilha Casquerinita, Santos, Brazil; June, 1913; H. Luderwaldt; 1 male, 1 female (47867).

Piassaguera, Santos, Brazil; June, 1913; H. Luderwaldt; 2 males (47859).

Iguape, São Paulo, Brazil; 1902; R. Krone; 1 male (47827).

SESARMA (HOLOMETOPUS) OCCIDENTALE Smith.

Sesarma occidentalis SMITH, Trans. Connecticut Acad. Arts and Sci., vol. 2, 1870, p. 158 (type-locality, Acajutla, west coast of Central America; male holotype in P. M. Y. U.).

Diagnosis.—Transverse. Front without deep median sinus. Hand cristate above.

Description.—Carapace narrower anteriorly than posteriorly, length less than anterior width. Supra-frontal lobes rough with coarse, sharp granules and separated by slight depressions, inner pair much wider than outer pair.

Front nearly perpendicular, three and one-half times as wide as high, its concave surface irregularly and coarsely granulous, lower margin curved forward and nearly straight.

Antero-lateral tooth acute, projecting well forward.

Chelipeds short, stout; anterior angle of merus sharp, dentate and raised in a thin crest at the distal end; carpus thickly beset with sharp granules; palm short. outer surface very granulous, upper margin a sharp crest.

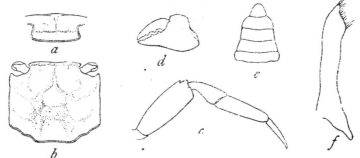

FIG. 148.—SESARMA (HOLOMETOPUS) OCCIDENTALE, MALE HOLOTYPE. *a*, ANTERIOR VIEW OF FRONT, × 1⅞; *b*, CARAPACE AND EYES, DORSAL VIEW, × 1⅞; *c*, THIRD LEG, × 1⅞; *d*, LEFT CHELA, OUTER VIEW, × 1⅞; *e*, LAST FIVE SEGMENTS OF ABDOMEN, × 1⅞; *f*, APPENDAGE OF FIRST SEGMENT OF ABDOMEN, × 7.

Third leg two and one-half times as long as carapace, its merus nearly three times as long as wide.

Measurements.—Male holotype, length of carapace 15.8, width of same 17.6, anterior width 16.9, width of front 9.4 mm.

Range.—Acajutla, Salvador; Panama (M. C. Z.); Tumaco, Colombia (Nobili).

Material examined.—Acajutla, Salvador; 1866; F. H. Bradley; 2 specimens of which 1 is type (Cat. No. 545, P. M. Y. U.).

SESARMA (HOLOMETOPUS) CINEREUM (Bosc).

WOOD CRAB; FRIENDLY CRAB; SQUARE-BACKED FIDDLER.

Plate 83.

Grapsus cinereus Bosc, Hist. Nat. Crust., vol. 1, an X [1801–1802], p. 204, pl. 5, fig. 1 (type-locality, "*la Caroline*"; type not extant).

Grapsus cinereus SAY, Journ. Acad. Nat. Sci. Philadelphia, vol. 1, 1818, p. 442 (not 1817, p. 99).

Sesarma cinerea MILNE EDWARDS, Hist. Nat. Crust., vol. 2, 1837, p. 75 (not all synonymy).

Diagnosis.—Transverse, nearly smooth. Front much more than half width of carapace. Merus of third leg more than two and one-half times as long as wide.

Description.—Carapace distinctly transverse, of nearly uniform width throughout, regions well marked, surface nearly smooth, punetate, toward the front rough with squamiform granules. Suprafrontal lobes well marked, the inner pair a little wider.

Front four times as wide as high, widening below, lower edge sinuous, somewhat four-lobed in dorsal view. Outer orbital angle acute.

Chelipeds heavy. Only the lower edge of the merus has a subdistal notch, upper edge sharp, inner edge irregularly dentate, with a triangular laminar expansion on the distal half. Inner angle of wrist rounded. Palm nearly twice as high as its superior length. The outer surface is covered with the same scabrous granules that form short transverse lines on arm and wrist; near the upper margin the granules are arranged in oblique parallel lines, but without forming the strong ridges which characterize the subgenus *Parasesarma;* inner face coarsely graunlous with a short prominent ridge near the distal end. Pollex elongate-triangular; dactylus thickened at base and somewhat dorsally flattened, a well marked concavity below at base. The fingers gape narrowly, and the largest tooth of their applied edges is at the middle of the pollex.

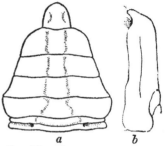

FIG. 149. — SESARMA (HOLOMETOPUS) CINEREUM, MALE (15072). *a*, ABDOMEN, × 3½; *b*, RIGHT APPENDAGE OF FIRST SEGMENT OF ABDOMEN, VENTRAL VIEW, × 7½.

Legs rather narrow, a sharp spine near the end of the merus; third pair over twice as long as carapace, its merus a little more than two and one-half times as long as wide.

Abdomen of male broadly triangular; the appendages of the first segment are hairy on the outer side of their extremities.

Measurements.—Male (15072), length of carapace 17.5, width of same 19.7, width of front above 11 mm.

Habits.—Lives under logs, drift, and roots, on wharves and piling. Found on muddy and sandy shores, near salt or brackish water, sometimes in mangrove swamps. In some situations makes burrows. Mr. J. D. Mitchell says of them: "They will go on board coasting boats and make trips of several days or a week, hiding under ropes or anywhere out of sight or reach. They prowl and feed mostly at night, but I have seen them come on deck and pick up scraps while the sailors were eating."

Range.—From Chesapeake Bay (Arundel-on-the-bay) to Tampico, Mexico.

Material examined.—

Arundel-on-the-bay, Chesapeake Bay, Maryland; on land near brackish water; W. P. Hay; 1 male, 5 females (22158).

Chesapeake Beach, Maryland: June 23, 1910; A. C. Weed and W. W. Wallis; 1 male, 1 female (40268). June 28, 1911; C. R. Shoemaker and W. D. Appel; 3 males, 1 female (43033).

Island Creek, Talbot County, Maryland; July 26, 1911; C. R. Shoemaker and W. D. Appel; 1 male, 2 females (1 ovig.) (45531).

Point Lookout, Maryland; Aug. 25, 1880; Wm. Palmer; 3 males, 1 female (15098).

Smiths Island, Northampton County, Virginia; Wm. Palmer: September, 1895; 1 male (18900). May 19, 1898; 1 male, 1 female (21623). Dec. 24, 1898; one specimen from the beach; the others from under roots and logs well up on the beach above high water; " they make holes in the sand "; 3 males, 5 females (22330).

Hampton, Virginia; June 28, 1892; *Grampus;* 20+ specimens (20117).

Winyah Bay, South Carolina; January, 1891; *Fish Hawk;* 7 specimens (17491).

Charleston Harbor, South Carolina; Mar. 18, 1880; M. McDonald; 2 females (3160).

West end Port Royal Island, South Carolina; Jan. 27, 1891; *Fish Hawk;* 1 specimen (17182).

Morgan River, South Carolina; around old palmetto tree; Feb. 24, 1891; *Fish Hawk;* 6 specimens (26138).

St. Johns River, Florida; G. Brown Goode; 3 specimens (P. M. Y. U.).

St. Augustine, Florida; H. S. Williams; 1 specimen (1022, P. M. Y. U.).

Mouth of Indian River, Florida; 1874; E. Palmer; 11 specimens (P. M. Y. U.).

Punta Rassa, Florida; mangrove swamp; February, 1884; H. Hemphill; 17 males, 13 females (15072).

Tampa Bay, Florida; opposite Port Tampa; shore; Jan. 26, 1898; *Fish Hawk;* 1 male, 4 females (26113).

Cedar Keys, Florida; December, 1883; H. Hemphill; 32 males, 49 females (6413).

St. Vincent Sound, Apalachicola, Florida; shore; Apr 2, 1915; *Fish Hawk;* 7 males, 5 females (49910).

Pensacola, Florida; July, 1893; J. E. Benedict; 20+ specimens. (17923).

Mobile Bay, Alabama; C. F. Baker; 1 female (21685).
Biloxi Bay, Mississippi; C. F. Baker; 1 male, 2 females (21686).
Oyster Bay, St. Bernard Parish, Louisiana; on muddy banks; Feb. 3, 1898; *Fish Hawk;* 3 specimens (26139).
Grand Isle, near New Orleans, Louisiana; G. Kohn; 3 specimens (2258).
Swan Lake, Galveston, Texas; Nov. 5-15, 1891; B. W. Evermann; 1 female (17111).
Tampico, Mexico; live in the soft mud of the river banks; June 1 1900; Edward Palmer; 4 females (2 soft shell) (45795).

SESARMA (HOLOMETOPUS) MIERSII Rathbun.

Plate 84.

Sesarma angustipes MIERS, Proc. Zool. Soc. London, 1881, p. 70; Rat Island, Monte Video. Not *S. angustipes* Dana. 1852.

Sesarma stimpsonii MIERS, *Challenger* Rept., Zool., vol. 17, 1886, p. 270 (type-locality, Monte Video; type in Brit. Mus.). Not *S. stimpsonii* Miers, 1881.

Sesarma (Holometopus) miersii RATHBUN, Proc. Biol. Soc. Washington, vol. 11, 1897, pp. 90 and 91 (type-locality, Abaco, Bahamas;[1] Cat. No. 11372, U.S.N.M.).

Diagnosis.—Little transverse. Front less than half width of carapace. Merus of third leg less than two and one-half times as long as wide.

Description.—Carapace a little broader than long, of equal width anteriorly and posteriorly. General appearance much like that of *S. cinereum*, punctae more numerous, the anterior granulated area is more extensive.

Front less than four times as wide as high, widening a little below. Of the superior frontal lobes those of the inner pair are very much wider than the outer pair.

The chelipeds differ as follows from those of *S. cinereum:* The arm has a low subdistal tooth on its upper margin; the laminar expansion of the inner margin is widest nearer the end of the segment. The palm is very coarsely granulate, being rough to the unaided eye.

Legs wider than in *S. cinereum;* the merus of the third pair is two and one-half or less than two and one-half times as long as wide.

Terminal segment of male abdomen large, as wide as long; the appendages of the first segment terminate in an oblique sinuous margin fringed with hair.

Measurements.—Male holotype, length of carapace 19.3, width of same 21.1, superior frontal width 11 mm.

---

[1] On page 92, Proc. Biol. Soc. Washington, vol. 11, I indicated Rat Island as the type-locality, but the specimen there described by me should be the type.

Habits.—On mangroves (Greeley). The following notes were furnished by Dr. E. A. Andrews, October 10, 1910:

Found in Montego River, crawling out on the stones and walls and rushing under water when disturbed. They were very abundant, though difficult to catch because extremely agile. Zoeas apparently belonging to this crab were taken in the tow-net in this river, and one is now alive here in fresh water, having passed through the megalops into the adult form. Others were hatched from eggs in fresh water.

Range.—From the Bahamas to Monte Video, Uruguay. Bermudas.

Material examined.—

Abaco, Bahamas; *Albatross;* 10 males, 10 females (11372).

Egg Island, Bahamas; Biol. Exped. State Univ. Iowa; specimen in Mus. S. U. I.

San Salvador, Bahamas; *Albatross;* 1 male, 4 females (11414).

Nueva Gerona, Isla de Pinos; July 11, 1900; Wm. Palmer and J. H. Riley; 1 female (23815).

Montego River, Montego Bay, Jamaica; Aug. 25, 1910; E. A. Andrews; 2, male and female (41525).

Swan Islands, Caribbean Sea; C. H. Townsend; 1 male, 3 females (14556).

Rio Parahyba do Norte, Cabedello, Brazil; on mangroves; June 20, 1899; Branner-Agassiz Exped.; 1 male (25712).

Rio de Janeiro, Brazil; Dr. Cunningham; 1 male (Cat. No. 68–82, Brit. Mus.).

Desterro, Brazil; Fr. Müller; 1 male (20312).

SESARMA (HOLOMETOPUS) MIERSII IHERINGI, new subspecies.

Plate 85.

Type-locality.—Bahia, Brazil, on the seashore; May, 1915; J. N. Rose, collector; from Carnegie Institution; male holotype, Cat. No. 48299, U.S.N.M.

Diagnosis.—Differs from typical *miersii* in having a little narrower carapace and longer legs.

Affinities.—Approaches very near *S. biolleyi* of the Pacific coast, which, however, has a higher front and differently shaped male abdomen.

Measurements.—Male holotype, length of carapace 20.7, width of same posteriorly 22, anteriorly 20.4, superior frontal width 11.7 mm.

Range.—Brazil, from Bahia to São Paulo.

Material examined.—Besides the type-specimen noted above, 1 male was collected at Iguape, São Paulo, Brazil, by R. Krone, and sent by H. von Ihering, Museu Paulista, to the National Museum (47830).

SESARMA (HOLOMETOPUS) MAGDALENENSE, new species.

Plate 86.

Type-locality.—Mangrove Island, Magdalena Bay, Lower California; Mar. 20, 1911; *Albatross;* holotype male, Cat. No. 45793, U.S.N.M.

Diagnosis.—Carapace broader than long, broadest anteriorly. Superior frontal lobes low and nearly smooth. Front three-fifths as wide as carapace.

Description.—Carapace distinctly broader than long, broadest at outer angles of orbit, diminishing posteriorly, a very shallow sinus in the lateral margins behind the antero-lateral angles. Surface for the most part smooth and shining, depressions moderately deep; pits of two sorts, a few large scattered ones visible to the naked eye, and numerous small ones, which become crowded on the anterior branchial region. On the anterior and antero-lateral regions there are a few scalelike granules. Antero-lateral angle a well-marked tooth.

Front about three-fifths as wide as carapace, surface nearly vertical, with the lower edge advanced; front widening below, lower margin arcuate, outer corners rounded; surface uneven, wrinkled, and unevenly granulate with fine, depressed granules; superior frontal lobes nearly smooth and feebly separated, the middle pair the wider.

Chelipeds of male massive; merus and carpus covered on the outer surface with short, granulated rugae; chelae high, swollen; immovable finger short, high, horizontal; dactylus strongly arched. Palm with lower margin very arcuate, its upper surface with several longitudinal broken lines of fine granules, its outer surface, as well as the upper surface of the proximal half of the dactylus, covered with fine scabrous granules; fingers punctate, gaping; basal half of the prehensile edge of the dactylus cut out in a deep sinus, into which projects a crenulated tooth of the immovable finger; both fingers irregularly dentate.

In the female both fingers are horizontal and longer than the immovable finger of the male; they do not gape and the teeth fit rather closely together. The chelae of the young male are intermediate in form between those of the female and of the adult male, and the gape is lacking.

Merus joints of legs rather short (in the fourth pair two and one-fourth times as wide as long), widening distally, and crossed by fine short rugae; dactyli slender, longer than their respective propodi, measured on the outer or anterior margin.

Abdomen of male broadly triangular; terminal segment as broad as long. Appendages of first segment rather slender. tips oblique.

Color.—Specimens in alcohol have a greenish-blue carapace mottled with purple; upper proximal half of chelae reddish-brown; upper surface of legs covered with a pattern of fine dots of dark purple on a light ground.

Measurements.—Holotype male, length of carapace 11.6, width between outer angles of orbits 14.2, width at postero-lateral angles 13.1 mm.

Material examined.—Eight males and eight females were taken at the type-locality.

SESARMA (HOLOMETOPUS) BIOLLEYI Rathbun.

Plate 87, figs. 2 and 3.

Sesarma (Holometopus) biolleyi RATHBUN, Proc. Biol. Soc. Washington, vol. 19, 1906, p. 100 (type-locality, Salinas de Caldera, Boca del Jesus Maria, Costa Rica; holotype male, Cat. No. 32490, U.S.N.M.).

Diagnosis.—Narrowed anteriorly. Front half width of carapace. Merus of third leg three times as long as wide.

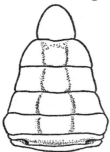

FIG. 150.—SESARMA (HOLOMETOPUS) BIOLLEYI, ABDOMEN OF MALE HOLOTYPE, × 3½.

Description.—Carapace a little broader than long and broader behind than before, very uneven, densely granulate anteriorly, punctate and wrinkled posteriorly. Front one-half width of carapace, vertical, widening below, lower edge projecting, convex in front view, slightly sinuous in dorsal view; superior lobes more transverse and more sharply marked than in the preceding species. Upper margin of orbit sinuous, very oblique, outer tooth acuminate.

Chelipeds rugose, the rugae becoming single tubercles or granules on the distal half of the palm; the latter much inflated, inner face sparingly granulous, a transverse row of granules near the distal end. Prehensile teeth of fingers fewer and stronger than in S. miersii. Inner laminar expansion of arm rather deeply cut into jagged teeth.

Legs long and narrow, third pair two and one-half times as long as carapace, its merus three times as long as wide.

Measurements.—Male holotype, length of carapace 19.1, anterior width of same 20.2, posterior width 21, width of front above 11.3 mm.

Habit.—Lives in mangroves.

Range.—Pacific coast of Costa Rica.

Material examined.—Costa Rica; P. Biolley and J. F. Tristan: Punta Arenas; Jan. 19, 1907; 2 males (39101). Feb., 1907; 4 males (39092).

Salinas de Caldera, Boca del Jesus Maria; on mangroves; Jan., 1906; 1 male, 1 female (male is holotype) (32490); 1 female (32316). Apr., 1905; 1 male (32317).

SESARMA (HOLOMETOPUS) TAMPICENSE Rathbun.

Plate 88.

Sesarma (*Holometopus*) *tampicense* RATHBUN, Proc. U. S. Nat. Mus., vol. 47, 1914, p. 124, pl. 8, text-fig. 4 (type-locality, Tampico, Mexico; holotype male, Cat. No. 45794, U.S.N.M.).

Diagnosis.—Sides of front parallel. Palm finely granulate. Fingers gaping. Merus joints of legs converging slightly from middle to distal end.

Description.—Carapace perceptibly wider than long, of nearly even width throughout, though widening slightly behind. Surface coarsely punctate at the middle, finely punctate on the branchial and intestinal regions, the punctae more or less connected by fine grooves; surface of frontal region and antero-lateral angles finely granulate; postero-lateral grooves fine.

Supra-frontal lobes deeply separated, the median groove larger than the lateral grooves; middle pair of lobes transversely arcuate; outer pair narrower, oblique, trending forward toward the orbit. Front relatively broad and low, about five times as wide as high, sides vertical, lower margin arcuate in front view, sinuous in dorsal view, surface concave in both directions. Upper margin of orbit nearly straight, up to the short, acute tooth at the outer angle of the orbit.

FIG. 151.—SESARMA (HOLOMETOPUS) TAMPICENSE, LEFT APPENDAGE OF FIRST SEGMENT OF ABDOMEN OF MALE HOLOTYPE, VENTRAL VIEW, × 8.

Merus of cheliped covered with granulated rugae on its outer face; lower outer margin with a well-marked subdistal tooth; tooth on upper margin nearly obsolete; inner margin denticulate, distally expanded and bearing a large tooth. Upper surface of carpus similar to outer surface of merus. Palm massive in the full-grown male, as high as its horizontal length, lower margin arcuate, upper margin marked by interrupted lines of granules; outer surface covered with fine, depressed granules, inner surface with much larger granules on the more elevated portion. Fingers rather long and slender, for the genus, gaping in the male except at the tip; prehensile edges dentate, with two or three teeth enlarged on each finger.

Legs of moderate length, the third pair about twice as long as the carapace; merus joints about two and three-fourths times as long as wide, converging slightly from the middle to the distal end.

Side margins of male abdomen sinuous. Appendages of first segment widened behind the tips.

Allied to *S. cinereum* and *S. miersii*. It differs from both in the front having parallel sides instead of widening below, in the fingers gaping, in the male appendages transversely, instead of obliquely, cut at the tip. From *S. cinereum* it differs also in its narrower carapace, without scattered tufts of hair, longer fingers, and narrower male abdomen; from *S. miersii* in the relatively smooth palms. those of *miersii* being coarsely granulate.

Measurements.—Male holotype, length of carapace 16.1, greatest width of same (at base of second leg) 17.3, width between outer angles of orbit 17.2, width of front 9.4 mm.

Habitat.—Lives in the soft mud of the river banks.

Range.—Known only from the type-locality.

Material examined.—Tampico, Mexico; June 1, 1910; Edward Palmer; 4 males, 1 of which is holotype (45794).

SESARMA (HOLOMETOPUS) RICORDI Milne Edwards.

BEACH CRAB.

Plate 89.

Sesarma ricordi MILNE EDWARDS, Ann. Sci. Nat., ser. 3, Zool., vol. 20. 1853, p. 183 [149], (type-locality, Haiti; type in Paris Mus.).

Sesarma guerini MILNE EDWARDS, Ann. Sci. Nat., ser. 3, Zool., vol. 20, 1853, p. 183 [149], ("*patrie inconnue*"; type in Paris Mus.).

Sesarma miniata SAUSSURE, Mém. Soc. Phys. Hist. Nat. Genève, vol. 14, 1858, p. 442 [26], (type-locality, Saint-Thomas; cotypes in Geneva and Paris Museums).

Sesarma angustipes STIMPSON, Ann. Lyc. Nat. Hist. New York, vol. 7, 1859, p. 66 (in part, at least). Not *S. angustipes* Dana, 1852.—SMITH, Trans. Connecticut Acad. Arts and Sci., vol. 2, 1870, p. 159.—DE MAN. Notes Leyden Mus., vol. 14, 1892, p. 253, pl. 10, fig. 5.

Sesarma stimpsonii MIERS, Proc. Zool. Soc. London, 1881, p. 70, (type-locality, Florida and the Tortugas; type in Brit. Mus.). Not *S. stimpsonii* Miers, 1886.

Sesarme cinerea HEILPRIN, Proc. Acad. Nat. Sci. Philadelphia, 1888, p. 320. Not *Sesarma cinerea* (Bosc).

Sesarma cinerea IVES, Proc. Acad. Nat. Sci. Philadelphia, 1891, p. 181.

Sesarma (Holometopus) ricordi RATHBUN, Proc. Biol. Soc. Washington, vol. 11, 1897, p. 91.

Diagnosis.—Carapace and chelipeds little rough. Third leg twice as long as carapace, merus three times as long as wide.

Description.—Carapace a little wider behind than in front, the length about equal to the anterior width. Surface finely punctate, but granulate only behind the orbits and sparingly on the suprafrontal lobes, and coarsely so on the deflexed surface of the front.

An irregular tubercle behind the orbit. Upper margin of orbit nearly straight, up to the subacute outer angle.

Supra-frontal lobes deeply separated; middle pair transverse, a little wider than the oblique outer pair. Front considerably wider below than above, inclined so that the greater part of its surface is visible in dorsal view, lower edge sinuous, the median sinus wide and deep.

The lower margin of the arm has the customary subdistal tooth, the inner margin has an arcuate laminar expansion which ends distally in a tooth. Chelae much swollen, slightly constricted at the base of the thumb, finely granulate outside, granules thickest on upper surface. Fingers with a wide aperture at base, the dactylus being much hollowed out; the largest prehensile tooth is near the middle of the pollex and is much thickened. Inner surface of palm coarsely granulate.

Legs narrow, third pair about twice as long as carapace, its merus three times as long as wide.

Extremity of appendage of first segment of male abdomen more transverse than in the allied species.

Color.—Orange or reddish-yellow, finely speckled by the black hairs of the carapace; feet often marbled (Saussure). Deep brown (Jarvis). Variable (A. E. Verrill).

Measurements.—Male (13798), length of carapace 19.5, greatest width of same 20.1, anterior width 19.1, width of front above 10.6 mm.

Habitat.—Along shore, under logs of driftwood, and among rocks and piles and dead seaweed, nearly up to high-water mark. Under log in pine woods about 50 yards from high-tide mark (Miller).

Range.—Southern Florida and Bahamas to Rio de Janeiro, Brazil; Bermudas.

Material examined.—

Smiths Place, south of South Bight, east side of Andros Island Bahamas; in pothole; May 3, 1912; Paul Bartsch; 4 males, 4 females (45626).

Near Lake Kissimmee, Florida; A. M. Reese; 11 males, 13 females (45529).

Key West, Florida; among rocks and dead seaweed, near high tide; 1885; H. Hemphill; 48 males, 95 females (13798).

Key West Harbor, Florida; 1884; Dr. E. Palmer; 1 female (13584).

Big Gasparilla Island, Florida; Mar. 6, 1889; *Grampus;* 1 male (15261).

Little Gasparilla Pass, Florida; *Grampus;* 1 male (15260).

Long Boat Key, Florida; Mar. 28, 1889; *Grampus;* 5 males, 9 females (15259).

Florida; 1901; J. E. Benedict; 1 y. (25576).

Biloxi, Mississippi; under log in pine woods about 50 yards from high-tide mark; Feb. 16, 1914; Gerrit S. Miller; 7 males, 5 females (46414).

Port of Silam, Yucatan; 1890; Ives, Mexican Exped.; specimens in Mus. Phila. Acad.

Belize, British Honduras; Rev. W. A. Stanton; 5 males, 9 females (22603).

Near Belize, British Honduras; Rev. W. A. Stanton; 8 specimens (21375).

Old Providence, West Indies; Apr. 4–9, 1884; *Albatross;* 1 female (18569).

Curaçao; Feb. 10–18, 1884; *Albatross;* 1 male, 2 females (18446).

Sabanilla, Colombia; Mar. 16–22, 1884; *Albatross;* 8 males, 30 females (17645).

Cabañas, Cuba; William Palmer and J. H. Riley; on rocks and piles along shore; June 3, 1900; 3 males, 1 female (23811); June 20, 1900; 1 female (23812).

Cuba; 1914; Henderson and Bartsch, *Tomas Barrera* Exped.: Los Arroyas; May 20; 7 males, 10 females (2 ovig.) (48573). Bahia Honda; June 7; 1 male (48581). Cabañas; on sand, shell, grass to mud bottom; June 8; 1 female (48616); 1 female y. (48617); 1 y. (48618).

Mariel, Cuba; June 10, 1900; William Palmer and J. H. Riley· 1 male (23813).

Point Vedado, Cuba; M. S. Roig; 1 male (46086)

Montego Bay, Jamaica; P. W. Jarvis; 1 male, 1 female (19678).

Kingston Harbor, Jamaica: Mar. 1–11, 1884; *Albatross;* 1 y. (19424). T. H. Morgan; 1 female (17227).

Jamaica; Mar. 1–11, 1884; *Albatross;* 12 males, 27 females (18558).

Haiti; Ricord, collector; type (Paris Mus.).

Porto Rico; 1899; *Fish Hawk:* Porto Real; Jan. 26; 1 female (24036). Boqueron Bay; Jan. 27; 1 male, 2 females, 1 y. (24016). Hucares; Feb. 14; 3 specimens (24035). Ensenada Honda, Culebra; Feb. 11; 9 males, 11 females, 6 y. (24047).

St. Thomas; Saussure, collector; 1 cotype of *S. miniata* (Paris Mus.); 2 males, 1 female, cotypes of *S. miniata* (Geneva Mus.).

St. Thomas; C. R. Shoemaker; 1915: Shore near town; June 30; 1 female (49817). Magens Bay; on shore in shade near small pool; July 4; 5 males, 4 females (49818). East side of harbor, on beach· July 11; 7 males, 8 females (49816). Buck Island; July 30; 1 female (49812).

St. Croix; 1 specimen (Copenhagen Mus.).

Martinique; M. A. Rousseau; 4 specimens (Paris Mus.).

Trinidad; Jan. 30–Feb. 2, 1884; *Albatross;* 3 males (18559).

Brazil; 1876-1877; R. Rathbun: Itaparica, Bahia; 1 female (40821). Bay of Rio de Janeiro; 1 female y. (40822).
Bermuda; July, 1888; Heilprin; specimens in Mus. Phila. Acad. Sci.
Locality unknown; 1 female, type of *S. guerini* (Paris Mus.).

SESARMA (HOLOMETOPUS) ANGUSTIPES Dana.

Plate 90.

Sesarma angustipes DANA, U. S. Expl. Exped., vol. 13, Crust., pt. 1, 1852, p. 353; atlas, 1855, pl. 22, fig. 7a–c; (type-locality, South America; type not extant).

Sesarma americana SAUSSURE, Mém. Soc. Phys. Hist. Nat. Genève, vol. 14, 1858, p. 441 [25] (type-locality, St. Thomas; cotypes in Geneva Mus.).

Diagnosis.—Sides of front parallel. Exorbital tooth small. Merus joints of legs diminishing slightly from middle to distal end.

Description.—Carapace considerably narrower anteriorly than posteriorly, length subequal to anterior width; surface pitted and wrinkled, anteriorly rough with granules arranged mostly in short rows.

Supra-frontal lobes deeply separated from one another, those of the inner pair not much wider than those of the outer pair. Front four and one-half times as wide as high, sides parallel, lower edge bilobed, with a broad and deep median sinus in dorsal view.

Outer angle of orbit acute but not prominent.

The laminar expansion on the inner margin of the arm is small and in the old male dentiform. The swollen palms are tuberculate or granulate inside and out; no definite marginal line above. Fingers smooth on outer and inner surface, gaping their whole length, very irregularly toothed, the largest tooth being one not far from the basal end of each finger.

Third leg a little more than twice as long as carapace, its merus about three times as long as wide. The merus joints converge slightly from the middle to the distal end.

Abdomen of male subtriangular; appendages narrow and somewhat club-shaped at the extremity.

Measurements.—Male (17295), length of carapace 24, width of same 25.7, anterior width 24.7, superior width of front 13.6 mm.

Habitat.—Fluviatile. Found in holes in perpendicular banks of river (Stanton). Has been transported in a bunch of bananas to Washington, District of Columbia.

Range.—West Indies. From Vera Cruz, Mexico, to Nicaragua. South America (probably Rio de Janeiro). West coast of Central America (?).

Material examined.—
Almendares River, "La Chorrere," near Vedado, Havana, Cuba;
M. S. Roig; 1 male, 1 female (46084, 46085).
Jamaica; 1910; E. A. Andrews: Montego Bay; Aug. 3; 8 males, 13 females (42876). Montego River; Mar. 20; 3 males, 2 females (42877). Montego River and Flint River; Aug. 29; 1 male, 20 females (42878).
St. Thomas; Saussure; 2 male and female, cotypes (Geneva Mus.)
Vera Cruz, Mexico; 2 males, 2 females (Brit. Mus.).
Belize, British Honduras; Mrs. F. E. D. Meehling; 2 females (3286).
Stann Creek; British Honduras; found in a hole in the perpendicular bank of the river; Rev. W. A. Stanton; 1 female (22596).
Ceiba, Honduras; Sept. 29, 1916; F. J. Dyer; 1 male, 1 female (49914).
Greytown, Nicaragua; Feb. 12, 1892; Chas. W. Richmond; 3 males, 2 females (19417).
"Central America," probably; from a bunch of bananas at Washington, District of Columbia; R. W. Brown; 1 female (16160).
"West coast of Central America," locality probably wrong; 2 males (17295).

SESARMA (HOLOMETOPUS) ROBERTI Milne Edwards.

BRACKISH-WATER CRAB.

Plate 91.

Sesarma roberti MILNE EDWARDS, Ann. Sci. Nat., ser. 3, Zool., vol. 20, 1853, p. 182 [148], (type-locality, Gorée; type in Paris Mus.).
Sesarma americana POCOCK, Ann. Mag. Nat. Hist., ser. 6, vol. 3, 1889, p. 7. Not *S. americana* Saussure, 1858.
Sesarma bromeliarum RATHBUN, Proc. U. S. Nat. Mus., vol. 19, 1896, p. 143 (type-locality, Rio Cobre (St. Catherine), Jamaica; male holotype, Cat. No. 19406, U.S.N M.).

*Diagnosis.—*Length greater than anterior width. Sides of front subparallel. Merus joints of legs of even width in their distal half.

*Description.—*Length of carapace exceeding its anterior width and nearly equaling its posterior width; surface rough with pits and wrinkles, and anteriorly rough with squamiform granules.

Supra-frontal lobes deeply separated, the middle pair not much wider than the outer pair which are very oblique. Front four times as wide as high, sides subparallel, lower edge sinuous, quadrilobate, median sinus very deep and wide.

Upper margin of orbit straight, up to the outer tooth, which is acute and well marked.

Inner margin of arm produced in a triangular lamina. Surface of cheliped very rough, the palm covered with coarse scaly granules, which are continued partly on the fingers. There is an oval gape at the base of the fingers.

Third leg twice as long as the carapace, its merus two and two-thirds times as long as wide. The merus joints do not diminish in width from the middle to the distal end.

Color.—Dull blackish; claws orange red; eyes green (A. H. Verrill).

Measurements.—Male holotype, length of carapace 26.5, greatest width of same 26.3, anterior width 24.7, width of front 14.1 mm.

Habitat.—Lives on steep muddy banks of rivers in small deep holes, which it burrows in the soft earth at the water's edge and into which it retreats when alarmed. Found in most of the rivers of Jamaica where the water is brackish (Jarvis); also on mangroves.

Range.—West Indies. Turbo, Panama (M. C. Z.). Also Gorée, Senegal.

Material examined.—

Baracoa, Cuba; Feb. 1, 1902; Wm. Palmer; 1 female (25550).

Rio Cobre (St. Catherine), Jamaica; P. W. Jarvis; 1 male (holotype of *S. bromeliarum*), 1 female (19406); 2 males, 2 females (19418).

Interior of Jamaica; E. A. Andrews; 1 female (17003).

Santo Domingo, West Indies; 1878; W. M. Gabb; 1 male (3193).

Porto Rico: Biol. Survey, U. S. Dept. Agriculture; Anasco and Bayamon, in bird stomachs; specimens in collection of Biol. Survey. Gundlach, collector (4801, Berlin Mus.).

Dominica, West Indies; A. H. Verrill; 1 male (32715).

Laion, Dominica; J. G. Ramage; 12 specimens (88.26, Brit. Mus.).

Martinique; M. A. Rousseau; 2 females (Paris Mus.).

Port Castries, St. Lucia; *Albatross;* 2 males, 1 female (22108).

Gorée, Senegambia; M. Robert; 1 male, 1 female (cotypes), (20311).

SESARMA (HOLOMETOPUS) FESTAE Nobili.

Sesarma (Holometopus) festae Nobili, Boll. Mus. Zool. Anat. Comp. R. Univ. Torino, vol. 16, No. 415, 1901, p. 42 (type-localities, Tumaco, Colombia, and Esmeraldas, Ecuador; types in Turin Mus.).

Diagnosis.—Near *angustum*. Carapace wider than long. Upper margin of hand with a small, denticulated crest.

Description.—Lateral margins of carapace almost a straight line, spreading a little posteriorly. Tooth at outer angle of orbit acute, directed forward and slightly inward. Front vertical, rather high, lower border with a deeply sinuous crest; frontal lobes subequal, those of median pair only a little wider than those of lateral pair; lobes prominent, as if arising from lower margin of front, armed with rather acute granules; deflexed part of front concave and granulous. Anterior part of carapace granulate; granules elongate, sub-squamiform, consisting of a linear series of two to five small granules, bearing now and then minute setae, visible only with a lens in the adult. Posterior part of carapace, except of branchial region, without granules but appearing punctate; in the young the punctae are beset with hair. Small transverse folds cross the posterior part of the branchial region.

General aspect of carapace rather flat, only feebly convex anteriorly. Mesogastric sulcus narrow, but reaching front; on each side a short sulcus runs at right angles toward the supero-internal angle of the orbit. Gastro-cardiac sulcus deep.

Chelipeds rather slender. Merus outwardly rugose, its upper angle acute and denticulate; lower angle also conspicuously denticulate; infero-inner angle dilated toward the apex and bearing a rather large tooth or denticulate projection. Carpus coarsely granulous, without inner tooth or prominence. Hand rather swollen, covered outside with rather large granules; granules of inner surface much depressed and less crowded; no trace of an inner crest; upper margin limited by a denticulate crest. Palm shorter than fingers; the latter are slightly dentate, leave a small gape when closed, the apices horny and slightly excavate.

Legs long and slender; penult pair notably longer than the others. Merus from two and one-half to three times as long as its width, the propodus one and one-half times as long as the dactylus. All the articles bear a small, denticulate crest above and are freely squamose on the posterior surface; the merus has at its extremity a stout, acute tooth directed forward. Dactylus and distal end of propodus bear a few short setae. (After Nobili.)

Measurements.—Male cotype, length of carapace 7, width of same at second ambulatory legs 8$\frac{1}{3}$, width between exorbital angles 8 width of front 4.5 mm.

Range.—Colombia (Pacific coast) and Ecuador.

SESARMA (HOLOMETOPUS) ANGUSTUM Smith.

Plate 92.

Sesarma angusta SMITH, Trans. Connecticut Acad. Arts and Sci., vol. 2, 1870, p. 159 (type-locality, Pearl Islands, Bay of Panama; male holotype, Cat. No. 659, P.M.Y.U.).

Diagnosis.—Longer than wide. Chelipeds small. **Merus joints of** legs diminishing slightly from middle to distal **end.**

Description.—Carapace a little longer than broad, and slightly narrower in front than behind; surface very rough with pits and wrinkles, and in the anterior third with granules which are coarse and scaly, except behind the orbits, where they are very fine.

Front equaling in width half the carapace, vertical; the projecting edge only is visible in dorsal view, sides parallel, median sinus extremely wide and deep, each lobe faintly bilobed; supra-frontal lobes subequal, deeply separated.

Upper margin of orbit slightly convex; outer angle small, acute.

Chelipeds small, very rough with sharp rugae on the arm, and sharp granules on wrist and swollen hand. Arm with a small distal expansion on the inner margin. Fingers little gaping, their prehensile edges not very unevenly dentate.

Legs long and narrow, third pair two and one-third times as long as carapace, its merus two and two-thirds times as long as wide. Merus joints narrowing from the middle toward the distal end.

Abdomen of male oblong-triangular, appendages of first segment much enlarged in their terminal third.

Measurements.—Male (32314), length of carapace 14.4, width of same 14, anterior width 13.9, width of front 7.5 mm.

Range.—From Costa Rica (west side) to Bay of Panama.

Habit.—Lives in mangroves (Tristan).

Material examined.—

Punta Arenas, Costa Rica; 1 male, 2 females (Copenhagen Mus.).

Boca del Jesus Maria, Costa Rica; Apr., 1905; P. Biolley and J. F. Tristan; 1 male, 1 female (32314).

Santo Domingo, Gulf of Dolce, Costa Rica; Apr., 1896; H. Pittier; 1 male, 1 female (19437).

Taboga Island, Bay of Panama: May 12-15, 1911; Meek and Hildebrand; 1 female ovig. (44175). June, 1914; J. Zetek; 1 male (48789).

Pearl Islands, Bay of Panama; 1866; F. H. Bradley; 1 male holotype (659, P.M.Y.U.).

SESARMA (HOLOMETOPUS) HANSENI Rathbun.

Plate 87, fig. 1.

Sesarma (Holometopus) hanseni RATHBUN, Proc. Biol. Soc. Washington, vol. 11, 1897, p. 92 (type-locality, West Indies; male holotype in Copenhagen Mus.).

Diagnosis.—Carapace transverse, broadest anteriorly, smooth. Hand with crest above.

Description.—Carapace much broader than long, broader anteriorly than posteriorly. Regions well marked; mesogastric very wide

behind, and with a curved sulcus parallel to its posterior margin.
Surface smooth, punctate and without granulation.

Superior margin of front uneven, the inner lobes sloping backward from the middle. Front more than four times as wide as high; lower margin projecting, thin, arcuate in front view, slightly sinuous in dorsal view.

Hand deep, covered with depressed tubercles: superior margin with a thin denticulate crest. Fingers irregularly toothed; the largest tooth of the dactylus is midway of its length and fits between the two largest teeth of the pollex.

Legs of moderate length, merus joints less than two and one-half times as long as wide.

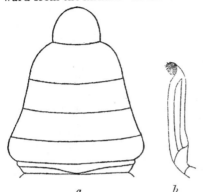

Fig. 152.—Sesarma (Holometopus) hanseni, male holotype a, abdomen, × 4⅔; b, appendage of first segment of abdomen, ventral view, × 5¼.

Measurements.—Male holotype, length of carapace 13.5, anterior width of same 16.6, posterior width of same 15.5, superior width of front 9.5 mm.

Range.—West Indies. Only one specimen known.

SESARMA (HOLOMETOPUS) BENEDICTI Rathbun.

Plate 93.

Sesarma recta de Man, Notes Leyden Mus., vol. 14, 1892, p. 249, pl. 10, figs. 4–4d; Surinam. Not *S. recta* Randall, 1840.

Sesarma (Holometopus) benedicti Rathbun, Proc. Biol. Soc. Washington, vol. 11, Apr. 26, 1897, p. 90 (type-locality, Surinam; type in Leyden Mus.).

Sesarma chiragra Ortmann, Zool. Jahrb., Syst., vol. 10, 1897, p. 331 (type-locality, Para; type in Mus. Phila. Acad. Sci.).—Calman, The Life of Crustacea, London, 1911, pl. 23 (facing p. 182), lower figure.

Diagnosis.—Movable finger remarkably enlarged at base. Carapace narrowed behind. Front very wide.

Description.—Carapace transverse, wider anteriorly than posteriorly, but not markedly so; mesogastric region deeply outlined; hepatic region depressed; surface rough with irregular pits and wrinkles, and on the anterior gastric region with small depressed squamiform granules.

Front wide, about two-thirds as wide as carapace, and five times as wide as high, increasing in width below; vertical, surface granulate, lower edge projecting, quadrilobate. Middle pair of superior lobes distinctly wider than outer pair.

Upper margin of orbit convex; outer tooth acute, rather prominent; behind it on the lateral margin there is a blunt, obtuse-angled rudiment of a tooth.

The arm has a triangular projection from the distal half of its inner margin. Palm very rough, with sharp granules which are continued the length of the upper surface of the dactylus and the lower surface of the pollex. Fingers remarkably modified in the male; the dactylus is depressed and flattened in its basal half, so that it is one and one-half times as wide as high; the dactylus is strongly swollen below, so that its lower margin is very convex, independent of the convexity of the palm. The fingers gape for their basal half and are irregularly toothed on the distal half.

Third leg a little less than twice as long as the carapace, its merus twice as long as wide.

Measurements.—Male (22838), length of carapace 18.5, anterior width of same 21.3, posterior width of same 20, width of front 13.3 mm.

Range.—Key West, Florida (M. C. Z.). Guiana to Rio de Janeiro, Brazil (M. C. Z.).

Material examined.

Demerara, British Guiana; 1 male, 1 female (Brit. Mus.).

Tajapouru, Brazil; Thayer Exped.; received from Mus. Comp. Zoöl.; 1 male, 1 female (22838).

Para, Brazil; Dr. T. B. Wilson; 2 males, 1 female, cotypes of *S. chiragra* (Mus. Phila. Acad. Sci.).

Genus METOPAULIAS Rathbun.

Metopaulias RATHBUN, Proc. U. S. Nat. Mus., vol. 19, 1896, p. 144; type, *M. depressus* Rathbun.

Carapace perfectly flat. Supra-frontal lobes subtriangular or obliquely truncated.

Differs from *Sesarma* in having the carapace perfectly flat except near the postero-lateral angles, where it is destitute of oblique ridges; the front deeply concave, both transversely and longitudinally, the upper and lower margins deeply divided, the supra-frontal lobes forming teeth which are subtriangular or obliquely truncated; the granules of the side walls arranged singly, not in pairs.

The antero-lateral margins are curved, unidentate, the extremities of the fingers very narrow, with tips curved, legs long and slender.

Known only from the type-species.

METOPAULIAS DEPRESSUS Rathbun.

PINE CRAB.

Plate 97, figs. 3 and 4.

Metopaulias depressus RATHBUN, Proc. U. S. Nat. Mus., vol. 19, 1896, p. 144 (type-locality, Newport (Manchester), Jamaica; female holotype, Cat. No. 19407, U.S.N.M.).

Diagnosis.—One lateral tooth. Legs slender.

Description.—Postero-lateral margins parallel, antero-lateral margins gently curved, so that the carapace is sensibly narrower in front; length equal to or greater than anterior width. Mesogastric and cardiac regions deeply delimited. Surface densely covered with punctae; toward the front there are sparsely distributed short lines of scabrous granules.

Front a little less than half as wide as carapace, about three and one-half times as wide as high, very concave in both directions, side margins vertical, surface nearly smooth; supra-frontal lobes flat above, strongly projecting, deeply separated, either subtriangular or obliquely truncated, those of the inner pair most produced at the inner angle, those of the outer pair at the outer angle; median sinus of lower margin much wider than that of the upper margin; with a projecting lobe either side, beyond which a shallow sinus extends to the rounded outer angle.

Upper margin of orbit concave; outer angle acuminate, directed forward. Lateral tooth subrectangular.

Chelipeds equal, of moderate size; outer surface of arm and wrist crossed by very short granulated lines; on the palm, the granules are arranged in short ridges at the proximal end, but elsewhere singly, except near the distal end of the outer surface which is smooth. Upper margin of arm granulate; lower and inner margins spinulous, the latter with a small triangular expansion on its distal half. Fingers long, rather narrow, prehensile edges irregularly toothed and nearly meeting. Dactylus granulate above for the proximal half.

Third leg two and one-half times as long as carapace, its merus from three and one-half to four and one-half times as long as wide. The merus joints are widest in the middle and diminish a little toward both ends; upper surface sparingly roughened, anterior margin acute at the end; subterminal tooth subrectangular with a short acute tip. Posterior margin of propodal joints and both margins of dactyli densely hairy, other margins sparingly so.

Measurements.—Female holotype, length of carapace in median line 17.8, greatest width of same 19.8, anterior width of same 17, width of front 8.5 mm.

Habitat.—Found on the large broad-leaved pine of a vivid green which grows abundantly on the tree trunks on the rugged limestone hills of Jamaica. The bases between the leaf bases of the pine form natural rain-water reservoirs, which do not fail even in dry weather, and these aquaria are tenanted by the crabs and other small creatures. The flat back of the crab (looking as if it has just passed through a mangle) enables it to crawl easily in the narrow spaces between the leaves. (Jarvis.)

Range.—Known only from Jamaica.

Material examined.—Jamaica, West Indies:

Near Montego Bay; from leaves of *Bromelia* from hill back of Snug Harbor; June 26, 1910; E. A. Andrews; 5 males, 4 females (42875).

Accompong; P. W. Jarvis; 3 y. (19426).

Newport, Manchester; P. W. Jarvis; 1 female holotype (19407).

Stony Hill; from water in crowns of Tillandsias, Bromelias, etc.; Mar. 9, 1877; H. G. Hubbard; 1 female, 13 y. (45519).

Holly Mount; found in water that collects in the wild pines on the trees; W. Harris; 1 female (32367).

Jamaica; P. W. Jarvis; 3 females (24939).

Additional record.—Ewarton, Jamaica, "the only place where it seems to be plentiful" (Jarvis).[1]

Genus METASESARMA Milne Edwards.

Metasesarma MILNE EDWARDS, Ann. Sci. Nat., ser. 3, Zool., vol. 20, 1853, p. 188 [154]; type, *M. rousseauxi* Milne Edwards.

Antenna excluded from orbit. Front vertical.

Differs from *Sesarma*, in having the tooth at the lower inner angle of the orbit meet the front near its lower angle, so as to exclude the antenna from the orbit. The reticulation of the under side of the carapace is finer, closer, and more confused.

Front vertically deflexed and deep. Supra-frontal tubercles feebly marked.

Distributed through the Indo-Pacific region and the east coast of Middle and South America.

METASESARMA RUBRIPES (Rathbun).

Plate 94.

Metopograpsus brasiliensis A. MILNE EDWARDS, label in Paris Mus., name unpublished.

Sesarma mülleri MIERS, Challenger Rept., Zool., vol. 17, 1886, p. 270, pl. 21, fig. 3; Bahia. Not *S. mullerii* A. Milne Edwards, 1869.

Sesarma (Holometopus) rubripes RATHBUN, Proc. Biol. Soc. Washington, vol. 11, p. 90, 1897 (type-locality, Bahia; type in Brit. Mus.).—MOREIRA, Arch. Mus. Nac. Rio de Janeiro, vol. 12, 1903, p. 112, pl. 1.

Sesarma benedicti MOREIRA, Arch. Mus. Nac. Rio de Janeiro, vol. 11, 1901, p. 40. Not *S. benedicti* Rathbun.

---

[1] The Jamaica Post, December, 1897.

Diagnosis.—Carapace very narrow behind. Legs very wide.

Description.—Carapace narrowing rapidly from before backward, about as long as the posterior width. Areolation much as in *Sesarma*, the mesogastric and cardiac regions well marked, lateral regions obliquely striated, surface densely punctate, anteriorly rough with short rows of granules.

Front about two-thirds as wide as the carapace and more than four times as wide as high, nearly vertical, widening below; upper edge nearly straight, median fissure deep and narrow, inner lobes narrower than outer lobes and separated from them by shallow emarginations; surface granulate; lower edge convex and lightly sinuous.

Outer angle of orbit blunt and short. No lateral tooth.

Outer surface of chelipeds coarsely roughened except the fingers. A triangular expansion of the distal half of the inner margin of the arm is cut into jagged teeth on its distal margin. Upper edge of palm subacute and finely granulate but not in a single line. Mobile finger very rough above, for two-thirds its length. Prehensile edges of fingers irregularly toothed, narrowly gaping, short-hairy at base.

Legs broad and flat; third pair less than twice as long as carapace, its merus twice as long as wide. Margins of last three segments hairy; the propodus is wider and more hairy in the first leg than in the other legs.

Measurements.—Male (19408), length of carapace 17.5, anterior width of carapace 20.7, posterior width of same 16.7, width of front 13.5 mm.

Range.—From Greytown, Nicaragua, to the Rio de la Plata.

Remarks.—This species has a suspicious resemblance to *Sesarma trapezium* Dana [1] said to inhabit the Hawaiian Islands, but not since found there.

Material examined.—

Greytown, Nicaragua; Feb. 8 and 12, 1892; Chas. W. Richmond; 1 male, 8 females (19420).

Santana, Paparo, Venezuela; H. Pittier; 1 male (46103).

Trinidad; Jan. 30–Feb. 2, 1884; shore; *Albatross;* 1 female (19421).

Cayenne, French Guiana; 2 males, 1 female (Paris Mus.).

S. Joao de Barra, Rio de Janeiro, Brazil; Dec., 1911; E. Garbe; 2 males, 1 female (47853).

Rio de Janeiro, Brazil (1331, Berlin Mus.).

Estasao Cubatoa, Santos, Brazil; December, 1895; Bisego, collector; 2 males, 1 female (47856).

Ilha Casquerinita, Santos, Brazil; June, 1913; H. Luderwaldt; 1 male, 1 female (47866).

---

[1] U. S. Expl. Exped., vol. 13, Crust., pt. 1, 1852, p. 354; atlas, 1855, pl. 22, fig. 8

Iguape, São Paulo, Brazil; 1901; R. Krone, collector; 1 male, 1 female (47865).
Desterro, Santa Catharina, Brazil; Müller, collector; 15 specimens (Paris Mus.).
Montevideo, Uruguay; 1885; W. E. Safford, collector; 2 males, 2 females (19408).
Rio de la Plata (Copenhagen Mus.).

Genus SARMATIUM Dana.

Sarmatium DANA, Amer. Journ. Sci., ser. 2, vol. 12, 1851, p. 288 [5]; Proc. Acad. Nat. Sci. Philadelphia, vol. 5, 1851 (1852), pp. 247 and 251; type, *S. crassum* Dana; U. S. Expl. Exped., vol. 13, Crust., pt. 1, 1852, p. 357.
Metagrapsus MILNE EDWARDS, Ann. Sci. Nat., ser. 3, vol. 20, 1853, p. 188 [154]; type, *M. curvatus* Milne Edwards.

Differs from *Sesarma* as follows:

(1) The front, instead of being abruptly and vertically deflexed, is gradually declivous and obliquely deflexed.

(2) The antero-lateral borders are usually arched instead of being in the same straight line with the postero-lateral borders.

(3) The abdomen of the male does not completely coincide with the breadth of the sternum at the level of the last pair of legs; and in the female the terminal segment is not deeply impacted in the penultimate segment.

Distribution.—West Indies; west coast of Africa; Indo-Pacific.

SARMATIUM CURVATUM Milne Edwards.

Plate 95.

Sesarma curvata MILNE EDWARDS, Hist. Nat. Crust., vol. 2, 1837, p. 75 (type-locality. Senegal; type in Paris Mus.).
Sesarma violacea HERKLOTS, Addit. Fauna Afr. Occ., 1851, p. 10, pl. 1, fig. 9 (type-locality, Boutry; type in Leyden Mus.).—DE MAN, Notes Leyden Mus., vol. 2, 1880, p. 31.
Metagrapsus curvatus MILNE EDWARDS, Ann. Sci. Nat., ser. 3, vol. 20, 1853, p. 189 [155].
Metagrapsus pectinatus MILNE EDWARDS, Ann. Sci. Nat., ser. 3, vol. 20, 1853, p. 189 [155] (type-locality, Martinique; type in Paris Mus.).
Sarmatium curvatum KINGSLEY, Proc. Acad. Nat. Sci. Philadelphia, 1880, p. 212.
Sesarma (Sarmatium) curvatum RATHBUN, Proc. U. S. Nat. Mus., vol. 22, 1900, p. 281.

Diagnosis.—Lateral margins tridentate. A pectinated ridge on upper-outer surface of hand in male.

Description.—Carapace deep, broader than long, coarsely punctate, regions plainly marked, 4 or 5 oblique striae on postero-lateral regions; of the 4 post-frontal lobes, the 2 middle ones are much wider and better marked than the outer ones.

Front more than half as wide as the whole anterior margin, its free edge concave at middle as seen from above, straight as seen from before. Antero-lateral margins of carapace distinctly arched and cut into three blunt teeth including the orbital tooth, the last tooth being much the smaller.

Chelipeds of male stout, hand above with a tuberculate edge and an obliquely longitudinal, pectinated ridge, outwardly a little roughened; movable finger crossed above by transversely oblique rugose ridges, and just outside these a longitudinal, milled ridge; carpus covered with fine, irregular granulated lines. In the female the fingers are nearly smooth.

Legs thick; merus joints with a subdistal spine above; dactyli stout, long-pointed, subequal to the propodal segments.

Color.—Violet.

Measurements.—Male (20309), length of carapace 23, width of same 28.5, width of front 13.8 mm.

Range.—Martinique, West Indies; west coast of Africa.

Material examined.—

Martinique; 1 male dried, type of *M. pectinatus* (Paris Mus.).

Senegal; 1 male type (Paris Mus.).

Congo; Dybowski coll.; 1896; 1 male, 1 female (20309); 9 specimens (Paris Mus.).

Remarks.—The type-specimen of *Metagrapsus pectinatus* in the Paris Museum has been compared with a typical specimen of *Sesarma curvata* from Senegal. The crest on the hand of the latter is less prominent than on the former, but the specimen is smaller.

Genus ARATUS Milne Edwards.

Aratus MILNE EDWARDS, Ann. Sci. Nat., ser. 3, Zool., vol. 20, 1853, p. 187 [153]; type, *A. pisonii* (Milne Edwards).

Very narrow behind. Antenna excluded from orbit. Abdomen of male subcircular.

Carapace trapezoidal, strongly narrowed behind. Front very broad, vertical, reaching nearly to buccal cavity, concealing the antennulae. Epistome very short, fore and aft. Antenna removed from the orbit by the broad lobe at the inner suborbital angle. Outer maxillipeds separated from each other by a small gap; merus oblong-oval, distinctly longer than the ischium. Dactyli of legs short. Abdomen of male, save for the last segment, subcircular. Otherwise much as in *Sesarma.*

Contains only one species.

ARATUS PISONII (Milne Edwards).

MANGROVE CRAB; TREE CRAB.

Plate 96.

Aratv pinima MARCGRAVE DE LIEDSTAD, Hist. Rer. Nat. Brasil, lib. 4, p. 185, fig. in text.
Sesarma pisonii MILNE EDWARDS, Hist. Nat. Crust., vol. 2, 1837, p. 76, pl. 19, figs. 4 and 5 (type-locality, Antilles; type in Paris Mus.).
Aratus pisonii MILNE EDWARDS, Ann. Sci. Nat., ser. 3, Zool., vol. 20, 1853, p. 187.

Diagnosis.—Sides entire. Claws bristly outside.

Description.—Carapace nearly as long as wide; width at the level of the base of the second foot three-fourths as great as width between outer angles of orbit. Sides acute, entire. Regions deeply marked; outer portion of branchial regions obliquely striated, these regions are also finely and densely punctate; gastric lobes and surface of front rough with fine sharp granules; otherwise the carapace is smooth and shining to the naked eye.

Upper margin of front concave, its four lobes well separated, margined by a line of granules, inner lobes narrower than outer; behind the outer pair, another narrower pair. Front nearly vertical, about four times as wide as high, sides parallel, lower edge with a broad median sinus.

Outer orbital tooth short, acute.

Chelipeds of moderate size, palms swollen. A spine on inner margin of ischium. Outer face of merus and carpus covered with granulated striae. Inner margin of arm spinulous, distal half expanded. Outer face of carpus narrow-oblong. An oval area on the outer surface of the chelae is covered with tufts of long black bristles. Palm tuberculate within and without. Fingers irregularly toothed, narrowly gaping.

Merus joints of legs very thin and flat, a terminal and a subterminal spine above, last two joints hairy on margins, propodi elongate, dactyli short. Third leg one and two-thirds times as long as carapace, its merus a little more than twice as long as wide.

Color.—Dark mottled green with reddish legs (Jarvis). Many-colored, brown, blue, white, red; chelae whitish-yellow at the extremities (Marcgrave).

Measurements.—Male (6434), length of carapace 21.2, anterior width of carapace 22.5, width at postero-lateral angles 17.3, width of front 14.8 mm.

Habits.—Found on mangroves, and sometimes alongshore, on rocks and piles, bushes and wharves; occurs near fresh, brackish, or salt water.

Ascends the mangrove trees and gnaws their leaves. By means of its short but remarkably acute claws (tips of the legs), which prick like pins when it runs over the hand, this crab climbs with the greatest agility up on the thinnest twigs (F. Müller).

Range.—From Florida (Tampa Bay on the west, and Miami on the east coast) and the Bahamas southward to São Paulo, Brazil. Corinto, west coast of Nicaragua (M. C. Z.) to Peru.

Material examined.—

Norris Cut, Miami, Florida; April, 1901; J. E. Benedict; 2 males, 1 female (25559).

Cape Florida, Florida; 1884; Edward Palmer; 1 male (13898).

Punta Rassa, Florida; mangrove swamp; February, 1884; Henry Hemphill; 24 males, 11 females (6434).

Little Gasparilla Pass, Florida; March 17, 1889; *Grampus;* 1 male (15263).

Long Boat Key, Florida; March 25, 1889; *Grampus;* 5 males, 8 females (15262).

Palma Sola, Florida; 1884; Edward Palmer; 1 male (15377).

Nassau, Bahamas; December 25, 1898; *Fish Hawk;* 1 female (24049).

Los Arroyas, Cuba; May 20, 1914; Henderson and Bartsch, *Tomas Barrera* Exped.; 10 males, 7+ females (48574).

Mariel, Cuba; on bushes and wharves; May 10, 1900; Wm. Palmer and J. H. Riley; 3 males, 9 females (23809).

Cabañas, Cuba; on rocks and piles along shore; June 3, 1900; Wm. Palmer and J. H. Riley; 3 males, 4 females (23810).

Coloma, Pinar del Rio, Cuba; on piles; Mar. 18, 1900; Wm. Palmer and J. H. Riley; 1 female (23633).

Mayaguez, Porto Rico; fresh water; Jan. 20, 1899; *Fish Hawk;* 2 females (24050).

Porto Real, Porto Rico; Jan. 27, 1899; *Fish Hawk;* 1 female (24048).

Boqueron Bay, Porto Rico; Jan. 27, 1899; *Fish Hawk;* 1 female (24013).

Hucares, Porto Rico; Feb. 14, 1899; *Fish Hawk;* 9 males, 19 females (24051).

Jamaica; Mar. 1–11, 1884; *Albatross;* 2 males, 4 females (18557).

Jamaica; Dr. T. H. Morgan; 2 males, 2 females (17228).

Bogue Islands, Montego Bay, Jamaica; June 20, 1910; C. B. Wilson; 1 male, 1 female (42879).

Near Belize, British Honduras; W. A. Stanton; 1 male, 1 female (21374).

Curaçao; Feb. 10–18, 1884; *Albatross;* 2 males, 8 females (18556).

Rio Parahyba do Norte, Cabedello, Brazil; on mangroves; June 20, 1899; Branner-Agassiz Exped.; 3 males, 6 females (25713).

Lagoa do Norte, Maceio, Brazil; on mangroves; July 27, 1899; Branner-Agassiz Exped.; 3 females (25714).

Mapelle, Bay of Bahia, Brazil; 1876–1877; Richard Rathbun; 1 male, 2 females (1 ovig.) (40604).

Estasao Piassaguera, Santos, Brazil; May, 1913; H. Luderwaldt; 2 females ovig. (47854).

Boca del Jesus Maria, Costa Rica; April, 1905; P. Biolley and J. F. Tristan; 1 male (32365).

Guayaquil, Ecuador; Dr. W. H. Jones, U. S. N.; 1 male (15109).

Near Capon, Peru; Feb. 2, 1906–1908; R. E. Coker; 1 female (40449).

Genus CYCLOGRAPSUS Milne Edwards.

Cyclograpsus MILNE EDWARDS, Hist. Nat. Crust., vol. 2, 1837, p. 77; type, *C. punctatus* Milne Edwards.

Gnathochasmus MACLEAY, Zool. S. Africa, Annulosa, 1838, p. 65; type, *G. barbatus* MacLeay, 1838=*C. punctatus* Milne Edwards, 1837.

Carapace with the front and antero-lateral margins forming a regular curve, postero-lateral margins subparallel, posterior margin long. No lateral teeth. Surface flat, except anterior third, which is deflexed; regions little marked and almost smooth.

The front and orbits occupy more than two-thirds the width of the carapace, front between one-third and one-half width of carapace. Lower margin of front nearly transverse. Orbits transversely oval, completely filled by the eyes.

The hairs of the side walls are stouter and the granules more irregular than in *Sesarma*. Antennules transverse; antennae short, lying in the orbital hiatus. Epistome short, fore and aft. Buccal cavern narrowing anteriorly. Maxillipeds widely gaping; merus elongate-quadrate, the palpus articulating on the anterior margin near the outer angle.

Chelipeds rather massive in the male, subequal, nearly smooth; palm swollen; fingers pointed. Legs narrow, of moderate length second pair longest.

Abdomen with seven segments in each sex, in the male not occupying the whole breadth of the sternum between the last pair of legs.

Tropical and subtropical America; tropical Pacific to Australia and New Zealand; East Atlantic.

Analogous species on opposite sides of the American continent: *integer* (Atlantic), *cinereus* (Pacific)

KEY TO THE AMERICAN SPECIES OF THE GENUS CYCLOGRAPSUS.

A¹. Carapace without a postorbital gutter below the antero-lateral margin.
integer, p. 326.
A². Carapace with a postorbital gutter below the antero-lateral margin.
  B¹. Front strongly deflexed. No tooth on merus joints of legs. *cinereus*, p. 327.
  B². Front slightly deflexed. An obtuse tooth on merus joints of legs.
punctatus, p. 328.

CYCLOGRAPSUS INTEGER (Milne Edwards).

MARSH CRABS (F. Müller).

Plate 97, figs. 1 and 2.

Grapsus integer LATREILLE, MS. in MILNE EDWARDS, Hist. Nat. Crust., vol. 2, 1837, p. 79.

Cyclograpsus integer MILNE EDWARDS, Hist. Nat. Crust., vol. 2, 1837, p. 79 (type-locality, Brazil; type in Paris Mus.).

Diagnosis.—Carapace granulate anteriorly. No deep postorbital sulcus. Tooth on merus joints of legs.

Description.—Carapace three-fourths as long as broad; gastrocardiac sulcus present; surface punctate and around the front and lateral margins finely granulate; margin a raised crenulate line. The front, in dorsal view, appears faintly bilobed; in front view, nearly straight. Upper margin of orbit slightly sinuous and sloping backward and outward. In lower border of orbit below outer angle there is a deep sinus from which a very shallow depression extends backward. Maxilliped broad, ischium and merus subequal in length.

Arm short and broad, outer surface finely rugose. Wrist bluntly angled, and a little rough along the inner margin. Hand smooth outside, higher than its superior length, margined above for a short distance at the proximal end; on inner face, a few granules near carpal cavity. Fingers moderately gaping, feebly toothed, tips horny.

Second pair of legs twice as long as carapace. Entrance to cavity between second and third legs densely fringed with hair. Merus joints partly rugose, with a low subterminal tooth above; last two joints pilose.

Abdomen of male narrow-triangular; appendages of first segment narrow.

Measurements.—Male (15071), length of carapace 12.6, width of same 15.4, fronto-orbital width 10.6 mm.

Habitat.—Among rocks and dead seaweed near high tide (Hemphill). In marshy places near the sea, where it forms burrows (F. Müller).

Range.—Southern Florida and Bahamas to Brazil (Milne Edwards). Bermudas.

Material examined.

Bermudas; 1876–1877; G. Brown Goode; 1 specimen (42672).

Watling Island, Bahamas; 1886; *Albatross;* 1 female (17676).

Near Lake Kissimmee, Florida; A. M. Reese; 3 males, 3 females (45521).

Key West, Florida: December, 1883; D. S. Jordan; 4 males, 16 females (5748). 1885; H. Hemphill; among rocks and dead seaweed, near high tide; 1 male, 1 female (15071).

Montego Bay, Jamaica; P. W. Jarvis; 1 male, 1 female (19064).
Port Henderson, Jamaica; P. W. Jarvis; 1 male, 1 female (19063).
Kingston Harbor, Jamaica: May-July, 1896; F. S. Conant; 2 y. (19603). T. H. Morgan; 1 male, 1 female (17229).
Ponce, Porto Rico; on reefs; Jan. 30, 1899; *Fish Hawk;* 2 males, 3 females (24038).
San Juan, Porto Rico; beach; Jan. 12, 1899; *Fish Hawk;* 4 males, 2 females (24012).
Hucares, Porto Rico; Feb. 4, 1899; *Fish Hawk;* 1 male (24039).
Santa Marta, Colombia; C. F. Baker; 1 male, 1 female (22552).
Sabanilla, Colombia; Mar. 16-22, 1884; *Albatross;* 13 males, 7 females (17669).

CYCLOGRAPSUS CINEREUS Dana.

Plate 98.

Cyclograpsus cinereus DANA, Proc. Acad. Nat. Sci. Philadelphia, 1851, p. 251 (type-locality, *Ad oras Chilenses;* type from Valparaiso in Mus. Comp. Zoöl.; female paratype, Cat. No. 2340, U.S.N.M.); U. S. Expl. Exped., vol. 13, Crust., pt. 1, 1852, p. 360; atlas, 1855, pl. 23, fig. 3*a-c.*
Cyclograpsus punctatus KINAHAN (not Milne Edwards), Journ. Roy. Soc. Dublin, vol. 1, 1857, p. 342.

Diagnosis.—Surface of front smooth. A deep postorbital sulcus. No tooth on merus joints of legs.

Description.—Narrower than *C. integer,* length about four-fifths of width. Surface of front smooth; granules only near the antero-lateral angle. A deep furrow extends backward from below the orbit. Merus of maxillipeds narrower than in *C. integer.*

Second leg less than twice as long as carapace. No tooth on merus joints of legs; last two joints sparingly hairy, the propodus hairy only on the distal half of lower margin. No unusual hairiness between bases of second and third legs.

Abdomen of male suboblong, except for the last segment; appendages very stout, narrowing at the extremity

Measurements.—Ovigerous female (13866), length of carapace 11.4, width of same 13.8, fronto-orbital width 9.1 mm.

Range.—Panama (M. C. Z.) to Lota, Chile. Hawaiian Islands (Dana); perhaps an error.

Material examined.—

San Lorenzo Island, Peru; Jan., 1884; Dr. W. H. Jones, U. S. Navy; 15 males, 17 females (13866). Dr. H. E. Ames, U. S. Navy; 1 male, 1 female (17689).

Chincha Islands, Peru; specimens in Copenhagen Mus.

Valparaiso, Chile; U. S. Expl. Exped.; 1 female paratype (2340).

Central Chile; from Province of Aconcagua to Province of Talca; received from C. E. Porter; 6 males, 3 females (1 ovig.) (45520).

Locality not known; U. S. Expl. Exped.; 3 males, 4 females (2437).

CYCLOGRAPSUS PUNCTATUS Milne Edwards.

Plate 90.

Cyclograpsus punctatus Milne Edwards, Hist. Nat. Crust., vol. 2, 1837, p. 78 (type-locality, Indian Ocean; type in Paris Mus.).
Gnathochasmus barbatus Macleay, Illus. Zool. S. Africa, 1838, p. 65, pl. 3 (type-locality, South Africa; type probably not extant).
Sesarma barbata Krauss, Sudafr. Crust., 1843, p. 45, pl. 3, fig. 3a–c.
Cyclograpsus minutus Jacquinot, in Hombron and Jacquinot, Voy. au Pole Sud, Zool., atlas, Crust., 1852, pl. 6, figs. 8 and H; vol. 3, Crust., 1853, p. 75 (type-locality, Talcahueno; type in Paris Mus.).

Diagnosis.—Edge of front plainly visible in dorsal view. A deep postorbital sulcus. Tooth on merus joints of legs. Dactyli of legs very stout.

Description.—Carapace larger and more uneven than in the two preceding species; length about four-fifths of width; on the anterior half, four large pits on each side; front and antero-lateral regions with uneven punctae. Front less deflexed than in *C. integer* or *cinereus*, its margin visible in dorsal view; width more than two-fifths width of carapace. Outer orbital angle blunt; lower margin very incomplete. A deep gutter leads backward from it under the antero-lateral margin.

Fig. 153.—CYCLOGRAPSUS PUNCTATUS, OUTER MAXILLIPED, ENLARGED. (AFTER MILNE EDWARDS.)

Ischium of outer maxilliped narrower behind than in *C. integer.*

Chelipeds very strong. Wrist and hand nearly smooth. A prominent longitudinal ridge inside palm. Prehensile teeth feeble, upper edge of pollex slightly convex.

Second leg one and three-fourths times as long as carapace. Merus joints with a feeble subterminal tooth; dactyli and only the distal end of the anterior edge of the propodi short-pilose; no great pubescence between bases of second and third legs; dactyli very stout, and last pair very short.

Abdomen of male tapering gradually from the third to the sixth segment, this part of the margin being concave.

Color.—Purplish-brown with black punctae (Stimpson). Greenish-yellow, carapace with dark-red spots or dots, claws unicolored or with dark red dots (Krauss).

Measurements.—Male (3240), length of carapace 23.5, width of same 28.5, fronto-orbital width 20.7 mm.

Habitat.—Under stones (Krauss). In holes in mud in brackish water at mouth of river (Krauss).

Range.—Chile: Valparaiso Bay (Valparaiso Mus.); Juan Fernandez (Lenz); Talcahuano (Jacquinot). Also South Africa. Indian

Ocean (M. Edw.). Hongkong (Stimpson). By many authors, this species is united with *C. audouinii* Milne Edwards[1] which inhabits New Guinea, Australia, New Zealand, etc., but the latter is more convex in both directions, the front is narrower and not visible in a dorsal view, the crest on the inner face of the hand is granulated, the prehensile edge of the thumb bears a prominent lobe, the male abdomen is much more triangular.

Material examined.—
Valparaiso Bay, Chile; C. E. Porter; 1 male, 1 female (Valparaiso Mus.).
Cape Town, South Africa; Aug. 12, 1877; I. Russell; 2 males. (3240).

Genus CHASMAGNATHUS de Haan.

Chasmagnathus DE HAAN, Fauna Japon., 1833, p. 5; 1835, p. 27; type, *C. convexus* de Haan.

Carapace transverse, sides more or less arcuate, dentate, postero-lateral regions inclined, front rounding down to the epistome. Front arcuate, its posterior width from one-half to one-third the width of the carapace. Orbits very open, a sulcus continued backward to their outer angle. Antennules obliquely folded, antennae standing in the orbital hiatus. Epistome deep, prominent beyond the front, bearing a transverse crest in line with the infero-orbital crest. Reticulation of the side walls fairly regular, the granules bead-like and arranged singly in rows, the hairs plumed at base. Merus of outer maxilliped larger than ischium, subtriangular, the palpus articulating on the broad anterior margin between the middle and the outer angle.

Chelipeds strong in the male; fingers pointed, tips hollowed. Legs of moderate length, second pair longest; dactyli styliform.

Abdomen with all segments separate, in the male not covering all the sternum between the last pair of legs.

*Distribution.—*Atlantic coast of South America, from Rio Janeiro southward. From Japan to Australia.

CHASMAGNATHUS GRANULATA Dana.

Plate 100; plate 159, fig. 9.

Chasmagnathus granulatus DANA, Proc. Acad. Nat. Sci. Philadelphia. vol. 5, 1851 (1852), p. 251 (type-locality, *palude juxta lacum "Peteninga" urbi " Rio Janciro " vicinum*; type not extant) ; U. S. Expl. Exped., vol. 13, Crust., pt. 1, 1852, p. 364; atlas, 1855, pl. 23, fig. 6a–d.

*Diagnosis.—*Carapace naked above. Palms granulate.
*Description.—*Carapace subquadrate, narrowest at the anterior angles, side margins slightly sinuous; convex in both directions, the

---
[1] Hist. Nat. Crust., vol. 2, 1837, p. 78.

antero-lateral margin upturned. Surface uneven, covered with granules, two oblique rows of granules on postero-lateral region. Regions well defined, hepatic area depressed. Front rounding downward, its posterior width nearly one-half width of carapace, sides forming an arch with the lower margin, a broad median longitudinal furrow. Upper margin of orbit sinuous, sloping backward from the front, outer angle acute, little advanced. Two lateral teeth acute, formed by narrow notches. Infero-orbital crest prominent and tuberculate, as is also the epistomial crest.

Outer surface of chelipeds rough with granules, arranged partly in rows on the arm, upper edge of arm denticulate, outer and inner edges tuberculate. Inner angle of wrist acute; palm high and swollen in the male, upper margin subacute, inner surface granulate at middle; fingers nearly meeting, prehensile edges irregularly dentate, a slight lobe on pollex, an enlarged tooth at base of dactylus.

Legs rather narrow, somewhat pilose, especially on the anterior (or upper) half of carpal and propodal joints; a blunt tooth on merus joints; dactyli long-pointed; second leg twice as long as carapace.

Abdomen of male narrow-triangular; last segment little longer than wide.

Small specimens are more quadrate, less rough, teeth more rectangular.

Measurements.—Male (22109), length of carapace 28.3, width of same 35, fronto-orbital width 28.9, posterior width of front 17.1 mm.

Habits.—Live in marshes and on the muddy shores of salt lakes, where they take refuge in holes. According to Moreira,[1] when the zone which they inhabit at Lake Maricá, Province of Rio de Janeiro, becomes submerged at high tide, the crabs come up on to the rocks.

Range.—Rio de Janeiro, Brazil (M. C. Z.), to San Matias Bay, Patagonia (M. C. Z.).

Material examined.—

Montevideo, Uruguay: 1885; Mr. Auchevaleta; 2 males, 1 female (12584). October, 1886; W. E. Safford; 5 males, 2 females (12580). 1888; *Albatross;* 7 males (22109). W. Sorensen; received from Copenhagen Mus.; 1 male (22304). 1897; Bisego, collector; received from H. von Ihering, Mus. S. Paulo; 1 male (47842). F. Felippone; 1 male (44686).

Rio Santa Lucia, Uruguay; F. Felippone; 1 male, received Dec., 1916, and returned to sender.

---

[1] Arch. Mus. Nac. Rio de Janeiro, vol. 11, 1901, p. 42.

Subfamily PLAGUSIINAE Dana.

Plagusiacaea Milne Edwards, Ann. Sci. Nat., Zool., ser. 3, vol. 20, 1853, p. 177 [143].
Plagusinae Dana, Amer. Journ. Sci., ser. 2, vol. 12, 1851, p. 288; U. S. Expl. Exped., vol. 13, Crust., pt. 1, 1852, pp. 129, 333, and 368.
Plagusiinae Alcock, Journ. Asiat. Soc. Bengal, vol. 69, 1900, pp. 289 and 297.

Front cut into lobes or teeth by the antennular fossae, which are visible in a dorsal view as deep clefts. The lower border of the orbit curves down into line with the prominent anterior border of the buccal cavern. The external maxillipeds do not completely close the buccal cavern but they do not leave a wide rhomboidal gap, they are without an oblique hairy crest, their palp articulates near the antero-external angle of the merus, and their slender exposed exognath has often no flagellum. Antennal flagella short. The male abdomen fills all the space between the last pair of legs.

Genus PLAGUSIA[1] Latreille.

Plagusia Latreille, Gen. Crust., vol. 1, 1806, p. 33; type, *P. depressa* (Fabricius).
Philyra de Haan, Fauna Japon., Crust., 1833, p. 5; 1835, p. 31; type, *P. depressa* (Fabricius). Not *Philyra* Leach, 1817, a genus of the Leucosiidae.

Carapace subcircular, depressed, antero-lateral borders toothed. Interorbital space about a third the breadth of the carapace. No true front; the antennular fossae in which the antennulae fold nearly vertically are visible in a dorsal view as deep clefts in the anterior border of the carapace. Inter-antennular septum broad. Orbits deep; the antennae stand in the wide orbital hiatus.

Epistome short; buccal cavern squarish, its anterior border crenate and projecting strongly in a horizontal direction. The narrow space between the external maxillipeds is closed by bristles; merus as broad as ischium.

Chelipeds and legs dorsally rugose. Chelipeds subequal; in the male they are more massive than the legs, and longer than those of first and last pairs; in the female they are shorter and slenderer than any of the legs; fingers stout, with rounded excavated tips.

Legs very stout, with broad massive meri and short, stout, thorny dactyli.

Abdomen of male triangular, covering all the sternum between the last pair of legs; the seven segments may be distinct or the third, fourth, and fifth fused. Abdomen of female broad, consisting of seven segments, but the third, fourth, and fifth do not move independently of one another.

---

[1] *Plagusia* Brown, 1756, a genus of fishes, was not revived till 1817 by Cuvier, R. Anim., vol. 2, p. 224 (*teste* Gill).

Habits.—Dodge about rocks that are awash at high tide and hide in crannies when pursued. Also at home on drift timber in the open sea, which accounts for the wide range of some of the species.

Distribution.—All warm seas; extends into the Mediterranean.

KEY TO THE AMERICAN SPECIES OF THE GENUS PLAGUSIA.

A^1. Carapace tuberculate. Merus joints of legs unispinose anteriorly.
   B^1. Tubercles of carapace well marked. Merus joints of legs with two fringes of hair on upper (or posterior) surface.
      C^1. Coxal joints of legs with a dentate lobe _____*depressa*, p. 332.
      C^2. Coxal joints of legs with an entire lobe____*depressa tuberculata*, p. 334.
   B^2. Tubercles of carapace depressed, ill-defined. Merus joints of legs with only a posterior fringe of hair on upper surface____*immaculata*, p. 335.
A^2. Carapace tomentose, nearly smooth. Merus joints of legs multispinose anteriorly_____*chabrus*, p. 336.

PLAGUSIA DEPRESSA (Fabricius).

Plate 101.

Cancer depressus FABRICIUS, Syst. Entom., 1775, p. 406 (type-locality, *in mari mediterraneo;* type in Kiel Mus.).[1]

Cancer squamosus HERBST, Natur. Krabben u. Krebse, vol. 1, 1790, p. 260, pl. 20, fig. 113 (type-locality, *Ostindien;* type in Berlin Mus.?).[2] Probably confused with *P. depressa tuberculata.*

Grapsus depressus BOSC, Hist. Nat. Crust., vol. 1, an X [1801–1802], p. 203.

Plagusia depressus SAY, Journ. Acad. Nat. Sci. Philadelphia, vol. 1, 1817, p. 100.

Plagusia sayi DE KAY, Zool. New York, Crust., 1844, p. 16 (type-locality, Gulf stream; type not extant).

Plagusia squamosa LATREILLE, Encyc. Méth., Hist. Nat., Entom., vol. 10, 1825, p. 145.

Plagusia gracilis SAUSSURE, Mém. Soc. Phys. Hist. Nat. Genève, vol. 14, 1858, p. 449 [33] (type-locality, Cuba; type in Geneva Mus.).

Diagnosis.—Carapace tuberculate. Coxal joints of legs with a dentate lobe; merus joints with one spine on anterior margin.

Description.—All the regions of the carapace are distinct, and the surface is covered with flat pearly or squamiform tubercles which are fringed anteriorly with small close-set bristles of uniform length. On the front of the gastric region is a series of about six prominent acute tubercles arranged in an arc concave forward. Antero-lateral border armed with four teeth (including orbital angle), which in the main decrease in size from before backward. Epistome prominent beyond the anterior border of the carapace and cut usually into five lobes.

Chelipeds of adult male massive and about as long as carapace, in the female slender and about three-fourths as long as carapace.

---

[1] The supposed type-specimen is labeled "*Alpheus depressus* F." No locality is given.
[2] I searched for the type but did not find it. Hilgendorf, in 1882 (SB. Ges. Naturf. Freunde Berlin, 1882, p. 24), says that the type-specimen of *Cancer squamosus* has the lobe at the base of the legs entire, but this is not the case in Herbst's fig. 113, pl. 20.

Inner angle of wrist coarsely dentiform. Tubercles on upper surface of palm and dactylus arranged in high relief in longitudinal rows, outer surface of the palm nearly smooth.

On the posterior edge of the dorsal surface of the coxae of the legs is an oblong lobe the upper edge of which on the second and third legs and sometimes on the first leg, is cut into two or three teeth. On the anterior border of the meropodites there is a single strong subterminal spine; the upper surface of the next three segments is traversed longitudinally by a dense strip of long hair; dactyli with two rows of strong spines on the concave side. The third or longest leg is about twice the length of the carapace.

Color.—Light reddish, dotted with blood-red, tubercles bordered with blackish ciliae with the extremity gray. Some blood-red spots on the legs. Under side of body yellowish. (Lareille.)

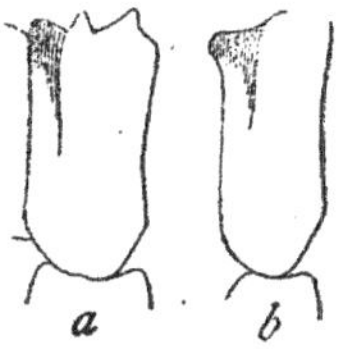

Measurements.—Female (18503), extreme length of carapace 45, width of same 48.6, fronto-orbital width 28.8 mm.

Range.—From Charleston Harbor, South Carolina, to Brazil. Gulf Stream. Bermudas. Eastern Atlantic. Amboina (de Man).

Material examined.—

Abaco, Bahamas; 1866; *Albatross;* 1 male (17647).

Fig. 154.—Coxa of third right leg of male of, *a*, PLAGUSIA DEPRESSA (40609), × 1⅔; *b*, P. DEPRESSA TUBERCULATA (18826), × 1½.

Spanish Wells, Bahamas; 1893; Biol. Exped. State Univ. Iowa; specimens in Mus. S.U.I.

Indian Key, Florida; among rocks, near low tide; H. Hemphill; 2 females (14056).

Dry Tortugas, Florida; June 5–8, 1893; Biol. Exped. State Univ. Iowa; 1 male, 1 female (Mus. S.U.I.).

Jamaica; *Albatross;* 1 male (18275).

Santo Domingo, West Indies; 1878; W. M. Gabb; 1 male (3190).

Arroyo, Porto Rico; February 4, 1899; *Fish Hawk;* 2 males (24040).

St. Thomas; shore near town; June 28 and 30, 1915; C. R. Shoemaker; 2 males (49806).

St. Croix; specimens in Copenhagen Mus.

Curaçao; *Albatross;* 1 female (18503).

Klein Bonaire; tide pools among coral rocks; July 11, 1905; J. Boeke; 1 male (42983).

Fernando Noronha; 1876–1877; R. Rathbun; 1 male, 1 female ovig. (40610).

Mamanguape, Brazil; stone reef; June 23, 1899; Branner-Agassiz Exped.; 1 female (25715).

Rio Goyanna, Brazil; stone reef; June 18, 1899; Branner-Agassiz Exped.; 3 females (25716).

Pernambuco, Brazil; 1876-1877; R. Rathbun; 1 male, 1 female (40609).

Hungry Bay, Bermudas; July-September; F. G. Gosling; 1 male (25438).

Fayal Harbor, Azores; found swimming beside hull of barkentine just in from Africa; December, 1892; Lewis Dexter; 2 males, 2 females (17711).

Madeira: N. Pacific Exploring Exped.; 3 specimens (2102). U. S. Exploring Exped.; 1 male, 1 female (2359).

Baya River, Elmina, Ashantee; Nov. 27, 1889; U. S. Eclipse Exped. to Africa, W. H. Brown; 1 male (14871).

Locality unknown; 1 male (17622).

PLAGUSIA DEPRESSA TUBERCULATA Lamarck.

Plate 102.

Plagusia squamosa LAMARCK, Hist. Nat. Anim. sans Vert., vol. 5, 1818, p. 246. Not *Cancer squamosus* Herbst, pl. 20, fig. 113.

Plagusia tuberculata LAMARCK, Hist. Nat. Anim. sans Vert., vol. 5, 1818, p. 247 (type-locality, *Ile de France;* type in Paris Mus.).

Plagusia orientalis STIMPSON, Proc. Acad. Nat. Sci. Philadelphia, vol. 10, 1858, p. 103 [49] (type-locality, Hong Kong; 2 males, cotypes, Cat. No. 2012, U.S.N.M.); Ann. Lyc. Nat. Hist. New York, vol. 7, 1860, p. 231.

Diagnosis.—Carapace tuberculate. Coxal joints of legs with an entire lobe; merus joints with one spine on anterior margin.

Description.—Differs very slightly from the preceding. The lobes on the coxae of the legs are rounded, not dentate (see fig. 154*b*). Usually no series of prominent tubercles on the gastric region. The tubercles of the carapace are variable, either prominent, or depressed, or almost obsolete on the convex portions of the carapace. Fringes of bristles also variable in extent, sometimes absent. The epistome has a tendency to be more divided than in *P. depressa*, into seven lobes, but this is not a constant character. Outer surface of palms rougher, especially in the male where the tubercles are well developed and arranged partly in longitudinal, and partly in transverse rows. The chelipeds in the adult male are half again as long as the carapace, in the female about as long as carapace.

Measurements.—Male (18826), extreme length of carapace 51.8, width of same 55.2, fronto-orbital width 32.1 mm.

Range.—Cape St. Lucas (Stimpson). Also throughout the Oriental Region, from the Hawaiian Islands westward and southward to Japan and the Arabian Sea.

Remarks.—This subspecies is not always well separated from typical *depressa*. See de Man, Notes Leyden Mus., vol. 5, 1883, p. 168.

Material examined.—
Hilo, Hawaii; 1901; U. S. Fish Comm.; 1 female (25306).
Kailua, Hawaii; 1901; U. S. Fish Comm.; 1 female (25305).
Maui, Hawaiian Islands; R. C. McGregor; 1 male y. (23694).
Honolulu, Oahu, Hawaiian Islands; June 8, 1901; U. S. Fish Comm.; 1 female (25304).
South coast, Oahu, Hawaiian Islands; Lae-o Ka Laau Light, N. 35°, W. 6.1'; 222–498 fathoms; Co. R. brk. Sh.; temp. 49.5° F.; Apr. 1, 1902; Sta. 3824; *Albatross;* 1 y. (29338).
Laysan, Hawaiian Islands; May, 1902; *Albatross;* 1 male, 3 females (29337).
Hawaiian Islands; W. H. Pease; 1 female (cotype of *P. orientalis* Stimp.) (2480).
Port Lloyd, Bonin Islands; U. S. S. *Tuscarora;* 1 male (5320).
Koneshine, Osumi, Japan; under rocks in shallow water; T. Urita; 1 female (48451).
Oho Sima, Tokaido coast, Japan; F. Sakamoto; 2 males (18826).
Hongkong, China; N. Pacific Exploring Exped.; 2 males (cotypes of *P. orientalis* Stimpson) (2012).
Anambas Islands, China Sea; 1899; Dr. W. L. Abbott; 1 male (23363).
Atjeh, Sumatra; Capt. Storm; 1 specimen (39157).
Great Nicobar, Indian Ocean; Mar. 16, 1901; Dr. W. L. Abbott; 1 male (25215).
Egmont Reef, W. Indian Ocean; 1905; H. M. S. *Sealark;* 1 male (41370).

PLAGUSIA IMMACULATA Lamarck.

Plate 103.

Plagusia immaculata LAMARCK, Hist. Nat. Anim. sans Vert., vol. 5, 1818, p. 247 (type-locality, *la Méditerranée? Je la crois de l'Occan Indien;* type in Paris Mus.).—MIERS, Ann. Mag. Nat. Hist., ser. 5, vol. 1, 1878, p. 150; *Challenger* Rept., Zool., vol. 17, 1886, p. 273, pl. 22, fig. 1.
Plagusia tuberculata RATHBUN (not Lamarck), Proc. U. S. Nat. Mus., vol. 21, 1898, p. 605.

*Diagnosis.—*Carapace very convex, feebly tuberculate; tubercles not fringed with hair. Merus joints of legs with only one fringe of hair, the posterior one, on upper surface.

*Remarks.—*Laurie[1] treats *immaculata* as a variety of *depressa.* The series in the United States National Museum seems to be specifically distinct. The lobe on the coxal joints of the legs is entire, as in *depressa tuberculata.*

*Color.—*Pale green, mottled with reddish-brown. Tarsi above dark purplish-brown, with small markings of very pale bluish green.

[1] Ceylon Pearl Oyster Fisheries, pt. 5, 1906, Suppl. Rept., No. 40, 1906, p. 429.

Carapace mottled and washed with pale, dirty green, dark reddish-brown, and straw-color, with a few orange dots. (Miers.)

Measurements.—Male (24758), length of carapace 33.3, width of same 35.7 mm.

Range.—From Costa Rica to Panama. Indo-Pacific region.

Material examined.—

Punta Arenas, Costa Rica; Jan. 19, 1907; P. Biolley; 1 female (39091).

Taboga Island, Panama; May 12, 1911; Meek and Hildebrand; 9 males, 8 females (44176).

Panama; Mar. 15, 1888; *Albatross;* 1 male y. (22110).

Pulo Midei, Natuna Islands, China Sea; May 23, 1900; Dr. W. L. Abbott; 1 male, 1 female (24758).

Indian Ocean; long. 90° E., on Equator; February, 1884; Capt. J. R. Lyon, British bark *Cashmere;* 3 males, 5 females (13911).

Voyage, Liverpool to Calcutta; Capt. J. R. Lyon, British bark *Cashmere;* November, 1883; 5 males, 4 females (13912).

PLAGUSIA CHABRUS (Linnaeus).

Plate 104.

Cancer chabrus LINNAEUS, Mus. Lud. Ulr., 1764, p. 438 (type-locality, *in Mari Indico;* type not extant).

Plagusia tomentosa MILNE EDWARDS, Hist. Nat. Crust., vol. 2, 1837, p. 92 (type-locality, Cape of Good Hope and Chile; type in Paris Mus.).—

KRAUSS, Südafr. Crust., 1843, p. 42, pl. 2, fig. 6.

Plagusia capensis DE HAAN, Fauna Japon., Crust., 1835, p. 58 (type-locality, Cape of Good Hope; type in Leyden Mus.).

Plagusia chabrus WHITE, Ann. Mag. Nat. Hist., vol. 17, 1846, p. 497.

Plagusia gaimardi MILNE EDWARDS, Ann. Sci. Nat., ser. 3, Zool., vol. 20, 1853, p. 178 (type-locality, Tongatabu; type in Paris Mus.).

Plagusia chabrus STEBBING, Ann. S. African Mus., vol. 6, 1910, p. 322.

Diagnosis.—Carapace tomentose, almost smooth. Basal joints of legs armed with a spine; merus joints with many spines or teeth on anterior margin.

Description.—Carapace, chelipeds, legs, and ventral surface of body covered with a short tomentum, and without tubercles, except one behind the orbit, and two or three elongated tubercles near the outer angle of the branchial region. Antero-lateral border armed with four strong, acute teeth, the two anterior larger than the two posterior. Front armed above with two small spines, and with a series of small tubercles or spines on its anterior margin. An acute tooth on the outer edge of the inner orbital lobe. Epistome trilobate, each lobe dentate. Exognath of outer maxilliped with a flagellum.

Chelipeds massive in the male, about one and two-thirds times as long as the carapace, coarsely tuberculate, the tubercles arranged in a longitudinal series on hand and dactylus. Lobe on basal joints of legs armed above with a spine; anterior margin of merus joints irregularly dentate or spinose.

Measurements.—Male (2476), extreme length of carapace 63.5, width of same 69.5, fronto-orbital width 39.5 mm.

Habitat.—In the holes excavated by the surf on the rocky coast of Table Bay (Krauss). Rather common about the rocks at half tide in Simons Bay (Stimpson).

Range.—Chile (Milne Edwards): Coquimbo and Island of Juan Fernandez (Porter). Also Cape of Good Hope; New South Wales; Tasmania; New Zealand; Tongatabu.

Material examined.—

Illawarra, New South Wales, Australia; U. S. Exploring Exped.; three specimens (one female) (2379).

Tasmania; received from Mus. Comp. Zoöl.; one male (2476)

Bay of Islands, New Zealand; U. S. Exploring Exped.; one male, one female (2360).

Genus PERCNON Gistel.

Acanthopus DE HAAN, Fauna Japon., Crust., 1833, p. 5; 1835, p. 29; type, *A. clavimana* de Haan, 1835=*planissimum* (Herbst, 1804). *Acanthopus* preoccupied by Klug, Illig. Mag., vol. 6, 1807, for genus of Hymenoptera.

Percnon GISTEL, Naturg. Thierreichs, 1848, p. 8; type, *P. planissimum* (Herbst).

Leiolophus MIERS, Cat. Crust. New Zealand, 1876, p. 46; type, *L. planissimus* (Herbst).

Carapace extremely flat and depressed, being quite disk-like. Interantennular septum narrow. Epistome almost linear. Merus of external maxillipeds very small, much narrower than the ischium, and disposed obliquely in repose; exognath extremely short and slender. Chelipeds and legs, though in places spiny, not rugose. Legs much slenderer, and though the meropodites are broad, they are very thin. The appendages of the first segment of the male abdomen end in a claw.

Distribution.—As *Plagusia*, but not in the Mediterranean.

PERCNON GIBBESI (Milne Edwards).

SPRAY CRAB (Jarvis).

Plate 103.

Acanthopus gibbesi MILNE EDWARDS, Ann. Sci. Nat., ser. 3, Zool., vol. 20, 1853, pp. 180 and 146 (type-locality, Antilles; type in Paris Mus.).

Leiolophus planissimus MIERS, Ann. Mag. Nat. Hist., ser. 5, vol. 1, 1878, p. 153 (part). Not *Cancer planissimus* Herbst, 1804.

Percnon planissimum RATHBUN, Proc. U. S. Nat. Mus., vol. 22, 1900, p. 281 (part).

Thus far I have considered the American and East-Atlantic form the same as that from the Indo-Pacific region, but have now come to the conclusion that they are different.

Diagnosis.—Tubercles of carapace flattened, depressed. No bristles on inner surface of palm. Merus joints of legs not especially elongated or narrowed.

Description.—Carapace ovate, thin, disk-like, longer than wide, covered with little short bristles, which, however, leave certain symmetrical raised linear patches bare; merus joints of legs clad in the same way, and with two long bare stripes.

Dorsal surface of carapace with a few low tubercles; one behind the inner angle of the orbit has a sharp point anteriorly.

Front, antennular and supra-orbital angles, and epistome all acutely spinose. Front, between the antennules, narrow, contained more than 6.5 times in the fronto-orbital distance; just as wide as long or very slightly wider than long, the length measured from the hinder end to the sinus of the posterior tooth; armed with four spines (two on each side), and a row of spinules, which are very inconspicuous, just within and parallel to each side-margin. Three spines on inner margin of orbit; the middle of the upper border is also more or less serrate. Eyes large and reniform. Antero-lateral border of carapace armed with four acute spines; the first is at the angle of the orbit; the other three are smaller and subequal to one another.

The chelipeds vary according to age and sex, but arm and wrist are always armed with spines; palm nearly smooth and nude, oval and somewhat compressed, very deep in the adult male; fingers short, blunt and hollowed at tip. Proximal upper surface of palm with a short, ill-defined furrow, not more than half as long as upper margin, and filled with pubescence. No patch of hair on inner surface of palm. Merus joints of legs not extraordinarily elongated or narrowed. Anterior border of merus joints armed along its whole length with large and even spines; posterior border ends in a spine; on the merus of the first two legs there is a second row of spinules parallel with the anterior border; on the third merus this row is indistinct; and on the fourth quite absent.

Color.—Carapace usually variegated or mottled with brown, pinkish flesh-color and salmon; there is generally a median longitudinal stripe of bright pale blue; the legs are banded with reddish-brown and light pink. Ventral side of body pale blue; of legs pale pink. (Verrill.)

Measurements.—Male (13828), total length of carapace 27.7, width of same 26, fronto-orbital width 17.1 mm.

Habitat.—Found on outlying rocks washed by the spray. Its body and spiny legs look as if they had been ironed out. (Jarvis.)

Range.—From Cape St. Lucas, Lower California, to Chile; southern Florida and Bahamas to Brazil; Bermudas; eastern Atlantic (Azores to Cape of Good Hope).

Material examined.—

Clarion Island, Mexico; from stomach of spotted grouper; Mar. 4, 1889; *Albatross;* 1 female (17451).

Sand Key Reef, Florida; May, 1911; J. B. Henderson; 1 female (45713).

Indian Key, Florida; along shore among rocks; H. Hemphill; 1 male, 1 female (14055).

Near Lake Kissimmee, Florida; A. M. Reese; 1 specimen (45353).

Key West, Florida; among rocks, low tide; H. Hemphill; 3 males, 1 female, 1 y. (13828).

Abaco, Bahamas; 1886; *Albatross;* 1 female (17670).

Andros, Bahamas; cave halfway between Smith's landing and lighthouse, south side of east end of South Bight, Long Bay Key District; May 14, 1912; Paul Bartsch; 1 male (45551).

Cuba; Henderson and Bartsch, *Tomas Barrera* Exped.: Ensenada de Cajon, off Cape Antonio; May 22–23, 1914; 2 females (48579). Cabañas; on reef; June 8–9, 1914; 1 male, 1 female (48592).

Kingston Harbor, Jamaica; T. H. Morgan; 3 females (17230).

Playa de Ponce Reef, Porto Rico; Feb. 1, 1899; *Fish Hawk;* 4 males, 2 females (24045).

Coral Reef, Colon, Panama; May 2, 1911; Meek and Hildebrand· 1 female ovig. (43992).

Hungry Bay, Bermudas; July–September; F. G. Gosling; 1 male, 1 female, 1 y. (25439).

Pastelleiro, Fayal, Azores; on rocky or stony beach; Feb. 1893; Lewis Dexter; 1 male (45552).

Madeira; U. S. Exploring Exped.; 1 male, 1 female (2349).

Family GECARCINIDAE Dana.

LAND CRABS.

Gecarciniens MILNE EDWARDS, Hist. Nat. Crust., vol. 2, 1837, p. 16.
Gecarcinacaea MILNE EDWARDS (part), Ann. Sci. Nat., Zool., ser. 3, vol. 20, 1853, p. 200 [166].
Gecarcinidae DANA (part), Amer. Journ. Sci., ser. 2, vol. 12, 1851, p. 289; U. S. Expl. Exped., vol. 13, Crust., pt. 1, 1852, p. 374.
Geocarcinidae MIERS, *Challenger* Rept., Zool., vol. 17, 1886, p. 216.—ALCOCK, Journ. Asiat. Soc. Bengal, vol. 69, 1900, pp. 283, 289, 297, and 440, and synonymy.

The palp of the external maxillipeds articulates either at the antero-external angle or at the middle of the anterior border of the merus, and is sometimes completely hidden behind the merus; the exognath is slender and inconspicuous, sometimes more or less concealed, and sometimes without a flagellum. Antennular fossae narrow. Front of moderate breadth, always strongly deflexed. Carapace transversely oval, antero-lateral borders strongly arched, fronto-orbital border very much less than the greatest breadth of the carapace. Male openings sternal.

KEY TO THE AMERICAN GENERA OF THE FAMILY GECARCINIDAE.

A^1. Fronto-orbital border more than half width of carapace. Buccal cavern elongate. Exognath of outer maxillipeds exposed and provided with a flagellum.
B^1. Outer maxillipeds gaping; anterior margin of merus emarginate. Dactyli of legs spinous_____Cardisoma, p. 340.
B^2. Outer maxillipeds meeting; merus quadrangular. Dactyli of legs not spinous_____Ucides, p. 346.
A^2. Fronto-orbital border less than half width of carapace. Buccal cavern rhomboidal. Exognath of outer maxillipeds concealed or nearly so and without flagellum.
B^1. Palpus of outer maxillipeds hidden_____Gecarcinus, p. 351.
B^2. Palpus of outer maxillipeds exposed_____Gecarcoidea, p. 362.

Genus CARDISOMA Latreille.

Cardisoma LATREILLE, Encyc. Méth., Hist. Nat., Entom., vol. 10, 1825, p. 685; type, *C. guanhumi* Latreille.
Discoplax A. MILNE EDWARDS, Ann. Soc. Entom. France, ser. 4, vol. 7. 1867, p. 284; type, *D. longipes* A. Milne Edwards=*rotundum* (Quoy and Gaimard, 1825).

Carapace deep, convex fore and aft, transversely oval or cordate, lateral borders tumid and strongly arched owing to the vault-like expansion of the gill chambers, pterygostomian regions densely tomentose.

Fronto-orbital border much more than half, and the deflexed and nearly straight front about a fourth, the greatest breadth of the carapace. Orbits deep, with the outer angle defined by a denticle, and with the tooth at the inner angle well developed but distant from the front; eyes very loose in the orbits.

Antennules obliquely folded, a good deal concealed by the front; interantennular septum very broad. The antennae lie in the orbital hiatus, which their broad basal joint nearly fills; flagellum very short.

Epistome short, prominent and well defined; buccal cavern elongate-squarish; the external maxillipeds have between them a rhomhoidal gape in which the mandibles are exposed. In the external maxillipeds, the merus is a longish joint, and carries the palp which is large and not at all concealed, at its antero-external angle; the exognath, which carries a flagellum, is exposed in much the greater part of its extent. The exognaths of the other maxillipeds are heavily fringed with coarse hair.

Chelipeds much more massive than the legs, markedly unequal in the American species, and differing little in the sexes. The larger cheliped alters considerably with age, the arm and fingers becoming elongated, and the whole hand increasing in size until it becomes longer than the carapace is broad and more than half as high as the carapace is long.

Legs stout; some of their joints are fringed with bristles, and their long strong dactyli are square in section and have a series of spines along all four edges.

The abdomen in both sexes consists of 7 separate segments and covers the whole width of the sternum between the last pair of legs.

The branchiae are 8 in number on either side; the gill chambers are vaulted and remarkably capacious, and they are lined by a thick vascular membrane folded to form a sort of pocket, and as in several other crabs—such as *Uca* and *Ocypode*—that spend most of their time out of water, a sort of "choroid process" of this membrane, shaped like a gill-plume, projects laterally over the pleura of the penultimate pair of legs.

Tropical America, Bermudas, Cape Verde Islands, west coast of Africa, Indo-Pacific from Madagascar to Hawaiian Islands and Chile.

Analogous species on opposite sides of the continent: *guanhumi* (Atlantic); *crassum* (Pacific), also on Atlantic side of Isthmus.

KEY TO THE AMERICAN SPECIES OF THE GENUS CARDISOMA

A^1. Extremity of appendages of first abdominal segment in male broad and blunt. Legs sparsely hairy .... *guanhumi*. p. 341.
A^2. Extremity of appendages of first abdominal segment in male with 2 slender processes. Legs very hairy .... *crassum*. p. 345.

CARDISOMA GUANHUMI Latreille.

GREAT LAND-CRAB; WHITE LAND-CRAB; MULATTO LAND-CRAB; JUEY (Porto Rico); TOURLOUROU; GUANHUMI (Brazil).

Plates 106 and 107.

Guanhumi MARCGRAVE DE LIEBSTAD, Hist. Rer. Natur. Brazil., 1648, p. 185, text-fig.; Brazil.

Cangrejos Terrestres PARRA, Descripcion de diferentes piezas de Historia Natural, Havana, 1787, p. 163, pl. 57.

Ocypode ruricola LATREILLE, Hist. Nat. Crust., vol. 6, 1803, p. 35, not pl. 24, fig. 2, copied from Herbst. Not *Cancer ruricola* Linnaeus.

Ocypode cordata LATREILLE. Gen. Crust., vol. 1, 1806, p. 31. Not *Cancer cordatus* Linnaeus.

Gecarcinus carnifex LATREILLE, Nouv. Dict. Hist. Nat., vol. 12, 1817, p. 511; St. Thomas. Not *Cancer carnifex* Herbst.

Cardisoma guanhumi LATREILLE, Encyc. Méth., Hist. Nat., Entom., vol. 10, 1825. p. 685 (type-locality, Brazil; type in Paris Mus.).

Ocypode (*cardisoma*) *cordata* DE HAAN, Fauna Japon., Crust., 1835, p. 27. Not *Cancer cordatus* Linnaeus.

Ocypode ruricola FREMINVILLE, Ann. Sci. Nat., ser. 2, Zool., vol. 3, 1835, p. 217. Not *Cancer ruricola* Linnaeus.

Ocypoda gigantea FREMINVILLE, Ann. Sci. Nat., ser. 2, Zool., vol. 3, 1835, p. 221.

Cardisoma guanhumi MILNE EDWARDS, Ann. Sci. Nat., ser. 3, Zool., vol. 20, 1853, p. 204 [170], pl. 9.

Cardisoma quadrata SAUSSURE, Mém. Soc. Phys. Hist. Nat. Genève, vol. 14, 1858, p. 438 [22], pl. 2, fig. 13 (type-locality, Haiti; type in Geneva Mus.).

Cardisoma diurnum GILL, Ann. Lyc. Nat. Hist. New York, vol. 7, 1859, p. 42 (type-localities, Barbados, Grenada, and St. Thomas; types not extant).

Cardisoma guanhumi SMITH, Trans. Connecticut Acad. Arts and Sci., vol. 2, 1870, p. 143, pl. 5, fig. 3.

Diagnosis.—First abdominal appendages in male blunt at tip. Legs sparsely hairy.

Description.—Carapace strongly convex fore and aft; the posterior gastric and the cardiaco-intestinal regions defined by grooves. Antero-lateral border defined by a raised line, which becomes indistinct with age and is not continuous with the outer margin of the orbital tooth, but starts at a small denticle behind that tooth. Sides of front very oblique. The sinuous upper border of the orbit runs very slightly backward to the base of the outer orbital tooth. The greatest height of the orbit is a little more than half its length. The basal joint of the antenna is large, touching the front. The breadth of the buccal cavern at its middle is about equal to its length in the middle line.

Chelipeds smooth except for a few small tubercles, wrinkles, denticles or granules along the edges of some of the joints and tubercles on the inner surface of palm and fingers; inner angle of the wrist dentiform, palm higher than its superior length, especially in the larger hand; the stout fingers meet only at tip and are much more gaping in the larger hand.

The size of the larger cheliped varies with age. In adults of moderate size it is about twice the length of the carapace, the ischium hardly projects beyond the carapace, and the length of the dactylus is about one and one-third times the height of the palm. In old specimens, especially in the male, it is about three times the length of the carapace, the ischium projects far beyond the carapace, and the length of the dactylus is twice the height of the palm.

Legs usually sparsely setose on the margins, though sometimes old specimens have conspicuous black bristles on the distal margin of the merus joints and on the margins and surfaces of the carpal and propodal joints with the exception of the posterior edge of the former.

The seventh segment of the male abdomen is less than half the length of the sixth, in old individuals much less than half. The first pair of appendages are triquetral, straight and stout, their tips rounded and slightly flattened laterally, and each is armed with a very small, scale-like appendage directed obliquely outward; and, on the upper edge, just above this appendage, there is a small process which is straight and does not reach beyond the rounded extremity of the thickened portion.

Color.—Adults, bluish-tinged ashy gray, occasionally dirty greenish or dirty white; young, violet.

Measurements.—Male, Jamaica, length of carapace 90, width of same 113, length of larger propodus 155 mm.

Habitat.—Live in great numbers in open fields, in forests, and on hills covered with bushes; also make deep burrows in the earth near the margin of swamps and ponds in which they remain during a part of the year throughout the greater part of the day, seeking their food chiefly at night. They are said to repair to the sea during the breeding season. On account of their swift movements it is almost impossible to catch them. For more detailed accounts of their habits, see Brooks,[1] Verrill,[2] and Henderson.[3]

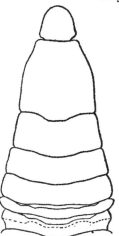

FIG. 155.—CARDISOMA GUANHUMI, ABDOMEN OF MALE (17987), NATURAL SIZE.

Range.—Bahamas, Southern Florida, West Indies, Texas to São Paulo, Brazil, and Bermudas. Also occurs in West Africa (Bouvier[4]).

The closely allied *C. carnifex* (Herbst) of the Indo-Pacific region has the orbits tapering more rapidly to the outer end; the merus and ischium of the outer maxillipeds broader; the chelipeds smoother; the legs more bushy-haired; the seventh segment of the abdomen longer.

Material examined.—

Little River, Florida; C. T. Simpson; cheliped of male (45963). " Very numerous."

Miami, Florida; G. M. Gray; 1 female (42144).

Norris Cut, near Miami, Florida; April, 1901; J. E. Benedict; 1 female (25558).

---

[1] Scribner's Magazine, vol. 14, July, 1893, p. 101.
[2] Trans. Connecticut Acad. Arts and Sci., vol. 13, 1908, p. 310.
[3] The Cruise of the Tomas Barrera, 1916, p. 179.
[4] Bull. Mus. Hist. Nat. Paris, 1901, p. 13. *C. guanhumi* Stimpson from Cape Verde Islands is a true *C. armatum* Herklots.

Cocoanut Grove, Florida; January, 1916; W. L. McAtee; 5 males, 3 females (48925).

Biscayne Key, Florida; G. Wurdemann; 2 specimens (2095).

Andros Island, Bahamas; received from Amer. Mus. Nat. Hist.; 1 male (31462).

Bahamas; May 11, 1912; Paul Bartsch; 1 specimen (46074).

St. Thomas; 1915; C. R. Shoemaker: July 17; 1 young male (49810). Magens Bay; July 27; 2 males, 2 females (50393).

Catano, Porto Rico; *Fish Hawk:* January 14, 1899; 2 females (24006). January 17, 1899; 2 males, 1 female (24007).

San Juan market, Porto Rico; *Fish Hawk;* January 15, 1899; 1 male, 1 female (24005).

Rio Bayamon, above Palo Seco, Porto Rico; *Fish Hawk;* January 16, 1899; 3 males, 1 female (24008).

Santo Domingo, West Indies; 1878; W. M. Gabb; 1 male, 1 female (3189).

Puerto Plata, Santo Domingo; Charles A. Fraser; 1 female y. (4140).

Jamaica; 1884; *Albatross;* 1 male (7340); 3 males, 3 females (7507).

Montego Bay, Jamaica; C. B. Wilson; from cocoanut groves: June 15, 1910; 2 males (42893). July 5, 1910; 1 female (42892).

Kingston Harbor, Jamaica; 1893; R. P. Bigelow; 4 males, 2 females (17987).

Mariel, Cuba; in swamp; May 10, 1900; William Palmer and J. H. Riley; 1 male (23802).

Cabañas, Cuba; May 27, 1900; William Palmer and J. H. Riley; 3 males (23801)

Cuba; May, 1914; Henderson and Bartsch, *Tomas Barrera* Exped.: Cayo Hutia; 1 specimen (48386). Ensenada de Santa Rosa; 1 to 3 fathoms; S. Sh. M., Sponge; 5 y. (48387). Santa Lucia; 1 male, 1 female (48393). Ensenada de Cajon, off Cape San Antonio; May 22; 1 female (48404).

Cuba; J. H. Hysell; 2 males, 1 female (24755).

Brazos de Santiago, Texas; G. Wurdemann; 1 specimen (2060).

East coast of Mexico; Edward Palmer; 1 cheliped (47107).

Mujeres Island, Yucatan; Mar. 24, 1901; Nelson and Goldman, Biol. Survey, U. S. Dept. of Agriculture; 1 female (24879).

Environs of Belize, British Honduras; W. A. Stanton; 2 males, 2 females (22594).

Patuca, Honduras; June 12 and Aug. 24, 1891; Harry W. Perry; 2 females (17714).

Bluefields, Nicaragua; April, 1892; Charles W. Richmond; 1 female (17642).

Old Providence Island; 1884; *Albatross;* 1 female (7331), 3 specimens (7532).

Colon (Aspinwall), Panama; J. Rowell; 3 specimens (2011).

Toro Point, Canal Zone, Panama; Meek & Hildebrand, Smithsonian Biol. Survey of the Panama Canal: April 4, 1911; 4 males, 1 female (44171). May 19, 1911; 2 males, 2 females (44170).

Sabanilla, Colombia; March 16–22, 1884; *Albatross;* 3 males, 5 females (7561).

Santa Marta, Colombia; C. F. Baker; 1 male (22549).

Curaçao; February 10–18, 1884; *Albatross;* 1 female (7581).

Macuto, Venezuela; August 5, 1900; Lyon and Robinson; 1 female (23796).

Guanta (near Barcelona), Venezuela; June 29, 1895; Lieut. Wirt Robinson; 1 male (18821).

Rio Formosa, Pernambuco, Brazil; 1876–1877; R. Rathbun, Hartt Explorations; 1 female ovig. (40577).

Ilha Sebastiao, São Paulo, Brazil; 1896; Bisego, collector; received from H. von Ihering, Mus. S. Paulo; 1 male (47849).

CARDISOMA CRASSUM Smith.

MOUTHLESS CRAB (Coker).

Plates 108 and 109.

Cardiosoma crassum SMITH, Trans. Connecticut Acad. Arts and Sci., vol. 2, 1870, p. 144, pl. 5 (type-locality, Gulf of Fonseca; type in Mus. Comp. Zoöl.).

Cardiosoma latimanus LOCKINGTON, Proc. California Acad. Sci., vol. 7, 1876 (1877), p. 151 [7], (type-locality, La Paz, Lower California; type not extant).

Diagnosis.—First abdominal appendages in male terminating in 2 slender appendages. Legs very hairy.

Description.—Differs little from the preceding.

Orbits, viewed from in front, less high as compared with their length, than in *guanhumi;* outer orbital tooth more advanced; epibranchial tooth more elevated.

The extremities of the first pair of male abdominal appendages are slightly flattened laterally, thickly clothed with hair on the outside and terminated by a long,

FIG. 156.—CARDISOMA CRASSUM, FRONT VIEW OF ORBITAL AND ANTENNAL REGION OF MALE (2137). × 1¼.

slender, hard, and horny tip, which curves outward for nearly half its length, then rapidly upward, and again outward at the end, forming thus about the third of a very elongated spiral. From the under edge, just below the base of this horny tip, there is a stout, straight process, which is soft and flexible, and clothed at the extremity with hair.

The legs are conspicuously hairy.

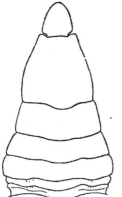

FIG. 157.—CARDISOMA CRASSUM, ABDOMEN OF MALE (48809), NATURAL SIZE.

Measurements.—Male (40501), length of carapace 100, width of same 128, length of propodus of larger chela 142, height of same 72 mm.

Range.—From San José and La Paz, Lower California, to Peru. Alcock gives "Chili" in the distribution of the genus *Cardisoma*. If this be correct, it is probably *C. crassum* that is found there.

Material examined.—

San José, Lower California (M. C. Z.).

La Paz, Lower California; L. Belding; 1 male, 1 female (4658).

Cape St. Lucas, Lower California; J. Xantus; 2 males (2137).

Acapulco, Mexico (M. C. Z.).

Panama Canal Zone; July 27, 1915; J. Zetek; 1 male (48809).

Cristobal, Panama; Mar. 13, 1912; Meek & Hildebrand, Smithsonian Biol. Survey of the Panama Canal; 1 female (46692).

Darien; Dr. Enrico Festa, collector; received from Turin Mus.; 1 male (20072).

Mouth of River Tumbes, Peru; Feb. 12; R. E. Coker; received from Peruvian Government; 1 male (40501).

Genus UCIDES Rathbun.

Uca LATREILLE, Nouv. Dict. Hist. Nat., vol. 35, 1819, p. 96; type, *U. uca* (Linnaeus, 1767)=*cordata* (Linnaeus, 1763).

Ucides RATHBUN, Proc. Biol. Soc. Washington, vol. 11, June 9, 1897, p. 154; type, *U. cordatus* (Linnaeus).

Œdipleura ORTMANN, Zool. Jahrb., Syst., vol. 10, 1897, p. 334; type, *O. cordata* (Linnaeus).

Carapace shaped much as in *Cardisoma*. Pterygostomian regions for the most part naked. Fronto-orbital border somewhat more than half, and the narrow, convex, deflexed front about one-sixth, the greatest breadth of the carapace. Orbits deep, the outer angle inconspicuous, the inner angle remote from the front but not dentiform. Orbits not very much larger than the eyes.

Antennules oblique, partly hidden by the front; interantennulary septum rather narrow, triangular. The basal antennal joint is longer than broad, and lies obliquely across the orbital hiatus; flagellum short.

Epistome small, prominent. Buccal cavern a little narrowed and arched anteriorly, completely closed by the outer maxillipeds. The merus of the latter is subrectangular, bearing at its external angle the palpus which is large and wholly exposed. The exognath is also visible and provided with a flagellum.

Chelipeds much stouter than the legs, more or less unequal in both sexes; while comparatively smooth outside, they are armed on the inner or prehensile surface with stout spines.

Legs stout, clothed in part with coarse bristles, dactyli unarmed.

The abdomen in the male has the fifth and sixth segments fused. The basal segments do not wholly cover the sternum between the last pair of legs.

Restricted to America. Analogous species on opposite sides of the continent: *cordatus* (Atlantic); *occidentalis* (Pacific).

KEY TO THE SPECIES OF THE GENUS UCIDES.

A^1. Carapace of male about 1¼ times as wide as long. Lower margin of palm convex_____*cordatus*, p. 347.
A^2. Carapace of male 1¼ times as wide as long. Lower margin of palm straight_____*occidentalis*, p. 350.

UCIDES CORDATUS (Linnaeus).

PAGURUS; KABURI.

Plates 110–113; plate 159, figs. 3 and 4.

Uca una Marcgrave de Liebstad, Hist. Rer. Natur. Brazil, 1648, p. 184, text-fig., male (type-locality, Brazil; type not extant).

Cancer Pagurus, hirsutus, Americanus, pronus Seba, Thesaurus, vol. 3, 1758, p. 51, pl. 20, fig. 4, male.

Cancer cordatus Linnaeus, Amœn. Acad., vol. 6, 1763, p. 414, male (type-locality, America; type not extant).

Cancer uca Linnaeus, Syst. Nat., ed. 12, vol. 1, pt. 2, 1767, p. 1041 (type-locality, America; type not extant).

Cancer uca Herbst, Natur. Krabben u. Krebse, vol. 1, 1783, p. 128.

Cancer cordatus Herbst, Natur. Krabben u. Krebse, vol. 1, 1783, p. 131, pl. 6, fig. 38, after Seba.

Cangrejos Ajaes de Manglar Parra, Descripcion de diferentes piezas de Historia Natural, Havana, 1787, p. 166, pl. 59 (male and female).

Ocypode cordata Latreille, Hist. Nat. Crust., vol. 6, an XI [1802–1803], p. 37, pl. 46, fig. 3, after Herbst.

Ocypode fossor Latreille, Hist. Nat. Crust., vol. 6, an XI [1802–1803], p. 38, female (type-locality, Cayenne; type in Paris Mus.).

Ocypode uca Latreille, Gen. Crust., vol. 1, 1806, p. 31.

Gecarcinus uca Lamarck, Hist. Nat. Anim. sans Vert., vol. 5, 1818, p. 251.

Gecarcinus fossor Desmarest, Consid. Gén. Crust., 1825, p. 114.

Uca uca Latreille, Cours d'Entom., 1831, p. 338.

Ocypode (uca) uca de Haan, Fauna Japon., Crust., p. 29, 1835, pl. c.

Uca una Guérin, Icon. Règne Anim. Cuvier, pl. 5, fig. 3, female.—Milne Edwards, Hist. Nat. Crust., vol. 2, 1837, p. 22, female; Ann Sci. Nat., ser. 3, Zool., vol. 20, 1853, p. 206 [172], pl. 10, fig. 2.—Young, Stalk-eyed Crustacea Brit., Guiana, 1900, p. 252, pl. 7 (female, colored plate).

Uca laevis MILNE EDWARDS, Hist. Nat. Crust., vol. 2, 1837, p. 22, male (type-locality, Antilles; type in Paris Mus.); *laevis* in index, vol. 3, 1840, p. 638.
Uca cordata WHITE, List Crust. Brit. Mus., 1847, p. 31.—SMITH, Trans. Connecticut Acad. Arts and Sci., vol. 2, 1869, pp. 13 and 36.—YOUNG, Stalk-eyed Crust. Brit. Guiana, 1900, p. 250, pl. 1 (male, colored plate).
Uca pilosipes GILL, Ann. Lyc. Nat. Hist. New York, vol. 7, 1859, p. 43 (type-locality, St. Thomas; type not extant).
Ucides cordatus RATHBUN, Ann. Inst. Jamaica, vol. 1, 1897, p. 25.
Oediplcura cordata ORTMANN, Zool. Jahrb., Syst., vol. 10, 1897, p. 336.

Diagnosis.—Carapace of male about one and one-fourth times as wide as long. Jugal region almost smooth. Lower margin of palm convex. First pair of legs of adult male the longest.

Description.—Carapace naked and smooth above, very broad, the greatest breadth much anterior to the middle, very convex fore and aft. Cervical suture very distinct. Gastric region broad and flattened in the middle, the protogastric subregions only partially separated from the mesogastric, the posterior part of which is rounded and slightly protuberant but still lower than the branchial regions. A deep furrow either side of the cardiac region. Branchial regions swollen, evenly rounded above, lateral margins very convex anteriorly, and indicated by a very slight denticulated ridge. Fronto-orbital border edged by a sharply raised rim; front lobiform, almost perpendicular. Orbit more than twice as long as it is high, with a broad and deep hiatus, over which projects the blunt outer angle; lower margin nearly straight and formed of two nearly parallel ridges, the lower of which is armed with a line of small tubercles, and the upper irregularly granulous. Inferior orbital region smooth, separated by a deep sulcus from the buccal area. Inferior lateral regions swollen and nearly smooth, only a few low granules near the orbital region. A ridge on each side of the buccal area is armed with a few small tubercles.

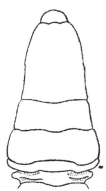

FIG. 158.—UCIDES CORDATUS, ABDOMEN OF MALE (17595), NATURAL SIZE.

Chelipeds very unequal, the greater one very large. Merus stout, the inferior angles armed with sharp spines, superior angle coarsely granulous. Inner margin of carpus rounded and armed with spines; distal margin below also armed with spines. Hand as long as, or longer than broad, spinous on the upper margin and inside, lower margin granulous, outer surface smooth. Fingers thick, curving toward each other, gaping, inner margins spinous. tips slightly excavate.

Legs smooth and naked above, but all the four long segments in the first three pairs are thickly clothed beneath, and the carpal and propodal segments of second and third pairs on the anterior side, with very long coarse hair; first pair longest in the adult, last pair much the shortest and slightly hairy. Dactyli of first two pairs very long and slightly curved downward; of the third and fourth pairs successively much shorter and curved backward as well as downward; the dactyli are six-sided with sharp granulated edges, except near the tips.

The female differs from the male in its much narrower carapace, stronger lateral margins, shorter chelipeds and legs, the latter being sparsely hairy.

Measurements.—Male (7671), length of carapace 70, width of same 89, length of propodus of large cheliped 97 mm. Female (17595), length of carapace 54.5, width of same 65 mm.

Habitat.—Lives not far from the sea in swampy ground where brackish water stands, and especially under mangroves.

Range.—West Indies to Rio de Janeiro.

Material examined.—

Playa de Marianao, Cuba; June 17, 1900; William Palmer and J. H. Riley; 1 male (23804).

Baracoa, E. Cuba; on tidal flat; Jan. 29, 1902; William Palmer; 1 female y. (25547).

Cape Cajon, Cuba; May, 1914; Bartsch and Henderson, *Tomas Barrera* Exped.; 1 male (48396).

Jamaica; Mar. 1-11, 1884; *Albatross;* 2 males, 1 female (7671).

Rio Bayamon, above Palo Seco, Porto Rico; Jan. 16, 1899; *Fish Hawk;* 3 females (24009).

St. Thomas; A. H. Riise; 1 female (2467).

Cristobal, Panama; Mar. 13, 1912; Meek and Hildebrand, Smithsonian Biol. Survey of Panama Canal; 1 female (46688).

Brazil; Lieut. F. E. Sawyer, U. S. Navy; 7 males, 1 female (17595).

Parahyba River, Cabedello, Brazil; in mangroves; June 20, 1899; A. W. Greeley, Branner-Agassiz Exped.; 2 females (25702).

Plataforma, Bahia, Brazil; 1876–1877; R. Rathbun, Hartt Explorations; 1 male. 1 female (40578).

Rio de Janeiro, Brazil; U. S. Exploring Exped.; 1 male (2126), 1 male (2361).

Terra de Masahe, Rio de Janeiro, Brazil; Jan., 1912; E. Barbe; recd. from H. von Ihering, Mus. S. Paulo; 2 males, 2 females (47860).

UCIDES OCCIDENTALIS (Ortmann).

Plates 114-116.

Uca una MILNE EDWARDS and LUCAS, d'Orbigny's Voy. l'Amér. Mérid., vol. 6, 1843, pt. 1, p. 23. Not *Uca una* Latreille.
Uca lævis MILNE EDWARDS, Arch. Mus. Hist. Nat. Paris, vol. 7, 1854, p. 185, pl. 16, figs. 1, 1a (type-locality, Guayaquil; type in Paris Mus.). Not *Uca laevis* (*lavis*) Milne Edwards, 1837.
Ocdiplcura occidentalis ORTMANN, Zool. Jahrb., Syst., vol. 10, 1897, p. 336. New name for *Uca laevis* Milne Edwards, 1854.

Diagnosis.—Carapace of male one and one-half times as wide as long. Jugal region tuberculous. Lower margin of palm straight. Second pair of legs of adult male the longest.

Description.—Carapace of male much wider than in *U. cordatus.* Lateral margin marked by a very faint tuberculate line. Orbits a little higher than in *U. cordatus.* Jugal regions tuberculous. Chelipeds not so strikingly unequal in this species, more elongate, especially the propodal segments. Palms distinctly longer than wide, their lower margins not dilated; the spines of the inner surface and the margins are smaller, more numerous, and more tuberculiform. Fingers very broad and flat, those of the smaller chela not gaping but beset on the inner surface with short, thick hair. Legs not very long, resembling those of the female of *U. cordatus.* The first three pairs are hairy beneath; the last pair is nearly naked. Ridges on the dactyli smooth.

FIG. 159.—UCIDES OCCIDENTALIS, ABDOMEN OF MALE (40490), NATURAL SIZE.

Color.—Carapace olive-green, margined with orange; claws, legs, and eyestalks deep red. (Coker.)

Measurements.—Male (2145), length of carapace 54, width of same 80, length of larger propodus 87 mm.

Habitat.—In the mud of mangrove swamps. They are taken at low tide by thrusting one's arm into the deep holes in the mud. The meat is of excellent flavor. (Coker.)

Range.—Lower California to Peru; Valparaiso (?).

Material examined.—
San Salvador, Central America; Capt. J. M. Dow; male (2145).
Guayaquil, Ecuador; four specimens (Paris Mus.).
Las Vacas, in the region of Capon, Peru; mangrove swamps; January 23, 1908; R. E. Coker, collector; received from Peruvian Government; 1 male, 1 female (40490).

Genus GECARCINUS Leach.

Gecarcinus LEACH, Edin. Encyc., vol. 7, 1814, p. 430; type, *G. ruricola* (Linnaeus).

Carapace shaped much as in *Cardisoma*. Pterygostomian regions nearly naked. Only the anterior part of the lateral border marked by a marginal line.

Fronto-orbital border half or less than half the greatest breadth of the carapace. Front strongly deflexed, edge horizontal, from one-fifth to one-eighth the greatest breadth of the carapace. Orbits deep, not much wider than high, outer angle obtuse and not prominent, inner angle a stout angular tooth which touches the front; next it a deep U-shaped sinus. Eyes nearly filling the orbits.

Antennules folded obliquely and in large part concealed by the front. Interantennular septum narrow. Antennae very short, their basal joint very small and not nearly reaching the front.

A strong oblique ridge runs outward from near the extremities of the epistome.

Epistome linear, sometimes overlapped by the outer maxillipeds. Buccal cavern subcircular or rhomboidal, widest at the middle; its margins with a dense fringe of hair; outer maxillipeds separated by a rhomboidal gape, the ischium and merus broad, of subequal length, the merus suboval, concealing the short, stout palp which is attached to the upper side of the middle of the distal extremity. Exognath concealed and devoid of a flagellum.

Chelipeds massive, equal or unequal, almost smooth.

Legs stout, second pair longest, last three joints armed with spines, dactyli with six or four rows.

The seven segments of the abdomen are in both sexes separate, short fringes of hair between the fourth and fifth segments of the sternum, and at the extremities of the first three segments of the abdomen conceal the outer end of an open channel or gutter which runs along the upper side of the second abdominal segment.

Tropical and warm north temperate America; Bermudas; Ascension Island; west and south Africa; Australasia.

KEY TO THE SPECIES OF THE GENUS GECARCINUS

A¹. Merus of maxilliped with entire margin.
  B¹. Merus of maxilliped distinctly longer than wide, covering the epistome and antennular cavities. Dactyl of ambulatory legs armed with six rows of coarse spines_____*ruricola*, p. 352.
  B². Merus of maxilliped about as long as wide, not covering epistome. Dactyli of ambulatory legs armed with four rows of medium-sized spinules, and sometimes an additional row of small spinules on the upper and lower surfaces_____*quadratus*, p. 358.

A². Merus of maxilliped with an inner distal emargination.
  B¹. Emargination of merus of maxilliped very shallow, rounded, not at all V-shaped _____ _lateralis, p. 355.
  B². Emargination of merus of maxilliped V-shaped, sometimes continued backward by a closed fissure.
    C¹. Antero-external angle of merus of maxilliped strongly produced beyond the emargination and fissure on the inner margin ___ lagostoma, p. 361.
    C². Antero-external angle of merus of maxilliped not produced much beyond antero-internal angle _____ planatus, p 359.

ANALOGOUS SPECIES ON OPPOSITE SIDES OF THE CONTINENT.

Atlantic.   Pacific.
ruricola.   }
lateralis. }   quadratus (also on the Atlantic side of the Isthmus).

SPECIES ERRONEOUSLY REFERRED TO THE GENUS GECARCINUS.

Gecarcinus barbiger Poeppig 1836=Paraxanthus barbiger (Poeppig).
Gecarcinus regius Poeppig 1836=Homalaspis plana (Milne Edwards, 1834).

GECARCINUS RURICOLA (Linnaeus).

BLACK OR MOUNTAIN CRAB; BLUE LAND-CRAB; RED TOURLOUROU.

Plates 117 and 118.

Cancer terrestris cuniculos sub terra agens SLOANE, Voy. Jamaica, vol. 1, 1707, pl. 2; vol. 2, 1725, p. 269; Jamaica.
Cancer sulcatus, terrestris, sive montanus, Americanus SEBA, Thesaurus, vol. 3, 1758, p. 51, pl. 20, fig. 5; Curassoa.
Cancer ruricola LINNAEUS, Syst. Nat., ed. 10, vol. 1, 1758, p. 626 (type-locality, America; type not extant).—HERBST, Natur. Krabben u. Krebse, vol. 1, 1783, p. 119, pl. 3, fig. 36, copied; pl. 4, fig. 37, copied; 1790, p. 262, pl. 20, fig. 116; vol. 3, 1799, p. 37, pl. 49, fig. 1.—BROWNE, Hist. Jamaica, index 3, 1789 (reference to p. 423).
Cangrejos Ajacs Terrestres PARRA, Descripcion de diferentes piezas de Historia Natural, Havana, 1787, p. 164, pl. 58 (color varieties).
Ocypode tourlourou LATREILLE, Hist. Nat. Crust., vol. 6, 1803, p. 36 (type-locality, Santo Domingo; type in Paris Mus.).
Gecarcinus ruricola LEACH, Trans. Linn. Soc. London, vol. 11, 1815, p. 322.
Ocypoda rubra FREMINVILLE, Ann. Sci. Nat., ser. 2, Zool., vol. 3, 1835, p. 222 (type-locality Antilles; type in Paris Mus.).
Gecarcinus agricola REICHENBACH, Zoologie oder Naturgeschichte des Thierreichs, vol. 2, 1836 (Zittau und Leipzig), p. 230; South America and Antilles.

Diagnosis.—Large. Fronto-orbital width much more than one-third width of carapace. Maxillipeds cover epistome and antennular cavities; merus suboval, subentire. Chelipeds equal. Dactyli of legs with six rows of spines.

Description.—Branchial regions greatly swollen sideways. Cervical suture very deep, ending anteriorly in a pit near the orbital angle. Equally deep is the median suture which runs nearly a third the length of the carapace; cardiac region less deeply outlined.

Surface closely covered with depressed granules; lateral regions with fine oblique granulated striae. A very short denticulated marginal line runs back from the obtuse orbital angle; this line is longer and more marked in young individuals.

Fronto-orbital width in the adult about two-fifths, in the half-grown about one-half, of the total width of the carapace. Front curved downward and backward, distinctly widening below, lower edge straight and horizontal, not visible from above, lateral edges sinuous. Height of orbit about two-thirds of its width; upper edge rimmed, subentire and continuous with the frontal margin; lower edge irregularly denticulate.

The lower surface of the carapace is granulate and for the most part crossed by broken lines of granules. The outer maxillipeds reach to the anterior of the antennular cavities, thus concealing the epistome and both pairs of antennae. The ischium measured on its outer margin is but little longer than the merus and not much longer than its own width. The articulation of the two segments is very oblique. The merus is not notched for the insertion of the palpus.

Fig. 160.—GECARCINUS RURICOLA, ABDOMEN OF MALE (7343), NATURAL SIZE.

Chelipeds equal, massive, but not extraordinarily enlarged, surface granulous like the carapace; lower edges of arm more or less spinulous or denticulate, upper edge rugose; inner edge of wrist rounded, and armed with a few spines on the distal half; hands broader than their superior length; fingers rough with fine horny, appressed spinules, prehensile margins irregularly dentate and narrowly gaping.

Legs of moderate length; merus joints roughened anteriorly with rugose lines; carpus with one or two anterior rows of spines; propodi with 4 rows of stout, horny-tipped spines; dactyli with 6 rows of the same. The propodi are considerably longer than wide, but shorter than the dactyli.

Terminal segment of male abdomen about half as wide as sixth segment.

Color.—A female of medium size (32722) was colored as follows: Carapace, eyes, wrists, and greater part of arms, also carpal and meral segments of ambulatories, almost black, with a purplish tinge. A small light yellowish spot near postero-lateral angle of carapace. Last two joints of legs red, distal part of spines yellow, tips horn color. Basal portion of legs and margins of arms mixed red and yellow. A red and yellow patch below orbit. Claws dark purplish proximally, rest of palm and fingers orange-yellow. A small yel-

strip around inner and proximal sides of ischiognath. Abdomen chiefly light yellowish mixed with violet, with irregular dark violet strip down the center.

Older specimens are said to be lighter in color.

Measurements.—Male (24756), length of carapace 63, width of same 89 mm.

Habits.—Freminville says of individuals of this species:

They live in the low and marshy ground of the savannas of the West Indies, not far from the shore. They hollow out burrows which are inclined obliquely and intersect one another in all directions. They stray but little except at night to seek their prey. During the day they stay like sentinels at the edge of the openings and at the least noise they enter precipitately. But in the season of abundant rain they spread over the country in a prodigious quantity, covering the prairies so that they appear all red. They run much more rapidly than the *Cardisoma* and are very difficult to catch.

Of the food value of this species, Browne writes:

When the black crab is fat and in a perfect state, it surpasses everything of the sort in flavour and delicacy, and frequently joins a little of the bitter with its native richness, which renders it not only the more agreeable in general, but makes it sit extremely easy upon the stomach. They are frequently boiled and served up whole, but are commonly stewed when served up at the more sumptuous tables.[1]

Calman says:

The migration to the sea takes place annually during the rainy season in May. The crabs come down from the hills in vast multitudes, clambering over any obstacles in their way, in their march toward the sea. The females enter the sea to wash off the eggs which they carry attached to their abdominal appendages, or rather, probably, to allow the young to hatch out. The crabs then return whence they came, and are followed later by the young, which having passed through their larval stages in the sea, leave the water, and are found in thousands clinging to the rocks on the shore.[2]

Range.—Bahamas; southern Florida; West Indies; Curaçao. Live specimens are not infrequently transported in bunches of bananas for long distances, as from Cuba to Washington, District of Columbia.

Material examined.—

Andros Island, Bahamas, received from Amer. Mus. Nat. Hist.; 1 female (31461).

Pothole, Smiths Place, south of South Bight, east side of Andros Island, Bahamas; May 3, 1912; Paul Bartsch, 1 female young (45624).

---

[1] Browne, Hist. Jamaica, 1789, p. 424.
[2] The Life of Crustacea, London, 1911, p. 190. See also Stebbing, A History of Crustacea (Recent Malacostraca); The International Scientific Series, No. 71, New York 1873, pp. 80–84.

Florida (M. C. Z.).

Cuba; brought to Washington in a bunch of bananas; A. Cerriglio; 1 female (32722).

Cuba; J. H. Hysell; 2 males, 3 females (24756).

Cuba; Bartsch and Henderson, *Tomas Barrera* Exped.: Ensenada de Cajon, off Cape San Antonio; May 22, 1914; 1 male, 3 y. (48405); 2 y. (49161). Cabañas; 2–5 fathoms; June 8, 1914; 1 male (48402).

Navassa Island; S. D. Nixon; 1 carapace (26378).

Hills back of Montego Bay, Jamaica; C. B. Wilson; 2 females (42901), 1 female (42902), 1 male y. (42903).

Montego Bay, Jamaica; E. A. Andrews; 1 male, 1 female (42917).

Jeremie and Grand Anse, Haiti (M. C. Z.).

Swan Island; C. H. Townsend; 1 male (17602).

Old Providence; 1884; *Albatross;* 9 males, 4 females (7343).

Barbados; Theodore N. Gill; 1 specimen (2065).

Found alive in Washington in bunch of bananas; John H. Semmes; 1 specimen (14851).

GECARCINUS LATERALIS (Freminville).

BLACK LAND-CRAB; COMMON LAND-CRAB.

Plates 119 and 120.

Ocypoda lateralis FREMINVILLE, Ann. Sci. Nat., ser. 2, Zool., vol. 3, 1835, p. 224 (type-localities, Martinique, Guadeloupe, Marie Galante, "la Desirade et les Saintes"; types in Paris Mus.).

Gecarcinus lateralis GUÉRIN, Icon. Règne Anim., pl. 5, fig. 1.—MILNE EDWARDS, Hist. Nat. Crust., vol. 2, 1837, p. 27, pl. 18, figs. 1–6.

Gecarcinus depressus SAUSSURE, Mém. Soc. Phys. Hist. Nat. Genève, vol. 14, 1858, p. 439 [23], pl. 2, fig. 14 (type-locality, Haiti; type in Geneva Mus.).

Diagnosis.—Small, narrow. Maxillipeds do not reach epistome; merus anteriorly with a shallow, rounded emargination. Chelipeds unequal in male. Dactyli of legs with four rows of spines.

Description.—Of smaller size than *G. ruricola;* sutures less deep; branchial regions less swollen anteriorly, the widest part of the carapace being in line with the antero-lateral angles of the mesogastric region, while in *G. ruricola* the widest part of the carapace is considerably in advance of that point.

Fronto-orbital width in the adult one-half or a little less than one-half of the total width. Front broader, less deep and less retreating than in the preceding species; it widens very little below, and its inferior margin is slightly arcuate. Fronto-orbital rim obscurely tuberculate.

The outer maxillipeds are smaller and do not attain the epistome, and their position is more longitudinal than in *G. ruricola;* the ischium is considerably longer than the merus and the latter has a shallow emargination on its anterior border.

Chelipeds very unequal in old males, less so in young males, equal or nearly so in females; the gape of the fingers being greater in the large cheliped than in the small one. In old specimens the spines of the carpus are obliterated.

Dactyli of legs with only four rows of spines, the ridge on the upper and the lower surface being unarmed, except that occasionally there are a few spines near the tip, or less frequently a whole row of very feeble spines.

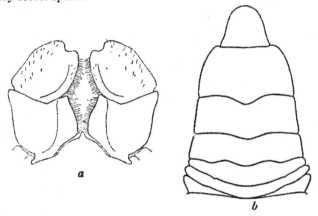

FIG. 161.—GECARCINUS LATERALIS, MALE (11368), ×2. *a,* OUTER MAXILLIPEDS, IN PLACE; *b,* ABDOMEN.

Color.—Commonly the carapace is mostly of a deep reddish brown or plum color; often this color is replaced posteriorly by a wide transverse band of lighter color spotted with yellow; this band extends forward, along each side, becoming narrower and darker, disappearing near the eye-sockets; a pair of small white spots close behind the eye-sockets and another pair in the cardiac region. Legs light grayish brown; chelipeds darker and more red; last joint bright orange. Underside white.[1]

Measurements.—Male, Green Cay, Bahamas, length of carapace 43.5, width of same 58 mm.

Habits.—Verrill says of its occurrence in Bermuda:

It is very common in sandy waste places on many of the smaller islands. It makes its deep burrows both near the shore and on the low hills, 20 to 30 feet

---

[1] Verrill, Trans. Connecticut Acad. Arts and Sci., vol. 13, 1908, pp. 308–9, fig. 2.

high, at some distance from the shore and where the shell-sand was nearly or quite dry. Its burrows are often very long and deep. Some of the young were exposed by turning over large flat stones under which they had burrowed.

Freminville says that individuals of this species never go into the water, but live constantly in the woods. Individuals plunged into sea water die in 2 or 3 minutes. His statements remain to be proved.

Range.—Bahamas; Florida Keys to Guiana; Bermudas; Ascension Island; Ecuador (Cano, Nobili).

Material examined.—

Loggerhead Key, Florida; Apr. 8, 1889; *Grampus;* 5 males, 1 female (15264).

Tortugas, Florida; 1893; Bahama Exped. S. U. I.; 4 males (Mus. S. U. I.).

Green Turtle Cay, Bahamas; E. A. Andrews; 1 female (20713).

Abaco, Bahamas; 1886; *Albatross;* 14 males, 6 females (11364).

Smith's place, south side of east end of South Bight, Long Bay Key District, Andros Island, Bahamas; Paul Bartsch: May 7, 1912; 9 males (45544). May 20, 1912; 9 males, 3 females (45543).

Green Cay, Bahamas; Apr. 12, 1886; *Albatross;* 1 male (11368).

Water Island, St. Thomas; on dry shore; June 28, 1915; C. R. Shoemaker; 3 males (49805).

Arroyo, Porto Rico; under logs on land, 20 feet above high water; Feb. 4, 1899; *Fish Hawk;* 5 y. (24010).

Santo Domingo, West Indies; 1878; W. M. Gabb; 3 males, 3 females (3204).

Haiti; 1 male, type of *G. depressus* (Geneva Mus.).

Jamaica; Mar. 1-11, 1884; *Albatross;* 1 male, 7 females (18539).

Mountains near Montego Bay, Jamaica; C. B. Wilson; 3 males, 2 females (42899); 1 male, 2 females (42900).

Snug Harbor, Jamaica; June 19, 1910; E. A. Andrews; 5 males, 3 females (42916).

Mariel, Cuba; between tides among rocks; May 10, 1900; William Palmer and J. H. Riley; 1 male, 1 female (23803).

Ensenada de Cajon, off Cape San Antonio, Cuba; May 22, 1914; Bartsch and Henderson, *Tomas Barrera* Exped.; 7 males, 6 females, 4 y. (48406).

Mujeres Island, Yucatan; Mar. 25, 1901; Nelson and Goldman; Biol. Survey, U. S. Dept. of Agriculture; 4 females (24880).

Cozumel Island, Yucatan: Jan. 28, 1885; *Albatross;* 2 specimens (9896). Apr. 24, 1901; Nelson and Goldman, Biol. Survey, U. S. Dept. of Agriculture; 1 male (24881).

Swan Islands; C. H. Townsend; 1 male (17619).

Near Belize, British Honduras; W. A. Stanton; 1 male (21383).

Ceiba, Honduras; under rotting log; Dec. 22, 1915; F. J. Dyer; 1 male (49907).

Toro Point, Canal Zone, Panama; Meek and Hildebrand, Smithsonian Biol. Survey of the Panama Canal: May 19, 1911; 2 males, 2 females (43993). June 24, 1912; 1 male (46687).

Sabanilla, Colombia; 1884; *Albatross;* 2 y. (17677), 27 males, 12 females (7558).

Santa Marta, Colombia; C. F. Baker; 1 male y. (22551).

Venezuela; 1900; Lyon and Robinson; 1 male y. (23799).

San Julian, Venezuela; July 17, 1900; Lyon and Robinson; 1 male (23798).

Macuto, Venezuela; July 29, 1900; Lyon and Robinson; 1 male (23797).

Guiana; 1 specimen (Paris Mus.).

Port Castries, St. Lucia; Dec. 3, 1887; *Albatross;* 9 males (40796)

Hungry Bay, Bermudas; F. G. Gosling; 5 males, 2 females (25435).

GECARCINUS QUADRATUS Saussure.

Plates 121 and 122.

Gecarcinus quadratus SAUSSURE, Rev. et Mag. de Zool., ser. 2, vol. 5, 1853, p. 360, pl. 12, fig. 2 (type-locality, Mazatlan; type in Geneva Mus.).

Diagnosis.—Small, narrow. Maxillipeds reach epistome or nearly so; merus not emarginate. Chelipeds unequal. Dactyli of legs with four rows of spines, besides often supernumerary spines above and below.

Description.—This species is intermediate between *G. ruricola* and *G. lateralis*, and still distinct from both. It has the size and shape of *lateralis*, and the front and orbits correspond to that species.

The outer maxillipeds just about reach the epistome; they are disposed as in *lateralis*, but the merus is rounded or a little truncate at the anterior end, which is not emarginate.

Chelipeds unequal in both sexes.

The dactyli of the legs bear sometimes 5 or 6 rows of spines, there being often a row of spines along upper and lower surfaces.

The abdomen of the male is narrower than in *G. lateralis*, excepting the seventh segment, which is distinctly more than half as wide as the sixth.

Measurements.—Male, Punta Arenas, length of carapace 44.7, width of same 55.8 mm.

Habitat.—Professor Tristan writes of this species that it is found frequently under the trunks of trees, six or eight together, and in the rainy season (June to October) is very abundant.

Range.—Along the Pacific coast, from Mexico to Colombia. Also at Turbo, on Atlantic side of Colombia.

Material examined.—
Corinto, Nicaragua (M. C. Z.).
Acapulco, Mexico; *Hassler* Exped.; recd. from Mus. Comp. Zoöl.; 3 males, 3 females (22819).
Mouth of Rio Jesus Maria, Costa Rica; April 22, 1905; P. Biolley and J. F. Tristan; 1 male (32311).
Pigres, Costa Rica; A. Zeledon; 1 female (46683).
Punta Arenas, Costa Rica; 2 males (Copenhagen Mus.).
Naios Id., Panama; G. M. Gray; 1 female (42141).
Turbo, Colombia (M. C. Z,).

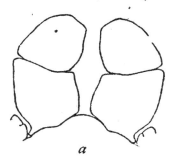 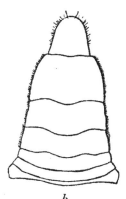

FIG. 162.—GECARCINUS QUADRATUS, MALE, × 2. *a*, OUTER MAXILLIPEDS, IN PLACE, OF SPECIMEN IN COPENHAGEN MUS.; *b*, ABDOMEN (32311).

GECARCINUS PLANATUS Stimpson.

Plates 123 and 124.

Gecarcinus planatus STIMPSON, Ann. Lyc. Nat. Hist. New York, vol. 7, 1860, p. 234 (type-locality, Todos Santos, near Cape St. Lucas; type not extant).—BOUVIER, Bull. Mus. Hist. Nat. Paris, vol. 4, 1898, p. 372.
Gecarcinus malpilensis FAXON, Bull. Mus. Comp. Zoöl., vol. 24, 1893, p. 157 (type-locality, Malpelo Island; type, Cat. No. 4492, Mus. Comp. Zoöl.); Mem. Mus. Comp. Zoöl., vol. 18, 1895, p. 28, pl. 4, figs. 2–2b.
Gecarcinus digueti BOUVIER, Bull. Mus. Hist. Nat. Paris, 1895, No. 1, p. 8 (type-locality, Lower California; type in Paris Mus.).—RATHBUN, North American Fauna, U. S. Department of Agriculture, No. 14, 1899, p. 73.

*Diagnosis.—*Large. Maxillipeds do not reach epistome; merus oblong, notched at extremity. Chelipeds unequal. Dactyli of legs with six rows of spines.

*Description.—*Of large size. Carapace narrower than in *G. ruricola*, flattened in the center and posteriorly. Outer gastric furrows obsolete anteriorly, a pit near the orbital angle; median furrow broad along the mesogastric region. The fine granulated marginal line in the male runs about one-third or less of the length of the lateral border; in the female and young this edge is longer, sharper, and denticulate.

Fronto-orbital width in the old male about one-third, in the adult female about nine-twentieths, of the width of the carapace. Front widening distinctly below, lower edge strongly projecting, granulate, slightly curved. Height of orbit about three-fourths of its width.

The outer maxillipeds may reach the epistome in old specimens, but fall short of it in smaller specimens; the gape is large, the line between ischium and merus almost transverse, the merus suboblong, with a -shaped notch in the rounded oblique distal extremity.

Chelipeds unequal in both sexes but most of all in the male. Wrists rather feebly spinulous or unarmed along the inner margin; larger palm of adult male much swollen, becoming elongate with age, fingers very rough, gaping.

Legs shorter, broader, and rougher than in *ruricola*, the merus joints distinctly margined above. Dactyli armed with 6 rows of spines.

Abdomen of male much wider than in *ruricola*, except the last segment, which is suboblong and half as wide as the preceding segment.

Measurements.—Male (20650), length of carapace 70, width of same 104 mm. Female (19646), length of carapace 57.4, width of same 75.3 mm.

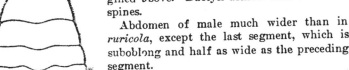

FIG. 163.—GECARCINUS PLANATUS, ABDOMEN OF MALE (20650), NATURAL SIZE.

Habitat.—On the Tres Marias we found them only on Maria Cleofa, where they were very numerous above high-water mark on the sandy beaches of the low eastern part of the island. They were also living very abundantly in burrows in the soft soil almost everywhere on the slopes of Isabel Island. They are nocturnal in habits and caused us some annoyance by walking over us at night while we were camped in their haunts. They began to come out of their burrows as soon as it became twilight in the evening. In both localities most of their burrows were found among the scrubby bushes. On Isabel Island they were often seen during the day sitting in the burrow a foot or so from the entrance, but scuttled back to a safe depth when I approached too near. (Nelson.)

Range.—West coast of Mexico and adjacent islands.

Material examined.—

Lower California; L. Diguet, collector; 1 male, holotype of *G. digueti* (Paris Mus.).

Maria Cleofa, Tres Marias Islands, Mexico; May 30, 1897; Nelson and Goldman, Biological Survey, U. S. Dept. Agriculture; 1 male (20650).
San Benedicto Island, Revilla Gigedo Islands, Mexico; A. W. Anthony; 1 male, 1 female (20690), 1 male (20697).
Socorro Island, Revilla Gigedo Islands, Mexico; A. W. Anthony; 2 males (20691).
Acapulco, Mexico (M. C. Z.).
Clipperton Island, North Pacific Ocean, west of Costa Rica; J. S. Arnheim; 1 male, 1 female (19646).
Malpelo Island, off Bay of Panama; *Albatross;* 1 male, holotype of *G. malpilensis* (4492, M. C. Z.).

GECARCINUS LAGOSTOMA Milne Edwards.

Plates 125 and 126.

Gecarcinus lagostoma MILNE EDWARDS, in Freminville, Ann. Sci. Nat., ser. 2, vol. 3, 1835, p. 218, footnote; Mauritius (*nomen nudum*); Hist. Nat. Crust., vol. 2, 1837, p. 27 (type-locality, Australasia;[1] type in Paris Mus.).—STEBBING, Hist. Crustacea, 1893, p. 84, pl. 2.

Gecarcinus ruricola GREEFF, Sitzsungsb. Ges. Marburg, 1882, p. 26; not *G. ruricola* (Linnaeus).—OSORIO, Jor. Sci. Lisboa, ser. 2, vol. 5, 1898, pp. 185 and 192; not *G. ruricola* (Linnaeus).

Gecarcinus lagostoma (?) MIERS, *Challenger* Rept., Zool., vol. 17, 1886, p. 218, pl. 18, figs. 2–2c.

Diagnosis.—Large. Maxillipeds cover epistome and antennular cavities; merus suboval, with an antero-internal fissure. Chelipeds subequal. Dactyli of legs with six rows of spines.

Description.—The dorsal aspect of the carapace is much like that of *G. ruricola.* The fronto-orbital distance is in the adult male three-eighths, in the half grown male less than half, the entire width of the carapace; the front is wider and the orbits narrower than in *ruricola.* The height of the orbit is about three-fourths of its width.

The outer maxillipeds touch the front, fitting snugly over the anten-

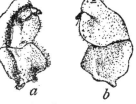

FIG. 164.— GECARCINUS LAGOSTOMA, OUTER MAXILLIPED, ASCENSION ISLAND, SLIGHTLY REDUCED. *a,* UPPER VIEW; *b,* LOWER VIEW. (AFTER MIERS.)

nular cavities; they are narrower, and at the middle project less outwardly than in *ruricola*; the merus is rounded at the end as in that species, but on the antero-internal edge has a deep fissure, closed by the overlapping of the proximal edge of the slit on the distal edge.

---

[1] Bouvier (Bull. Mus. Hist. Nat. Paris, 1906, p. 498) considers this locality erroneous.

Chelipeds subequal, almost unarmed in the adult male; carpal spinules few and inconspicuous; chelae of moderate size, and fingers little gaping, as in *ruricola*.

Legs broad, very rough; merus joints strongly serrulate above; propodal joints not much longer (on the upper edge) than their width; dactyli with six rows of spines.

Color.—Greeff says: The color of the carapace in most of these crabs is a deep violet, but in some cases dark red; the sides are al ways violet, the middle grayish-yellow. Chelae violet and also the bases of the legs; these become lighter toward the extremity, the dactyls being reddish-yellow with yellow spines. The color, however, varies in different individuals.

Measurements.—Male (14869), length of carapace 69.5, width of same 94.2 mm.

Habitat.—Greeff says of the occurrence of *G. lagostoma* at the islands of St. Thomas and Rolas, West Africa:

FIG. 165.—GECARCINUS LAGOSTOMA, ABDOMEN OF MALE (14869), NATURAL SIZE.

From the forests of the coast region, their ordinary places of abode, they undertake excursions toward the interior of the islands and into the mountains. Here one meets with them in the open much more rarely than in the lowlands, but they occur also in holes in the earth, under stones, damp foliage, and rotten trees. They are without doubt very adept at climbing, as they were occasionally seen sitting high up in the trees.

Range.—Trinidad, West Indies; Fernando de Noronha, Brazil. Also recorded from Australasia; Mauritius; Madagascar;[1] South and West Africa, including Ascension Island, South Trinidad Island, South Atlantic Ocean (Baylis), and from Bermuda with doubt. The writer has examined a subfossil specimen of a movable finger from the Bermudas.

Material examined.—

Trinidad, West Indies; Ensign Roger Welles, Jr., U. S. N.; 1 male (17603).

Fernando de Noronha, Brazil; 1876–1877; R. Rathbun, Hartt Explorations; 2 males, 2 females (40574).

Ascension Island; March 21, 1890; W. H. Brown, U. S. Eclipse Exped. to Africa; 1 male (14869).

Genus GECARCOIDEA Milne Edwards.

Gecarcinus DE HAAN, Fauna Japon., Crust., p. 30, 1835, pl. C. Not *Gecarcinus* Leach, 1814.

Gecarcoidea MILNE EDWARDS, Hist. Nat. Crust., vol. 2, 1837, p. 25; type, *G. lalandii* Milne Edwards.
Pelocarcinus MILNE EDWARDS, Ann. Sci. Nat., Zool., ser. 3, vol. 20, 1853, p. 203 [169]; type, *G. lalandii* Milne Edwards.
Hylaeocarcinus WOOD-MASON, Ann. Mag. Nat. Hist., ser. 4, vol. 14, 1874, p. 189; type, *H. humei* Wood-Mason.
Limnocarcinus DE MAN, Notes Leyden Mus., vol. 1, 1879, p. 65; type, *L. intermedius* de Man.

Carapace transversely oval, somewhat depressed, with the lateral borders tumid and strongly arched owing to the vault-like expansion of the gill-chambers; the gastric region particularly well defined.

The extent of the fronto-orbital border is less than half the greatest breadth of the carapace, that of the strongly deflexed and nearly straight front is from one-sixth to one-seventh the greatest breadth of the carapace.

Orbits deep, broadly oval, demarcated dorsally by a sharpish, slightly raised border, their outer angle not defined, a wide gap in their lower border; at the inner angle there is a strong tooth which may or may not meet the front: if it does so, the antennae, which are much reduced in size, are excluded from the orbit.

The antennules fold obliquely beneath the front, and the interantennular septum is of moderate width.

Epistome sunken, hairy posteriorly so as to appear ill defined from the palate. Buccal cavern rounded anteriorly, not nearly closed by the external maxillipeds, which leave between them a wide rhomboidal gap in which the mandibles are exposed.

The external maxillipeds are rather short; their merus lies obliquely, and its anterior edge is excavated for the insertion of the palp, which is short and coarse and is completely exposed; their exognath is very short and almost entirely concealed and is without a flagellum. The exognaths of the other maxillipeds are heavily fringed with hair.

Chelipeds much more massive than the legs, equal or nearly so in both sexes, though larger and longer in the male than in the female.

Legs stout; in all, the anterior border of the carpus and all the borders of the propodite and dactylus are spiny, there being 6 rows of spines on the dactylus.

The abdomen in both sexes consists of 7 separate segments and in the male its base covers all the breadth of the sternum between the last pair of legs.

The gill-chamber and its lining membrane, and the number of branchiae, are as in *Cardisoma*.

Only one species known, recorded from Brazil (?); Andamans and Nicobars, Christmas Island, Celebes, Philippines, New Guinea, Loyalty Islands.

? GECARCOIDEA LALANDII Milne Edwards.

Plate 160, figs. 7 and 8.

Gecarcoidea lalandii MILNE EDWARDS, Hist. Nat. Crust., vol. 2, 1837, p. 25, (type-locality, Brazil; type destroyed).—ORTMANN, Zool. Jahrb., Syst., vol. 7, 1893, p. 738.—CALMAN, Proc. Zool. Soc. London, 1909, p. 710.
Pelocarcinus lalandei MILNE EDWARDS, Ann. Sci. Nat., ser. 3, Zool., vol. 20, 1853, p. 203; Arch. Mus. Hist. Nat. Paris, vol. 7, 1854, p. 183, pl. 15, figs. 2, 2a.
Hylaeocarcinus humei WOOD-MASON, Journ. Asiat. Soc. Bengal, vol. 42, pt. 2, 1873, p. 260, pls. 15 and 16 (type-locality, Nicobar; type in Indian Mus.).
Limnocarcinus intermedius DE MAN, Notes Leyden Mus., vol. 1, 1879, p. 65 (type-locality, near the bay of Gorontalo, Celebes; type in Leyden Mus.).
Hylaeocarcinus natalis POCOCK, Proc. Zool. Soc. London, 1888, p. 561 (type-locality, Christmas Island; type in Brit. Mus.).
Pelocarcinus marchei A. MILNE EDWARDS, Nouv. Arch. Mus. Hist. Nat. Paris, ser. 3, vol. 2, 1890, p. 173, pl. 12 (type-locality, Philippines: Île des Deux-Sœurs; type in Paris Mus.).
Pelocarcinus cailloti A. MILNE EDWARDS, Nouv. Arch. Mus. Hist. Nat. Paris, ser. 3, vol. 2, 1890, p. 174, pl. 13 (type-locality, Loyalty Islands; type in Paris Mus.).
Pelocarcinus humei ALCOCK, Journ. Asiat. Soc. Bengal, vol. 69, 1900, p. 449.
Gecarcinus lagostomus ANDREWS, Monogr. Christmas Island, 1900, p. 163; not *G. lagostoma* Milne Edwards.

Diagnosis.—Fronto-orbital border less than half width of carapace. Buccal cavern rhomboidal. Outer maxillipeds gaping; merus distally excavate, bearing at its summit the palpus exposed to view; exognath partly concealed and without flagellum.

Description.—Median furrow very pronounced. Branchial regions very large and very swollen. The cardiac lobe is prolonged strongly backward between the bases of the posterior feet.

Front small, strongly curved downward, and terminated by a prominent and straight border.

Antennular cavities almost circular, small and plainly visible; septum between them narrow. Antennae almost rudimentary and wholly under the front.

Orbits small, oval, deep, bordered by a small sharp rim; upper margin a little sinuous and directed forward and outward, outer angle not prominent; outer suborbital lobe forms a rounded tooth which joins the angle of the front in such a way as to exclude the antennae from the orbit.

Chelipeds strong and almost equal; the arms surpass much the lateral border of the carapace and their anterior border is roughened with obtuse denticles; hands rounded; fingers elongate, cylindrical and almost meeting.

Legs strongly dentate along the last three joints; 6 rows of spines on the dactyli.

Color.—Brownish-red.

Habitat.—Lives in moist land or mud. Migrates to the sea during the rainy season to hatch the eggs. Vast quantities of the megalopa-larva have been found in the sea shortly after migration; also, at a later date, the young crab.[1]

Measurements.—Male type (according to the figure), length of carapace 61, width of same 86.5 mm.

Range.—Brazil (Milne Edwards); South America (White, Brit. Mus.). Both these localities are doubtful. It is likely that the species inhabits only Indo-Pacific Islands.

Family OCYPODIDAE Ortmann.

Ocypodiens MILNE EDWARDS, Hist. Nat. Crust., vol. 2, 1837, p. 39.
Ocypodinæ MILNE EDWARDS, Ann. Sci. Nat., ser. 3, Zool., vol. 18, 1852, p. 140 [104] (part).
Macrophthalmidæ DANA, U. S. Expl. Exped., vol. 13, Crust., pt. 1, 1852, pp. 308 and 312.
Ocypodidæ ORTMANN, Zool. Jahrb., Syst., vol. 7, 1894, pp. 700 and 741.—
ALCOCK, Journ. Asiat. Soc. Bengal, vol. 69, 1900. pp. 283, 290, and 294.

Palp of external maxillipeds coarse, articulating at or near the antero-external angle of the merus; exognath generally slender and often more or less concealed. Front usually of no great breadth and often a narrow lobe more or less deflexed. The orbits occupy the whole anterior border of the carapace outside the front, and their outer wall is often defective.

Buccal cavern usually large and a little narrower in front than behind, the external maxillipeds often, but not always, completely close it. Abdomen of male narrow. Male openings sternal.

Amphibious littoral and estuarine crabs, burrowing and commonly gregarious.

KEY TO THE AMERICAN SUBFAMILIES AND GENERA OF THE FAMILY OCYPODIDAE.

A[1]. An orifice or recess, the edge of which is thickly fringed with hair, between the bases of the second and third pairs of ambulatory legs. Antennular flagella small or rudimentary, folding obliquely or almost vertically, the interantennular septum broad. Chelipeds very unequal either in both sexes or in the male only_____Subfamily *Ocypodinae*, p. 366.
B[1]. Antennular flagella rudimentary, completely hidden beneath the front; antennae small, almost rudimentary. Eyes very large, occupying the greater part of the ventral surface of the eyestalks. Chelipeds very unequal in both sexes_____*Ocypode*, p. 366.

---

[1] See Andrews, Monogr. Christmas Island, 1900, p. 163.

male one cheliped is enormously enlarged, and the other very small, as are both chelipeds in the female_____*Uca*, p. 374.

A². No special recess between the bases of any of the legs. Antennular flagella well developed, folding transversely, the interantennular septum very narrow. Chelipeds usually subequal__Subfamily *Macrophthalminae*, p. 423.

Euplax, p. 423.

Subfamily OCYPODINAE Dana.

Ocypodiacés ordinaires MILNE EDWARDS, Ann. Sci. Nat., Zool., ser. 3, vol. 18, 1852, p. 140 [104].
Ocypodinæ DANA, Amer. Journ. Sci., ser. 2, vol. 12, 1851, p. 286; U. S. Expl. Exped., vol. 13, Crust., vol. 1, 1852, p. 312 (part).—ALCOCK, Journ. Asiat. Soc. Bengal, vol. 69, 1900, p. 290.

Carapace deep, subquadrilateral, the regions seldom well defined; front narrow, deflexed, commonly a mere lobe between the long eye stalks; antennular flagellum small, folding obliquely or almost vertically, the interantennular septum broad; the external maxillipeds completely close the buccal cavern, their exognath is inconspicuous but is not, or not entirely, concealed, and may either have, or be destitute of, a flagellum; chelipeds remarkably unequal either in both sexes or in the male only. There is an orifice or recess, the edge of which is thickly fringed with hair, between the bases of the second and third pairs of true legs.[1]

Genus OCYPODE Fabricius.

Ocypode FABRICIUS, Entom. Syst., Suppl., 1798, p. 312; type, *O. ceratophthalma* (Pallas).
Monolepis SAY, Journ. Acad. Nat. Sci. Philadelphia, vol. 1, 1817, p. 155; type. *M. inermis* Say, larval stage of *O. albicans* Bosc.
Ceratophthalma MACLEAY, Zool. S. Africa, Annul., 1838, p. 64; type *C. cursor* (Herbst, 1782) = *ceratophthalma* (Pallas, 1772).
Parocypoda NEUMANN, Cat. Podoph. Crust. Heidelberger Mus., 1878, p. 26; type, *P. ceratophthalma* (Pallas).

Carapace deep, subquadrilateral, somewhat broader than long, moderately convex, strongly declivous anteriorly, dorsal surface closely granular, regions indistinctly and incompletely defined. Front a narrow deflexed lobe, from one-seventh to one-ninth the greatest breadth of the carapace.

Orbits large, divided into 2 chambers. The eye occupies the ventral surface of the eyestalk and is often prolonged in a horn or style.

Only the basal joint of the antennulae is visible; antennae small and rudimentary.

---

[1] For a description of this connection with the branchial cavity, see Fritz Müller's Facts and Arguments for Darwin, translated by W. S. Dallas, London, 1869, pp. 33–36.

merus and ischium of which are subquadrilateral, the palpus articulating at the external angle of the former.

Chelipeds shorter than legs, very unequal, the larger one stouter than the legs. Chelae flattened; palm short and high and bearing on the inner surface a stridulating ridge which can be scraped against a corresponding ridge on the ischium.

Legs broad, the fourth pair much shorter and slenderer than the others, which are nearly equal. Between the basal joints of the second and third pairs, is an orifice, thickly protected by hairs, leading toward the branchial cavity.

Abdomen composed of seven segments in both sexes.

Tropical and subtropical coasts, from the American Atlantic (Rhode Island to Brazil), through the Mediterranean and Red Seas, to the American Pacific (Lower California to Chile).

Analogous species on opposite sides of the continent: *albicans* (Atlantic); *occidentalis* (Pacific).

KEY TO THE AMERICAN SPECIES OF THE GENUS OCYPODE.

A¹. Eyestalks not prolonged beyond the eyes in the form of a style. Fingers pointed.
  B¹. Orbital angles distinctly less advanced than the front. Merus of ambulatory legs hairy_____*albicans*, p. 367.
  B². Orbital angles as advanced or nearly as advanced as the front. Merus of ambulatory legs naked_____*occidentalis*, p. 372.
A². Eyestalks prolonged beyond the eyes in the form of a style. Fingers truncate_____*gaudichaudii*, p. 373.

SPECIES ERRONEOUSLY REPORTED AS AMERICAN.

O. macrocera MILNE EDWARDS. Recorded from Brazil in Hist. Nat. Crust., vol. 2, 1837, p. 49, but not in later work. Ann. Sci. Nat., Zool., ser. 3, vol. 18, 1852, p. 142 [106].

OCYPODE ALBICANS Bosc.

SAND CRAB; GHOST CRAB.

Plates 127 and 128.

Cancer arenarius CATESBY, Nat. Hist. Carolina, vol. 2, 1743, p. 35, pl. 35, "Ilathera, and many others of the Bahama Islands."
Cancer vocans LINNAEUS, Syst. Nat., ed. 10, vol. 1, 1758, p. 626 (part).
Cancer quadratus FABRICIUS, Mant. Ins., vol. 1, 1787, p. 315 (type-locality, Jamaica; type examined in Copenhagen Mus.); Ent. Sys., auct. et emend., vol. 2, 1793, p. 439; not *C. quadrata* Meuschen, 1781.
Ocypode quadrata FABRICIUS, Ent. Syst., Suppl., 1798, p. 347.
? *Ocypode rhombea* FABRICIUS, Ent. Syst., Suppl., 1798, p. 348; locality not known.
Ocypoda albicans Bosc, Hist. Nat. Crust., vol. 1, an X [1801–1802], p. 196, not pl. 4, fig. 1 (type-locality, *la Caroline;* type not extant).

Ocypode albicans LATREILLE, Hist. Nat. Crust., vol. 6, an XI [1802–1803], p. 48.

Ocypode arenarius SAY, Journ. Acad. Nat. Sci. Philadelphia, vol. 1, 1817, p. 69 (type-locality, "Inhabits sandy beaches of the sea, in holes of considerable depth"; a specimen (perhaps cotype) was presented by Say to the Brit. Mus.).

Monolepis inermis SAY, Journ. Acad. Nat. Sci. Philadelphia, vol. 1, 1817, p. 157 (type-locality, eastern shore of Maryland; type not extant); larval stage.

Ocypoda arenaria MILNE EDWARDS, Hist. Nat. Crust., vol. 2, 1837, p. 44, pl. 19, figs. 13 and 14.—MIERS, Ann. Mag. Nat. Hist., ser. 5, vol. 10, 1882, p. 384, pl. 17, figs. 7–7b.

Ocypode albicans RATHBUN, Proc. Washington Acad. Sci., vol. 2, 1900, p. 134.

Diagnosis.—Outer angle of orbit less advanced than the front. Eyes without style. Fingers pointed.

Description.—Greatest breadth, which is a fifth more than greatest length, is at the anterior third of the carapace. Lateral borders anteriorly convex, but posteriorly slightly concave or straight. Front and side margins beaded or serrulate. The H-form depression in the center of the carapace, the anterior half of the cervical suture, and the lobe at the inner angle of the branchial region are well marked. Granulation of surface coarser on the sides and along the front than elsewhere.

Upper border of orbit transverse and very sinuous, the outer angle acute, less advanced than the front; lower border notched at the middle, and with a triangular sinus below the outer angle. The eyes do not reach the outer angle.

Chelipeds rough, merus and carpus moderately so, merus serrulate above, toothed on lower margins; carpus with a sharp spine at inner angle. Chelae coarsely scabro-tuberculate; margins of palm and fingers dentate; fingers pointed. Stridulating ridge of larger palm less than half greatest width of palm and composed of 14 or 15 small tubercles. It plays against a smooth ridge along the distal half of the upper margin of the lower surface of the ischium.

The ambulatory legs are almost smooth, fringed with long yellow hair; the second pair the longest, about two and two-thirds times length of carapace; the third pair a little shorter; the last pair much the shortest, reaching only to distal fourth of propodus of third pair. Merus of first three pairs broadened; propodus of same pairs with longitudinal brushes of hair on anterior surface. Dactyli of all fluted, depressions hairy.

Color.—Pepper-and-salt, pale yellow, straw-color or yellowish-white, imitating the color of the beaches (Verrill). Sometimes light amber and often iridescent.

Measurements.—Length of carapace of male (11367) 44, width of same 50.2 mm.

Habitat.—On sandy beaches in deep burrows, near or above high tide. When pursued they run very rapidly, often suddenly stopping and squatting so closely in the sand that they could easily be overlooked (Verrill). They are partly nocturnal, usually remaining in their burrows during the daytime, and coming out at dusk or later to feed on animal refuse.[1]

Range.—From Block Island, Rhode Island, to the State of Santa Catharina, Brazil. Rare east of New Jersey. Bermudas. Megalops stage in Vineyard Sound, Massachusetts, and Narragansett Bay, Rhode Island.

Material examined.—

Woods Hole, Massachusetts; U. S. Fish Comm.: 4 megalops (11175). August, 1885; electric light; 6 megalops (10995).

Off Delaware, lat. 38° 25′ N.; long. 72° 40′ W.; surface; Sept. 21, 1885; U. S. Fish Comm.: 1 megalops (11128).

Bethany Beach, Delaware; July 4–10, 1912; W. D. Appel; 6 males (44571).

Wallops Island, Virginia; May 26, 1913; U. S. Dept. Agri.; 1 male (45962).

Smiths Island, Northampton County, Virginia: May 16–25, 1914; Chas. W. Richmond; 3 males, 1 female, 5 y. (18287). Sept. 28, 1897; Chas. W. Richmond and Wm. Palmer; 5 y. (20589). August, 1909; J. H. Riley and E. J. Brown; 15+ specimens (39860). 1910; J. H. Riley; 2 y. (41019). June, 1912: George Hitchens; 23 specimens (44524).

Willoughby Point, Virginia; May, 1880; Earll & McDonald; 3 males, 3 females (3879).

Buckroe Beach, Virginia; May 5, 1898; M. C. Marsh; 5+ specimens (32251).

Fort Macon, North Carolina; Oct. 7, 1904; B. W. Evermann; 1 male (48294).

St. Augustine, Florida; J. C. H. Smith; 4 specimens (2008).

East Peninsula, opposite Micco, Florida; O. Bangs; 10+ specimens (18738).

Indian River Inlet, Florida; Jan. 23, 1896; U. S. Fish Comm.; 1 female (20101).

Loggerhead Key, Florida; Apr. 8, 1889; U. S. Fish Comm.; 2 males (20098).

Marquesas Key, Florida; 1889; *Grampus;* 2 males (20100).

[1] For detailed observations on the habits of this species at Tortugas, Florida, see R. P. Cowles, Habits, Reactions, and Associations in *Ocypoda arenaria*. Papers from Tortugas Laboratory of Carnegie Institution of Washington, vol. 2, 1908, pp. 1–41, text-figs. 1–10, pls. 1–4 (1 colored).

Key West, Florida; Jan. 27, 1901; B. A. Bean & W. H. King; 1 y. (24839).

Garden Key, Tortugas, Florida; beach; Dec. 25, 1912; Bur. Fisheries; 7 young (49913).

Little Gasparilla Pass, Florida; Mar. 17, 1889; U. S. Fish Comm.; 1 male (20099).

Long Key, Florida; Jan. 29, 1903; *Fish Hawk;* 1 female (43290).

Clearwater, Florida; July 14, 1879; S. T. Walker; 1 female (3260).

Santa Rosa Island, Pensacola, Florida; outer beach; July 25, 1893; J. E. Benedict; 1 male (17925).

Cameron, Louisiana; R. P. Cowles; 1 male (30568).

Matagorda peninsula near *Fish Hawk* anchorage, Texas; Jan. 10, 1905; *Fish Hawk;* 1 specimen (33034).

Padre Island, Texas; Aug. 27, 1891; Wm. Lloyd; 1 male (17705).

Texas; Aug. 16, 1891; Wm. Lloyd; 1 male (17706).

Tampico, Mexico; A. Dugès; 1 male (18688).

La Barra, near Tampico, Mexico; Feb. 7, 1910; E. Palmer; 9 y. (43291).

Mujeres Island, Yucatan; April, 1901; Nelson and Goldman; 2 males (25235).

Cozumel, Yucatan; seine; Jan. 29, 1885; *Albatross;* 1 male (9894).

Near Belize, British Honduras; W. A. Stanton; 1 male (21372).

British Honduras; W. A. Stanton; 2 males (22599).

Swan Islands, Caribbean Sea; February, 1887; C. H. Townsend; 3 males, 1 female (17649).

Old Providence; Apr. 4–9, 1889; *Albatross;* 2 males, 4 females, 2 y. (7531).

Greytown, Nicaragua; Feb. 10, 1892; C. W. Richmond; 5 males (17651).

Toro Point, Canal Zone, Panama; May 19, 1911; Meek and Hildebrand; 1 male (43991).

Sabanilla, Colombia; March 16–22, 1884; *Albatross;* 7 males (17652).

Santa Marta, Colombia; C. F. Baker; 1 female (22550).

Pernambuco, Brazil; 1876–1877; R. Rathbun; 1 male, 2 females (40600). Pernambuco?; 1876–1877; R. Rathbun; 1 y. (40599).

Fernando de Noronha, Brazil; 1876–1877; R. Rathbun; 1 male, 1 female (40602).

Maceio, Alagoas, Brazil; July 25, 1899; Branner-Agassiz Exped.; on sand beach; 1 male (25696); on coral reef; 1 male (25696).

Bahia, Brazil; July 12, 1874; I. Russell; 1 female (13858).

Bahia, Brazil; Dec. 23, 1887; *Albatross;* 1 male (22075).

Itaparica, Bahia, Brazil; April, 1876; R. Rathbun; 1 y. (40597).

Mar Grande, Bahia, Brazil; 1876-1877; R. Rathbun; 1 male, 1 female (40598).

Caravellas, Bahia, Brazil; 1876-1877; R. Rathbun; 1 male (40601).

Abrolhos Islands, Brazil; Dec. 27, 1887; *Albatross;* 1 male (22076).

Rio de Janeiro, Brazil; U. S. Exploring Exped.; 2 males, 1 female (2362).

Ilha Grande, Rio de Janeiro; July, 1915; J. H. Rose; 1 female (48300).

Matuba, São Paulo, Brazil; 1905; E. Garbe; 1 male (47837).

Dog Island, Nassau, Bahamas; June 20, 1903; B. A. Bean; 1 female (31043).

Green Cay, Bahamas; Apr. 12, 1886; *Albatross;* 3 males (11367).

Watling Island, Bahamas; 1886; *Albatross;* 1 male (17679).

Cabañas, Cuba; June 5, 1900; Wm. Palmer and J. H. Riley; 2 y. (23820).

Mariel, Cuba; sandy shore; May 10, 1900; Wm. Palmer and J. H. Riley; 3 females (23819).

Cuba; 1914; Bartsch and Henderson, *Tomas Barrera* Exped.: On reef flat between Cayo Hutia and little Cayo, N. E. of Light; May; 11 males, 3 females (48385). Ensenada de Santa Rosa; 1-3 fathoms; S. Sh. M. Sponge; May; 1 male (48391). Reef Lavesos Italienos, off Cayo Lavesos; 2-3 fathoms; co. S. R.; June 2; 1 male (48397). Bahia Honda; June 7; 1 male (49159).

Montego Bay, Jamaica; C. B. Wilson; small brackish pond near shore; June 23, 1910; 1 male (42904); shore; 2 males, 1 female (42905).

Kingston Harbor, Jamaica: *Albatross;* Mar. 1-11, 1884; 2 y. (7695). T. H. Morgan; 1 male, 1 female (17221).

Porto Rico; 1899; *Fish Hawk:* Aguadilla; Jan. 18; 1 y. (24527). Playa de Ponce; Jan. 31; 1 y. (24529). San Juan; Jan. 11 and 12; 1 male, 2 y. (24526). Hucares; Feb. 14; 1 y. (24531). Ensenada Honda, Culebra; Feb. 10; 2 males (24530).

Santo Domingo, West Indies: 1877; Chas. A. Fraser; 1 female (3295). 1878; W. M. Gabb; 1 female y. (3203).

St. Thomas; 1915; C. R. Shoemaker: Magens Bay; beach, lives in burrows in sand; July 4; 2 males (49803). East side St. Thomas Harbor; beach; July 11; 4 young (49840). Drift Bay, Water Island; beach; July 17; 1 male, 2 females (49804). Drift Bay; beach; 1 male (49802).

Port Castries, St. Lucia; 1887; *Albatross:* Dec. 2; 1 female, 1 y. (22074). Dec. 3; 8 males, 3 females (40797).

Hungry Bay, Bermudas; July-Sept.; F. G. Gosling; 1 male, 1 female (25434).

Locality not given; 1 male (49251).

OCYPODE OCCIDENTALIS Stimpson.

Plate 129, figs. 2 and 3.

Ocypoda occidentalis STIMPSON, Ann. Lyc. Nat. Hist. New York, vol. 7, 1860, p. 229 (type-locality, Cape St. Lucas; cotypes, Cat. No. 2016, U.S.N.M.; cotypes also in M. C. Z. and, probably, in Brit. Mus.).

Ocypoda gaudichaudii ? LOCKINGTON, not Milne Edwards and Lucas, Proc. California Acad. Sci., vol. 7, 1876 (1877), p. 145.

Ocypoda kuhlii, var. *occidentalis* MIERS, Ann. Mag. Nat. Hist., ser. 5, vol. 10, 1882, p. 386.

? *Ocypoda urvillei* DOFLEIN, Sitzungsber. k. bayer. Akad. Wiss., math.-phys. Cl., vol. 29, 1899, p. 189, Mollendo, Peru; probably not *O. urvillei* Milne Edwards.[1]

Diagnosis.—Outer angle of orbit about as advanced as the front. Eyes without style. Fingers pointed.

Description.—Closely allied to *O. albicans*, from which it is distinguished by the following characters:

The carapace is narrower, the greatest breadth being only a sixth more than its greatest length.

Granulation of carapace more uniform and coarser.

Outer angles of orbit more produced, being nearly or quite as advanced as the front.

Arm and wrist rougher; teeth on lower margin of propodus more numerous but less prominent. Stridulating ridge longer, its tubercles from 17 to 20.

Ambulatory legs shorter, second pair not two and one-half times as long as carapace, fourth pair reaching to middle of propodus of third pair. Merus of all the legs as well as carpus and propodus of last pair naked. Hair of carpus and propodus of first three pairs short and furry, and lacking on the lower margins. Margins of merus joints denticulate.

Color.—Darkish, and can be distinguished very well on the dry sand, but its color is rather confused with the sand when wet.

Measurements.—Length of carapace of male cotype 43.2, width of same 50.5 mm.

Habits.—When pursued they run to and fro till they find themselves on soft sand, when they hide themselves quickly, diving a little into the sand and confounding themselves with it to such an extent that it is difficult to distinguish them. (Tristan, in letter.)

Range.—From Turtle Bay, west coast Lower California, to Peru.

Material examined.—

Turtle Bay, Lower California; July 30, 1896; A. W. Anthony; 4 males, 4 females (19511).

---

[1] Doflein had but a young specimen, and the young of *occidentalis* resembles that of *ceratophthalma*—*urvillei*.

Cape St. Lucas, Lower California: John Xantus; 33 specimens (cotypes) (2016). Wm. Stimpson; 1 male (2406).
San Jose del Cabo, Lower California; 1 male, 2 y. (20701).
West coast, Lower California; A. W. Anthony; 1 male (20702).
Tres Marias Islands, Mexico; 1897; Nelson and Goldman: Maria Magdalena; May 28; 1 female (20648). Maria Cleofa; May 30; 1 male (20649).
Acapulco, Mexico (M. C. Z.).
Salina Cruz, Mexico; on sand beach; December 25, 1898; Chas. A. Deam; 2 specimens (32252).
Boca del Rio Jesus Maria, Costa Rica; April, 1905; P. Biolley and J. F. Tristan; 1 male (32313).
Pigres, Costa Rica; A. Zeledon; 18 y. (43294); 1 male (43293).
Puntarenas (Estero-side), Costa Rica; February, 1905; J. F. Tristan; 1 male (32283).
Ancon, Peru; Geo. Keifer; 2 males (9236).

OCYPODE GAUDICHAUDII Milne Edwards and Lucas.

CARRETERO OR CART-DRIVER (Coker).

Plate 129, fig. 1; plate 130, fig. 1.

Ocypoda gaudichaudii MILNE EDWARDS and LUCAS, d'Orbigny's Voy. dans l'Amér. Mérid., vol. 6, 1843, Crust., p. 26; vol. 9, atlas, 1847, pl. 11, figs. 4–4b (type-locality, Chile; type in Paris Mus.).

Diagnosis.—Eyestalks forming a style beyond the eyes. Fingers truncate.

Description.—Greatest breadth at antero-lateral angles, two-sevenths more than greatest length. Lateral borders sinuous. Grannlation coarser on the sides and along the front. Upper margin of orbit sinuous at inner end, then sloping straight backward, its outer angle directed outward, not at all forward. A notch in outer sinus of lower margin. Eyes prolonged beyond the orbit in a blunt-pointed style of variable length, but not exceeding the length of the eye.

Chelipeds decidedly roughened throughout. Tubercles of palm rather small, numerous. Fingers broadly truncated, terminal edge horny. The stridulating ridge occupies two-thirds the width of the palm; its upper half is composed of small tubercles, gradually changing to striae which are very crowded on lower half. There are about 12 round tubercles, and about 45 striae. The complementary ridge occupies two-thirds the length of the ischium. Length of ambulatory legs about as in *O. occidentalis;* surface rather finely rugose, edges denticulate, the lower edge of the merus joints most distinctly so. Merus and carpus naked. Dactyli and upper half of anterior surface of propodi short-hairy.

In other respects resembling *O. albicans.*

Color.—Yellowish.

Measurements.—Length of carapace of male (17297) 31.7; width of same 40.5 mm.

Habits.—The following notes were furnished by Mr. J. Fid Tristan of Costa Rica:

They live on the seashore where they dig holes in the sand. Their yellowish color is more or less confused with the color of the sand. The rising tide entirely covers the holes with sand; at low tide the crabs begin to come out sideways, remaining on the edge of the holes and hiding very quickly at the slightest noise. If anything falls near them they jump at it with extraordinary rapidity, as a spider in its web, and try to secure it with their claws so as to carry it off quickly to their holes.

Range.—Gulf of Fonseca, Salvador, to Chile; Galapagos Islands. San Pablo, California (Mus. Comp. Zoöl.). Also said by Cano to occur at Honolulu.

Material examined.

Gulf of Fonseca, Salvador (M. C. Z.).

Corinto, Nicaragua (M. C. Z.).

Boca del Rio Jesus Maria, Costa Rica; April 1905; P. Biolley and J. F. Tristan; 1 female (32312).

Pigres, Costa Rica; Mar. 3, 1905; A. Zeledon; 3 females (43292).

Taboga Island, Panama; June, 1914; J. Zetek; 5 males (48790).

Panama: Mar. 15, 1888; *Albatross;* 1 female (22077). H. A. Ward; 2 males (17297).

Bahia, Ecuador; beach; Aug. 26, 1884; Dr. W. H. Jones, U. S. Navy; 9 males (13871).

Manta, Ecuador; sandy beach; Aug. 28, 1884; Dr. W. H. Jones. U. S. Navy; 1 male (13872).

Guayaquil, Ecuador; Dr. W. H. Jones, U. S. Navy; 1 male, 1 female (14094).

Peru; 1906–1908; R. E. Coker: Las Vacas, near Capon; on beach; Jan. 23, 1908; 3 males, 1 female (40457). Lobos de Afuera; beach; Mar. 30; 1 male, 1 female (40456). Chimbote; Feb. 27; sand beach; 2 males, 1 female (40458).

Chatham Island, Galapagos Islands; *Albatross:* Apr. 4, 1888; 1 female (22078). Jan. 8, 1905; 2 females (33163).

Valparaiso, Chile; 2 young (M. C. Z.).

Locality?; *Albatross;* 7 males, 6 females (22079).

Genus UCA Leach.

FIDDLER CRABS; CALLING CRABS.

Uca LEACH, Edin. Encyc., vol. 7, 1814. p. 430; type, *U. una* Leach, 1814= *U. heterochelos* (Lamarck), 1801.—RATHBUN, Proc. Biol. Soc. Washington, vol. 11, 1897, p. 154.

Gelasimus LATREILLE, Nouv. Dict. Hist. Nat., vol. 12, 1817, p. 517; type *G. maracoani* (Latreille).

Acanthoplax MILNE EDWARDS, Ann. Sci. Nat., ser. 3, Zool., vol. 18, 1852, p. 151 [115]; type, *A. insignis* Milne Edwards.
Eurychelus L. AGASSIZ, MS. label; type, *E. monilifer* L. Agassiz, MS.

Carapace deep, subquadrilateral, with the antero-lateral angles more or less produced, and the lateral borders more or less convergent posteriorly, occasionally subhexagonal, broader than long, the regions never very strongly defined, surface usually smooth, sometimes very convex. The female is usually narrower than the male and sometimes much rougher. The antero-lateral angles are very variable in individuals of the same species and in the two sides of the same individual. This irregularity is most noticeable in the males, the side of the carapace which bears the large claw being most strongly developed.

Front a deflexed lobe varying in width from one-twentieth to one-third of the width of the carapace and in form from narrow spatulate to broad arcuate.

The orbits occupy the remaining width of the carapace, and are deep, more or less sinuous, and oblique; lower margin cut into numerous small lobes more or less truncate, upper margin double, the intervening space broad, shallow, more or less triangular; this surface I have called the eyebrow.

Antennulae very small; antennae of good size.

Epistome short, distinct. Buccal cavity broader than long, side margins convex, completely closed, except for a chink anteriorly, by the outer maxillipeds, of which the merus is somewhat oblique and bears the palpus at its antero-external angle.

The chelipeds of the male are extraordinarily unequal, of the female small and equal. The large cheliped of the male is heavy, the merus projects beyond the body, the carpus is subquadrilateral in its external outline, the propodus and dactylus elongate and often enormously developed. The small cheliped of the male and both chelipeds of the female are similar to one another and shorter than the ambulatory legs and bear spoon-shaped fingers. Legs flattened, more or less hairy, the last pair smaller than the rest. The merus joints are sometimes wider in the female than in the male. A hairy-edged pouch leads into the branchial cavity between the bases of the second and third pairs.

Abdomen of male usually, of female always, composed of seven distinct segments. In the male three or four segments may be coalesced.[1]

---

[1] For accounts of the behavior of various species of fiddlers, see Fritz Müller, Facts and Arguments for Darwin (translation), London, 1869, pp. 19, 36, etc.; also Kosmos, vol. 8, 1831, p. 472 (color change); J. S. Kingsley, Something about Crabs, in Amer. Nat., vol. 22, 1888, pp. 888-896; A. S. Pearse, Habits of Fiddler Crabs, in Philippine Jour. Sci., vol. 7, 1912, pp. 113-133, also Smithsonian Report for 1913 (1914), pp. 415-428.

Distributed from Atlantic America (Boston to Uruguay) eastward, including the Mediterranean, to Pacific America (San Diego, California, to Chile).

KEY TO THE AMERICAN SPECIES OF THE GENUS UCA.

A^1. Front between the eyes narrow and spatuliform, less than one-tenth width of carapace.
  B^1. Eyes without terminal stylets.
    C^1. Lateral margins either granulate or unarmed.
      D^1. Dactylus of large claw of male wider in the middle than at the proximal end.
        E^1. Surface of fingers rough _____ _maracoani_, p. 378
        E^2. Surface of fingers smooth _____ _monilifera_, p. 380.
      D^2. Dactylus of large claw of male more slender, and tapering from the proximal end to the tip (except for the tooth at the middle of its prehensile edge).
        E^1. Fingers of large claw of male gaping to the extremity. Carapace broad, width one and four-fifths times its length.
                heterochelos, p. 381.
        E^2. Fingers of large claw of male partially in contact in their terminal half. Carapace narrower, width less than one and four-fifths times its length _____ _princeps_, p. 382.
    C^2. Lateral margins armed with large spiniform tubercles in both sexes. Merus joints of legs armed with tubercles or spines___ _insignis_, p. 385.
  B^2. Eyes with terminal stylets.
    C^1. Oblique ridge inside palm continued to upper margin. Ocular stylet longer than peduncle _____ _stylifera_, p. 383.
    C^2. Oblique ridge inside palm ending at carpal cavity. Ocular stylet as long as cornea _____ _heteropleura_, p. 385.
A^2. Front between the eyes wider, increasing in width from below upward, posteriorly more than one-tenth width of carapace.
  B^1. Carapace not semicylindrical.
    C^1. Anterior part of side margins convex and curving gradually backward.
      D^1. An oblique tuberculate ridge on inner surface of larger palm of male extending upward from lower margin.
        E^1. Front less than one-fifth width of carapace.
          F^1. Carapace coarsely granulate or tuberculate. Immovable finger flat, lower edge acute _____ _tangeri_, p. 387.
          F^2. Carapace nearly smooth. Immovable finger convex, lower edge rounded.
            G^1. Carapace strongly convergent posteriorly _____ _thayeri_, p. 406.
            G^2. Carapace moderately convergent posteriorly __ _rectilatus_, p. 405.
        E^2. Front more than one-fifth width of carapace.
          F^1. Front wide, at least one-third of fronto-orbital width.
            G^1. Oblique ridge inside palm a single row of tubercles; the two rows of tubercles at base of dactylus are parallel and near together. Eyebrow nearly vertical and scarcely visible in a dorsal view _____ _minax_, p. 389.
            G^2. Oblique ridge several tubercles in width; the two rows of tubercles at base of dactylus diverge from below upward and leave a considerable space between them. Eyebrow inclined and visible in a dorsal view _____ _mordax_, p. 391.

F^2. Front narrower, less than one-third of fronto-orbital width.
    G^1. Oblique tuberculate ridge on inner surface of larger palm of male terminating at carpal cavity.
        H^1. Eyebrow nearly vertical, scarcely visible in a dorsal view. Space between carpal cavity and dactylus coarsely granulate_____*pugnax*, p. 395.
        H^2. Eyebrow inclined, visible in a dorsal view. Space between carpal cavity and dactylus finely granulate.
                                                      pugnax rapax, p. 397.
    G^2. Oblique tuberculate ridge not terminating at carpal cavity, but continued upward with an angular turn.
        H^1. Side margins strongly convergent posteriorly.
                                                macrodactylus, p. 404.
        H^2. Side margins moderately convergent.
            J^1. Merus joints of legs not enlarged_____*brevifrons*. p. 393.
            J^2. Merus joints of legs enlarged_____*galapagensis*, p. 403.
  D^2. No oblique tuberculate ridge on inner surface of palm, extending upward from lower margin_____*pugilator*, p. 400.
C^2. Anterior part of side margins straight, or nearly so, and continued thence backward with an angular turn.
  D^1. An oblique tuberculate ridge on inner surface of palm extending upward from lower margin.
    E^1. Oblique ridge terminating at carpal cavity_____*uruguayensis*, p. 413.
    E^2. Oblique ridge not terminating at carpal cavity but continued upward with an angular turn.
        F^1. A spine or tooth on inner surface of carpus. Oblique ridge very prominent_____*spinicarpa*, p. 411.
        F^2. No spine on inner surface of carpus. Oblique ridge moderately prominent.
            G^1. Oblique ridge and proximal ridge at base of dactylus meeting above_____*oerstedi*, p. 414.
            G^2. Oblique ridge and proximal ridge at base of dactylus not meeting.
                H^1. Upper surface of palm turned abruptly at right angles with outer surface and forming a flat or concave surface.
                                          coloradensis, p. 410.
                H^2. Upper surface of palm rounding gradually into outer surface.
                    J^1. Upper margin of palm with a carina.
                        K^1. Proximal ridge at base of dactylus almost at right angles to lower margin_____*crenulata*, p. 409.
                        K^2. Proximal ridge at base of dactylus very oblique to lower margin _____*speciosa*, p. 408.
                    J^2. Upper margin of palm without carina_____*helleri*. p. 415.
  D^2. No oblique tuberculate ridge on inner surface of palm extending upward from lower margin_____*panamensis*, p. 412.
B^2. Carapace very convex, semicylindrical.
  C^1. An oblique tuberculate ridge on inner surface of palm extending upward from lower margin.
    D^1. Third to sixth abdominal segments in male more or less fused.
        E^1. Carapace little narrowed behind.
            F^1. An oblique stridulating ridge near proximal lower corner of inner surface of large palm of male. Upper margin of orbit little oblique_____*musica*, p. 417.

F². No oblique stridulating ridge on inner surface of palm. Upper margin of orbit more obl ... _____ *stenodactylus*, p. 416.
E². Carapace strongly narrowed behind_____ ___*leptodactyla*, p. 420.
D². No abdominal segments fused. Fingers of large cheliped three or four times as long as palm_____*festae*, p. 420.
C². No oblique tuberculate ridge on inner surface of palm extending upward from lower margin.
D¹. Abdomen of male very wide, no segments fused. Antero-lateral angles obtuse and blunt_____ _____*subcylindrica*, p. 419.
D². Abdomen of male moderately wide, third to sixth segments fused. Antero-lateral angles either acute or right angles.

latimanus, p. 422.

ANALOGOUS SPECIES ON OPPOSITE SIDES OF THE CONTINENT.

Atlantic.	Pacific.
maracoani.	*monilifera.*
heterochelos.	*princeps.*
speciosa.	*crenulata.*
subcylindrica.	*stenodactylus.*

Species on both sides of the continent: *mordax.*
Species not determinable:
"*Gelasimus pugilator*" Cano,[1] not (Bosc). Panama.
"*Gelasimus vocator*" Cano.[1] Gulf of Panama.
Uca vocator Doflein.[2] Guayaquil, Ecuador.
Uca vocator, var., Nobili.[3] Esmeraldas, Ecuador.

UCA MARACOANI (Latreille).

Plate 130, figs. 2 and 3; plate 131, fig. 3.

Maracoani MARCGRAVE DE LIEBSTAD, Hist. Rer. Natur. Brazil, 1648, p. 184, text-fig.; Brazil.
Cancer palustris cuniculos sub terra agens SLOANE, Nat. Hist. Jamaica, vol. 2, 1725, p. 260; Jamaica.
Ocypode maracoani LATREILLE, Hist. Nat. Crust., vol. 6 (an 11) 1802–1803, p. 46; Cayenne.
Ocypode heterochelos OLIVIER, Encyc. Méth., Insectes, vol. 8, 1811, p. 417 (part).
Gelasima maracoani LATREILLE, Nouv. Dict. Hist. Nat., vol. 12, 1817, p. 519; Tabl. Encyc. Méth., Crust., vol. 24, 1818, pl. 296, fig. 1.
Gonoplax maracoani LAMARCK, Hist. Nat. Anim. sans Vert., vol. 5, 1818, p. 254.
Gelasimus maracoani MILNE EDWARDS, Ann. Sci. Nat., ser. 3, vol. 18, 1852, p. 144 [108]. pl. 3. figs. 1–1*b*.—KINGSLEY, Proc. Acad. Nat. Sci. Philadelphia, 1880, p. 136, pl. 9, fig. 1 (part); not *G. armatus* Smith, nor *G. natalensis* Wilson, MS.

Diagnosis.—Front very narrow, spatuliform. Movable finger of large claw widest in distal half. Fingers rough. Carapace with dorsal side-margins.

---

[1] Boll. Soc. Nat. Napoli, ser. 1, vol. 3. 1889, p. 233.
[2] Sitzungsb. k. bayer. Akad. Wiss., math.-phys. Cl., vol. 29, 1899, p. 191.
[3] Boll. Mus. Zool. Anat. comp. R. Univ. Torino, vol. 16, No. 415, 1901, p. 49.

Description.—Width of carapace at the antero-lateral angles one and one-half times its greatest length. Carapace strongly convex antero-posteriorly, inter-regional grooves rather deep. Posterior border a little more than half of anterior. True lateral borders moderately convergent. The raised line that bounds the dorsal plane on each side is sinuous and terminates a little in front of the postero-lateral margin and above the base of the last foot. In the male it is finely granulate and ends in a tubercle; in the female it is coarsely bead-granulate throughout. Surface in the male very finely granulate; in the female coarsely so, especially at the sides.

The greatest breadth at the front is one-nineteenth of the width of the carapace; the upper surface of the front, or the space inclosed between the margins of the orbits, is linear.

Orbits almost transverse, edges sinuous, the lower one crenated.

In the large cheliped of the adult male the upper border of the merus is smoothly rounded; inner border armed with blunt, well-separated teeth increasing distally; outer border with a few granules. Wrist almost smooth, very finely wrinkled, a few granules. The propodus is one and three-fourths times as long as the width of the carapace, and three times as long as wide; basal portion covered and margined with coarse tubercles; digital portion covered with pits filled with short hair, and bordered by a raised margin, entire below and granulate above. Inner face of palm with a transverse tuberculate ridge bent at a right angle at the middle. Dactylus broadest at distal third, margins granulate, a tubercle near distal third of prehensile edge; upper margin convex, lower nearly straight. Both fingers are smooth inside and have an inner granulate ridge on their occludent margins. The immovable finger has a third or intermediate row of granules on this margin which projects in a broad lobe at the proximal third. When closed the fingers gape for their basal third, and overlap distally, their tips spiniform and crossing each other. In the male the ambulatory legs are nearly smooth. In the female the merus joints are somewhat tuberculate, their upper margin is distinctly denticulate, and the lower margin is tuberculo-dentate; in the last pair the margins are furred with short hair.

Measurements.—Length of carapace of male (17657) 22, width of same 34 mm.; female (17657), length of carapace 20.5, width of same 29 mm.

Habitat.—Salt marshes (Sloane).

Range.—Jamaica (Sloane). Cayenne to Rio de Janeiro.

Material examined.—

Cayenne. French Guiana; specimens in Copenhagen Mus.

Maranhão, Brazil: Lieutenant F. E. Sawyer, U. S. Navy; 13 males, 5 females (17657). Thayer Exped.; 2 males, 2 females (22190).

Natal, Rio Grande do Norte, Brazil; June 29, 1899; Branner-Agassiz Exped.; 6 males, 1 female (25697).
Plataforma, Bahia, Brazil; 1876–1877; R. Rathbun; 11 males, 5 females (40613, 40614).
Porto Seguro, Brazil; Thayer Exped.; 2 males, 1 female (22191).

UCA MONILIFERA Rathbun.

Plate 132.

Eurychelus monilifer L. AGASSIZ, MS. label.
Uca monilifera RATHBUN, Proc. U. S. Nat. Mus., vol. 47, 1914, p. 126, pl. 9 (type-locality, Guaymas, Mexico; holotype male, Cat. No. 1578, M. C. Z.).

Diagnosis.—Front very narrow, spatuliform. Movable finger of large claw widest at middle. Fingers smooth. Carapace without dorsal side-margins.

Description.—Considerably larger than *maracoani*. There is no raised or granulated line bounding the dorsal plane on either side, but the dorsal rounds smoothly into the lateral surface. As in *maracoani* the anterior margin of the carapace or superior margin of the orbit is transversely sinuous, forming a triangular tooth at the antero-lateral angles, and the front between the eyes is extremely narrow and spatuliform, its median furrow linear and not reaching the broadest part of the spatula.

The lower margin of the orbit is deeply crenated or turreted throughout.

The large cheliped of the male is much smoother than in the Atlantic form. The inner border of the arm has a large laminar expansion directed upward, edge arcuate, denticulate. The wrist is more elongate than in *maracoani*. The tubercles of the palm are few and indistinct. Fingers quite smooth except at the margins and at the base of the dactylus, where there are a few depressed tubercles. On the immovable finger there is, as in *maracoani*, a raised line just above the lower margin and continued backward on the palm; there is, however, no broad lobe or tooth on the proximal half of the prehensile edge. The movable finger has a different shape from that of the allied species; while the upper margin is a regular and moderate curve, the prehensile edge is concave for its basal three-eighths, then straight to near the tip, forming a small tooth at the meeting of the two lines; this brings the widest part of the finger just proximal to its middle. In *maracoani*, the widest part of the finger is near its distal third, and the whole of the distal half is much wider than in this species. The spiniform finger tips are not strongly bent in *monilifera*.

The ambulatory legs are almost bare except the dactyli.

The chief difference in the form of the male abdomen of the two species lies in the penultimate segment: in *monilifera* it is less than twice as wide as it is long; in *maracoani* just twice as wide as long.

Measurements.—Holotype male, length of carapace 28.7, width of same at antero-lateral angles 45.4 mm.

Range.—West coast of Mexico.

Material examined.—

Montague Island, mouth of Colorado River, Mexico; May 18, 1915; E. A. Preble, Biol. Survey, U. S. Dept. of Agriculture; 1 male (48829).

Guaymas, Mexico; 1859; Capt. C. P. Stone, U. S. Navy; 1 male holotype (1578, M. C. Z.), 1 male paratype (22180).

UCA HETEROCHELOS (Lamarck).

Plate 131, figs. 1 and 2.

Cancer Uka una, Brasiliensis SEBA, Thesaurus, vol. 3, 1758, p. 44, pl. 18, fig. 8 (inaccurate); Brazil.

Cancer vocans major HERBST, Natur. Krabben u. Krebse, p. 83, 1782, pl. 1, fig. 11 (after Seba); Bahamas.

Ocypoda heterochelos LAMARCK, Syst. Anim. sans Vert., 1801, p. 150 (no locality given; type in Paris Mus.?); refers to figures of Seba and Herbst.

Ocypoda heterochelos BOSC, Hist. Nat. Crust., vol. 1, an 10 [1801–1802], p. 197, "*l'Amérique méridionale.*"

Cancer uka SHAW, Natur. Misc., vol. 14, 1802, pl. 588, colored (after Seba) and accompanying text (type-locality, Brazil); *Cancer uca* in index. Not *Cancer vca* (*uca*) Linnaeus, 1767.

Ocypode heterochelos OLIVIER, Encyc. Méth., Insectes, vol. 8, 1811, p. 417 (part).

Uca una LEACH, Edin. Encyc., vol. 7, 1814, p. 430.

Gelasimus maracoani DESMAREST, Consid. Crust., 1825, p. 123 (part).

Gelasimus platydactylus MILNE EDWARDS, Hist. Nat. Crust., vol. 2, 1837, p. 51 (type-locality, Cayenne; type in Paris Mus.); Ann. Sci. Nat., ser. 3, Zool., vol. 18, 1852, p. 144 [108], pl. 3, figs. 2–2b.

Gelasimus grangeri DESBONNE, in Desbonne und Schramm, Crust. de la Guadeloupe, 1867, p. 45 (type-locality, Guadeloupe, in the Canal; type probably not extant).

Gelasimus heterochelcs KINGSLEY, Proc. Acad. Nat. Sci. Philadelphia, 1880, p. 137, pl. 9, fig. 2 (part).

Uca platydactyla ORTMANN, Zool. Jahrb., Syst., vol. 10, 1897, p. 346 (part).

Diagnosis.—Front very narrow, spatuliform. Movable finger of large claw tapering. Fingers gaping throughout. Wrist and hand granulate or finely tuberculate.

Description.—Allied to *U. maracoani*. Carapace wide, width of male one and four-fifths times length. The raised line that bounds the dorsal plane on each side is bent transversely inward at the posterior end, and is more coarsely granulated in the female than in the male; the latter is true also of the dorsal surface.

Front one-twentieth of the width of the carapace. Orbits more oblique than in *U. maracoani;* lower margin turreted.

Upper border of arm of large cheliped of adult male rounded, sparsely granulate, inner border limbed, limb dentate at the distal end; outer border tuberculo-dentate. Wrist coarsely granulate, inner edge dentate. The propodus is one and one-half times as long as the width of the carapace, and nearly three times as long as wide; palmar portion covered with low tubercles or coarse granules, which become reduced on the fingers to granules not distinguishable by the naked eye. The pollex has a submarginal, besides a marginal, row of granules, which extends half the length of the palm; width of pollex uniform for more than its basal half, then tapering to the slightly curved tip; the middle row of prehensile tubercles is replaced by teeth on this terminal portion, the first tooth the largest. The dactylus is regularly tapering and has a tooth just posterior to the middle of its prehensile edge. Fingers gaping, gape diminishing to the overlapping tips. The ambulatory legs of the female differ from those of the male in having the surface and margins rougher and in the upper and lower margins of the merus joint and the upper margin of the carpus and half the propodus of the last pair furred with short hair.

Measurements.—Male (11375), length of carapace 19, width of same 34 mm.; female (11375), length of carapace 13.5, width of same 23.3 mm.

Habitat.—In mangrove marshes (Desbonne).

Range.—Bahamas and Cuba to Brazil.

Material examined.—

Watling Island, Bahamas; 1886; *Albatross;* 6 males, 1 female (11375).

Baracoa, E. Cuba; tidal flat; Jan. 29, 1902; William Palmer; 6 males, 1 female (25549).

Jamaica; E. A. Andrews; 1 male (42918)

Other records.—Mexico (Kingsley); Guadeloupe (Desbonne); Cayenne (Milne Edwards); Brazil (Seba, Shaw).

UCA PRINCEPS (Smith).
MAESTRO-SASTRE OR MASTER-TAILOR (Coker).
Plate 133; plate 160, fig. 6.

Gelasimus princeps SMITH, Trans. Connecticut Acad. Arts and Sci., vol. 2, 1870, p. 120, pl. 2, fig. 10, pl. 3, figs. 3–3c (type-locality, Corinto, west coast of Nicaragua; type in Mus. Comp. Zoöl.).—LOCKINGTON, Proc. California Acad. Sci., vol. 7, 1876 (1877), p. 146 [2].

?*Gelasimus platydactylus* SAUSSURE, Rev. et Mag. Zool., ser. 2, vol. 5, 1853, p. 362.

Uca platydactyla ORTMANN, Zool. Jahrb., Syst., vol. 10, 1897, p. 346 (part).

Gelasimus heterochelos KINGSLEY, Proc. Acad. Nat. Sci. Philadelphia, 1880, p. 137 (part).

Diagnosis.—Front very narrow, spatuliform. Movable finger of large claw tapering. Fingers gaping in basal half. Wrist and hand coarsely tuberculate.

Description.—A larger species than *U. heterochelos*. Carapace narrower, width less than one and four-fifths times length. Upper margin of orbit more sinuous, but less oblique.

Inner border of arm of large cheliped of adult male strongly dentate. Outer surface of wrist and hand coarsely tuberculate, much more so than in *U. heterochelos*. While the surface of the pollex is smooth, except on the margins, the dactylus is more or less tuberculate on its upper half, the tubercles becoming smaller distally. When the fingers are closed, they gape for only about their basal half.

The upper margin of the merus, carpus, and two-thirds of the propodus of the last leg of the female is furred with hair.

Measurements.—Length of carapace of male (20689) 25.5, width of same 42.5 mm. Length of carapace of female (20689) 24.2, width of same 38 mm.

Habitat.—Inhabits salt marshes; also found under rocks at low tide.

Range.—From San Bartolome Bay, Lower California, to Peru.

Material examined.—

Abreojos Point, Lower California; salt marsh; April 18, 1897; A. W. Anthony; 6 males, 3 females (20689).

La Paz, Lower California; L. Belding; 1 male, 2 females, 1 y. (14826).

San Blas, Tepic, Mexico; June 14, 1897; Nelson and Goldman; Biol. Survey, U. S. Dept. Agriculture; 1 male, 3 females (20654).

Corinto, Nicaragua; J. A. McNiel; 12 cotypes (M. C. Z.).

Punta Arenas (ocean side), Costa Rica; February, 1907; P. Biolley; 1 male, y. (39099).

Santo Domingo, Gulf of Dulce, Costa Rica; April, 1896; H. Pittier; 1 male (19440).

Puerto Grande on the Rio Zarumilla (2 leagues from Capon), Peru; on salt flats; February 2, 1908; R. E. Coker; 1 male, received from Peruvian Government (40468).

Back of Chulliyache (on Bay of Sechura), Peru; salt marshes; R. E. Coker; 3 males, received from Peruvian Government (40467).

UCA STYLIFERA (Milne Edwards).

Plate 134, figs. 1 and 2.

Gelasimus platydactylus MILNE EDWARDS, Règne Anim. Cuvier, disciples ed., Crust., pl. 18, fig. 1*a* (not Milne Edwards, 1837).

Gelasimus styliferus MILNE EDWARDS, Ann. Sci. Nat., ser. 3, Zool., vol. 18, 1852, p. 145 [109], pl. 3, fig. 3 (type-locality, Guayaquil; type in Paris Mus.).

Gelasimus heterophthalmus SMITH, Trans. Connecticut Acad. Arts and Sci., vol. 2, 1870, p. 116, pl. 2, figs. 6 and 6a; pl. 3, figs. 1–1b (type-locality, Gulf of Fonseca; type in Mus. Comp. Zoöl.).

Uca platydactyla, var. *stylifera* ORTMANN, Zool. Jahrb., Syst., vol. 10, 1897, p. 347 (part).

Diagnosis.—Front very narrow, spatuliform. Ocular stylet present, longer than peduncle. Oblique ridge on inner surface of palm continued by a right angle to upper margin. Pollex with strong tooth on basal half.

Description.—Width of carapace from one and two-fifths to one and three-fourths times its length. Dorsal surface smooth and shining, convex longitudinally but not at all laterally. Sulci shallow. Superior border of orbit subentire, strongly arcuate in the middle, outer angle strongly produced laterally on the side of the large cheliped but not on the other. Only the posterior part of the lateral margin is indicated by a faint, granulous line. Front spatulate; the upper margins of the orbits form a narrow, triangular median interspace. Inferior border of the orbit denticulate with shallow, flattened, truncate teeth. The ocular peduncle on the side of the large cheliped is terminated beyond the cornea by a very slender filiform stylet much longer than the peduncle itself, and slightly flattened and expanded at the tip. No trace of a terminal stylet on the peduncle of the other side.

Inner margin of merus of larger cheliped armed with a low crest, which may be slightly dentate at its distal extremity; outer margin sharp and denticulate. Carpus tuberculous. Basal portion of propodus coarsely and densely tuberculous, margins thin and dentate; inner surface smooth, with a rectangular, transverse ridge. Fingers long, compressed and high, smooth inside, prehensile edges evenly tuberculated and each armed with a single stout submedian tooth, that on the pollex considerably within that on the dactylus. Fingers gaping in basal half. Outer surface of pollex rough with irregular shallow punctures, lower edge granulous, with a submarginal granulous line. Dactylus smooth outside except near the base, upper edge subentire.

Ambulatory legs smooth and unarmed.

Measurements.—Male (cotype of *heterophthalmus*), length of carapace 18.7, width of same 32.2 mm.

Range.—Gulf of Fonseca, Salvador, to Guayaquil, Ecuador.

Material examined.—

Gulf of Fonseca, Salvador; J. A. McNiel; 4 males, cotypes of *heterophthalmus* (M. C. Z.).

Puntarenas (Pacific Ocean or Estero side), Costa Rica; Feb., 1905; J. F. Tristan; 2 males (32325)

UCA HETEROPLEURA (Smith).

Plate 161, figs. 1–4.

Gelasimus heteropleurus SMITH, Trans. Connecticut Acad. Arts and Sci., vol. 2, 1870, p. 118, pl. 2, fig. 7; pl. 3, figs. 2–2*b* (type-locality, Gulf of Fonseca; type in Mus. Comp. Zoöl.).

Uca platydactyla, var. *stylifera* ORTMANN, Zool. Jahrb., Syst., vol. 10, 1897, p. 347 (part).

Diagnosis.—Front very narrow, spatuliform. Ocular stylet present, as long as cornea. Oblique ridge on inner surface of palm ending at carpal cavity. Pollex without strong tooth on basal half.

Description (after Smith).—Allied to *U. stylifera*. Carapace slightly granulous. Branchial regions separated by deep sulci from gastric and cardiac regions. The antero-lateral angle on the side of the larger cheliped projects outwardly as a prominent obtuse tooth. Lateral margins armed with a very marked line of sharp granules. The ocular peduncle on the side of the larger cheliped terminates in a slender, flattened stylet about as long as the cornea.

Inner and outer margins of merus of large cheliped of male minutely denticulate. Carpus flattened and granulous. Basal portion of propodus coarsely and thickly tuberculous outside, the tubercles larger above; inner surface armed only with the oblique tubercular crest running from the lower margin and terminating below the middle. Fingers but little longer than palm; prehensile edges armed with a tooth a little way from the tip and nearly straight, gaping as far as the teeth. Outer surface of fingers granulous or minutely tuberculous, upper edge of dactylus tuberculous or denticulate.

Abdomen of male more narrowed toward the tip, edges slightly concave.

Otherwise as in *U. stylifera*.

Measurements.—Length of carapace of a cotype male 15.8, width of same 25 mm.

Range.—Known only from the type-locality.

Material examined.—Gulf of Fonseca, Salvador; J. A. McNiel; 1 male cotype (M. C. Z.).

UCA INSIGNIS (Milne Edwards).

Plate 161, figs. 5–15.

Acanthoplax insignis MILNE EDWARDS, Ann. Sci. Nat., ser. 3, Zool., vol. 18, 1852, p. 151, pl. 4, fig. 23 (type-locality, Chili; type in Paris Mus.); Arch. Mus. Hist. Nat. Paris, vol. 7, 1854, p. 162, pl. 11, figs. 1–16.

Gelasimus (Acanthoplax) excellens GERSTAECKER, Arch. f. Naturg., vol. 22, 1856, pt. 1, p. 138 (type-locality, Veragua; type in Berlin Mus.).

Gelasimus armatus SMITH, Trans. Connecticut Acad. Arts and Sci., vol. 2, 1870, p. 123, pl. 2, fig. 5; pl. 3, figs. 4–4*d*, male (type-locality, Gulf of Fonseca; type in Mus. Comp. Zoöl.).

Gelasimus ornatus SMITH, Trans. Connecticut Acad. Arts and Sci., vol. 2, 1870, p. 125, pl. 2, figs. 9–9a; pl. 3, figs. 5–5c, female (type-locality, west coast of Central America; type in Mus. Comp. Zoöl.).

Gelasimus insignis SMITH, Trans. Connecticut Acad. Arts and Sci., vol. 2, 1870, p. 126.

Diagnosis.—Carapace less than one and one-half times as wide as long; in the female tuberculous or granulous; in the male smooth. Side margins of female tuberculous or dentate. Merus of legs tuberculous or spinous.

Description.—Male, carapace one and two-fifths times as wide as long, slightly convex, very little narrowed posteriorly, surface naked and deeply areolated. Cardiac region large and prominent. Branchial regions covered by raised veinlike markings branching from a central trunk. Front small, spatulate. Upper margin of orbit raised, slightly sinuous, antero-lateral angles prominent, the one on the side of the large hand directed strongly outward, the other directed forward. Anterior part of lateral margin longitudinal, so that the carapace is scarcely wider between the antero-lateral angles than a short distance behind them. At the posterior end of this longitudinal portion are two marginal tubercles from which a grannlated line extends to the bases of the posterior legs where there is another small tubercle.

Lower margin of orbit armed with from 15 to 18 slender, compressed and truncated teeth. Jugal regions veined somewhat as the regions above.

Merus of large cheliped smooth and rounded above; inner border armed with slender tubercles, and rising into a low crest distally, outer border armed with scattered tubercles. Carpus with several low tubercles on outer surface, and one or two small tubercles at proximal end of inner margin. Propodus high and lamellar, and exceeding in length twice the length of the carapace. Basal portion short, covered outside with large, depressed, well separated tubercles; inferior margin thin, armed with dentiform tubercles; inner surface with a prominent tubercle near the middle, from which an oblique line of obscure tubercles extends to the lower margin. Digital portion very broad at base, coarsely punctate outside, inferior edge denticulate and margined on the outside, prehensile edge straight except a slight excavation at base, armed with small tubercles and a high, tubercular median ridge, and at the tip with a slender tooth. Dactylus broadest in distal half, upper edge arcuate, denticulate, prehensile edge straight, closing against the pollex except at the base, and armed with three rows of tubercles, and a sharp terminal tooth, tubercle at basal third the largest.

Merus of last three pairs of legs armed on posterior edge with from three to seven spines.

Female.—Carapace narrower than in the male, widest at antero-lateral angles, surface tuberculous, some tubercles on the branchial region large and depressed. Upper border of orbit regularly arcuate, denticulate, lateral angle a slender tooth projecting forward and outward. Curved portion of lateral margin ornamented with 8 to 10 bead-like tubercles. Posterior margin tuberculous. Ambulatory legs each armed with a spine on the ischium, and from one to five spines on the merus.

Measurements.—Male (type of *G. armatus*), length of carapace 25.2, width of same 35.5 mm. Female (type of *G. ornatus*), length of carapace 26.6, width of same 36 mm.

Habitat.—On salt marshes.

Range.—Gulf of Fonseca, Salvador, to Chile.

Material examined.—

Gulf of Fonseca, Salvador; J. A. McNiel; 1 male, holotype of *armatus* (M. C. Z.).

West coast of Central America (probably same locality as the foregoing); J. A. McNeil; 1 female, holotype of *ornatus* (M. C. Z.).

Punta Arenas, Costa Rica; 1 male (Copenhagen Mus.).

Veragua, Panama; 1 male, holotype of *excellens* (Berlin Mus.).

Back of Chulliyache (on Bay of Sechura), Peru; on salt marshes; R. E. Coker; 1 male, received from Peruvian Government (40489).

UCA TANGERI (Eydoux).

Plates 135 and 136.

Gelasimus tangeri Eydoux, Mag. de Zool., 1835, Cl. 7, pl. 14, colored (type-locality, Tanger; type in Mus. Phila. Acad. Sci.).

Gelasimus perlatus Herklots, Addit. Fauna Afr. Occ., 1851, p. 6, pl. 1, fig. 3 (type-locality, near Boutry; type in Leyden Mus.).

Gelasimus cimatodus Rochebrune, Bull. Soc. Philom. Paris, ser. 7, vol. 7, 1882, p. 171 (type-locality, les deux Mamelles & côte des Maringouins pointe des Chameaux; type in Paris Mus.).

Uca tangieri Ortmann, Zool. Jahrb., Syst., vol. 10, 1897, p. 356.

Uca tangeri Rathbun, Proc. U. S. Nat. Mus., vol. 22, 1900, p. 276.

Diagnosis.—Front narrow, but wider than long. Carapace coarsely granulate. Fingers of large cheliped of male thin and flat.

Description.—Regions of carapace marked by deep furrows; surface covered with fine granules mixed with coarse granules or tubercles, most striking on branchial and cardiac regions; surface of front smooth. Front tapering anteriorly, subtruncate, a little rounded at extremity. Lateral margins nearly straight and longitudinal or a little curved for a short distance behind antero-lateral angles, then curving rather abruptly inward and backward; margins formed of alternately large and small granules. Upper margin of orbit very sinuous, marked with small, distant granules; lower margin of eyebrow ill defined; lower margin of orbit deeply turreted. Sides of carapace nearly smooth, under part coarsely tuberculate.

Large cheliped of male very strong; merus, carpus, and manus unevenly tuberculate and granulate; antero-distal angle of merus furnished with a tuft of hair; upper margin of palm armed with a few blunt spines or teeth. Fingers over one and one-half times as long as palm, measured through the middle; immovable finger wider and smoother than dactylus, and widest in the middle, its granules growing regularly larger toward end of finger; outer surface of dactylus a little convex especially proximally, and with two longitudinal rows of coarser granules, one on upper edge and the other above the middle. Fingers with a wide gape in basal half. Inner surface smooth, except for a strong, right-angled ridge at middle of palm and two diverging ridges at base of dactylus.

Merus joints of legs roughened, upper margin serrulate, lower tuberculate; propodi and dactyli with fringed margins.

Carapace of female narrower, sides less converging, lower surface more hairy.

Measurements.—Male (14874), length of carapace 21.5, width of same 31.3 mm. Female (21387), length of carapace 18, width of same 24.1 mm.

Range.—From Portugal to north and west coasts of Africa, as far as Algiers and Angola. The only records of occurrence in America are by Miers,[1] "West Indies (Frazer) in the British Museum collection," and by Kingsley,[2] "(?) Bahia! E. Wilson (Phila. Acad.)."

Material examined.

Dakar, West Africa; May 3, 1892; O. F. Cook; 2 males, 1 female (21387).

Rock Spring, Monrovia; April, 1894; O. F. Cook and G. N. Collins; 4 males, 2 females (20575).

Mouth of Mesurado River, Monrovia; O. F. Cook; 3 males (20669).

Monrovia; April, 1894; O. F. Cook and G. N. Collins; 1 male (20576).

Baya River, Elmina, Ashantee; Nov. 27, 1889; W. H. Brown, U. S. Eclipse Exped. to Africa; 3 males (14872), 2 males, 4 females (14873).

Chinchoxo; Falkenstein, collector; received from Berlin Mus.; 1 male (19551).

St. Paul de Loanda; Dec. 11, 1889; W. H. Brown, U. S. Eclipse Exped. to Africa; 7 males, 4 females (14874), 1 male (14875).

---

[1] Ann. Mag. Nat. Hist., ser. 5, vol. 8, 1881, p. 262.
[2] Proc. Acad. Nat. Sci. Philadelphia, 1880, p. 153.

Mouth of the Kwilu River, West Africa; Falkenstein, collector; received from Berlin Mus.; 1 male (19557).

UCA MINAX (LeConte).

RED-JOINTED FIDDLER CRAB.

Plate 137.

Gelasimus minax LeConte, Proc. Acad. Nat. Sci. Philadelphia, vol. 7, 1855, p. 403 (type-locality, Beesley's Point, N. J.; type in Mus. Phila. Acad. Sci.).
Gelasimus palustris Stimpson, Ann. Lyc. Nat. Hist. New York, vol. 7, 1859, p. 62 (part).
Gelasimus minax Smith, Trans. Connecticut Acad. Arts and Sci., vol. 2, 1870, p. 128, pl. 2, fig. 4; pl. 4, figs. 1–1b.
Uca vocator, var. *minax* Ortmann, Zool. Jahrb., Syst., vol. 10, 1897, p. 353.

Diagnosis.—Oblique ridge inside hand terminating at carpal cavity. Front very broad and shallow. Eyebrow nearly vertical. Red spots at articulations of large cheliped. Hand very rough.

Description.—Carapace widest behind the antero-lateral angles; greatest width less than one and one-half times greatest length. Surface convex in both directions, very finely granulate, except near the antero-lateral angle where it is coarsely so. H-form depression shallow, a horizontal depression behind the orbit; lateral portions of the dorsal plane covered with a pubescence easily rubbed off. Front considerably more than one-third the fronto-orbital width, very broadly convex. Upper margin of orbit sloping strongly backward with a slight sinuosity to the obtuse antero-lateral angle.

Superior lateral margins strongly converging in a very sinuous line to the level of the middle of the cardiac region. Posterior and parallel to this line is a similar but short ridge.

The eyebrow is about 5 times as wide as it is deep, and is nearly vertical. Lower margin of orbit armed with subtruncate, separated teeth.

Upper surface of arm of large cheliped sparingly granulate; inner edge denticulate, outer edge granulate. Wrist tuberculate. Outer face of palm tuberculate on the upper two-thirds, the tubercles diminishing to granules below; in shape subtriangular, as the upper portion is bent over almost horizontally. Inner surface with an oblique single row of tubercles from the lower margin to the carpal cavity; also a short parallel row leading down from the ridge on the proximal half of the upper margin; between the ridges a tuberculated area. The two rows of tubercles at the base of the dactylus are parallel and near together. Fingers one and a half times as long as palm, smooth to naked eye, prehensile edges armed with three rows of irregular tubercles, the largest one at the middle of the pollex.

Dactylus much longer than the pollex and curving downward past the tip of the latter, leaving a wide gape throughout the entire length.

Ambulatory legs long-hairy, merus-joints wrinkled and in the first three pairs enlarged toward the middle.

Female narrower than male.

Color.—The three articulations of the large joints of the greater claw have a red spot in life.

Measurements.—Male (12781), length of carapace 25.2, width of same 38 mm.

Habits.—Lives in marshes of estuaries of rivers and brooks, in brackish or even quite fresh water. Abundant in the mangroves (Colombia, Pearse). According to Fowler [1]—

It is a vegetarian, feeding on the algae which grows in muddy salt-marshes. This is often a minute-green algoid plant covering the surface of the mud. The male uses its small claw exclusively in obtaining its food and conveying it to the mouth. The female uses either of her small ones indifferently. In enlarging its burrows the crabs were seen to scrape off the mud from the inside by means of the claws of the ambulatory legs, and having formed the mud into a pellet, pushed it up out of the hole by means of the elbow joint at the base of the great claw, when this is folded down. The crab was also found to construct a regular oven-like arch of mud over the mouth of its burrow. This archway is horizontal and large enough to contain the crab, who quietly sits in this curious doorway on the outlook for his enemies of all kinds.

Range.—From Wareham, Massachusetts, to Texas. Colombia (Pearse).

Material examined.—

Wareham, Massachusetts; July 21, 1887; U. S. Fish Comm.; 3 males (12781).

Seekonk River, Rhode Island; F. P. Drowne; low tide near brackish water under stones; 2 males, 1 female (19569); burrows on shore of river, burrows in groups and located near eel grass; 2 males, 1 female (19570).

Newport, Rhode Island; 1880; U. S. Fish Comm.; 3 males (4909).

Chesapeake Beach, Maryland; found in marsh grass; July 4, 1914; C. R. Shoemaker; 2 females (1 ovig.) (49106).

Point Lookout, Maryland; J. E. Benedict; 3 males, 1 female (16159).

St. Georges Island, Maryland; June 30–July 6, 1890; H. M. Smith; 1 male (22807).

North Carolina; John W. Hays, jr.; 1 male (13890).

Winyah Bay, South Carolina; Jan., 1891; *Fish Hawk;* 5 males, 1 female (17487).

Bulls Island, South Carolina; fresh pools near beach; Mar. 21, 1891; 9 males (17183).

---

[1] Ann. Rept. New Jersey State Museum, 1911 (1912), p. 453.

Charleston, South Carolina; market; 1881; C. C. Leslie; 1 female (4089).
South Carolina; Mar. 22, 1894; *Fish Hawk;* 5 males, 3 females (22279).
Savannah, Georgia; 7 males, 5 females (19497).
Jacksonville, Florida; Apr. 1, 1869; J. A. Allen; 3 males, 3 females (22188).
Sarasota Bay, Florida; Union College Collection; 1 female (42618).
Arlington, Florida; F. C. Goode; 1 female (3178).
Mobile Bay, Alabama; C. F. Baker; 1 male (21081).
Biloxi Bay, Mississippi; July 23, 1898; C. F. Baker; 1 male (21845).
Matagorda Bay, Texas; Mar. 23, 1905; T. E. B. Pope; 13 males, 1 female (33035).

UCA MORDAX (Smith).

Plate 134, figs. 3 and 4.

Gelasimus vocator von MARTENS, Archiv f. Naturg., vol. 35, 1869, pt. 1, p. 6. Probably not *Cancer vocator* Herbst.
Gelasimus mordax SMITH, Trans. Connecticut Acad. Arts and Sci., vol. 2, 1870, p. 135, pl. 2, fig. 3; pl. 4, figs. 4 and 4*a* (type-locality, Canals at Para; type in Mus. Comp. Zoöl.).
Gelasimus vocator KINGSLEY, Proc. Acad. Nat. Sci. Philadelphia, 1880, p. 147 (part).
?*Gelasimus affinis* STREETS, Proc. Acad. Nat. Sci. Philadelphia, 1872, p. 131 (type-locality, Island of St. Martin, West Indies; type not extant).
Uca vocator ORTMANN, Zool. Jahrb., Syst., vol. 10, 1897, p. 352 (part).
Uca minax RATHBUN, not LeConte, Ann. Inst. Jamaica, vol. 1, 1897, p. 27.
? *Uca amazonensis* DOFLEIN, Sitzb. k. bay. Akad. Wiss. München, math.-phys. Cl., vol. 29, 1899, p. 193 (type-locality, Teffé, Amazon River; type in Munich Mus.).
Uca mordax RATHBUN, Proc. U. S. Nat. Mus., vol. 22, 1900, p. 276.

Diagnosis.—Palm elongate, proximally narrow, the upper part turned inward to form an upper surface. Front very shallow and broad. Eyebrow obliquely inclined. Hand slightly roughened.

Description.—Carapace widest behind the antero-lateral angles; greatest width about one and one-half times greatest length. Surface very convex antero-posteriorly, slightly so transversely; antero-lateral regions punctate, punctae filled with hair. H-form depression deep; as is also the semicircular depression which begins behind the orbit and is continued along the antero-lateral margin.

Front at its base one-third of the fronto-orbital width; rounding strongly downward; lower edge very broad and subtruncate. Upper margin of orbit sinuous and almost transverse; antero-lateral angle directed forward, very broadly obtuse. Lateral margins and lower margin of orbit much as in *U. minax.* Eyebrow oblique.

Large cheliped of male rather smooth. Merus with rugose lines on upper and outer surfaces, and fine granules on lower half of inner surface; edges of this surface tuberculo-denticulate. Wrist finely granulate-rugose. Palm elongate, its outer aspect considerably narrowed proximally on account of the palm being strongly bent over to form a broad upper surface; outer upper surface granulate, granules varying in size, but on the whole coarser on the upper half. The oblique ridge on the inner face which leads up to the carpal cavity is broad and usually tuberculated, being several tubercles in width; sometimes the tubercles are obsolete except at the proximal end. Normally this ridge ends at the cavity but sometimes it is continued upward by an angular turn. A more vertical ridge is situated on the upper half and is continued in a curved line to the proximal end of the upper margin. The two rows of tubercles at the base of the dactylus diverge from below upward and leave a considerable space between them. Inner surface granulate, granules coarser at the middle. Fingers not much longer than palm, widely gaping, dactylus longer than pollex and curving past it; pollex obliquely truncate at tip; surface almost smooth, shining, prehensile edges irregularly tuberculate, with the largest tubercle at the middle of the pollex.

FIG. 166.—UCA MORDAX, INNER SURFACE OF LARGE CHELA OF MALE FROM PARA, SLIGHTLY REDUCED. (AFTER SMITH.)

Ambulatory legs hairy; merus joints enlarged (save in last pair) and slightly rugose.

The females are narrower than the males and the carapace is more granulous.

Measurements.—Male (21373), length 21.5, width 32.2 mm.

Remarks.—This species is extremely variable as to the ornamentation of the inside of the palm, but the characters of the carapace, outside of the palm, etc., serve to distinguish it with certainty.

Habitat.—" Common in holes in the clay flat at the mouth of the Manzanares River [Colombia] and at Punta Gruesa in holes in sand, under logs, and among mangrove roots." (Pearse.)

Range.—From the Bahamas and Gulf of Mexico to Rio de Janeiro. West coast of Mexico. Liberia (?).

Material examined.—

Cameron, Louisiana; R. P. Cowles; 1 male (30570).

Tampico, Mexico: Live in the soft mud of the river banks; June 1, 1910; Edward Palmer; 8 males, 1 y. (43353). A. Dugès; 1 male (18689).

Belize, British Honduras; Rev. W. A. Stanton; 11 males, 4 females (21373 and 22604).

Swan Island, Caribbean Sea; C. H. Townsend; 9 males (18430).
Greytown, Nicaragua; Chas. W. Richmond: February 8–9, 1892; 8 males, 2 females (18433). April 9, 1892; 1 male (18434).
Rum Cay, Bahamas; 1886; *Albatross;* about 1500 males, 500 females (11357).
Marianao Playa, Cuba; C. F. Baker; 2 males (31891).
Kingston Harbor, Jamaica; 1893; R. P. Bigelow; 11 males, 10 females (18431).
Fresh water, mouth of Rio Cobre, Jamaica; 1893; R. P. Bigelow; 2 males (18432).
Jamaica: Mar. 1–11, 1884; *Albatross;* 1 male y., 2 females (18553). T. H. Morgan; 3 males, 4 females (22307).
Santo Domingo, West Indies; 1878; W. M. Gabb; 2 males, 1 female (3199).
Porto Rico; 1899; *Fish Hawk:* Mayaguez; fresh water; Jan. 20; 1 male, 1 female (24544). Rio Bayamon above Palo Seco; Jan. 16; 2 females (24543). Hucares; Feb. 14; 4 males (24545).
St. Thomas, West Indies: Jan. 17–24, 1884; *Albatross;* 1 female (18564). East side of harbor; beach; July 11, 1915; C. R. Shoemaker; 7 males, 3 females (49811).
St. Croix, West Indies; Riise, collector; 2 males (19713).
Pernambuco, Brazil; on mangroves; Aug. 1, 1899; Branner-Agassiz Exped.; 1 male (25698).
Itabapuana, Brazil; Thayer Exped.; 4 males, 4 females (22186).
Rio de Janeiro, Brazil; 1876–1877; R. Rathbun; 2 males, 2 females (19971).
Terra de Masabe, State of Rio Janeiro, Brazil; January, 1912; E. Garbe; 2 males (47834).
Liberia; O. F. Cook; 1 female (21847). Identification not certain.
San Blas, Tepic Territory, Mexico; Nelson & Goldman; 4 males (22306).
Acapulco, Mexico; *Hassler* Exped.; 1 specimen (M. C. Z.).

UCA BREVIFRONS (Stimpson).

Plate 138.

Gelasimus brevifrons STIMPSON, Ann. Lyc. Nat. Hist. New York, vol. 7, 1860, p. 292 (type-locality, lagoon at Todos Santos near Cape St. Lucas; type not extant).—LOCKINGTON, Proc. California Acad. Sci., vol. 7, 1876 (1877), p. 147 [3].

Gelasimus vocator KINGSLEY, Proc. Acad. Nat. Sci. Philadelphia, 1880, p. 147 (part).

Uca vocator ORTMANN, Zool. Jahrb., Syst., vol. 10, 1897, p. 352 (part).

Uca brevifrons HOLMES, Proc. California Acad. Sci., ser. 3, vol. 3, 1904, p. 308, pl. 35, figs. 1–5.

Diagnosis.—Oblique ridge inside hand continued toward upper margin. Front very broad and shallow. Merus joints of legs not enlarged.

Description.—Carapace convex, minutely punctate and granulate, branchial regions more or less plainly marked with small tubercles near the outer edge; -shaped impression deep, a faint longitudinal groove on branchial regions; groove behind orbital margins. Front broad, about one-third width of carapace, little produced. Lower edge of eyebrow with a strong convexity midway between the front and lateral angles. These last are almost right angles and slightly or not at all produced forward. Lateral margins of carapace for a short distance nearly parallel, arching slightly outward, and then converging quite strongly to the posterior margin.

Merus of large cheliped of male marked outside with transverse, sparingly setose striae; carpus long, roughened with small blunt tubercles; hand large and stout, nearly twice as long as width of carapace, upper surface with small tubercles which gradually become smaller further down on the outer surface; lower side of the palm marked with a fine raised line which fades out on the base of the pollex. Inner surface of palm marked with a short, very prominent, tuberculated ridge which runs from near the lower margin toward the supero-proximal portion of the hand; from the upper end of this ridge there is a short tuberculated ridge extending toward the base of the dactyl; the prominent lower margin of the carpal groove extends backward from the angle where these ridges meet, in the same direction as the upper ridge; the meeting point of the 3 ridges forms the apex of a large, high, three-sided pyramid. Pollex gently curved, prehensile edge with a few teeth on distal half, tip truncate; dactyl curving from near the middle, overreaching the pollex and bearing a few prehensile teeth. Ambulatory legs slender and hairy. (Abridged from Holmes.)

Color.—Carapace in alcohol dark olivaceous (Holmes).

Measurements.—Male (43352), length of carapace 19.5, width of same 28.6 mm.

Habitat.—Very abundant in the woods 2 kilometers from the seashore (Tristan). Found also in fresh water and at Todos Santos in a lagoon.

Range.—Lower California to Panama.

Material examined.—

Cape St. Lucas, Lower California; John Xantus; 1 female (M. C. Z.).

Acapulco, Mexico; *Hassler* Exped.; 3 males (M. C. Z.).

Boca del Jesus Maria, Costa Rica: Apr. 22, 1905; P. Biolley and J. F. Tristan; 1 male (32323). J. F. Tristan; 1 male, 1 female (32324).

Pigres, Costa Rica; A. Zeledon; 9 males, 1 female (43352).
Quebrada Chavarria, Golfito, Gulf of Dolce, Costa Rica; April, 1896; H. Pittier; 3 males, 4 females (19435).
La Capitana, Canal Zone; H. Pittier; 4 males (43848 and 44320).
Rio Calabre, Panama; Mar. 18, 1911; Meek and Hildebrand; 4 males (43988).
Marraganti, Panama; June 2, 1912; E. A. Goldman; 1 male (48277).
Panama; Sternbergh, collector; 1 male (22185).
Additional records.—Magdalena Bay, Lower California (Lockington). San José del Cabo, Lower California (Holmes).

UCA PUGNAX (Smith).

Plate 139.

Gelasimus vocans GOULD, Rept. Invert. Massachusetts, 1841, p. 325 (part). Not *Cancer vocans* Linnaeus.
Gelasimus vocans, var. A, DE KAY, Nat. Hist. New York, 1844, Crust., p. 14, pl. 6, fig. 10.
Gelasimus pugilator LECONTE, Proc. Acad. Nat. Sci. Philadelphia, vol. 7, 1855, p. 403 (not Bosc).
Gelasimus palustris SMITH, Amer. Nat., vol. 3, 1870, p. 557; male with nearly equal hands.
Gelasimus pugnax SMITH, Trans. Connecticut Acad. Arts and Sci., vol. 2, 1870, p. 131, pl. 2, fig. 1; pl. 4, figs. 2–2d (type-locality, New Haven, locality of specimens figured; type in P. M. Y. U.).
Gelasimus vocator KINGSLEY, Proc. Acad. Nat. Sci. Philadelphia, 1880, p. 147 (part).
Uca vocator ORTMANN, Zool. Jahrb., Syst., vol. 10, 1897, p. 352 (part).
Uca pugnax RATHBUN, Amer. Nat., vol. 34, 1900, p. 585.

Diagnosis.—Oblique ridge inside hand terminating at carpal cavity. Inner surface of hand coarsely granulate. Eyebrow nearly vertical.

Description.—Carapace so convex fore and aft that in a dorsal view the lower edge of the front and the eyebrow are invisible. The most conspicuous depressions are the H-form depression, a pit on the branchial region in line with the gastro-cardiac sulcus, and a pit behind the middle of the orbit. Front about two-sevenths of the fronto-orbital width, margin regularly arched. Upper margin of orbit sinuous and oblique. Antero-lateral angle obtuse, not produced.

Large cheliped of male rough. Merus with granulated, rugose lines outside, lower margin granulate. Carpus and palm tuberculous outside. Oblique ridge leading from lower margin of inner surface of palm tuberculate, but less prominent than in *U. mordax;* space between crests coarsely granulate, or tuberculate. Fingers one and one-half times as long as palm, disposed much as in *U. mordax*, but without a large tubercle at middle of pollex.

Ambulatory legs especially hairy on the carpal and propodal joints.

Color.—Dorsal surface of carapace of male very dark greenish olive, the middle and anterior portion mottled with grayish white; the front, between and above the bases of the ocular peduncles, light blue, varying somewhat in intensity in different specimens, and the anterior margin tinged with brown. Large cheliped lighter than carapace, marked with pale brownish yellow at the articulations and along upper edge of dactylus, and both fingers nearly white along prehensile edges. Exposed portions of ocular peduncles and eyes like dorsal surface of carapace. Smaller cheliped and legs somewhat translucent and thickly mottled and specked with dark grayish olive. Sternum and abdomen mottled ashy gray. The females differ from the males in having dorsal surface of carapace less distinctly mottled with whitish and in wanting the blue on the front. (Smith).[1]

Measurements.—Male (3836), length of carapace 14.8, width of same 22.5 mm.

Habits.—Makes its burrows only upon salt marshes, but is often seen in great companies wandering out upon muddy or sandy flats, or even upon the beaches of the bays and sounds (Verrill and Smith).[2] Usually digs its holes in mud (Pearse).[3] Lives on a muddy substratum which is well shaded by a dense vegetation and hence continually moist (Schwartz and Safir).[4]

Range.—From Provincetown and Barnstable, Massachusetts, to Louisiana.

Material examined.—

Provincetown, Massachusetts; sand; April 14, 1879; U. S. Fish Comm.; 18 specimens (3828)

Barnstable, Massachusetts; 1875; U. S. Fish Comm.; 1 male, 1 female (4163).

Cataumet, Massachusetts; July 22; U. S. Fish Comm.; 9 males, 2 females (40793).

Woods Hole, Massachusetts: U. S. Fish Comm.; 4 males (40795). 1911; U. S. Bur. Fisheries; 15 males, 1 female (43156).

Vineyard Sound, Woods Hole, Massachusetts; 1911; U. S. Bur. Fisheries; 1 specimen (43192).

Naushon Island, Massachusetts; 1882; U. S. Fish Comm.: 158 males, 18 females (5864). August 15; 7 males, 1 female (40794).

---

[1] Trans. Connecticut Acad. Arts and Sci., vol. 2, 1870, p. 132.
[2] Rept. U. S. Commr. Fisheries, vol. 1, for 1871-72 (1873), p. 545 [251].
[3] Ann. Rept. Smithson. Inst. for 1913 (1914), p. 416.
[4] For a comparison of *U. pugnax* with *U. pugilator*, see Schwartz and Safir, The Natural History and Behavior of the Fiddler Crab. Cold Spring Harbor Monographs, No. VIII. Brooklyn Inst. Arts and Sci., 1915, pp. 1-24.

Newport, Rhode Island; shore; 1880; U. S. Fish Comm.; 8 males (4123). September 2; 4 specimens (36957).

Prudence Island, Narragansett Bay, Rhode Island; August 30, 1901; E. A. Mearns; 18 males (25867).

New Haven, Connecticut; shore; 1879; R. Rathbun; 17 specimens (3836).

The Gut, Cold Spring Harbor, Long Island; July 23, 1898; Biological Laboratory; 24 specimens (21661).

Great South Bay, Long Island; 1898; T. H. Bean; 4 males, 1 female (43360).

Atlantic City, New Jersey; April 26, 1900; E. P. Miller; 23 specimens (23635).

Chincoteague, Virginia; July, 1913; Henderson and Bartsch; 5 specimens (46285).

Smiths Island, Northampton County, Virgina; May 16–25, 1894; Charles W. Richmond; 4 males (18285).

Mainland off Smiths Island, Virginia; E. A. Mearns; 24 males, 7 females (41017).

Cherrystone, Virginia; Aug. 27, 1881; M. McDonald; 1 male, 1 female (3475).

Wallops Island, Virginia; May 26, 1913; Department of Agriculture; 2 males (45959).

Mill Creek Shore at Fort Monroe, Virginia; May 5, 1898; M. C. Marsh; 8 males (43359).

Winyah Bay, South Carolina; January, 1891; *Fish Hawk;* 2 males (17488).

Charleston Harbor, South Carolina; Mar. 18, 1880; M. McDonald; 11 males, 4 females (15056).

Port Royal, South Carolina; Feb. 1, 1891; *Fish Hawk;* 11 males, 1 female (17184).

Myrtle Bush Creek, South Carolina; Feb. 7, 1891; *Fish Hawk;* 2 males, 3 females (17185).

South Carolina; Mar. 22, 1894; *Fish Hawk;* 28 males, 7 females (22278).

Grand Isle, near New Orleans, Louisiana; G. Kohn; 3 specimens, not typical (2259).

UCA PUGNAX RAPAX (Smith).

Plate 140.

?*Gelasimus palustris* MILNE EDWARDS, Ann. Sci. Nat., ser. 3, Zool., vol. 18, 1852, p. 148 [112], pl. 4, fig. 13 (type-locality, Antilles; type in Paris Mus.).

Gelasimus rapax SMITH, Trans. Connecticut Acad. Arts and Sci., vol. 2, 1870, p. 134, pl. 2, fig. 2; pl. 4, fig. 3 (type-locality, Aspinwall; type in P. M. Y. U.).

Gelasimus vocator KINGSLEY, Proc. Acad. Nat. Sci. Philadelphia, 1880, p. 147 (part).
Uca vocator ORTMANN, Zool. Jahrb., Syst., vol. 10, 1897, p. 352 (part).
Uca pugnax rapax RATHBUN, Bull. U. S. Fish Comm., for 1900, vol. 2, 1901, p. 7.

Diagnosis.—Oblique ridge inside hand terminating at carpal cavity. Inner surface of hand finely granulate. Eyebrow visible in dorsal view.

Description.—Differs from typical *U. pugnax* in having the surface of the eyebrow more oblique and visible in a dorsal view, and in the inner surface of the palm almost smooth to the naked eye, finely granulate under the lens.

The carapace is often narrow posteriorly, and the lateral margins more sinuous, but these characters are not invariable.

Measurements.—Male (18551), length of carapace 15.5, width of same 25.5 mm.

Habits.—Frequents salt marshes and mangrove swamps.

Mr. P. W. Jarvis, Kingston, Jamaica, has written of this species that it occurs in large numbers near the coasts on the sides of ditches and sometimes in thousands on the edges of swamps. When alarmed the male opens the large claw which it holds erect in readiness for the attack and at the same time rapidly scuttles away to its burrow in the mud, down which it quickly descends. The female leaves her burrow in search of food only at dusk, but the male is out and about most of the day. There is a local superstition that this crab can cure deafness and earache, and in consequence it is known to the dwellers on the coast as the "Deaf-ear crab." The method consists in crushing the living crab and pouring the juice into the ear.

Range.—Gulf of Mexico and Florida Keys to Rio de Janeiro.

Variations.—Some specimens, from Key West and Havana, show characters intermediate between *pugnax* and *pugnax rapax*.

Material examined.—

Miami, Florida; April, 1898; *Fish Hawk;* 1 male (43362).

Key West, Florida: D. S. Jordan; December, 1883; 3 males, 1 female (6369). H. Hemphill; high tide, among rocks or burrowing in sand; 8 specimens (13813); 1885; 1 male, 5 females (15055). *Albatross;* Apr. 15-27, 1884; 26 males, 4 females (7515).

Sarasota Bay, Florida; Union College collection; 1 male (42616).

Near Mobile, Alabama; Jan. 29, 1883; C. L. Herrick; 1 female (15057).

Matagorda Bay, Texas; Mar. 22, 1905; T. E. B. Pope; 1 male (33031).

Drain near Matagorda Bay, Texas; fresh water; J. D. Mitchell· 5 males, 4 females (25033).

Cuba; 1900; Wm. Palmer and J. H. Riley: Cabañas; June 2; 13 males, 4 females (23824). June 3; on rocks and piles along shore; 1 male (23825). Mariel; May 10; from tidal swamp; 20 males, 3 females (23823). June 10; 2 males (23822). Nueva Gerona, Isla de Pinos; July 11; 13 males, 3 females (23826).

Cuba; May, 1914; Henderson and Bartsch, *Tomas Barrera* Exped.: Dimas Bay; M. plants; May 17; 1 male (49182). Ensenada de Santa Rosa; 1-3 fathoms; S. Sh. M. Sponge; 50 males, 3 females (49390). Santa Lucia; S. Sh. M. Sponge; 46 males, 1 female (48392). Los Arroyas; May 20; 20 males, 30 females (48410).

Montego Bay, Jamaica: P. W. Jarvis; 1 male, 1 female (19056); E. A. Andrews; July 7, 1910; 1 male (42921).

Snug Harbor, Montego Bay, Jamaica; June 19, 1910; E. A. Andrews (42919).

Bogue Islands, Montego Bay, Jamaica: June 20, 1910; C. B. Wilson; 6 males (42906). June 24, 1910; E. A. Andrews; 6 males, 1 female (42920).

Kingston, Jamaica; P. W. Jarvis; 1 male (19059).

Kingston Harbor, Jamaica: 1893; R. P. Bigelow; 5 males, 2 females (19477). May-July, 1896; F. S. Conant; 1 female ovig. (19599). T. H. Morgan; 2 males (17220). *Albatross;* Mar. 1-11; 1884; 6 males, 1 female (18550).

Haiti; 1917; Henderson and Bartsch: Thomazeau; 6 males, 1 female (50489). Petit Guave; 28 males (50490). Port-au-Prince; 11 males, 1 female (50491).

Mayaguez, Porto Rico; fresh water; Jan. 20, 1899; *Fish Hawk;* 4 males, 2 females (24536).

Mangrove swamp, San Juan Bay, Porto Rico; Jan. 18, 1899; Paul Beckwith; 3 males (22794).

Catano, Porto Rico; Jan. 4, 1899; *Fish Hawk;* 14 males, 1 female (24533).

Hucares, Porto Rico; Feb. 14, 1899; *Fish Hawk;* 48 males, 22 females (24539).

Ensenada Honda, Culebra, Porto Rico; Feb. 9, 1899; *Fish Hawk;* 1 male (24538).

St. Thomas, West Indies: *Albatross;* Jan. 17-24; 4 males (18549). Riise, collector; 3 males, 1 female (19712). Gregerie Bay; brought in by fishermen; July 22, 1915; C. R. Shoemaker; 20 males, 2 females, mostly young (49813).

West Indies; 1 male (15058).

Old Providence, West Indies; Apr. 4-9, 1884; 6 males (7546).

Coral Reef, Colon, Panama; May 2, 1911; Meek and Hildebrand; 2 males (43989).

Sabanilla, Colombia; Mar. 16-22, 1884; *Albatross;* 5 males (18436).

St. Joris Bay, Curaçao; Apr. 3, 1905; J. Boeke; 1 male (42957).

Curaçao; Feb. 10-18, 1884; *Albatross;* 14 males, 3 females (18551).
Puerto Cabello, Venezuela: Schibbye, collector; 1 male (19714). Oct., 1916; J. N. Rose; 7 males (49632).
Trinidad; shore; Jan. 30–Feb. 2, 1884; *Albatross;* 11 males, 3 females (7641).
Pernambuco, Brazil; 1876–1877; R. Rathbun; 1 male (40623).
Plataforma, Bahia, Brazil; 1876–1877; R. Rathbun; 1 male, 1 female (40620).
Salt Lagoa, Caravellas, Bahia, Brazil; 1876–1877; R. Rathbun; 1 male (40621).
Maruim, Brazil; Sept. 4; Hygom, collector; 1 male (19715).
Itabapuana, Brazil; Thayer Exped.; 3 males, 1 female (22187).
Brazil; 1876–1877; R. Rathbun; 1 female (40622).

UCA PUGILATOR (Bosc).

Plate 141; plate 160, fig. 2.

Ocypoda pugilator Bosc, Hist. Nat. Crust., vol. 1, an X [1801–1802], p. 197 (type-locality, Caroline; type not extant).
Gelasima pugillator LATREILLE, Nouv. Dict. Hist. Nat., vol. 12, 1817, p. 519.
Gelasimus vocans GOULD, Rept. Invert. Massachusetts, 1841, p. 325 (part). Not *Cancer vocans* Linnaeus.
Gelasimus vocans DEKAY, Nat. Hist. New York, Crust., 1844, p. 14 (part), pl. 6, fig. 9.
Gelasimus pugilator MILNE EDWARDS, Ann. Sci Nat., ser. 3, Zool., vol. 18, 1852, p. 149 [113], pl. 4, fig. 14.
Gelasimus pugilator SMITH, Trans. Connecticut Acad. Arts and Sci., vol. 2, 1870, p. 136, pl. 4, fig. 7.
Gelasimus pugilator KINGSLEY, Proc. Acad. Nat. Sci. Philadelphia, 1880, p. 150.
Uca pugilator ORTMANN, Zool. Jahrb., Syst., vol. 10, 1897, p. 352 (part).

Diagnosis.—No oblique ridge on inner surface of palm. Carapace broad behind. Ambulatory legs narrow.

Description.—Carapace very convex, smooth, less uneven than in *U. pugnax.* Front more than one-third of fronto-orbital width, broadly rounded below. Eyebrow deeper than in *U. pugnax*, its lower part visible in dorsal view. Upper margin of orbit slightly sinuous; antero-lateral angle little prominent. Lateral margin strongly curved outward behind the orbit.

Large cheliped of male shaped much as in *U. pugnax*, although the fingers do not attain so great a length. A ridge runs along the pollex from the tip backward a slight distance upon the palm. The largest prehensile tubercles are just behind the middle of the pollex and near the tip. Inner surface of palm granulate, the granules coarser on the thickest part of the palm; no oblique ridge across the middle.

Color.—Carapace of male a dull light purplish or grayish-blue, with a large patch of deep purplish-blue on the anterior half. Large cheliped dull light blue with white tubercles; articulations yellowish; fingers mostly white. Small cheliped and legs buff ground with blue or brown speckles, articulations yellowish.

Measurements.—Male (43356), length of carapace 17.3 mm., width of same 26 mm.

Habits.—Lives upon muddy and sandy flats and beaches (Verrill and Smith). Usually digs its holes in sand (Pearse). Lives in areas which have a consistent sandy substratum the surface of which may become dry through evaporation (Schwartz and Safir). Mr. J. D. Mitchell has furnished the following notes on the occurrence of *U. pugilator* in Texas:

This fiddler crab inhabits all salt-water marsh land, but thrives best on the bay shores next to the Gulf, such as the immense marshes on the north side of the Matagorda Peninsula and the chain of islands on our coast. They dig holes in the ground near water, about half an inch in diameter, the holes connecting under the ground in many instances. They are, next to the mullet, the greatest food product of the coast. Nearly everything in the water and on the land near the water eats fiddlers. They are the main food of the raccoon, and are also eaten by skunks, possums, wild cats, and wolves, and by almost every fish that bites at a hook. The blue crabs catch them if they come near the bay shore and sand crabs eat them if they wander near the Gulf shore. They are a source of endless amusement to the small boy, who makes a pen by setting 12-inch planks edgewise and drives them into it, playing they are cattle; once cut off from their holes they huddle together like sheep and can be driven anywhere.

Range.—From Boston Harbor, Massachusetts, to Galveston, Texas. Jeremie, Haiti; Dr. D. F. Weinland (M. C. Z.).

Material examined.—

Corn Hill and Truro, Massachusetts; F. W. True; 1 male (43355).
Harwich, Massachusetts; U. S. Bur. Fisheries; 1 male (43358).
Cataumet, Massachusetts; July 22; U. S. Fish Comm.; 5 males, 4 females (36873).
Woods Hole, Massachusetts; U. S. Fish Comm.; 85 specimens (3212); 1882; 1 male (14402).
Vicinity of Woods Hole, Massachusetts; M. J. Rathbun; 2 males 1 female (32481).
Naushon Island, Massachusetts; low water, gutters; U. S. Fish Comm.; 9 males, 5 females (3347).
Sipaquisset, Vineyard Sound, Massachusetts; Sept. 15, 1875; U. S. Fish Comm.; 2 males, 3 females (2556).
Vineyard Sound; 1875; U. S. Fish Comm.; 2 specimens (35304).
Gardiners Bay, Long Island; J. F. Fowler; 1 male (3771).
The Gut, Cold Spring Harbor, Long Island; July 23, 1898; Biological Laboratory; 6 specimens (21662).

Great South Bay, Long Island; 1898; T. H. Bean; 1 male (43356).
Smiths Island, Northampton County, Virginia: May 16–25, 1894; Chas. W. Richmond; 1 male (18286). May 19, 1898; Wm. Palmer; 1 male, 4 females (21622).
Mainland opposite Smiths Island, Virginia; E. A. Mearns; 6 males, 20 females (41018).
Revels Island, Virginia; Oct. 13, 1915; W. L. McAtee; 1 male, 2 females (49102).
Mill Creek Shore at Fort Monroe, Virginia; May 5, 1898; M. C. Marsh; 6 males, 1 female (43357).
Willoughby Point, Virginia; May, 1880; McDonald & Earll; 4 males, 3 females (3146).
Morehead City, North Carolina; Feb. 30, 1863; A. S. Bickmore; 3 males, 3 females (22184).
Porchers Bluff, Charleston County, South Carolina; May, 1911; E. A. Mearns; 1 male (43361).
Port Royal, South Carolina; Feb. 1, 1891; *Fish Hawk;* 5 males, 5 females (17188).
Port Royal Island, west end, South Carolina; Jan. 27, 1891; *Fish Hawk;* 23 males, 8 females (17186).
Cat Island Creek, Port Royal, South Carolina; Feb. 2, 1891; *Fish Hawk;* 4 males, 2 females (17187).
Myrtle Bush Creek, South Carolina; Feb. 7, 1891; *Fish Hawk;* 1 male (17189).
South Carolina: Kurtz and Stimpson; 65 males, 62 females (2061). *Fish Hawk;* Mar. 22, 1894; 9 males, 5 females (22280).
St. Augustine, Florida; L. W. Ledyard; 2 males (2195).
Ponce Park, Florida; Nov. 24, 1908; B. A. Bean; 1 male (39193).
Near Lake Kissimmee, Florida; A. M. Reese; 4 males, 1 female (43354).
Cards Sound, Florida; Feb. 13, 1889; *Grampus;* 1 male, 1 female (15255).
Pine Key, Florida; January, 1884; H. Hemphill; 79 males, 7 females (6440).
Key West, Florida: H. Hemphill; 1 male, 1 female (17688). Apr. 15–27, 1884; *Albatross;* 2 males (18552).
Cape Sable Creek, Florida; Feb. 20, 1889; *Grampus;* 13 males, 18 females (15254).
Marco, Florida: May, 1884; H. Hemphill; 21 males, 8 females (6964). Feb. 25, 1889; *Grampus;* 2 males, 1 female (15252).
Punta Rassa, Florida: April, 1883; C. W. Ward; 48 males, 1 female (5871). February, 1884; H. Hemphill; 33 males, 10 females (6435). H. Hemphill; 1 male (17687).
Big Gasparilla Pass, Florida; Mar. 5–16, 1889; *Grampus;* 26 males, 23 females (15253).

Tampa Bay, Florida; Mar. 17, 1885; *Albatross;* 23 males, 3 females (9908).
Clearwater, Florida; July 14, 1879; S. T. Walker; 7 males, 11 females (3276).
Cedar Keys, Florida; December, 1883; H. Hemphill; 109 males, 41 females, 3 y. (6412).
Cocoanut Grove, Florida; January, 1916; W. L. McAtee; 7 males 4 females (48924).
Florida; L. E. Daniels; 1 male (22984).
Near Mobile, Alabama; 1883; C. L. Herrick; 1 male (5781).
Biloxi Bay, Mississippi; shore; C. F. Baker; 1 male (21683).
Biloxi, Mississippi; P. R. Hoy; 1 male (45958).
Galveston, Texas; 1871; Boll, collector; 20 specimens (M.C.Z.).

UCA GALAPAGENSIS Rathbun.

Plate 142.

Uca galapagensis RATHBUN, Proc. Washington Acad. Sci., vol. 4, 1902, p. 275, pl. 12, figs. 1 and 2 (type-locality, Indefatigable Island; type, Cat. No. 22319, U.S.N.M.).

Diagnosis.—Oblique ridge inside palm continued to upper margin. Orbital margin very oblique. Merus of ambulatory legs enlarged.

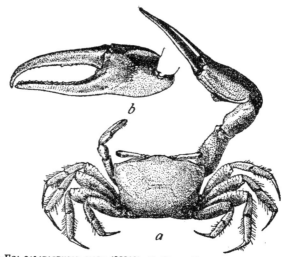

FIG. 107.—UCA GALAPAGENSIS, MALE (22319), × 1½. *a*, DORSAL VIEW; *b*, INNER SURFACE OF LARGER CHELA.

Description.—Carapace very convex in both directions, surface smooth and shining, with small distant punctae and microscopically granulate. H-form depression very shallow. Front about one-third the anterior width of the carapace, edge scarcely visible in dorsal view, eyebrow partly visible. Upper margin of orbit very sinuous and oblique. Antero-lateral angles almost rectangular. Lateral margins for the most part convex, slightly sinuous posteriorly; typically they slope outward a little from the orbital angle, so that the carapace is widest behind that angle, but such is not always the case, as one side may slope out and the other in.

Arm of large cheliped of male granulate-rugose; wrist and hand coarsely granulate, granules on the latter increasing in size above. Surface of the palm strongly bent over above, a longitudinal sulcus at the bend. Inner face with a tuberculated zigzag ridge across the middle; a small area of fine granules on the upper half. Fingers one and one-third times as long as palm; prehensile tubercles of pollex coarser than those of dactylus, which has, however, several large tubercles. For the rest, much as in *U. pugnax*.

Measurements.—Male (22319), length of carapace 13.6, width of same 19.8 mm.

Habitat.—Lives on salt flats.

Range.—Galapagos Islands; Peru.

Material examined.—

Galapagos Islands: James Island; Apr 11, 1888; *Albatross*; 1 male (22320). Indefatigable Island; Apr 12, 1888; *Albatross;* 6 males (types) (22319). South Seymour Island; May 3, 1899; Stanford University; 2 males, 1 female (25665).

Puerto Grande, Rio Zarumilla (2 leagues from Capon), Peru; salt flats; Feb. 2, 1908; R. E. Coker; 5 males (40488).

UCA MACRODACTYLUS (Milne Edwards and Lucas).

Plate 143.

Gelasimus macrodactylus MILNE EDWARDS and LUCAS, in d'Orbigny's Voy. dans l'Amér. Mérid., vol. 6, 1843, Crust., p. 27; vol. 9, atlas, 1847, pl. 11, fig. 3 (type-locality, *Côtes du Valparaiso;* type in Paris Mus.).

Gelasimus annulipes KINGSLEY, Proc. Acad. Nat. Sci. Philadelphia, 1880, p. 148 (part). (Not Milne Edwards.)

Uca macrodactyla NOBILI, Boll. Mus. Zool. Anat. comp. R. Univ. Torino, vol. 16, 1902, No. 415, p. 49.

Diagnosis.—Carapace strongly narrowed behind.. Oblique ridge inside palm continued to upper margin. Short sulcus outside palm at base of thumb.

Description.—Convexity moderate. H-form depression shallow; sulcus behind the orbit deep.

Front about one-third width of carapace. Orbits very oblique, slightly sinuous. Sides strongly converging, sinuous; carapace widest at antero-lateral angles, which are subacute.

Merus of large cheliped of male marked with very fine granulated rugae. Carpus and palm coarsely granulate, the granules continued on the fingers, becoming very fine toward the tips; granulation coarser on the upper part of the chela than on the lower. Besides the sulcus following the lower margin of the palm there is another very short longitudinal sulcus at the distal end of the palm, and opposite the middle of the pollex.

The oblique, tuberculate ridge leading up from the lower margin of the inner surface is continued in a zigzag line to upper margin, surface distal to it granulate. Fingers about as long as palm, narrowly gaping; prehensible tubercles coarse, a larger one at middle of pollex.

Measurements.—Male (19441), length 8, width 11.5 mm.

Habits.—Extremely abundant in the mud of the mangroves. In some places it is quite impossible to walk along without hurting many of them. In the stomach of a kind of owl (*Ciccaba nigrolineata* Sclater) I found many carapaces of the little species. (Note by J. F. Tristan.)

Range.—Mexico to Chile.

Material examined.—

Guaymas, Mexico; Capt. C. P. Stone; 1 male without large cheliped (M. C. Z.).

Corinto, Nicaragua; J. A. McNiel; 2 males, 1 female (M.C.Z.).

Boca del Rio Jesus Maria, Costa Rica; April, 1905; J. F. Tristan; 4 males, 3 females (32320).

Santo Domingo, Gulf of Dulce, Costa Rica; April, 1896; H. Pittier; 2 males, 3 females (19441).

Chile; 2 males, cotypes (Paris Mus.).

Additional records.—Colombia: Tumaco (Nobili). Ecuador: Puntilla di S. Elena (Nobili).

UCA RECTILATUS (Lockington).

Gelasimus rectilatus LOCKINGTON, Proc. California Acad. Sci., vol. 7, 1876 (1877), p. 148 (type-locality, west coast of Lower California; type not extant, Cat. No. 3112, California Acad. Sci.).

Gelasimus annulipes KINGSLEY, Proc. Acad. Nat. Sci. Philadelphia, 1880, p. 148 (part). (Not Milne Edwards.)

Uca rectilata HOLMES, Occas. Papers California Acad. Sci., vol. 7, 1900, p. 76, pl. 1, figs. 10-14.

Diagnosis.—Side margins strongly convergent and almost straight lines. Front narrow.

Description.—Carapace wide, nearly flat transversely, but longitudinally strongly convex. Front about one-sixth width of carapace, arched strongly forward. Upper edge of eyebrow much curved, lower edge close to upper but distinct. Postorbital angle prominent, acute, and directed obliquely outward. Lateral margins strongly convergent, the lower margins straight, but the upper margins, or those which bound the dorsal plane, are slightly sinuous and are prolonged almost to the posterior margin.

Merus of larger cheliped of male slightly rugose; carpus lightly granulated; hand similar to that of *U. crenulata;* outer surface of palm finely granulated and lower edge margined; an oblique, granu-

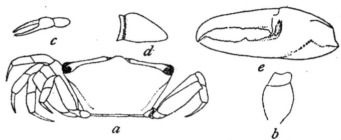

FIG. 168.—UCA RECTILATUS, COTYPE, ENLARGED, AFTER HOLMES. *a*, DORSAL VIEW OF CARAPACE, EYES, AND LEGS; *b*, ISCHIUM AND MERUS OF RIGHT OUTER MAXILLIPED; *c*, CARPUS AND CHELA OF SMALLER CHELIPED; *d*, MERUS OF RIGHT OR LARGER CHELIPED, INNER FACE; *e*, LARGER CHELA, INNER FACE.

lated ridge on inner surface, beginning at lower margin and ending at carpal groove. Pollex tapering, a tooth near the middle, extremity slightly excavated; dactyl curved more strongly toward the tip, which slightly overreaches the pollex. Ambulatory legs slender. (Abridged from Holmes.)

Measurements.—Male, type, length of carapace 8.75, width of same 13.5 mm.

Range.—West coast of Lower California. Known only from the type specimens, which were destroyed in the San Francisco fire.

UCA THAYERI Rathbun.

Plate 144.

Ciecie Ete MARCGRAVE DE LIEBSTAD, Hist. Rer. Natur. Brazil, 1648, p. 183, text fig.; Brazil.

Uca thayeri RATHBUN, Proc. Washington Acad. Sci., vol. 2, 1900, p. 134, pl. 8, figs. 1 and 2 (type-locality, Rio Parahyba do Norte at Cabedello; type, Cat. No. 23733, U.S.N.M.).

Diagnosis.—Front narrow. Sides strongly convergent. Oblique ridge inside palm not ending at carpal cavity but continued partly upward.

Description.—Carapace strongly narrowed behind, widest either at the lateral angles or a little farther back; sides sinuous. Interregional depressions deep; surface finely granulose and tomentose.

Front narrower than usual in this group, less than one-fifth width of carapace, its sides longer than the slightly convex lower margin. Orbital margin moderately oblique and sinuous.

Merus of large cheliped of male crossed with short granulate rugae; carpus scabrous. Upper part of outer surface of hand coarsely granulate or **tuberculate**, very finely granulate below, as on the fingers also. Fingers twice as long as palm, narrowly gaping; prehensile edges with one or more larger tubercles; one of these is at the basal fourth of the dactylus. Inner face of palm with a single row of tubercles on the oblique **ridge**, which turns at a right angle at the middle, after which it is continued but a short distance. Surface between ridges smooth.

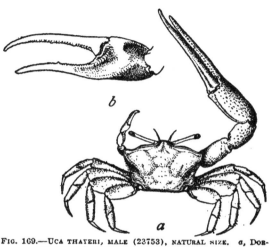

FIG. 169.—UCA THAYERI, MALE (23753), NATURAL SIZE. a, DORSAL VIEW; b, INNER SURFACE OF LARGER CHELA.

Merus of ambulatory legs much dilated.

Measurements.—Male (22192), length of carapace 17.2, width of same 27.5 mm.

Range.—Jamaica and Porto Rico to Brazil.

Material examined.—

Jamaica; Mar. 1–11, 1884; *Albatross;* 1 male (22321).

Rio Bayamon, above Palo Seco, Porto Rico; Jan. 16, 1899; *Fish Hawk;* 2 males, 1 female (24541).

Fajardo, Porto Rico; Feb. 17, 1899; *Fish Hawk;* 1 male, 1 female (24542).

Victoria, Brazil; Thayer Exped.; 2 females (22193); 5 males, 19 females (M.C.Z.).

Natal, Rio Grande do Norte, Brazil; June 29, 1899; Branner-Agassiz Exped.; 1 male (25699).

Cabedello, Rio Parahyba do Norte, Brazil; June 20, 1899; Branner-Agassiz Exped.; 7 males, 1 female (types) (23753).

Rio Parahyba do Norte, Brazil; Thayer Exped.; 1 male (22192), 4 males, 1 female (M.C.Z.).

Plataforma, Bahia, Brazil; 1876–1877; R. Rathbun; 5 males, 5 females (40619).

Sao Matheos, Brazil; Thayer Exped.; 4 males, 9 females (M.C.Z.).

UCA SPECIOSA (Ives).

Plate 145.

Gelasimus speciosus IVES, Proc. Acad. Nat. Sci. Philadelphia, 1891, p. 179, pl. 5, figs. 5 and 6 (type-locality, Port of Silam, Yucatan; type in Mus. Phila. Acad. Sci.).

Uca stenodactyla ORTMANN, Zool. Jahrb., Syst., vol. 10, 1897, p. 356 (part).

Diagnosis.—Oblique ridge inside palm continued to upper margin, bending at a right angle. Lateral margins of carapace angled.

Description.—A rather small species. H-form depression deep; sulcus behind the orbits shallower; surface smooth. Carapace widest at antero-lateral angles. Anterior part of lateral margins straight and convergent; the margin then turns at an obtuse angle and is continued in a sinuous line to a point opposite the middle of the cardiac region. The front is nearly one-third the width of the carapace; its lower margin, as also the surface of the eyebrow, is visible in a dorsal view. Upper margin of orbit oblique, sinuous, anterolateral angle acute, directed slightly forward.

Large cheliped of male with arm rugose, wrist and hand granulous outside, fingers smooth. On the inner surface of the palm the oblique, tuberculated ridge leading up from the lower margin turns at the middle at a right angle and is continued to near the upper margin. The tuberculated ridge parallel to the distal end of the palm is very oblique to the lower margin and nearly reaches the upper margin. Space between ridges smooth. Fingers one and two-thirds times as long as palm; fingers slender, gaping; of the prehensile tubercles, there is a larger one at middle of pollex and at basal fourth of dactylus.

Measurements.—Male (18435), length of carapace 10, width of same 14.4 mm.

Range.—South and west Florida; West Indies.

Material examined.—

Cocoanut Grove, Florida; Jan. 1916; W. L. McAtee; 3 males, 1 female (48980).

Cards Sound, Florida; Feb. 15, 1889; *Grampus*; 4 males, 2 females (15256).

Key Vaccas, Florida; H. Hemphill; 1 male (18435).

Key West, Florida; high-tide among stones or burrowing in sand; H. Hemphill; 8 males, 3 females (22309).

Sarasota Bay, Florida; Union College collection; 1 male, 2 females (42617).

Florida; 1879; J. W. Milner; 6 males, 3 females (18422).
Curaçao; *Albatross;* 7 males, 1 female (22310).

UCA CRENULATA (Lockington).

Plate 146.

Gelasimus crenulatus LOCKINGTON, Proc. California Acad. Sci., vol. 7, 1876 (1877), p. 149 (type-locality, Todos Santos Bay; type not extant)
Gelasimus vocator KINGSLEY, Proc. Acad. Nat. Sci. Philadelphia, 1880, p. 147 (part).
Gelasimus gracilis RATHBUN, Proc. U. S. Nat. Mus., vol. 16, 1893, p. 244 (type-locality, La Paz; male type, Cat. No. 4622, U.S.N.M.).
Gelasimus macrodactylus BOUVIER, Bull. Mus. Hist. Nat. Paris, No 1, p. 8, 1895; not Milne Edwards and Lucas.
Uca vocator ORTMANN, Zool. Jahrb., Syst., vol. 10, 1897, p. 352 (part).
Uca stenodactyla ORTMANN, Zool. Jahrb., Syst., vol. 10, 1897, p. 356, part: San Diego; not *G. stenodactylus* Milne Edwards and Lucas.
Uca crenulata HOLMES, Occas. Papers California Acad. Sci., vol. 7, 1900, p. 75, pl. 1, figs. 7-9.

Diagnosis.—Oblique ridge inside palm continued to upper margin, and bending at an obtuse angle. Lateral margins of carapace angled.

Description.—The Pacific analogue of *U. speciosa*. The anterior margin of the carapace is a little more horizontal; and the lateral margins less convergent. The anterior, or straight portion of the lateral margin may be longitudinal, or slope inward, or outward, and is likely to be different on the two sides. The palm in the large cheliped of the male appears narrower, as a part of it is bent over to form the upper surface. Fingers more evidently granulous. The oblique ridge on the inner face of the palm is more nearly perpendicular to the lower margin and bends at an obtuse angle in continuing toward the upper margin.

Measurements.—Length of carapace of male (17504) 10, width of same 15 mm.

Malformation.—A curious malformation occurs in one claw among a lot collected at San Felipe Bay by Nelson and Goldman, 1905. The dactylus has an accessory branch, which is half as long as the dactylus itself and arises near its base and is directed downward at almost a right angle; it is subcylindrical and there are rudiments of teeth on the distal edge. It rubs against the inner face of the pollex and prevents the two fingers from closing together. The tip of the branch is broken off.

Range.—From San Diego, California, to Gulf of California, as far north as San Felipe Bay, Lower California; Guaymas and Mazatlan, Mexico.

Material examined.—
San Diego, California; 1872; H. Hemphill; 5 males (17504).

Todos Santos Bay, Lower California: H. Hemphill; 4 males (17576). C. R. Orcutt; 25 specimens (19033).

Santo Domingo, Lower California; Sept. 26, 1905; Nelson and Goldman; 15 males (33415).

Mangrove Island, Magdalena Bay, Lower California; Mar. 20, 1911; *Albatross;* 2 males (44580).

La Paz, Gulf of California; L. Belding; 34 males, 6 females (types of *Gelasimus gracilis* Rathbun) (4622).

Pichilinque Bay, Gulf of California; Apr. 29, 1888; *Albatross;* 23 males, 7 females (22080).

San Luis Gonzales Bay, Gulf of California; Mar. 27, 1889; *Albatross;* 4 males (17458).

San Felipe Bay, Gulf of California; June 23, 1905; Nelson and Goldman; 13 males, 1 female (32630).

Lower California; M. Diguet; 1 male, 1 female (20301).

Guaymas, Mexico; Capt. C. P. Stone; 13 males, 3 females (M. C. Z.).

Mazatlan, Mexico; H. Edwards; received from Peabody Acad. Sci., Salem; 1 female y. (M. C. Z.).

UCA COLORADENSIS (Rathbun).

Plate 147.

Gelasimus coloradensis RATHBUN, Proc. U. S. Nat. Mus., vol. 16, 1893, p. 246 (type-locality, Horseshoe Bend, Colorado River; male holotype, Cat. No. 17459, U.S.N.M.).

Uca coloradensis HOLMES, Occas. Papers California Acad. Sci., vol. 7, 1900, p. 76.

Diagnosis.—Oblique ridge inside palm continued to upper margin. Upper surface of palm at right angles to outer surface. Lateral margins of carapace angled.

Description.—Allied to the two preceding. Carapace widest behind the antero-lateral angles, the side margins being straight anteriorly, then turning abruptly inward at the widest point. Dorsal sulci very deep; a long groove with lateral branches on the branchial region. Front one-fourth the extreme width of the carapace.

The outer surface of the palm is abruptly bent over above, forming an almost horizontal upper surface; granulation coarse along the bend. Fingers about twice as long as palm.

Oblique tuberculate ridge on inner surface of palm bent at a right angle at the middle.

Otherwise much as in *U. crenulata.*

Measurements.—Male (17459), length of carapace 12.5; width of same 20 mm.

Range.—Mexico: Head of Gulf of California; Guaymas.

Material examined.—
Colorado River, opposite mouth of "Hardy's Colorado" River, Sonora, Mexico; Mar. 27, 1914; Dr. E. A. Mearns, U. S. Army; 32 males, 14 females (18292).
Horseshoe Bend, Colorado River, Lower California; *Albatross;* 1 male holotype (17459).
Guaymas, Mexico; Capt. C. P. Stone; 7 males, 4 females (M. C. Z.).

UCA SPINICARPA Rathbun.

Plate 148.

Uca spinicarpa RATHBUN, Amer. Nat., vol. 34, 1900, p. 586 (type-locality, Galveston; male type, Cat. No. 22183, U.S.N.M.).

*Diagnosis.—*Oblique ridge inside palm continued to upper margin and very prominent. Tooth on inner face of carpus. Lateral margins of carapace angled.

*Description.—*Allied to *U. speciosa.* Orbital margin less oblique, sides less converging. Front about two-sevenths width of carapace, lower edge subtruncate. Anterior part of lateral margins straight and subparallel. Antero-lateral angle a right angle.

Merus of large cheliped of male granulate-rugose. Carpus scabrous-granulate on outer surface; inside there is a sharp spine or tooth at the middle of oblique ridge. The palm is granulate outside, granules increasing in size above, surface bent over almost horizontally above. Inner surface with a strongly protruding tuberculate ridge, which bends at a right angle at the middle and is continued to upper margin; space between ridges smooth. Fingers one and one-third times as long as palm, finely granulate; prehensile edges irregularly tuberculate, a slightly larger tubercle near base of dactylus and at middle of pollex.

*Measurements.—*Male (22183), length of carapace 9.5, width of same 13.8 mm.

*Habitat.—*Mr. J. D. Mitchell says of the habitat of this species:

Drain runs half a mile into the land from Matagorda Bay; in the head of this drain there are several shallow wells of fresh water for cattle, walled on three sides. These crabs were all around these wells.

*Range.—*Alabama to Mexico. Jamaica. Brazil.
Material examined.
?Florida; U. S. Fish Comm.; 1 male (26111).
Near Mobile, Alabama; Jan. 29, 1883; C. L. Herrick; 3 males, 1 female (22312).
Shores of Biloxi Bay, Mississippi; C. F. Baker; 1 male (21684).
Galveston, Texas; 1871; Boll, collector; 2 males, 1 female (types) (22183).
Drain near Matagorda Bay, Texas; fresh water; J. D. Mitchell; 1 male (25034).

Maron, Lagoon Madre, Mexico; Apr. 10, 1910; E. Palmer; 1 female y. (43364).

Tampico, Mexico; A. Dugès; 1 male (22311).

Kingston Harbor, Jamaica; 1893; R. P. Bigelow; 1 male (22313).

Mamanguape, Brazil; stone reef; June 20, 1899; Branner-Agassiz Exped.; 1 male (25700).

UCA PANAMENSIS (Stimpson).

Plate 149.

Gelasimus panamensis STIMPSON, Ann. Lyc. Nat. Hist. New York, vol. 7, 1859, p. 63 (type-locality, Panama; type not extant).
Gelasimus panamensis SMITH, Trans. Connecticut Acad. Arts and Sci., vol. 2, 1870, p. 139, pl. 4, fig. 5.
Uca panamensis NOBILI, Boll. Mus. Zool. Anat. comp. R. Univ. Torino, vol. 16, 1901, p. 49, No. 415.

Diagnosis.—Antero-lateral angles prominent. Proximal end of palm thick; inner surface smooth, without tuberculate ridge. Lateral margins of carapace angled.

Description.—Surface anteriorly uneven; narrow part of mesogastric region well marked; two frontal lobes high; a prominent tubercle behind base of eyestalk. Carapace broadest at the antero-lateral angles which are strongly produced forward in an obtuse tooth, making the orbital margins very sinuous. Side margins slightly concave or straight anteriorly, continued backward at a very oblique angle; the slight oblique or curved ridge at the posterior extremity of the lateral margin is prominent but short. Front prominent, broad, and rounded below as in *U. pugilator.*

Large cheliped of male almost smooth. Arm and wrist sparingly granulous. Supero-proximal portion of hand strongly bent over horizontally; infero-proximal angle thick, very prominently projecting backward so that the hand can not be bent outward further than at a right angle to wrist and arm. Fingers flat, tapering; gape narrow, uniform; prehensile tubercles low, a large one near middle of pollex. Inner surface of palm almost smooth, oblique ridge smooth and subhorizontal.

Measurements.—Male (22181), length of carapace 12.4, width of same 18 mm.

Range.—Gulf of Fonseca, Salvador, to Peru.

Material examined.—

Gulf of Fonseca, Salvador; J. A. McNiel; 1 specimen (M.C.Z.).

Punta Arenas, Costa Rica; February, 1907; P. Biolley and J. F. Tristan; 1 male (39100).

Taboga Island, Panama; J. Zetek: June, 1914; 1 male (48784). Sept. 12, 1914; 2 females (48785).

Panama: 1 male (20277). Dr. Maack; 2 males (22181). A. Agassiz, 2 males (22182).
?Panama; 1 male (17766).
Payta, Peru; Dr. W. H. Jones, U.S.N.; 5 specimens (M.C.Z.).

UCA URUGUAYENSIS Nobili.

Plate 150.

Uca uruguayensis NOBILI, Boll. Mus. Zool. Anat. comp. R. Univ. Torino, vol. 16, 1901, No. 402, p. 14 (type-locality, La Sierra; type in Turin Mus.).

Diagnosis.—Fourth to sixth abdominal segments in male fused. Oblique ridge inside of palm terminating at carpal cavity. Carapace widest at antero-lateral angles.

Description.—This species resembles the group which includes *U. speciosa*, *U. crenulata* and *U. coloradensis*, but differs from all in having the fourth, fifth, and sixth segments of the male abdomen fused.

Carapace very convex in both directions, sulci of moderate depth, surface smooth. Width of front less than one-fourth width of carapace. Upper margin of orbit oblique. Side margins of carapace straight and sloping backward and inward, then turning at an angle as in the allied species.

Arm and wrist sparsely roughened. Palm granulate outside, granules increasing in size and prominence from below upward. The palm curves over to upper margin much as in *U. speciosa*. Ridges of inner surface coarsely tuberculate; those at base of dactyl are at right angles to lower margin; the oblique ridge ends at middle of palm; between this ridge and the others the surface is very coarsely granulate. Fingers about one and one-fourth times as long as palm (measured horizontally). Both fingers have several larger tubercles scattered along their prehensile edges.

Although the fourth, fifth, and sixth segments are fused, there is a faint sulcus between the fourth and fifth.

Color.—In formalin, uniform red-brown on the carapace and coralline red on chelipeds and legs.

Measurements.—Male (22194), length of carapace 10, width of same 16.5 mm.

Range.—Rio de Janeiro to Uruguay.

Material examined.—

Rio de Janeiro, Brazil; 1876–1877; R. Rathbun; 4 males, 2 females (40624).

E. Piassaguera, Santos, Brazil; June, 1913; H. Luderwaldt; 1 male (47870).

Maldonado, Uruguay; J. G. Cary; 1 male, 1 female (22194); 19 males, 5 females (M.C.Z.).

Uruguay; F. Felippone; 4 males (39120).

UCA OERSTEDI Rathbun.

Plate 152, figs. 1 and 2.

Uca oerstedi RATHBUN, Proc. Biol. Soc. Washington, vol. 17, 1904, p. 161 (type-locality, Punta Arenas, Costa Rica; male holotype in Copen. Mus.).

Diagnosis.—Carapace deeply areolated; sides strongly angulated; front narrow; oblique and distal ridges on inner side of palm meeting.

Description.—Surface uneven; a deep groove on outer side of gastric and cardiac regions is continued anteriorly in a transverse groove behind orbits, and posteriorly toward postero-lateral angle; a second longitudinal groove outside the first divides the branchial regions unequally.

Front posteriorly one-sixth as wide as distance between antero-lateral angles, gradually narrowing to a rounded end. Superior orbital margin very oblique and sinuous; eyebrow largely visible from above. Antero-lateral angle little more than a right angle; anterior portion of side margin straight or nearly so, and directed backward and a little outward; the margin then turns abruptly inward at an oblique angle and is very concave almost to its posterior end; the isolated bit of margin above the base of the last foot is very short and convex.

Large cheliped of male stout. The merus has a laminate and denticulate upper margin; outer face with a few scattered scabrous granules. The carpus has similar but more numerous granules; the oblique ridge on its inner face has a few tubercles at the proximal extremity. Palm covered with coarse tubercles outside; a beaded ridge on lower margin. Fingers broad, flat, almost smooth; margins raised; a triangular gape between their basal halves; prehensile tubercles bead-like, a tooth at middle of pollex, and sometimes a larger tubercle at middle of dactylus. Inner surface smooth, except for the tuberculate ridges; the oblique ridge starting from the lower margin extends to a point above the middle, then turns at a very prominent right angle toward the supero-distal end of the palm where it joins the longer and more proximal of the two ridges parallel to the base of the dactylus.

Merus of legs broad, margins convex.

Measurements.—Male, holotype, length of carapace 12, width of same 16.8 mm.

Range.—Costa Rica.
Material examined.—Punta Arenas, Costa Rica; Oersted, collector; 1 male holotype, 1 female (Copen. Mus.); 1 male paratype (31506).

UCA HELLERI Rathbun.

Plate 151.

Uca helleri RATHBUN, Proc. Washington Acad. Sci., vol. 4, 1902, p. 277, pl. 12, figs. 3 and 4 (type-locality, Mangrove Point, Narboro Island; male holotype, Cat. No. 24829, U.S.N.M.).

Diagnosis.—Orbits very oblique. Upper margin of palm without carina; oblique ridge continued to upper margin.

Description.—A small species. Carapace moderately convex, depressions shallow. Surface microscopically granulate, with more distant punctae. Front less than one-third width of carapace; margin broadly rounded. Upper margin of orbit very oblique, scarcely sinuous. Carapace widest at antero-lateral angles, which are rectangular. Sides strongly convergent posteriorly, moderately sinuous.

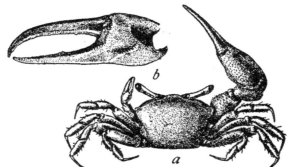

FIG. 170.—UCA HELLERI, MALE, × 3. *a*, DORSAL VIEW OF HOLOTYPE; *b*, INNER SURFACE OF LARGER CHELA (25666).

Lower margin of orbit as well as greater part of eyebrow visible in dorsal view.

Outer surface of merus and carpus of large cheliped of male crossed by short transverse lines of fine granules. Outer surface of manus covered with granules larger above and very fine on outer portion, somewhat reticulated, leaving small smooth patches. An irregular pit behind union of fingers. Oblique ridge of inner surface at angle of 45° to lower margin, turning at middle of palm at a right angle or somewhat obtuse angle, and continued in an uneven line to upper margin. Surface between ridges almost smooth. Fingers long and slender, with a broad gape. Of the prehensile tubercles, there is a larger one near middle of pollex, one at basal third or fourth of dactylus, and sometimes one at distal third of same. Merus of first three legs dilated.

Measurements.—Male (24829), length of carapace 5.6, width of same 8.1 mm.

Range.—Galapagos Islands.

Material examined.—Mangrove Point, Narborough Island, Galapagos Islands; March, 1899; from Stanford Univ.; 3 males (1 is holotype), 1 female (24829). Black Bight, Albemarle Island, Galapagos Islands; Jan. 9, 1899; from Stanford Univ.; 1 male (25666).

UCA STENODACTYLUS (Milne Edwards and Lucas).

Plate 152, fig. 3; plate 153.

Gelasimus stenodactylus MILNE EDWARDS and LUCAS, in d'Orbigny's Voy. dans l'Amér. Mérid., vol. 6, 1844, Crust., p. 26; vol. 9, atlas, 1847, pl. 11, fig. 2 (type-locality, Valparaiso; type in Paris Mus.).

Gelasimus gibbosus SMITH, Trans. Connecticut Acad. Arts and Sci., vol. 2, 1870, p. 140, pl. 2, fig. 11; pl. 4, fig. 8 (type-locality, Gulf of Fonseca; type in M.C.Z.).

Uca stenodactyla ORTMANN, Zool. Jahrb., Syst., vol. 10, 1897, p. 356 (part).

Uca gibbosa HOLMES, Occas. Papers California Acad. Sci., vol. 7, 1900, p. 77.

Diagnosis.—Third to sixth abdominal segments in male incompletely fused. Oblique ridge inside palm forming a right angle and continued to upper margin. Carapace subcylindrical.

Description.—Small species.

Carapace semicylindrical, narrowing little behind; surface smooth; interregional depressions deep; antero-lateral region curving abruptly downward at the sides.

Front less than one-quarter width of carapace, the sharp marginal carina being obliterated on the anterior border. Upper margin of orbit very oblique, slightly sinuous. Eyebrow very deep. The carapace may be widest at the subrectangular antero-lateral angles, or a little behind that point. The forward part of the side margins is concave; the whole margin is sinuous.

The merus of the large cheliped of the male is longer than the carapace; its short, transverse, granulated lines are inconspicuous. Wrist and palm granulate; granules on lower half of palm smaller and more flattened than those on upper half. A deeply depressd area at the union of the palm with the immovable finger. Fingers very slender and tapering, widely gaping; of the prehensile tubercles there is a larger one near the base of the dactyl, another near middle of pollex. The transverse ridge across the inner surface of the palm is bent at a right angle at the center, armed with a single row of large tubercles below and a band of smaller ones above. Space smooth between this ridge and the one at the distal end of the palm.

Ambulatory legs narrow; merus of first three pairs somewhat dilated proximally.

Abdomen of male with third to sixth segments fused.

Measurements.—Male (32322), length of carapace 8.5, width of same 13 mm.

Range.—Gulf of Fonseca, Central America, to Valparaiso. Brazil (Milne Edwards).

Material examined.—

Gulf of Fonseca, Salvador; J. A. McNiel; 1 male, holotype of *Gelasimus gibbosus* (M. C. Z.).

Boca del Rio Jesus Maria, Costa Rica; mangrove swamps; April, 1905; J. F. Tristan; 1 male (32321).

Puntarenas, Costa Rica; February, 1907; P. Biolley and J. F. Tristan; 3 males (39098).

Puntarenas, Estero side, Costa Rica; February, 1905; J. F. Tristan; 3 males (32322).

Panama City, Panama; Dec. 12, 1913; J. Zetek; 1 specimen (48795).

Valparaiso, Chile; 1 male, 1 female, cotypes (Paris Mus.).

UCA MUSICA Rathbun.

Plate 154.

? *Gelasimus annulipes* BATE, in Lord's Naturalist in Vancouver Island and British Columbia, vol. 2, 1866, p. 271.—? KINGSLEY, Proc. Acad. Nat. Sci. Philadelphia, 1880, p. 148 (part: Vancouver).

Gelasimus gibbosus LOCKINGTON, Proc. California Acad. Sci., vol. 7, 1876 (1877), p. 150 [6].—STREETS, Bull. U. S. Nat. Mus., vol. 7, 1877, p. 113. Not *G. gibbosus* Smith.

Uca stenodactyla ORTMANN, Zool. Jahrb., Syst., vol. 10, 1897, p. 356 (part).

Uca stenodactylus RATHBUN, Proc. U. S. Nat. Mus., vol. 21, 1898, p. 603 (not synonymy).

Uca musica RATHBUN, Proc. U. S. Nat. Mus., vol. 47, 1914, text-fig. 5, pl. 10 (type-locality, Pichilinque Bay, Gulf of California; holotype male, Cat. No. 22081, U.S.N.M.).

Diagnosis.—Third to sixth abdominal segments in male almost completely fused. Oblique ridge inside palm bent at an obtuse angle; an oblique stridulating ridge near proximal, lower angle. Carapace subcylindrical.

Description.—Very like *U. stenodactylus* but differs as follows: The upper margin of the orbit is much less oblique; the lateral angle of the carapace, marking its greatest width, is farther back. The granules of the palm are of more uniform size. The palm is scarcely depressed near the immovable finger. The dactylus is more strongly arched. The transverse ridge across the inner surface of the palm is very prominent, is bent at an obtuse and rounded angle and is armed for nearly its whole length with a row of large tubercles. Near the proximal lower corner of the inner surface there is a longitudinally oblique stridulating ridge extending from the articulation with the carpus to the lower marginal line of the palm almost below the angle of the transverse ridge. The stridulating ridge is made up of closely placed parallel lines oblique to the axis of the ridge and subparallel to the lower margin of the palm. When the cheliped is flexed the ridge plays against the line of granules on the lower or anterior surface of the first ambulatory leg; this line extends nearly the whole length of the carpal segment and part way along the merus. The third to sixth abdominal segments of the male are more completely fused than in *stenodactylus*.

Color.—Prevailing tint of carapace and limbs (in spirit) blue, of varying intensity, shading in parts into greenish and on the fingers of the chelipeds into white (Lockington).

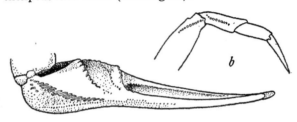

FIG. 171.—UCA MUSICA, MALE HOLOTYPE, × 3½. *a*, LOWER VIEW OF LARGER, LEFT CHELA, SHOWING STRIDULATING RIDGE; *b*, ANTERIOR (LOWER) VIEW OF PORTION OF FIRST LEFT LEG, SHOWING GRANULES WHICH PLAY AGAINST STRIDULATING RIDGE.

Measurements.—Male holotype, length of carapace 8, width of same 12.9 mm.

Range.—From San Diego, California, to Mazatlan, Mexico; occasionally farther north.

Material examined.—

Vancouver Island, British Columbia; photographs of a male, received from C. F. Newcombe (U.S.N.M.).

Seattle, Washington; D'Arcy W. Thompson, collector; photographs of a large chela (U.S.N.M.).

Pichilinque Bay, Gulf of California, Lower California; Apr. 29, 1888; *Albatross;* 1 male holotype (22081).

La Paz, Gulf of California, Lower California; Dr. Thos. H. Streets, U. S. Navy; 6 males (2294).

Guaymas Bay, Gulf of California, Mexico; shore; Feb. 20, 1904; Wm. Palmer; 1 male (31512).

Mazatlan, Mexico; 1881; C. H. Gilbert; 2 males (5054).
Additional records.—San Diego (Ortmann). San Bartolome Bay, Lower California (Lockington).

UCA SUBCYLINDRICA (Stimpson).

Plate 155; plate 160, fig. 5.

Gelasimus subcylindricus STIMPSON, Ann. Lyc. Nat. Hist. New York, vol. 7, 1859, p. 63 (type-locality, Matamoras, Rio Grande; cotypes in M.C.Z.).
Gelasimus subcylindricus SMITH, Trans. Connecticut Acad. Arts and Sci., vol. 2, 1870, p. 137, pl. 4, figs. 6–6b, 1870.
Uca pugilator ORTMANN, Zool. Jahrb., Syst.. vol. 10, 1897, p. 352 (part).

Diagnosis.—Inner surface of palm without oblique ridge. Abdomen of male very wide. Carapace subcylindrical.

Description.—A larger species than *U. stenodactylus*. Carapace longer in proportion to the width. Depressions shallow or obsolete. Antero-lateral region less strongly curved downward. Front nearly one-third width of carapace at antero-lateral angles, its marginal carina well marked. Eyebrow shallower, invisible in dorsal view. Carapace widest behind the blunt and obtuse-angled antero-lateral angles; side margins anteriorly straight and oblique; then faintly marked by broken granulate lines.

Large cheliped of male stouter than in *U. stenodactylus*. Merus shorter than carapace; granules of carpus few and scabrous; those of palm very coarse everywhere. Fingers little if any longer than palm, moderately narrow, a longitudinal ridge runs along middle of the pollex and is continued upon the palm; several enlarged tubercles on each finger. No oblique ridge across inner surface of palm, but the surface distal to the carpal cavity is coarsely granulate. Merus of first three pairs of ambulatory legs rather broad. Abdomen of male very broad, all segments distinct.

Measurements.—Male (23655), length of carapace 12.5, width of same 19 mm.

Range.—Texas and Mexico.

Material examined.—

Corpus Christi, Texas; 1900; V. Bailey, Biol. Survey, U. S. Dept. Agriculture; 1 male (23655).

Near Santa Rosa, Cameron Co., Texas; 1891; William Lloyd, collector; from U. S. Dept. of Agriculture; 1 male (17807).

Matamoros, Mexico, on the Rio Grande; "L. B.," meaning perhaps "left bank"; M. Berlandier, collector; deposited by Lieut. Couch, U. S. Army; 1 male, 3 females, cotypes (M. C. Z.).

Mexico; received from William Stimpson; 2 males (Paris Mus.) Perhaps these also are cotypes.

UCA FESTAE Nobili.

Uca festae NOBILI, Boll. Mus. Zool. Anat. comp. R. Univ. Torino, vol. 16, 1902, No. 415, p. 51 (type-locality, Rio Daule Inferiore; type in Turin Mus.).

Diagnosis.—Oblique ridge inside palm continued to upper margin. Fingers very long. Carapace subcylindrical.

Description (after Nobili).—Carapace very wide, usually (but not always) wider behind than between, the external orbital angles. Sulci very deep. Orbital angle, a right angle and variable.

Chelipeds enormously developed, four times width of carapace. Merus and carpus rugose-granulose, the latter bearing inside two or three large dentiform tubercles at its base. Hand very long, narrow, rather cuneiform, smooth below, granulate above, the granules near the articulation of the finger much larger and more spherical. On the inner surface, an angled crest analogous to that in *U. speciosa* and its allies. Surface between crests smooth. Fingers 3 or 4 times as long as palm, very slender, twisted and terminating in a point. Dactyl longer than pollex, denticulate in proximal third only, pollex denticulate throughout.

Merus joints of legs of moderate width.

Measurements.—Male, Rio Daule, length of carapace 12, width of same 18 mm.

Locality.—Ecuador: Rio Daule Inferiore, in brackish water.

Affinity.—Differs from *U. stenodactylus*, in having the palm narrower, less narrowed toward the base, fingers longer and more curved, abdominal segments all free.

UCA LEPTODACTYLA Rathbun.

Plate 156.

Gelasimus leptodactylus GUÉRIN, MS.
Gelasimus stenodactylus KINGSLEY, Proc. Acad. Nat. Sci. Philadelphia, 1880, p. 154 (part).
?*Gelasimus poeyi* GUÉRIN MS. (See Kingsley, Proc. Acad. Nat. Sci. Philadelphia, 1880, p. 154.)
Uca stenodactyla ORTMANN, Zool. Jahrb., Syst., vol. 10, 1897, p. 356 (part).
Uca leptodactyla RATHBUN, in Rankin, Ann. New York Acad. Sci., vol. 11, 1898, No. 12, p. 227 (type-locality, near Ft. Montagu, Nassau, N. P.; type, Cat. No. 22315, U.S.N.M.).
?*Uca gibbosa* NOBILI, Boll. Mus. Zool. Anat. comp. R. Univ. Torino, vol. 14, 1899, No. 355, p. 5.

Diagnosis.—Carapace widest at antero-lateral angles, subcylindrical. Oblique ridge inside palm continued to upper margin. Fingers slender, one and one-half times as long as palm. Third to sixth abdominal segments in male fused.

Description.—A small, semicylindrical species. Carapace strongly narrowed behind; surface smooth, depressions shallow. Front about one-fourth width of carapace, its anterior border subtruncate. Upper margin of orbit very oblique, slightly sinuous. Eyebrow partly visible from above. Carapace widest at the antero-lateral angles, which are prominent and acutangular or rectangular. Side margins straight or a little concave in front; then after a very obtuse angle they form a sinuous line.

The merus of the large cheliped of the male is longer than the carapace and bears short and fine, granulated rugae. The carpus and manus are coarsely granulate above, the granules of the palm becoming smaller on the outside. The palm turns over rather abruptly above. Fingers one and one-half times as long as palm, narrow; a large tubercle at middle of dactylus, another at its basal fourth; between these two there is a low tooth on the pollex, and another subterminal. A coarsely tuberculate ridge on inner side of palm, upper half transverse, lower half oblique as usual. Surface distal to the ridge smooth.

Merus joints of ambulatory legs more dilated than in *U. stenodactylus*. Legs crossed by transverse bands of dark spots. Third to sixth abdominal segments in male more or less fused.

Measurements.—Male (24546), length of carapace 6, width of same 9.8 mm.

Range.—West coast of Florida; Bahamas; West Indies; Mexico to Santos, Brazil.

Material examined.—

West Florida; J. E. Benedict; 1901; 1 female (25557).

Green Turtle Cay, Bahamas; E. A. Andrews; 4 males (22314).

Near Fort Montagu, Nassau, New Providence, Bahamas; holes in sand between tides about 5-6 inches deep; Jan. 28, 1890; J. I. Northrup; 1 male, 1 female (22315).

New Providence, Bahamas; *Albatross;* 3 males (11384).

Baracoa, Cuba; tidal flat; Jan. 29, 1902; Wm. Palmer; 2 males, 2 females (25548).

Jamaica; T. H. Morgan; 4 males, 1 female (22316).

San Antonio Bridge, San Juan, Porto Rico; Jan. 14, 1899; *Fish Hawk;* 13 males, 9 females (24546).

Fajardo, Porto Rico; Feb. 17, 1899; *Fish Hawk;* 1 female (24549).

Ensenada Honda, Culebra, Porto Rico; Feb. 9, 1899; *Fish Hawk;* 1 female (24548).

Mexico; T. B. Wilson; 3 males, labeled by Guérin (Mus. Phila. Acad.).

Amarracão, State of Paiuhy, Brazil; 1913; Fr. Iglezias; 1 male (48894).

Parahyba River, Brazil; mangroves; June 21, 1899; Branner-Agassiz Exped.; 1 male (25701).
Pernambuco, Brazil; 1876-1877; R. Rathbun; 14 males, 3 females (40617).
? Pernambuco, Brazil; 1876-1877; R. Rathbun; 12 males, 7 females (40618).
Bahia, Brazil; sea shore; May, 1915; J. N. Rose; 3 males, 3 females (48297).
Plataforma, Bahia, Brazil; 1876-1877; R. Rathbun; 17 males, 9 females (40615-40616).
Porto Seguro, Brazil; Thayer Exped.; 3 males, 1 female (22189).
Rio de Janeiro, Brazil; Dr. Naegeli; 1 male, 1 female (22317).
Maruim, Brazil; Sept. 9, 1863; Hygom, collector; 4 males (22318); many specimens in Copenhagen Mus., labeled " *G. Hygomi* Ltk."
S. Sebastiao, São Paulo, Brazil; 1896; Bisego, collector; 1 male (47850).
Santos, Brazil; 2 specimens (Copenhagen Mus.).

UCA LATIMANUS (Rathbun).

Plate 157.

Gelasimus stenodactylus ? LOCKINGTON, Proc. California Acad. Sci., vol. 7, 1876 (1877), p. 148 [4].

Gelasimus latimanus RATHBUN, Proc. U. S. Nat. Mus., vol. 16, 1893, p. 245 (type-locality, La Paz; type, Cat. No. 17500, U.S.N.M.).

Uca latimana NOBILI, Boll. Mus. Zool. Anat. comp. R. Univ. Torino, vol. 16, 1901, No. 415, p. 52.

Diagnosis.—Inner surface of palm granulate and without oblique ridge. Fingers shorter than palm. Third to sixth abdominal segments in male fused. Carapace subcylindrical.

Description.—A small, subcylindrical species. Carapace moderately narrowed behind; surface finely granulate, depressions well marked. Front nearly one-third width of carapace, its anterior border slightly curved. Upper margin of orbit slightly oblique, sinuous. Eyebrow deepest at its middle. Antero-lateral angles acute and prominent, projecting forward. Carapace widest behind those angles, the anterior part of the side margins being concave or straight.

Large cheliped of male short and stout. Merus shorter than carapace, its lower margin sharply serrulate, outer surface granulate-rugose. Carpus granulate. Palm not much longer than broad, coarsely granulate, as is also the upper surface of the dactylus and lower surface of pollex. Fingers broad and flat, two-thirds as long as palm, regularly tapering, slightly curved, blunt tipped, granulate; interspace very narrow and regularly diminishing. Prehensile tubercles low and molariform; a prominent submarginal row of granules above and below. Inner surface of palm without oblique transverse ridge; the subtriangular area forming the lower proximal portion is granulate.

Merus of first three pairs of legs much enlarged. The third to sixth segments of the abdomen of the male are fused, though vestiges of the sutures remain.

Measurements.—Male, holotype, length of carapace, 6.3; width of same, 10 mm.

Range.—From La Paz, Lower California, to Tumaco, Colombia, on the frontier of Ecuador (Nobili).

Material examined.—La Paz, Lower California; L. Belding; 1 male, holotype (17500). Santo Domingo, Gulf of Dulce, Costa Rica; April, 1896; H. Pittier; 1 male (19442).

Subfamily MACROPHTHALMINAE Dana.

Gonoplacés vigils MILNE EDWARDS, Ann. Sci. Nat., ser. 3, Zool., vol. 18, 1852, p. 155 [119] (part).

Macrophthalminæ DANA, Amer. Journ. Sci., ser. 2, vol. 12, 1851, p. 286; U. S. Expl. Exped., vol. 13, Crust., pt. 1, 1852, p. 312.—ALCOCK, Journ. Asiat. Soc. Bengal, vol. 69, 1900, p. 290.

Carapace usually quadrilateral, broader than long (sometimes more than twice as broad as long), flattish and not very deep, the regions usually well defined; front variable, but never very broad; antennules with a well-developed flagellum that folds transversely, interantennular septum very narrow; eyestalks usually elongate; the external maxillipeds do not always meet across the buccal cavern, though the gap between them is never very wide, their exognath is not, or not entirely, concealed and has a flagellum; chelipeds usually subequal. No special recess between the bases of any of the legs.

Genus EUPLAX Milne Edwards.

Euplax MILNE EDWARDS, Ann. Sci. Nat., Zool., ser. 3, vol. 18, 1852, p. 160 [124]; type, *E. leptophthalmus* Milne Edwards.

Chaenostoma STIMPSON, Proc. Acad. Nat. Sci. Philadelphia, vol. 10, 1858, p. 97 [43]; type, *C. orientale* Stimpson, 1858=*E. bosci* (Audouin, 1825).

Carapace not much wider than long; lateral margins dentate. Eyestalks of moderate length, not overreaching the orbital angles, which do not form a prominent tooth. Outer maxillipeds with a rhomboidal gape.

Distributed through the Indo-Pacific, Australia, and Chile.

EUPLAX LEPTOPHTHALMA Milne Edwards.

Euplax leptophthalmus MILNE EDWARDS, Ann. Sci. Nat., ser. 3, Zool., vol. 18, 1852, p. 160 [124] (type-locality, Chile; type in Paris Mus.).

Eyestalks very slender and much shorter than the orbits. Front of medium width. Three lateral teeth, wide and elevated. (After Milne Edwards.)

Range.—Chile.

FIG. 172.—PREHISTORIC BOWL UNEARTHED IN COSTA RICA, THE BASE REPRESENTING A FEMALE LAND-CRAB, CARDISOMA GUANHUMI, WITH THE CLAWS FOLDED UNDER THE EYES; × ABOUT ⅔. ORIGINAL IN POSSESSION OF MRS. ZELEDON; CAST IN U. S. NATIONAL MUSEUM.

EXPLANATION OF PLATES.

PLATE 1.

Trizocarcinus dentatus, male holotype, ×1½.

FIG. 1. Antero-ventral view.
2. Dorsal view.
3. Posterior view.

PLATE 2.

Bathyplax typhla, ×1⅜.

FIG. 1. Male (9724), ventral view, to show maxilliped and abdomen.
2. Female (9729), dorsal view.
3. Male (9724), ventral view, to show chelae.

PLATE 3.

Pilumnoplax elata, ×3.

FIG. 1. Male (11407), dorsal view, with 2 detached legs.
2. Male (11407), ventral view, and right cheliped detached.
3. Female (19880), dorsal view.

PLATE 4.

FIG. 1. *Goneplax barbata*, male (4116, M.C.Z.), dorsal view, ×3⅞.
2. *Goneplax sigsbei*, female (4117, M.C.Z.), dorsal view, ×3⅞.
3. *Goneplax barbata*, male (4116, M.C.Z.), ventral view, ×3⅞.
4. *Goneplax sigsbei*, female (4117, M.C.Z.), ventral view, ×3⅞.

PLATE 5.

Goneplax barbata, male (46309), ×4.

FIG. 1. Anterior view.
2. Dorsal view.
3. Posterior view.
4. Ventral view.

PLATE 6.

FIG. 1. *Prionoplax atlantica*, male (15272), anterior view, ×3.
2. Same, dorsal view, with one leg and part of right cheliped detached, ×3.
3. *Tetraplax quadridentata*, male (24564), dorsal view, ×3.
4. Same, ventral view, ×3.

PLATE 7.

Euryplax nitida, male (45824), × about 3.

FIG. 1. Anterior view.
2. Dorsal view.
3. Ventral view.

PLATE 8.

Speocarcinus carolinensis, female (45951), ×1¼.

FIG. 1. Dorsal view.
2. Ventral view.

PLATE 9.

Speocarcinus granulimanus, male (17461), ×1⅜.

FIG. 1. Dorsal view.
2. Ventral view.

PLATE 10.

FIG. 1. *Speocarcinus ostrearicola*, male holotype, dorsal view, ×2.
2. *Speocarcinus californiensis*, male (45581), dorsal view, ×1½.
3. Same, ventral view, ×1½.

PLATE 11.

Cyrtoplax spinidentata, male holotype, ×1½.

FIG. 1. Antero-ventral view.
2. Dorsal view.
3. Posterior view.

PLATE 12.

FIG. 1. *Panoplax depressa*, male (24556), dorsal view, ×3.
2. Same, ventral view, ×3.
3. *Eucratopsis crassimanus*, male (45952), dorsal view, ×2.

PLATE 13.

FIG. 1. *Oediplax granulata*, female holotype, dorsal view, ×1.
2. Same, ventral view, ×1.
3. *Glyptoplax smithii*, male (19796), ventral view, ×4.
4. Same, female (18269), dorsal view, ×4.

PLATE 14.

FIG. 1. *Chasmocarcinus obliquus*, male holotype, dorsal view, ×5.
2. Same, ventral view of body, ×5.
3. *Pseudorhombila octodentata*, male holotype, dorsal view, ×1.

PLATE 15.

FIG. 1. *Pinnotheres holmesi*, female holotype, dorsal view, ×2 7/10.
2. Same, ventral view, ×2 7/10.
3. *Pinnotheres ostreum*, female (49208), dorsal view, ×1⅜.
4. Same, ventral view, ×1⅜.
5. *Pinnotheres ostreum*, male (49209), ventral view, ×6½.
6. Same, dorsal view, ×6½.

PLATE 16.

FIG. 1. *Pinnotheres geddesi*, female (23439), ventral view, ×2 7/10.
2. Same, dorsal view, ×2 7/10.
3. *Pinnotheres geddesi*, female (cotype, Brit. Mus.), ventral view, ×2 7/10.
4. Same, dorsal view, ×2 7/10.
5. *Pinnotheres angelicus*, female (17467), ventral view, ×2 7/10.
6. Same, dorsal view, ×2 7/10.

PLATE 17.

FIG. 1. *Pinnotheres depressus*, male (48594), ventral view, × nearly 7.
2. Same, dorsal view, × nearly 7.
3. *Pinnotheres maculatus*, male (18014), dorsal view, × nearly 2.
4. Same, female (3818), dorsal view, × nearly 2.
5. Same, male (18014), ventral view, × nearly 2.
6. Same, female (3818), ventral view, × nearly 2.
7. *Pinnotheres pugettensis*, female (39131), ventral view, × nearly 3.
8. Same, dorsal view, × nearly 3.

PLATE 18.

FIG. 1. *Pinnotheres pugettensis*, female (40396), infero-external view of right chela, and superior view of four left legs, $\times 4\frac{4}{5}$.
2. *Pinnotheres muliniarum*, male holotype, dorsal view, $\times 10\frac{1}{2}$.
3. Same, ventral view, $\times 10\frac{1}{2}$.

PLATE 19.

FIG. 1. *Pinnotheres serrei*, female (48571), dorsal view, × nearly 2.
2. Same, ventral view, × nearly 2.
3. *Pinnotheres serrei*, female (49214), ventral view, × nearly 3.
4. Same, dorsal view, × nearly 3.
5. *Pinnotheres serrei*, male holotype, ventral view, × nearly $3\frac{3}{4}$.
6. *Pinnotheres serrei*, male (49213), dorsal view, × nearly 7.
7. Same, ventral view, × nearly 7.
8. *Pinnotheres barbatus*, male (23435), dorsal view, × nearly 2.
9. Same, female (23435), dorsal view, × nearly 2.
10. Same, male (23435), ventral view, × nearly 2.
11. Same, female (23435), ventral view, × nearly 2.

PLATE 20.

FIG. 1. *Pinnotheres strombi*, female holotype, ventral view, × nearly 2.
2. Same, dorsal view, × nearly 2.
3. *Pinnotheres concharum*, female (45611), dorsal view, × nearly 9.
4. Same, ventral view, × nearly 9.
5. *Pinnotheres concharum*, female (18410), dorsal view, × nearly 4.
6. Same, ventral view, × nearly 4.

PLATE 21.

FIG. 1. *Pinnotheres reticulatus*, female (18217), ventral view, $\times 1\frac{9}{10}$.
2. Same, dorsal view, $\times 1\frac{9}{10}$.
3. *Pinnotheres moseri*, female (23440), ventral view, $\times 2\frac{4}{5}$ (maxillipeds removed).
4. Same, dorsal view, $\times 2\frac{4}{5}$.
5. *Pinnotheres taylori*, male (40397), ventral view, $\times 3\frac{3}{4}$.
6. Same, dorsal view, $\times 3\frac{3}{4}$.
7. *Pinnotheres taylori*, female (40397), ventral view, $\times 4\frac{2}{3}$.
8. Same, dorsal view, $\times 4\frac{2}{3}$.

Plate 22.

Fig. 1. *Pinnotheres shoemakeri*, female (49220), dorsal view, $\times 6$.
2. Same, male holotype, dorsal view, $\times 6\frac{1}{8}$.
3. Same, female (49220), ventral view, $\times 6$.
4. Same, male holotype, ventral view, $\times 6\frac{1}{8}$.
5. *Pinnotheres orcutti*, male holotype, ventral view, $\times 6$.
6. Same, dorsal view, $\times 6$.

Plate 23.

Fig. 1. *Pinnotheres hemphilli*, male holotype, dorsal view, $\times 13$.
2. Same, ventral view, $\times 13$.

Plate 24.

Fig. 1. *Fabia subquadrata*, female (23928), dorsal view, $\times 1\frac{2}{3}$.
2. *Fabia lowei*, female (45583), dorsal view, $\times 1\frac{2}{3}$.
3. *Fabia subquadrata*, female (23928), ventral view, $\times 1\frac{2}{3}$
4. *Fabia lowei*, female (45583), ventral view, $\times 1\frac{2}{3}$.
5. *Fabia canfieldi*, female holotype, dorsal view, $\times 2\frac{2}{10}$.
6. *Fabia byssomiae*, female (48595), dorsal view, $\times 3\frac{1}{3}$.
7. *Fabia canfieldi*, female holotype, ventral view, $\times 2\frac{2}{10}$.
8. *Fabia byssomiae*, female (48595), ventral view, $\times 3\frac{1}{3}$.

Plate 25.

Fig. 1. *Pinnaxodes meinerti*, female (5760, M.C.Z.), ventral view, $\times 1\frac{1}{8}$.
2. Same, dorsal view, $\times 1\frac{1}{8}$.
3. *Pinnaxodes meinerti*, male holotype, dorsal view, $\times 1\frac{3}{10}$.
4. *Parapinnixa bouvieri*, male (5744, M.C.Z.), legs of left side, $\times 6$.
5. Same, ventral view, $\times 6$.
6. Same, dorsal view, $\times 6$.
7. *Parapinnixa bouvieri*, female (holotype), dorsal view, $\times 6$.
8. Same, ventral view, $\times 6$.
9. Same, right chela, $\times 6$.
10. Same, left chela and right legs, $\times 6$.

Plate 26.

Fig. 1. *Parapinnixa hendersoni*, male holotype, anterior view, $\times 2\frac{3}{4}$.
2. Same, dorsal view, $\times 2\frac{3}{4}$.
3. Same, ventral view, $\times 2\frac{3}{4}$.
4. *Parapinnixa hendersoni*, female (48711), ventral view, $\times 2\frac{3}{4}$.
5. Same, antero-dorsal view, $\times 2\frac{3}{4}$.
6. *Dissodactylus nitidus*, female (22113), ventral view, $\times 2\frac{3}{4}$.
7. Same, dorsal view, $\times 2\frac{3}{4}$.

Plate 27.

Fig. 1. *Dissodactylus encopei*, male (23430), ventral view, $\times 3$.
2. Same, dorsal view, $\times 3$.
3. *Dissodactylus encopei*, female (23431), ventral view, $\times 3$.
4. Same, dorsal view, $\times 3$.
5. *Dissodactylus borradailei*, male (23790), ventral view, $\times 4$.
6. Same, dorsal view, $\times 4$.
7. *Dissodactylus borradailei*, female holotype, ventral view, $\times 3$.
8. Same, dorsal view, $\times 3$.

PLATE 28.

FIG. 1. *Dissodactylus stebbingi*, male holotype, ventral view, ×4.
2. Same, dorsal view, ×4.
3. *Dissodactylus alcocki*, male (23447), ventral view, together with right leg, ×4.
4. Same, dorsal view, together with right leg, ×4.
. *Dissodactylus calmani*, female holotype, ventral view, ×3.
5. Same, dorsal view, ×3.
7. *Dissodactylus mellitae*, male (40271), ventral view, ×7.
8. Same, dorsal view, together with left leg, ×7.

PLATE 29.

FIG. 1. *Pinnixa transversalis*, female (20625), dorsal view, × nearly 2.
2. *Pinnixa transversalis*, male (20625), ventral view, × nearly 2.
3. Same, dorsal view, × nearly 2.
4. *Pinnixa faxoni*, male (7639), ventral view, × nearly 2.
5. Same, dorsal view, × nearly 2.
6. *Pinnixa faxoni*, female (7639), ventral view, × nearly 2.
7. Same, dorsal view, × nearly 2.
8. *Pinnixa cristata*, female (42817), ventral view, × nearly 2.
9. Same, dorsal view, also separately, left chela, third leg and fourth leg, × nearly 2.

PLATE 30.

FIG. 1. *Pinnixa patagoniensis*, male holotype, anterior view. ×2.
2. Same, dorsal view, ×2.
3. Same, ventral view, ×2.
4. *Pinnixa floridana*, female holotype, ventral view. ×3.
5. Same, dorsal view, ×3.
6. *Pinnixa floridana*, male (49249), dorsal view, together with legs of right side, ×4.
7. Same, ventral view, ×4.
8. *Pinnixa tomentosa*, female (29948), dorsal view, ×2.

PLATE 31.

FIG. 1. *Pinnixa faba*, female (31599), dorsal view, ×1½.
2. Same, male (31599), dorsal view, ×1½.
3. Same, female (31599), ventral view, ×1½.
4. Same, male (31599), ventral view, ×1½.
5. *Pinnixa littoralis*, female (31600), dorsal view, ×1½.
6. Same, male (31600), dorsal view, ×1½.
7. Same, female (31600), ventral view, ×1½.
8. Same, male (31600), ventral view, ×1½.

PLATE 32.

Pinnixa barnharti.

FIG. 1. Male (5742, M.C.Z.), ventral view, ×2.
2. Female holotype, dorsal view, ×2.
3. Same, ventral view, ×2.
4. Male (5742, M.C.Z.), abdomen, ×6.

PLATE 33.

FIG. 1. *Pinnixa valdiviensis*, male (5740, M.C.Z.), dorsal view, × nearly 2.
2. Same, ventral view, × nearly 2.
3. *Pinnixa chaetopterana*, female (26105), dorsal view, × nearly 2.
4. Same, male (5824), dorsal view, × nearly 2.
5. Same, female (26105), ventral view, × nearly 2.
6. Same, male (5824), ventral view, × nearly 2.

PLATE 34.

FIG. 1. *Pinnixa occidentalis*, male (17470), dorsal view, nat. size.
2. *Pinnixa sayana*, male (48438), dorsal view, ×3.
3. *Pinnixa sayana*, female (34017), ventral view, ×4.
4. Same, dorsal view, ×4.
5. *Pinnixa valdiviensis*, male cotype, ventral view, ×2.
6. Same, dorsal view, ×2.

PLATE 35.

FIG. 1. *Pinnixa franciscana*, male (48445), ventral view, ×3.
2. Same, dorsal view, ×3.
3. *Pinnixa franciscana*, female (48450), dorsal view, ×3.
4. Same, ventral view, ×3.
5. *Pinnixa cylindrica*, male (17952), dorsal view, ×2.
6. *Pinnixa schmitti*, female holotype, dorsal view, ×3.
7. Same, male (25850), dorsal view, ×3.
8. *Pinnixa cylindrica*, male (17952), ventral view, ×2.
9. *Pinnixa schmitti*, female holotype, ventral view, ×3.

PLATE 36.

FIG. 1. *Pinnixa hiatus*, female (29949), dorsal view, × nearly 4.
2. Same, ventral view, × nearly 4.
3. Same, two legs of left side, × nearly 4.
4. Same, chela and four legs of right side, × nearly 4.
5. *Pinnixa tubicola*, female (20860), dorsal view, × nearly 2.
6. Same, third leg of right side, × nearly 2.
7. Same, ventral view, × nearly 2.
8. *Pinnixa tubicola*, male, Trinidad (Stanford Univ.), dorsal view, × nearly 3.
9. *Pinnixa weymouthi*, male holotype, dorsal view, together with cheliped and two legs of right side, × nearly 4.
10. Same, ventral view, × nearly 4.

PLATE 37.

FIG. 1. *Scleroplax granulata*, male (49247), dorsal view, ×3⅔.
2. Same, ventral view, ×3⅔.
3. *Scleroplax granulata*, female paratype, dorsal view, ×2.
4. *Opisthopus transversus*, male (3446), ventral view, ×1⅜.
5. Same, dorsal view, ×1⅜.

Plate 38.

Pinnaxodes chilensis.

Fig. 1. Male (5737), dorsal view, × nearly 3.
2. Same, ventral view, × nearly 3.
3. Same, inner view of left chela, × nearly 4.
4. Same, outer view of left chela, × nearly 4.
5. Same, leg, × nearly 4.
6. Female (49238), ventral view, × nearly 2.
7. Same, dorsal view, × nearly 2.

Plate 39.

Fig. 1. *Pinnotherelia laevigata*, male (40445), anterior view, × nearly 2.
2. Same, dorsal view, × nearly 2.
3. Same, ventral view, × nearly 2.
4. *Tetrias scabripes*, female (21595), ventral view, × nearly 2.
5. Same, dorsal view, × nearly 2.

Plate 40.

Fig. 1. *Pinnotherelia laevigata*, female (2435), dorsal view, ×4.
2. Same, ventral view, ×4.
3. *Cymopolia sica*, male (46063), dorsal view, ×2¾.
4. Same, ventral view, ×2¾.

Plate 41.

Fig. 1. *Pinnixa retinens*, male holotype, dorsal view, ×4.
2. Same, ventral view, ×4.
3. *Cymopolia floridana*, female holotype, dorsal view, ×4.
4. Same, ventral view, ×4.

Plate 42.

Cymopolia alternata.

Fig. 1. Male (19840), dorsal view, ×2¾.
2. Female (49190), ventral view, ×2¾.
3. Same, dorsal view, ×2¾.

Plate 43.

Cymopolia alternata.

Fig. 1. Male (19840), ventral view, ×2¾.
2. Female, variety (15282), dorsal view, ×2¾.
3. Same, ventral view, ×2¾.

Plate 44.

Fig. 1. *Cymopolia lucasii*, male holotype, dorsal view, ×2.
2. Same, ventral view, ×2.
3. *Cymopolia zonata*, male (22071), dorsal view, ×2.

PLATE 45.

FIG. 1. *Cymopolia zonata*, male (22071), ventral view, ×2.
2. *Cymopolia faxoni*, female holotype, dorsal view, ×3.
3. Same, ventral view, together with right chela and last left leg, detached, ×3.

PLATE 46.

Cymopolia affinis.

FIG. 1. Male (24515), dorsal view, ×2.
2. Same, ventral view, ×2.
3. Male (49196), dorsal view, ×3.

PLATE 47.

FIG. 1. *Cymopolia bahamensis*, male (22301), dorsal view, ×3.
2. *Cymopolia bahamensis*, female (22301), ventral view, ×3.
3. *Cymopolia affinis*, male (49196, second specimen), dorsal view, ×3.

PLATE 48.

FIG. 1. *Cymopolia rathbuni*, female immature (49191), dorsal view, ×5.
2. Same, ventral view, ×5.
3. *Cymopolia isthmia*, female holotype, dorsal view, ×5.
4. Same, ventral view, ×5.

PLATE 49.

Cymopolia obesa.

FIG. 1. Male (17895), dorsal view and left chela, ×2.
2. Male (17894), ventral view, ×2.
3. Same, dorsal view, ×2.

PLATE 50.

Cymopolia gracilis.

FIG. 1. Male (49192), ventral view, ×2⅔.
2. Same, dorsal view, ×2⅔.
3. Female (11411), ventral view, ×1⅔.

PLATE 51.

FIG. 1. *Cymopolia gracilis*, female (11411), dorsal view, ×1$\frac{7}{10}$.
2. *Cymopolia fragilis*, male (20620), dorsal view, ×2⅔.
3. Same, ventral view, ×2⅔.

PLATE 52.

FIG. 1. *Cymopolia cursor*, female (7818), dorsal view, with detached legs, ×1¾.
2. Same, left chela, ×5¼.
3. *Cymopolia gracilipes*, male (11395), dorsal view, with detached legs, ×4½.
4. Same, ventral view, ×2⅔.

PLATE 53.

Grapsus grapsus, male (22086), dorsal view, ×$\frac{21}{25}$.

PLATE 54.

Grapsus grapsus, male (48601).

FIG. 1. Ventral view, ×$\frac{4}{5}$.
2. Anterior view of front, ×$\frac{4}{5}$.

PLATE 55.

Geograpsus lividus.

FIG. 1. Male (24041), ventral view, ×$\frac{9}{10}$.
2. Male (47857), dorsal view, ×$\frac{9}{10}$.

PLATE 56.

Leptograpsus variegatus, male (49057), ×$\frac{4}{5}$.

FIG. 1. Dorsal view.
2. Ventral view.

PLATE 57

Goniopsis cruentata, male (23805), ×$\frac{2}{3}$.

FIG. 1. Ventral view.
2. Dorsal view.
3. Anterior view.

PLATE 58.

Goniopsis pulchra, male (17294), ×1.

FIG. 1. Ventral view.
2. Dorsal view.

PLATE 59.

Pachygrapsus crassipes, male (32234), ×$\frac{9}{10}$.

FIG. 1. Ventral view.
2. Dorsal view.

PLATE 60.

FIG. 1. *Pachygrapsus maurus*, male (44529), ventral view, ×3.
2. *Pachygrapsus maurus*, female (44529), dorsal view, ×3.
3. *Pachygrapsus gracilis*, male (49254), ventral view, ×2.

PLATE 61.

FIG. 1. *Pachygrapsus gracilis*, male (49254), dorsal view, ×1$\frac{2}{5}$.
2. *Pachygrapsus transversus*, male (40825), dorsal view, ×1$\frac{2}{5}$.
3. Same, ventral view, ×1$\frac{2}{5}$.

PLATE 62.

Pachygrapsus marmoratus, male (14497), ×1½.

FIG. 1. Dorsal view.
2. Ventral view.

PLATE 63.

Planes minutus, male (20695), ×2.

FIG. 1. Dorsal view.
2. Ventral view.

PLATE 64.

Planes marinus, male holotype, ×1 1/16.

FIG. 1. Antero-ventral view.
2. Dorsal view.
3. Ventral view.

PLATE 65.

Cyrtograpsus angulatus, male (18624), ×1½.

FIG. 1. Dorsal view.
2. Ventral view.

PLATE 66.

Cyrtograpsus altimanus, male, ×1½.

FIG. 1. Ventral view of holotype.
2. Dorsal view of same.
3. Antero-ventral view of paratype.

PLATE 67.

Hemigrapsus affinis, male (22104), ×5.

FIG. 1. Ventral view.
2. Dorsal view.
3. Ventral view, to show chelae.

PLATE 68.

Hemigrapsus crenulatus, male (22102), ×1½.

FIG. 1. Dorsal view.
2. Ventral view.

PLATE 69.

Hemigrapsus nudus, male (32233), ×¾.

FIG. 1. Anterior view.
2. Ventral view.
3. Dorsal view.

PLATE 70.

Hemigrapsus oregonensis, male (48818), ×1¼.

FIG. 1. Ventral view.
2. Dorsal view.
3. Anterior view.

PLATE 71.

Tetragrapsus jouyi, male (17496), ×2⅔.

FIG. 1. Ventral view.
2. Dorsal view.
3. Ventral view, to show chelae.

PLATE 72.

FIG. 1. *Glyptograpsus impressus*, male (44174), dorsal view, × nearly 2.
2. Same, ventral view, × nearly 2.
3. *Glyptograpsus jamaicensis*, male holotype, dorsal view, slightly reduced.

PLATE 73.

Platychirograpsus typicus, male holotype, dorsal view, slightly reduced.

PLATE 74.

FIG. 1. *Euchirograpsus americanus*, male (44675), dorsal view, ×2½.
2. Same, ventral view, ×2½.

PLATE 75.

Sesarma (Chiromantes) africanum, male (14870), slightly enlarged.

FIG. 1. Anterior view.
2. Dorsal view.
3. Ventral view.

PLATE 76.

Sesarma (Sesarma) verleyi, female holotype. ×1¼.

FIG. 1. Ventral view.
2. Dorsal view.
3. Anterior view.

PLATE 77.

Sesarma (Sesarma) reticulatum, male (45530), ×1½.

FIG. 1. Anterior view.
2. Dorsal view.
3. Ventral view.

PLATE 78.

FIG. 1. *Sesarma (Sesarma) curacaoense*, male (17678), anterior view, ×1⅜.
2. Same, dorsal view, ×1⅜.
3. *Sesarma (Sesarma) sulcatum*, male (4631), chela, ×⅝.
4. Same, dorsal view, ×⅝.

PLATE 79.

Sesarma (Sesarma) rhizophorae, male (32491), ×2.

FIG. 1. Anterior view.
2. Dorsal view.
3. Ventral view.

PLATE 80.

Sesarma (Sesarma) bidentatum, male (32285), ×1½.

FIG. 1. Anterior view.
2. Dorsal view.
3. Ventral view.

PLATE 81.

Sesarma (Sesarma) jarvisi, male holotype, ×2.

FIG. 1. Anterior view.
2. Dorsal view.
3. Ventral view.

PLATE 82.

Sesarma (Holometopus) rectum, male (47862), slightly enlarged.

FIG. 1. Anterior view.
2. Dorsal view.
3. Ventral view.

PLATE 83.

Sesarma (Holometopus) cinereum, male (15072), ×1⅜.

FIG. 1. Antero-dorsal view.
2. Dorsal view.
3. Ventral view.

PLATE 84.

Sesarma (Holometopus) miersii, male holotype, ×1½.

FIG. 1. Anterior view.
. Dorsal view.
3. Ventral view.

PLATE 85.

Sesarma (Holometopus) miersii iheringi, male (48299), ×1⅜.

FIG. 1. Ventral view.
2. Dorsal view.

PLATE 86.

Sesarma (Holometopus) magdalenense, male (45793), ×2½.

FIG. 1. Anterior view.
2. Dorsal view.
3. Ventral view.

PLATE 87.

FIG. 1. *Sesarma* (*Holometopus*) *hanseni*, male holotype, slightly reduced.
    2. *Sesarma* (*Holometopus*) *biolleyi*, male holotype, dorsal view, slightly reduced.
    3. Same, ventral view, slightly reduced.

PLATE 88.

Sesarma (*Holometopus*) *tampicense*, male holotype, $\times 1\frac{7}{10}$.

FIG. 1. Anterior view.
    2. Dorsal view.
    3. Ventral view.

PLATE 89.

Sesarma (*Holometopus*) *ricordi*, male (13798), $\times 1\frac{2}{3}$.

FIG. 1. Anterior view.
    2. Dorsal view.
    3. Ventral view.

PLATE 90.

Sesarma (*Holometopus*) *angustipes*.

FIG. 1. Female (42876), dorsal view, $\times 1\frac{1}{2}$.
    2. Male (46084), ventral view, $\times 1\frac{1}{2}$.
    3. Same, anterior view, $\times 1\frac{1}{2}$.

PLATE 91.

Sesarma (*Holometopus*) *roberti*, male (22108), $\times 1\frac{3}{20}$.

FIG. 1. Anterior view.
    2. Dorsal view.
    3. Ventral view.

PLATE 92.

Sesarma (*Holometopus*) *angustum*, male (32314), $\times 2$.

FIG. 1. Anterior view.
    2. Dorsal view.
    3. Ventral view.

PLATE 93.

Sesarma (*Holometopus*) *benedicti*, male (22838), $\times 1\frac{2}{3}$.

FIG. 1. Anterior view.
    2. Dorsal view.
    3. Ventral view.

PLATE 94.

Metasesarma rubripes, male (19408), $\times 2$

FIG. 1. Anterior view.
    2. Dorsal view.
    3. Ventral view.

PLATE 95.

Sarmatium curvatum, male (20309), ×1½.

FIG. 1. Anterior view.
2. Dorsal view.
3. Ventral view.

PLATE 96.

Aratus pisonii, male (25559), ×1½.

FIG. 1. Ventral view.
2. Dorsal view.

PLATE 97.

FIG. 1. *Cyclograpsus integer*, male (17669), dorsal view, ×1⅔.
2. Same, ventral view, ×1⅔.
3. *Metopaulias depressus*, female (42875), dorsal view, ×1⅔.
4. Same, anterior view, ×1⅔.

PLATE 98.

Cyclograpsus cinereus, male (45520), ×2⅔.

FIG. 1. Anterior view.
2. Dorsal view.
3 Ventral view.

PLATE 99.

Cyclograpsus punctatus, male (3240), ×1½.

FIG. 1. Dorsal view.
2. Ventral view.

PLATE 100.

Chasmagnathus granulata, male (22109), ×1 7/10.

FIG. 1. Dorsal view.
2. Ventral view.

PLATE 101.

Plagusia depressa, male (40609), ×⅔.

FIG. 1. Dorsal view.
2. Ventral view.

PLATE 102.

Plagusia depressa tuberculata, male (18826), ×7/10.

FIG. 1. Dorsal view.
2. Ventral view.

PLATE 103.

Plagusia immaculata, male (24758), ×⅞.

FIG. 1. Dorsal view.
2. Ventral view.
3. Maxillipeds.

PLATE 104.

Plagusia chabrus, male (2476), dorsal view, ×$\frac{87}{100}$.

PLATE 105.

Percnon gibbesi, male (48592), ×⅘.

FIG. 1. Dorsal view.
2. Ventral view.

PLATE 106.

Cardisoma guanhumi, male (17987), dorsal view, ×$\frac{44}{100}$.

PLATE 107.

Cardisoma guanhumi, female (42892), ventral view, ×$\frac{44}{100}$.

PLATE 108.

Cardisoma crassum, male (4658), dorsal view, ×$\frac{44}{100}$.

PLATE 109.

Cardisoma crassum, male (4658), ventral view, ×$\frac{44}{100}$.

PLATE 110.

Ucides cordatus, male (17595), dorsal view, ×⅘.

PLATE 111.

Ucides cordatus, male (17595), ventral view, ×$\frac{44}{100}$.

PLATE 112.

Ucides cordatus, female (17595), dorsal view, ×1¼.

PLATE 113.

Ucides cordatus, female (17595), ventral view, slightly enlarged.

PLATE 114.

Ucides occidentalis, male (2145), dorsal view, ×⅔.

PLATE 115.

Ucides occidentalis, male (2145), ventral view, $\times \frac{2}{10}$.

PLATE 116.

Ucides occidentalis, male (2145), antero-dorsal view, $\times \frac{2}{10}$.

PLATE 117.

Gecarcinus ruricola, female (42901), dorsal view, slightly enlarged.

PLATE 118.

Gecarcinus ruricola, female (42901), ventral view, $\times \frac{4}{5}$.

PLATE 119.

Gecarcinus lateralis, male (11368), dorsal view, $\times 1\frac{1}{5}$.

PLATE 120.

Gecarcinus lateralis, male (11368), ventral view, $\times 1\frac{1}{5}$.

PLATE 121.

Gecarcinus quadratus, male (32311), dorsal view, $\times 1\frac{1}{5}$.

PLATE 122.

Gecarcinus quadratus, male (32311), $\times \frac{5}{6}$.

FIG. 1. Ventral view.
2. Ventral view showing under side of chelae.

PLATE 123.

Gecarcinus planatus, male (20650), dorsal view, $\times \frac{3}{4}$.

PLATE 124.

Gecarcinus planatus, male (20650), ventral view, $\times \frac{4}{5}$.

PLATE 125.

Gecarcinus lagostoma, male (14869), dorsal view, $\times \frac{5}{6}$.

PLATE 126.

Gecarcinus lagostoma, male (14869), ventral view, $\times \frac{3}{4}$.

PLATE 127.

Ocypode albicans, male (49251), dorsal view, $\times 1\frac{1}{3}$.

THE GRAPSOID CRABS OF AMERICA.   441

PLATE 128.

Ocypode albicans, male (49251), ventral view, ×1½.

PLATE 129.

Fig. 1. *Ocypode gaudichaudii*, male (17297), dorsal view, ×⅔.
  2. *Ocypode occidentalis*, male (2016), ventral view, ×⅔.
  3. Same, dorsal view, ×⅔.

PLATE 130.

Fig. 1. *Ocypode gaudichaudii*, male (17297), ventral view, ×$\frac{1}{10}$.
  2. *Uca maracoani*, male (17657), dorsal view, ×$\frac{1}{10}$.
  3. Same, female (17657), dorsal view, ×1⅔.

PLATE 131.

Fig. 1. *Uca heterochelos*, male (11375), anterior view, ×1.
  2. Same, dorsal view, ×1.
  3. *Uca maracoani*, male (17657), anterior view, ×1.

PLATE 132.

Uca monilifera, male holotype, ×⅔.

Fig. 1. Anterior view.
  2. Dorsal view.
  3. Ventral view.

PLATE 133.

Uca princeps.

Fig. 1. Female (20689), dorsal view, ×1.
  2. Male (20689), anterior view, ×1.
  3. Same, dorsal view, ×1.

PLATE 134.

Fig. 1. *Uca stylifera*, male (32325), anterior view, ×1.
  2. Same, dorsal view, ×1.
  3. *Uca mordax*, male (21373), anterior view, ×1.
  4. Same, dorsal view, ×1.

PLATE 135.

Uca tangeri, male (14874), dorsal view, ×1½.

PLATE 136.

Uca tangeri, male (14874), ×1½.

Fig. 1. Anterior view.
  2. Inner face of chela.

PLATE 137.

Uca minax, male (17183), ×1½.

Fig. 1. Anterior view.
  2. Inner face of chela.
  3. Dorsal view.

PLATE 138.

Uca brevifrons, male (32323), ×1⅓.

FIG. 1. Outer face of chela.
2. Inner face of chela.
3. Dorsal view.

PLATE 139.

Uca pugnax, male (3828), ×1⅓.

FIG. 1. Anterior view.
2. Inner face of chela.
3. Dorsal view.

PLATE 140.

Uca pugnax rapax, male (19477), ×1⅓.

FIG. 1. Anterior view.
2. Inner face of chela.
3. Dorsal view.

PLATE 141.

Uca pugilator, male (5871), ×1½.

FIG. 1. Anterior view.
2. Inner face of chela.
3. Dorsal view.

PLATE 142.

Uca galapagensis, male (22319), ×1⅔.

FIG. 1. Anterior view.
2. Inner face of chela.
3. Dorsal view.

PLATE 143.

Uca macrodactylus, male (32320), ×2.

FIG. 1. Anterior view.
2. Inner face of chela.
3. Dorsal view.

PLATE 144.

Uca thayeri.

FIG. 1. Female (23753), dorsal view, ×2.
2. Male (23753), outer face of chela, ×2.

PLATE 145.

Uca speciosa, male (15256), ×2⅔.

FIG. 1. Dorsal view.
2. Outer face of chela.
3. Inner face of chela.

PLATE 146.

Uca crenulata.

FIG. 1. Male (44580), antero-dorsal view, ×2.
    2. Same, dorsal view, ×2.
    3. Male (33415), outer face of chela, ×2.
    4. Same, inner face of chela, ×2.

PLATE 147.

Uca coloradensis, male (18292), ×2.

FIG. 1. Anterior view.
    2. Inner face of chela.
    3. Dorsal view.

PLATE 148.

Uca spinicarpa, male (21684), ×2.

FIG. 1. Anterior view.
    2. Inner face of chela.
    3. Dorsal view.

PLATE 149.

Uca panamensis, male (48784), ×2.

FIG. 1. Outer face of chela.
    2. Lower face of chela.
    3. Inner face of chela.
    4. Dorsal view.

PLATE 150.

Uca uruguayensis, male (22194), ×2.

FIG. 1. Anterior view.
    2. Inner face of chela.
    3. Dorsal view.

PLATE 151.

Uca helleri, male (24829), ×3.

FIG. 1. Anterior view.
    2. Inner face of chela.
    3. Dorsal view.
    4. Antero-dorsal view.

PLATE 152.

FIG. 1. *Uca oerstedi,* male holotype, anterior view, ×1⅜.
    2. Same, dorsal view, ×1⅜.
    3. *Uca stenodactylus,* male (32322), antero-dorsal view, ×2¼.

PLATE 153.

Uca stenodactylus, male (32322), ×2¼.

FIG. 1. Anterior view.
    2. Inner face of chela.
    3. Dorsal view.

PLATE 154.

Uca musica, male holotype, ×1⅞.

FIG. 1. Anterior view.
2. Antero-dorsal view.
3. Dorsal view.
4. Ventral view.

PLATE 155.

Uca subcylindrica, male (23655), ×2.

FIG. 1. Dorsal view.
2. Ventral view.

PLATE 156.

Uca leptodactyla, male (47850), ×2⅞.

FIG. 1. Anterior view.
2. Inner face of chela.
3. Antero-dorsal view.
4. Dorsal view.

PLATE 157.

Uca latimanus, male (17500), ×3.

FIG. 1. Inner face of chela.
2. Outer face of chela.
3. Antero-dorsal view.
4. Dorsal view.

PLATE 158.

(After A. Milne Edwards.)

FIG. 1. *Glyptoplax pugnax*, male, Panama, dorsal view, × about 3.
2. Same, antennal and buccal regions, enlarged.
3. Same, abdomen and last sternal segments, enlarged.
4. Same, right chela, outer view, enlarged.
5. Same, left chela, outer view, enlarged.
6. Same, outer maxilliped, enlarged.
7. *Glyptoplax smithii*, male, Florida, right chela, outer view, enlarged.
8. Same, left chela, outer view, enlarged.
9. Same, dorsal view, × about 4½.
10. Same, abdomen and last sternal segments, enlarged.

PLATE 159.

FIG. 1. *Eucratopsis crassimanus*, male, abdomen and sternum, slightly reduced (after Dana).
2. Same, right chela, slightly reduced (after Dana).
3. *Ucides cordatus*, buccal cavity and outer maxillipeds, slightly reduced (after de Haan).
4. Same, right outer maxilliped, slightly reduced (after de Haan).
5. *Pinnotheres politus*, female, dorsal view, ×1½ (after Lenz).
6. *Speocarcinus carolinensis*, male, portion of abdomen and sternum, slightly reduced (after Stimpson).

Fig. 7. *Cyrtograpsus angulatus*, male, left chela, slightly reduced (after Dana).
8. Same, ventral view in part, slightly reduced (after Dana).
9. *Chasmagnathus granulata*, male, ventral view in part, ×1¾ (after Dana).
10. *Pinnotheres bipunctatus*, male, dorsal view, ×4 (after Nicolet).
11. Same, outer maxilliped, enlarged (after Nicolet).
12. Same, abdomen, enlarged (after Nicolet).

PLATE 160.

Fig. 1. *Pachygrapsus pubescens*, female, dorsal view, × about ⅔ (after Heller).
2. *Uca pugilator*, male, abdomen; × about ⅔ (after Smith).
3. *Sesarma (Sesarma) curacaoense*, male holotype, abdomen, × nearly 2 (after de Man).
4. *Pachygrapsus corrugatus*, dorsal view, ×⅕ (after von Martens).
5. *Uca subcylindrica*, male, abdomen, × about ⅔ (after Smith).
6. *Uca princeps*, male, abdomen, × about ⅔ (after Smith).
7. *Gecarcoidea lalandii*, male, dorsal view, about ½ natural size (after Milne Edwards).
8. Same, ventral view of anterior portion, reduced (after Milne Edwards).

PLATE 161.

Figures (except fig. 8) about ⅔ natural size and after Smith.

Fig. 1. *Uca heteropleura*, male, dorsal view of carapace and eyes.
2. Same, right and larger chela, outer view.
3. Same, left and larger chela of another specimen, inner view.
4. Same, merus of same cheliped as fig. 3, anterior surface.
Uca insignis, female, type of *Gelasimus ornatus* Smith, anterior view.
Same specimen, dorsal view of carapace and eyes.
5. *Uca insignis*, male, type of *Gelasimus armatus* Smith, dorsal view of carapace and eyes.
8. *Uca insignis*, female, right outer maxilliped, enlarged (after Milne Edwards).
9. *Uca insignis*, male, type of *G. armatus*, left and larger chela, outer view.
10. Same specimen, posterior right leg.
11. Same specimen, third right leg.
12. Same specimen, second right leg.
13. *Uca insignis*, female, type of *G. ornatus*, posterior right leg.
14. Same specimen, third right leg.
15. Same specimen, second right leg.

TRIZOCARCINUS DENTATUS. (PAGE 18.)

FOR EXPLANATION OF PLATE SEE PAGE 425.

BATHYPLAX TYPHLA. (PAGE 19.)
FOR EXPLANATION OF PLATE SEE PAGE 425.

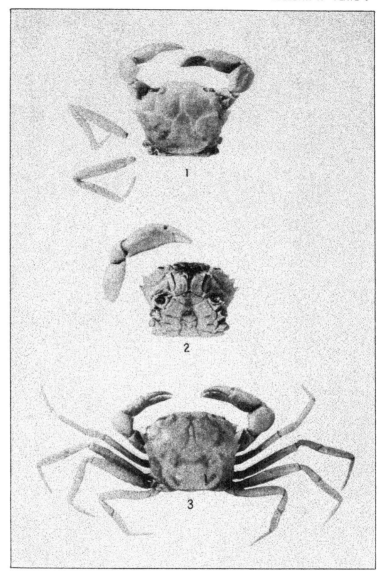

PILUMNOPLAX ELATA. (PAGE 23.)
FOR EXPLANATION OF PLATE SEE PAGE 425.

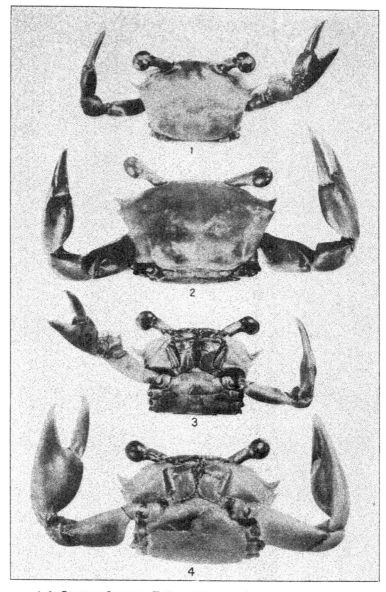

1, 3. GONEPLAX BARBATA. (PAGE 26.) 2, 4. G. SIGSBEI. (PAGE 26.)

FOR EXPLANATION OF PLATE SEE PAGE 425.

M.B.L. LIBRARY - WOODS HOLE, MASS.

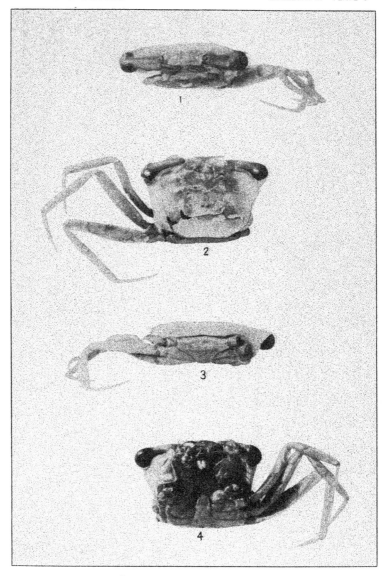

GONEPLAX BARBATA. (PAGE 26.)

FOR EXPLANATION OF PLATE SEE PAGE 425.

M R L LIBRARY - WOODS

2. PRIONOPLAX ATLANTICA. (PAGE 30.) 3, 4. TETRAPLAX QUADRIDENTATA. (PAGE 32.)

For explanation of plate see page 425.

M.B.L. LIBRARY - WOODS HOLE, MASS.

EURYPLAX NITIDA. (PAGE 34.)
FOR EXPLANATION OF PLATE SEE PAGE 425.

M. B. L. LIBRARY - WOODS HOLE, MASS.

SPEOCARCINUS CAROLINENSIS. (PAGE 39.)

FOR EXPLANATION OF PLATE SEE PAGE 426.

M. B. L. LIBRARY, WOODS HOLE, MASS.

SPEOCARCINUS GRANULIMANUS. (PAGE 40.)

FOR EXPLANATION OF PLATE SEE PAGE 426.

M. B. I. LIBRARY - WOODS HOLE, MASS.

1. SPEOCARCINUS OSTREARICOLA. (PAGE 41.) 2, 3. S. CALIFORNIENSIS. (PAGE 42.)

FOR EXPLANATION OF PLATE SEE PAGE 426.

M B L LIBRARY - WOODS HOLE, MASS.

CYRTOPLAX SPINIDENTATA. (PAGE 46.)

FOR EXPLANATION OF PLATE SEE PAGE 426.

M B L LIBRARY - WOODS HOLE MASS.

1, 2. PANOPLAX DEPRESSA. (PAGE 47.) 3. EUCRATOPSIS CRASSIMANUS. (PAGE 52.)
FOR EXPLANATION OF PLATE SEE PAGE 426.

M. B. L. LIBRARY WOODS HOLE, MASS.

1, 2. OEDIPLAX GRANULATA. (PAGE 44.)   3, 4. GLYPTOPLAX SMITHII. (PAGE 51.)

For explanation of plate see page 426.

M B L LIBRARY WOODS HOLE, MASS.

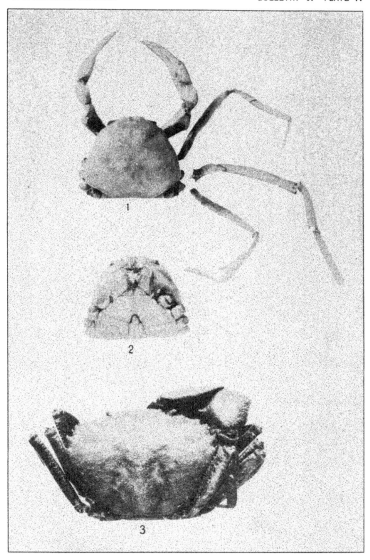

1, 2. CHASMOCARCINUS OBLIQUUS. (PAGE 58.) 3. PSEUDORHOMBILA OCTODENTATA. (PAGE 43.)

FOR EXPLANATION OF PLATE SEE PAGE 426.

1, 2. PINNOTHERES HOLMESI. (PAGE 68.) 3–6. P. OSTREUM. (PAGE 66.)

FOR EXPLANATION OF PLATE SEE PAGE 426.

1–4. PINNOTHERES GEDDESI. (PAGE 70.) 5, 6. P. ANGELICUS. (PAGE 72.)

FOR EXPLANATION OF PLATE SEE PAGE 426.

M B L LIBRARY WOODS HOLE, MASS.

1, 2. PINNOTHERES DEPRESSUS. (PAGE 79.) 3-6. P. MACULATUS. (PAGE 74.)
7, 8. P. PUGETTENSIS. (PAGE 82.)

FOR EXPLANATION OF PLATE SEE PAGE 427.

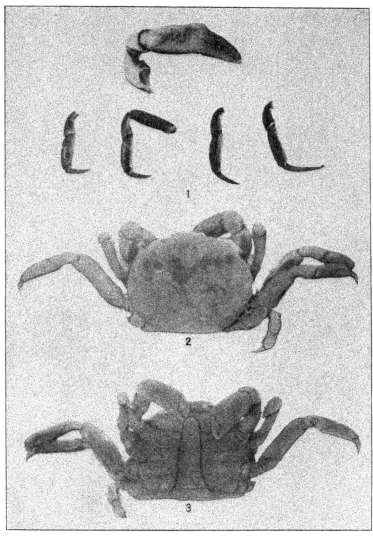

1. PINNOTHERES PUGETTENSIS. (PAGE 82.)   2, 3. P. MULINIARUM. (PAGE 81.)

FOR EXPLANATION OF PLATE SEE PAGE 427.

1-7. PINNOTHERES SERREI. (PAGE 84.) 8-11. P. BARBATUS. (PAGE 88.)

FOR EXPLANATION OF PLATE SEE PAGE 427.

附录

1, 2. PINNOTHERES STROMBI. (PAGE 90.) 3-6. P. CONCHARUM. (PAGE 86.)

FOR EXPLANATION OF PLATE SEE PAGE 427.

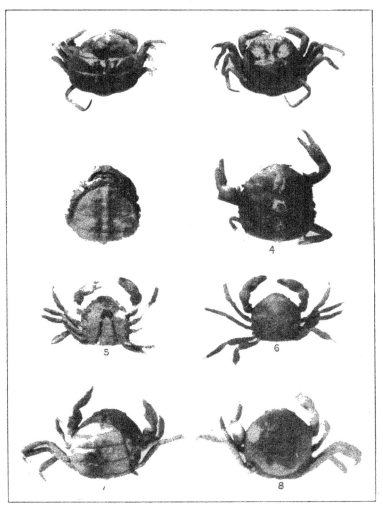

1, 2. PINNOTHERES RETICULATUS. (PAGE 93.) 3, 4. P. MOSERI. (PAGE 94.)
5-8. P. TAYLORI. (PAGE 97.)

FOR EXPLANATION OF PLATE SEE PAGE 427.

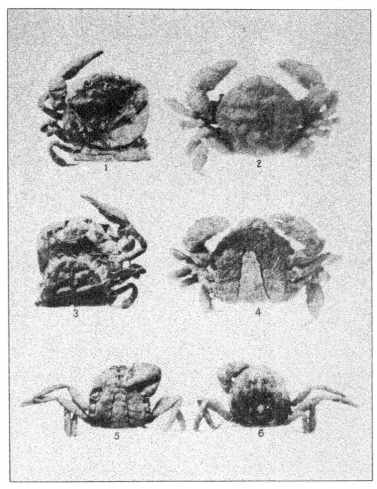

1-4. PINNOTHERES SHOEMAKERI. (PAGE 95.) 5, 6. P. ORCUTTI. (PAGE 98.)

FOR EXPLANATION OF PLATE SEE PAGE 428.

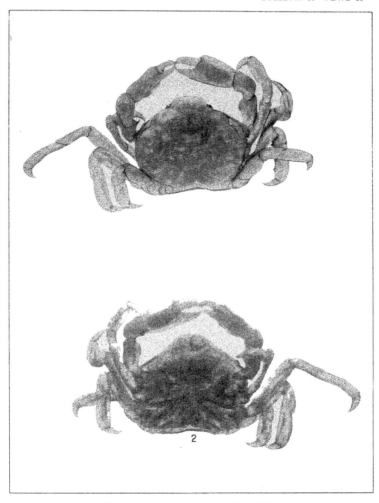

PINNOTHERES HEMPHILLI. (PAGE 99.)
FOR EXPLANATION OF PLATE SEE PAGE 428.

III.

1, 3. FABIA SUBQUADRATA. (PAGE 102.) 2, 4. F. LOWEI. (PAGE 104.) 5, 7. F. CANFIELDI. (PAGE 106.) 6, 8. F. BYSSOMIAE. (PAGE 105.)

FOR EXPLANATION OF PLATE SEE PAGE 428.

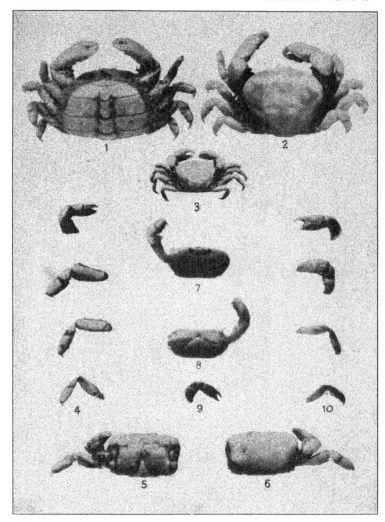

1-3. PINNAXODES MEINERTI. (PAGE 177.) 4-10. PARAPINNIXA BOUVIERI. (PAGE 111.)

FOR EXPLANATION OF PLATE SEE PAGE 428.

1–5. PARAPINNIXA HENDERSONI. (PAGE 109.) 6, 7. DISSODACTYLUS NITIDUS. (PAGE 116.)

FOR EXPLANATION OF PLATE SEE PAGE 428.

1–4. DISSODACTYLUS ENCOPEI. (PAGE 119.) 5–8. D. BORRADAILEI. (PAGE 121.)

FOR EXPLANATION OF PLATE SEE PAGE 428.

1, 2. DISSODACTYLUS STEBBINGI. (PAGE 123.) 3, 4. D. ALCOCKI. (PAGE 124.)
5, 6. D. CALMANI. (PAGE 125.) 7, 8. D. MELLITAE. (PAGE 117.)

FOR EXPLANATION OF PLATE SEE PAGE 429.

1-3. PINNIXA TRANSVERSALIS. (PAGE 131.) 4-7. P. FAXONI. (PAGE 133.)
8, 9. P. CRISTATA. (PAGE 134.)

For explanation of plate see page 429.

1–3. PINNIXA PATAGONIENSIS. (PAGE 135.) 4–7. P. FLORIDANA. (PAGE 138.) 8. P. TOMENTOSA. (PAGE 141.)

FOR EXPLANATION OF PLATE SEE PAGE 429.

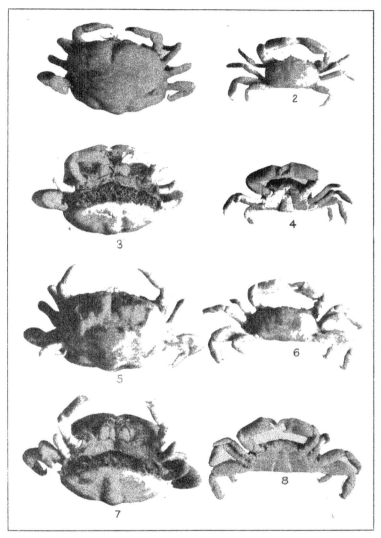

1-4, PINNIXA FABA. (PAGE 142.) 5-8, P. LITTORALIS. (PAGE 145.)
FOR EXPLANATION OF PLATE SEE PAGE 429.

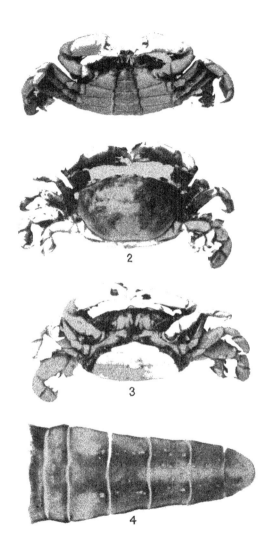

PINNIXA BARNHARTI. (PAGE 149.)
FOR EXPLANATION OF PLATE SEE PAGE 429.

1, 2. PINNIXA VALDIVIENSIS. (PAGE 154.) 3-6. P. CHAETOPTERANA. (PAGE 151.)

FOR EXPLANATION OF PLATE SEE PAGE 430.

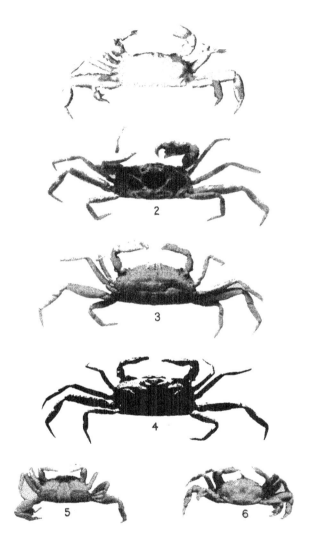

PINNIXA OCCIDENTALIS. (PAGE 155.) 2–4. P. SAYANA. (PAGE 156.) 5, 6. P. VALDIVIENSIS. (PAGE 154.)

FOR EXPLANATION OF PLATE SEE PAGE 430.

1-4. PINNIXA FRANCISCANA. (PAGE 161.) 5, 8. P. CYLINDRICA. (PAGE 159.)
6, 7, 9. P. SCHMITTI. (PAGE 162.)

FOR EXPLANATION OF PLATE SEE PAGE 430.

1-4. PINNIXA HIATUS. (PAGE 164.) 5-8. P. TUBICOLA. (PAGE 165.)
9, 10. P. WEYMOUTHI. (PAGE 166.)

FOR EXPLANATION OF PLATE SEE PAGE 430.

1–3. SCLEROPLAX GRANULATA. (PAGE 171.) 4, 5. OPISTHOPUS TRANSVERSUS. (PAGE 173.)

FOR EXPLANATION OF PLATE SEE PAGE 430.

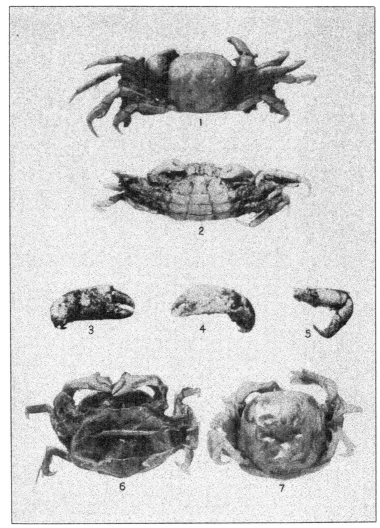

PINNAXODES CHILENSIS. (PAGE 175.)

FOR EXPLANATION OF PLATE SEE PAGE 431.

1–3. PINNOTHERELIA LAEVIGATA. (PAGE 181.) 4, 5. TETRIAS SCABRIPES. (PAGE 179.)

FOR EXPLANATION OF PLATE SEE PAGE 431.

1, 2. PINNOTHERELIA LAEVIGATA. (PAGE 181.) 3, 4. CYMOPOLIA SICA. (PAGE 208.)

FOR EXPLANATION OF PLATE SEE PAGE 431.

1, 2. PINNIXA RETINENS. (PAGE 139.) 3, 4. CYMOPOLIA FLORIDANA. (PAGE 220.)

FOR EXPLANATION OF PLATE SEE PAGE 431.

CYMOPOLIA ALTERNATA. (PAGE 188.)

FOR EXPLANATION OF PLATE SEE PAGE 431.

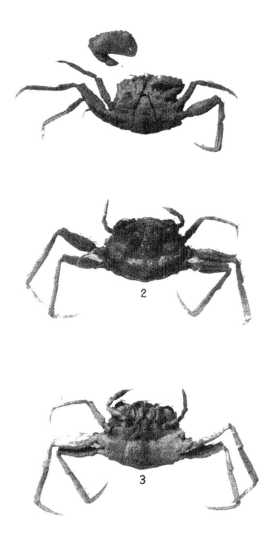

CYMOPOLIA ALTERNATA. (PAGE 188.)

FOR EXPLANATION OF PLATE SEE PAGE 431.

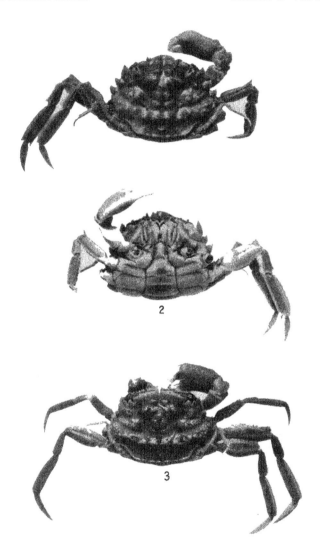

2. CYMOPOLIA LUCASII. (PAGE 193.) 3. C. ZONATA. (PAGE 190.)

FOR EXPLANATION OF PLATE SEE PAGE 431.

1. CYMOPOLIA ZONATA. (PAGE 190.) 2, 3. C. FAXONI. (PAGE 194.)

FOR EXPLANATION OF PLATE SEE PAGE 432.

CYMOPOLIA AFFINIS. (PAGE 196.)

For explanation of plate see page 432.

1, 2. CYMOPOLIA BAHAMENSIS. (PAGE 200.) 3. C. AFFINIS. (PAGE 196.)

For explanation of plate see page 432.

1, 2. CYMOPOLIA RATHBUNI. (PAGE 198.) 3, 4. C. ISTHMIA. (PAGE 201.)

FOR EXPLANATION OF PLATE SEE PAGE 432.

CYMOPOLIA OBESA. (PAGE 205.)
FOR EXPLANATION OF PLATE SEE PAGE 432.

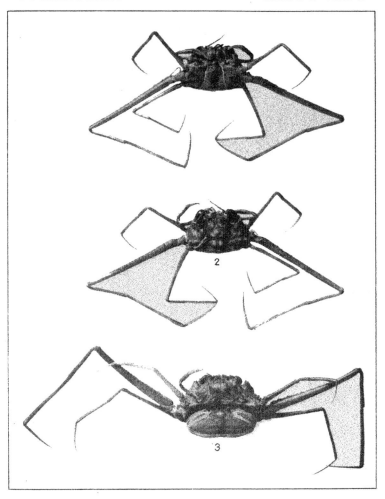

CYMOPOLIA GRACILIS. (PAGE 218.)
FOR EXPLANATION OF PLATE SEE PAGE 432.

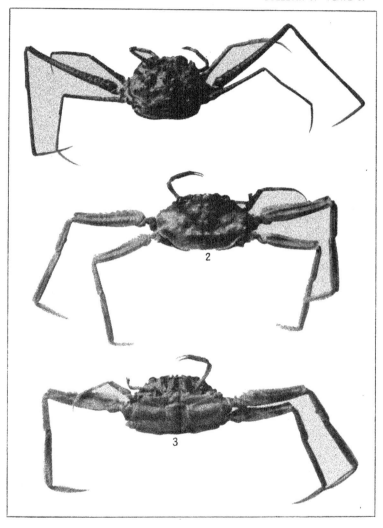

1. CYMOPOLIA GRACILIS. (PAGE 218.) 2, 3. C. FRAGILIS. (PAGE 213.)
FOR EXPLANATION OF PLATE SEE PAGE 432

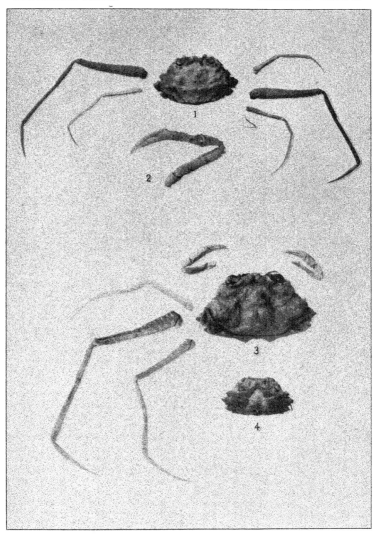

1, 2. CYMOPOLIA CURSOR. (PAGE 215.) 3, 4. C. GRACILIPES. (PAGE 221.)
FOR EXPLANATION OF PLATE SEE PAGE 432.

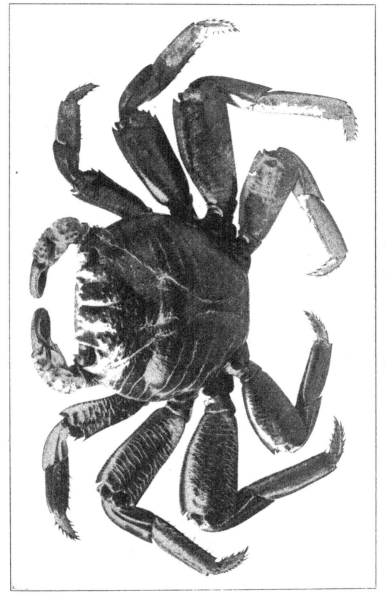

GRAPSUS GRAPSUS. (PAGE 2)
FOR EXPLANATION OF PLATE SEE PAGE 433.

GRAPSUS

FOR EXPLANATION OF PLATE SEE PAGE 341.

GEOGRAPSUS LIVIDUS. (PAGE 232.)

FOR EXPLANATION OF PLATE SEE PAGE 433.

LEPTOGRAPSUS VARIEGATUS. (PAGE 234.)

FOR EXPLANATION OF PLATE SEE PAGE 433.

GONIOPSIS ORUENTATA. (PAGE 237.)

For explanation of plate see page 433.

GONIOPSIS PULCHRA. (PAGE 239.)

FOR EXPLANATION OF PLATE SEE PAGE 433.

PACHYGRAPSUS CRASSIPES. (PAGE 241.)

FOR EXPLANATION OF PLATE SEE PAGE 433.

1, 2. PACHYGRAPSUS MAURUS. (PAGE 244.) 3. P. GRACILIS. (PAGE 249.)

FOR EXPLANATION OF PLATE SEE PAGE 433.

1. PACHYGRAPSUS GRACILIS. (PAGE 249.) 2, 3. P. TRANSVERSUS. (PAGE 244.)
FOR EXPLANATION OF PLATE SEE PAGE 433.

PACHYGRAPSUS MARMORATUS. (PAGE 250.)

FOR EXPLANATION OF PLATE SEE PAGE 434.

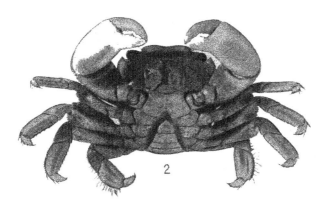

PLANES MINUTUS. (PAGE 253.)
FOR EXPLANATION OF PLATE SEE PAGE 434.

PLANES MARINUS. (PAGE 258.)
FOR EXPLANATION OF PLATE SEE PAGE 434.

CYRTOGRAPSUS ANGULATUS. (PAGE 261.)

FOR EXPLANATION OF PLATE SEE PAGE 434.

CYRTOGRAPSUS ALTIMANUS. (PAGE 262.)

FOR EXPLANATION OF PLATE SEE PAGE 434.

HEMIGRAPSUS AFFINIS. (PAGE 264.)
FOR EXPLANATION OF PLATE SEE PAGE 434.

HEMIGRAPSUS CRENULATUS. (PAGE 266.)

FOR EXPLANATION OF PLATE SEE PAGE 434.

HEMIGRAPSUS NUDUS. (PAGE 267.)
FOR EXPLANATION OF PLATE SEE PAGE 434.

HEMIGRAPSUS OREGONENSIS. (PAGE 270.)

FOR EXPLANATION OF PLATE SEE PAGE 435.

TETRAGRAPSUS JOUYI. (PAGE 273.)

FOR EXPLANATION OF PLATE SEE PAGE 435.

1, 2. GLYPTOGRAPSUS IMPRESSUS. (PAGE 275.) 3. G. JAMAICENSIS. (PAGE 277.)

FOR EXPLANATION OF PLATE SEE PAGE 435.

PLATYCHIROGRAPSUS TYPICUS. (PAGE 81.)

FOR EXPLANATION OF PLATE SEE PAGE 435.

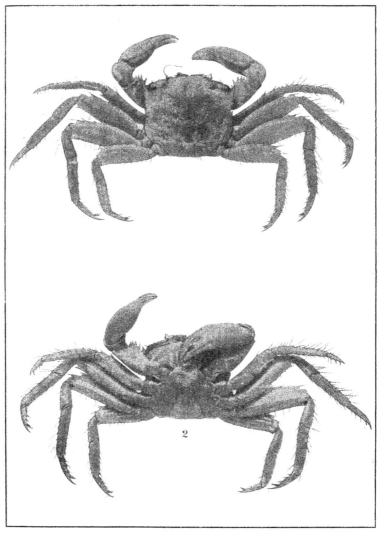

EUCHIROGRAPSUS AMERICANUS. (PAGE 282.)

For explanation of plate see page 435.

SESARMA (CHIROMANTES) AFRICANUM. (PAGE 287.)

For explanation of plate see page 435.

SESARMA (SESARMA) VERLEYI. (PAGE 288.)

FOR EXPLANATION OF PLATE SEE PAGE 435

SESARMA (SESARMA) RETICULATUM. (PAGE 290.)
FOR EXPLANATION OF PLATE SEE PAGE 435.

1, 2. SESARMA (SESARMA) CURACAOENSE. (PAGE 293.) 3, 4. S. (S.) SULCATUM. (PAGE 289.)

FOR EXPLANATION OF PLATE SEE PAGE 435.

SESARMA (SESARMA) RHIZOPHORAE. (PAGE 294.)
FOR EXPLANATION OF PLATE SEE PAGE 436.

SESARMA (SESARMA) BIDENTATUM. (PAGE 295.)

FOR EXPLANATION OF PLATE SEE PAGE 436.

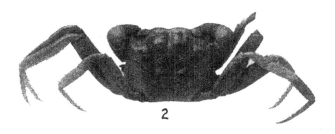

SESARMA (SESARMA) JARVISI. (PAGE 296.)

FOR EXPLANATION OF PLATE SEE PAGE 436.

SESARMA (HOLOMETOPUS) RECTUM. (PAGE 298.)

FOR EXPLANATION OF PLATE SEE PAGE 436.

SESARMA (HOLOMETOPUS) CINEREUM. (PAGE 300.)
FOR EXPLANATION OF PLATE SEE PAGE 436.

SESARMA (HOLOMETOPUS) MIERSII. (PAGE 303.)

FOR EXPLANATION OF PLATE SEE PAGE 436.

SESARMA (HOLOMETOPUS) MIERSII IHERINGI. (PAGE 304.)

For explanation of plate see page 436.

SESARMA (HOLOMETOPUS) MAGDALENENSE. (PAGE 305.)

FOR EXPLANATION OF PLATE SEE PAGE 436.

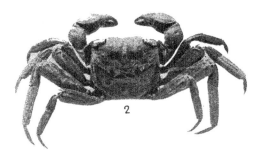

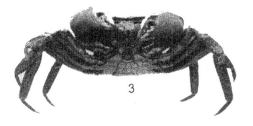

SESARMA (HOLOMETOPUS) HANSENI. (PAGE 315.) 2, 3. S. (H.) BIOLLEYI. (PAGE 306.)

FOR EXPLANATION OF PLATE SEE PAGE 437.

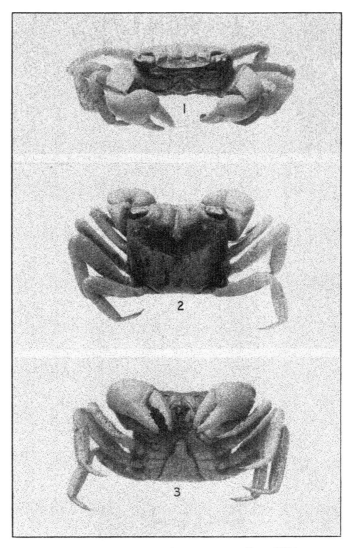

SESARMA (HOLOMETOPUS) TAMPICENSE. (PAGE 307.)

FOR EXPLANATION OF PLATE SEE PAGE 437.

SESARMA (HOLOMETOPUS) RICORDI. (PAGE 308.)
FOR EXPLANATION OF PLATE SEE PAGE 437.

SESARMA (HOLOMETOPUS) ANGUSTIPES. (PAGE 311.)

FOR EXPLANATION OF PLATE SEE PAGE 437.

SESARMA (HOLOMETOPUS) ROBERTI. (PAGE 312.)

FOR EXPLANATION OF PLATE SEE PAGE 437.

SESARMA (HOLOMETOPUS) ANGUSTUM. (PAGE 314.)

FOR EXPLANATION OF PLATE SEE PAGE 437.

SESARMA (HOLOMETOPUS) BENEDICTI. (PAGE 316.)

FOR EXPLANATION OF PLATE SEE PAGE 437.

METASESARMA RUBRIPES. (PAGE 319.)

For explanation of plate see page 437.

SARMATIUM OURVATUM. (PAGE 321.)

FOR EXPLANATION OF PLATE SEE PAGE 438.

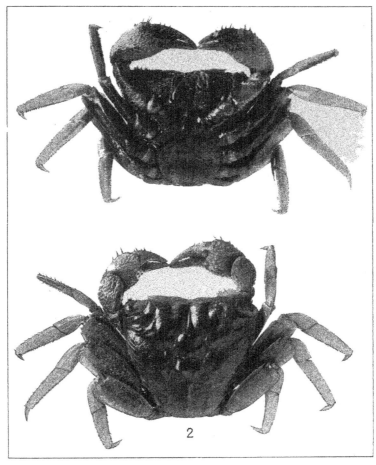

ARATUS PISONII. PAGE 323.)
FOR EXPLANATION OF PLATE SEE PAGE 438.

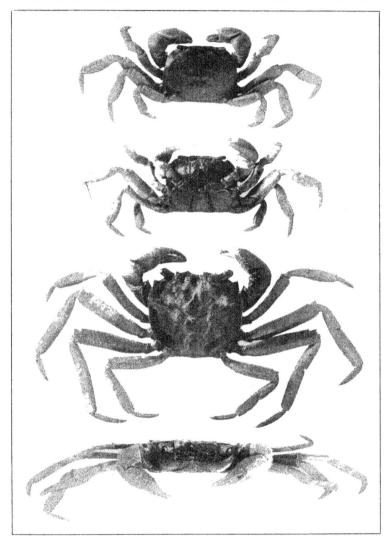

1, 2. CYCLOGRAPSUS INTEGER. (PAGE 326.) 3, 4. METOPAULIAS DEPRESSUS. (PAGE 318.)

FOR EXPLANATION OF PLATE SEE PAGE 438.

CYCLOGRAPSUS CINEREUS. (PAGE 327.)

FOR EXPLANATION OF PLATE SEE PAGE 438.

CYCLOGRAPSUS PUNCTATUS. (PAGE 328.)

FOR EXPLANATION OF PLATE SEE PAGE 438.

CYCLOGRAPSUS PUNCTATUS. (PAGE 328.)

FOR EXPLANATION OF PLATE SEE PAGE 438.

CHASMAGNATHUS GRANULATA. (PAGE 329.)

FOR EXPLANATION OF PLATE SEE PAGE 438.

PLAGUSIA DEPRESSA. (PAGE 332.)

FOR EXPLANATION OF PLATE SEE PAGE 438.

PLAGUSIA DEPRESSA TUBERCULATA. (PAGE 334.)

FOR EXPLANATION OF PLATE SEE PAGE 438.

PLAGUSIA IMMACULATA. (PAGE 335.)
FOR EXPLANATION OF PLATE SEE PAGE 439.

PLAGUSIA CHABRUS. (PAGE 8.)

FOR EXPLANATION OF PLATE SEE PAGE 8.

PERCNON GIBBESI. (PAGE 337.)

FOR EXPLANATION OF PLATE SEE PAGE 439.

CARDISOMA GUANHUMI. PAGE 341.

CARDISOMA QUANHUMI. (PAGE 341.)

For explanation of plate see page 6.

CARDISOMA CRASSUM. (PAGE 8.)

FOR EXPLANATION OF PLATE SEE PAGE 8.

CARDISOMA CRASSUM. (Page 8.)

For explanation of plate see page 8.

UCIDES CORDATUS. PAGE 339.

FOR EXPLANATION OF PLATE SEE PAGE 439.

UCIDES CORDATUS. (PAGE 347.)
FOR EXPLANATION OF PLATE SEE PAGE 439.

Ucides cordatus. (Page 347.)
For explanation of plate see page 6.

UCIDES CORDATUS.
For explanation of plate see page 439.

UCIDES OCCIDENTALIS. (PAGE 350.)

FOR EXPLANATION OF PLATE SEE PAGE 439

UCIDES OCCIDENTALIS. (PAGE 350.)

FOR EXPLANATION OF PLATE SEE PAGE 440.

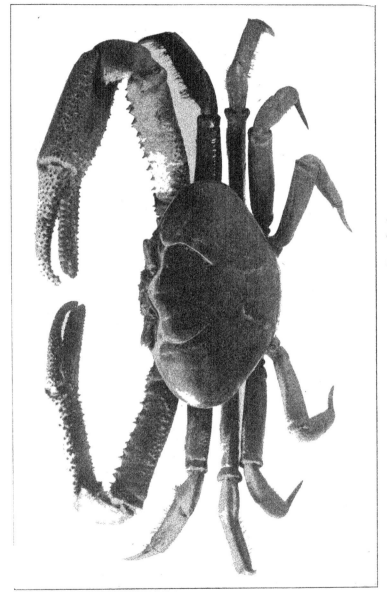

UCIDES OCCIDENTALIS. (PAGE 8.)

FOR EXPLANATION OF PLATE SEE PAGE 4.

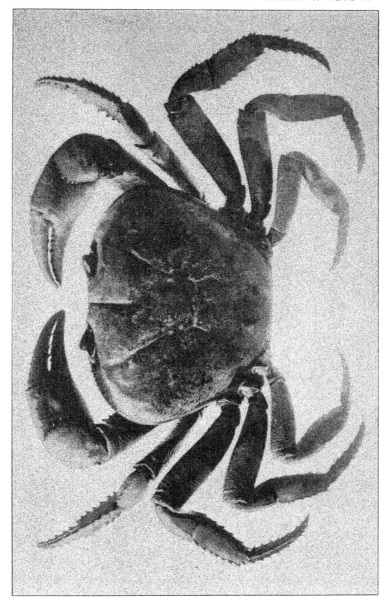

For explanation of plate see page 8.

GECARCINUS RURICOLA. PAGE 8.
FOR EXPLANATION OF PLATE SEE PAGE 44.

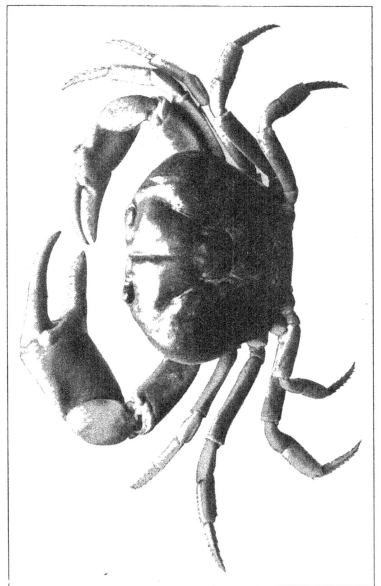

GECARCINUS LATERALIS. PAGE 355.

FOR EXPLANATION OF PLATE SEE PAGE 40.

GECARCINUS LATERALIS. PAGE 8.

For explanation of plate see page 8.

GECARCINUS QUADRATUS. (PAGE 8.

FOR EXPLANATION OF PLATE SEE PAGE 4.

GECARCINUS QUADRATUS. (PAGE 358.)

FOR EXPLANATION OF PLATE SEE PAGE 440.

GECARCINUS PLANATUS. (PAGE 8.)

FOR EXPLANATION OF PLATE SEE PAGE 440.

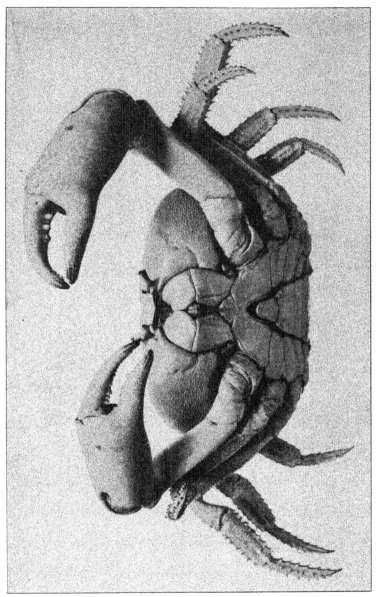

GECARCINUS PLANATUS. (PAGE 58.
FOR EXPLANATION OF PLATE SEE PAGE 440.

GECARCINUS LAGOSTOMA. PAGE 361.

FOR EXPLANATION OF PLATE SEE PAGE 8.

GECARCINUS LAGOSTOMA. (PAGE 361.)

FOR EXPLANATION OF PLATE SEE PAGE 4.

OCYPODE ALBICANS. (PAGE 367.)

FOR EXPLANATION OF PLATE SEE PAGE 440.

OCYPODE ALBICANS. (PAGE 367.)

FOR EXPLANATION OF PLATE SEE PAGE 44.

1. OCYPODE GAUDICHAUDII. (PAGE 373.) 2, 3. O. OCCIDENTALIS. (PAGE 372.)

FOR EXPLANATION OF PLATE SEE PAGE 441.

1. OCYPODE GAUDICHAUDII. (PAGE 373.) 2, 3. UCA MARACOANI. (PAGE 378.)

FOR EXPLANATION OF PLATE SEE PAGE 441.

1, 2. UCA HETEROCHELOS. (PAGE 381.) 3. U. MARACOANI. (PAGE 378.)
FOR EXPLANATION OF PLATE SEE PAGE 441.

UCA MONILIFERA. (PAGE 380.)
FOR EXPLANATION OF PLATE SEE PAGE 441.

UCA PRINCEPS. (PAGE 382.)
FOR EXPLANATION OF PLATE SEE PAGE 441.

1, 2. UCA STYLIFERA. (PAGE 383.) 3, 4. U. MORDAX. (PAGE 391.)
FOR EXPLANATION OF PLATE SEE PAGE 441.

UCA TANGERI (PAGE 387.)

For explanation of plate see page 44

UCA TANGERI. (PAGE 387)

FOR EXPLANATION OF PLATE SEE PAGE 44

UCA MINAX. (PAGE 389.)
FOR EXPLANATION OF PLATE SEE PAGE 441.

UCA BREVIFRONS. (PAGE 393.)
FOR EXPLANATION OF PLATE SEE PAGE 442.

UCA PUGNAX. (PAGE 395.)

FOR EXPLANATION OF PLATE SEE PAGE 442.

UCA PUGNAX RAPAX. (PAGE 397.)

FOR EXPLANATION OF PLATE SEE PAGE 442.

UCA PUGILATOR. (PAGE 400.)
FOR EXPLANATION OF PLATE SEE PAGE 442.

UCA GALAPAGENSIS. (PAGE 403.)
FOR EXPLANATION OF PLATE SEE PAGE 442.

UCA MACRODACTYLUS. (PAGE 404.)

FOR EXPLANATION OF PLATE SEE PAGE 442.

Uca thayeri. (Page 406.)
For explanation of plate see page 442.

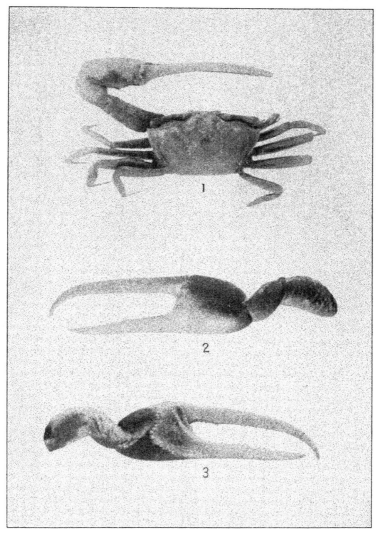

UCA SPECIOSA. (PAGE 408.)
FOR EXPLANATION OF PLATE SEE PAGE 442.

UCA CRENULATA. (PAGE 409.)
FOR EXPLANATION OF PLATE SEE PAGE 443.

UCA COLORADENSIS. (PAGE 410.)

FOR EXPLANATION OF PLATE SEE PAGE 443.

UCA SPINICARPA. (PAGE 411.)
FOR EXPLANATION OF PLATE SEE PAGE 443.

UCA PANAMENSIS. (PAGE 412.)
FOR EXPLANATION OF PLATE SEE PAGE 443.

UCA URUGUAYENSIS. (PAGE 413.)
FOR EXPLANATION OF PLATE SEE PAGE 443.

UCA HELLERI. (PAGE 415.)

FOR EXPLANATION OF PLATE SEE PAGE 443.

1, 2. Uca oerstedi. (Page 414.) 3. Uca stenodactylus. (Page 416.)

For explanation of plate see page 443.

UCA STENODACTYLUS. (PAGE 416.)

FOR EXPLANATION OF PLATE SEE PAGE 443.

UCA MUSICA. (PAGE 417.)
FOR EXPLANATION OF PLATE SEE PAGE 444.

UCA SUBCYLINDRICA. (PAGE 419.)
FOR EXPLANATION OF PLATE SEE PAGE 44.

UCA LEPTODACTYLA. (PAGE 420.)
FOR EXPLANATION OF PLATE SEE PAGE 444.

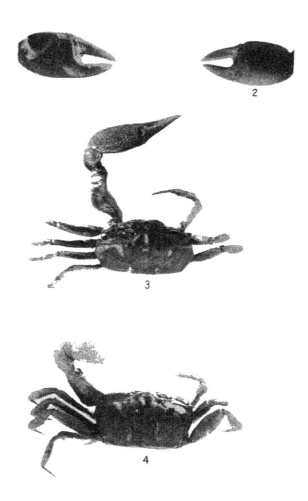

UCA LATIMANUS. (PAGE 422.)

FOR EXPLANATION OF PLATE SEE PAGE 444.

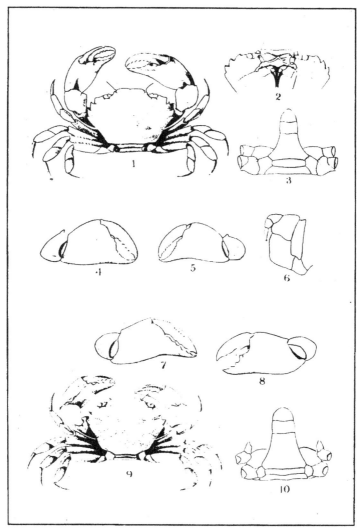

1-6. GLYPTOPLAX PUGNAX. (PAGE 50.) 7-10. G. SMITHII. (PAGE 51.)

FOR EXPLANATION OF PLATE SEE PAGE 444.

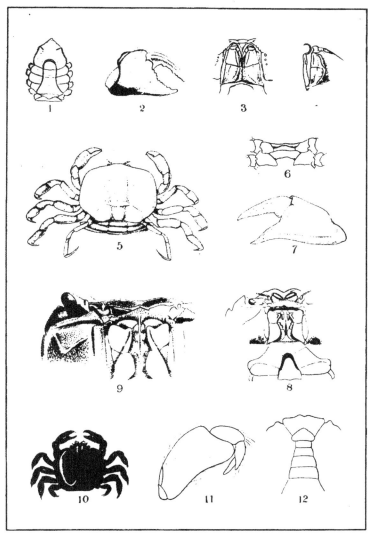

1, 2. EUCRATOPSIS CRASSIMANUS. (PAGE 52.) 3, 4. UCIDES CORDATUS. (PAGE 347.) 5. PINNOTHERES POLITUS. (PAGE 71.) 6. SPEOCARCINUS CAROLINENSIS. (PAGE 39.) 7, 8. CYRTOGRAPSUS ANGULATUS. (PAGE 261.) 9. CHASMAGNATHUS GRANULATA. (PAGE 329.) 10–12. PINNOTHERES BIPUNCTATUS. (PAGE 78.)

FOR EXPLANATION OF PLATE SEE PAGES 444–445.

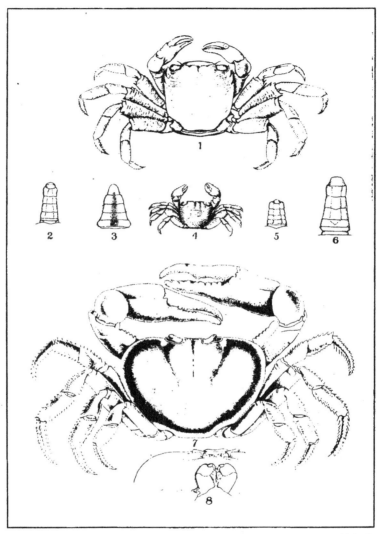

1. PACHYGRAPSUS PUBESCENS. (PAGE 252.) 2. UCA PUGILATOR. (PAGE 400.) 3. SESARMA (SESARMA) CURACAOENSE. (PAGE 293.) 4. P. CORRUGATUS. (PAGE 252.) 5. U. SUBCYLINDRICA. (PAGE 419.) 6. U. PRINCEPS. (PAGE 382.) 7, 8. GECARCOIDEA LALANDII. (PAGE 364.)

FOR EXPLANATION OF PLATE SEE PAGE 445.

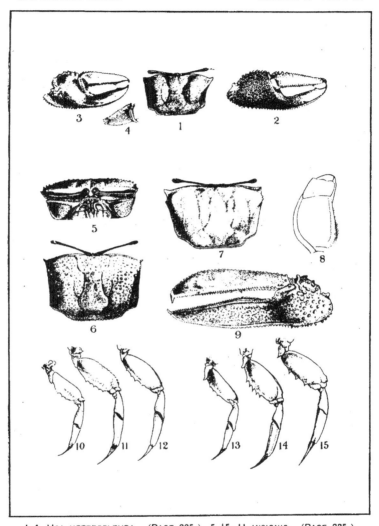

1-4. UCA HETEROPLEURA. (PAGE 385.) 5-15. U. INSIGNIS. (PAGE 385.)

FOR EXPLANATION OF PLATE SEE PAGE 445.

INDEX.

A.

	Page.
Acanthoplax	375
insignis	375, 385
Acanthopus	337
clavimana	337
gibbesi	337
acutifrons, Cymopolia	184, 186, 223
Palicus	223
advena, Pachygrapsus	245
aequatoriale, Sesarma	13, 285, 287
Sesarma (Sesarma)	292
aequatorialis, Sesarma	292
affinis, Cymopolia	184, 185, 196, 432
Gelasimus	391
Hemigrapsus	264, 266, 434
Palicus	196
Parapinnixa	12, 107, 111
Pinnixa	12, 130, 131, 168
africana, Sesarma	287
africanum, Sesarma	285
Sesarma (Chiromantes)	287, 435
Sesarma (Perisesarma)	287
agassizi, Palicus	196, 197
agassizii, Palicus	196
agricola, Gecarcinus	352
albeola, Charitonetta	172
albicans, Ocypode	13, 366, 367, 368, 372, 373, 440, 441
albus, Loxechinus	176, 177
alcocki, Dissodactylus	115, 124, 429
Alcyonium	105
algosus, Mytilus	72
Alpheus depressus	332
alternata, Cymopolia	12, 184, 185, 188, 191, 194, 196, 431
alternatus, Palicus	188, 196, 198
altifrons, Grapsus	227, 230
altimanus, Cyrtograpsus	261, 262, 434
amazonensis, Uca	391
americana, Pilumnoplax	21
Sesarma	311, 312
americanus, Euchirograpsus	282, 283, 435
Pilumnoplax	21
Amphitrite	166
ornata	152
angelica, Pinnotheres	72
angelicus, Pinnotheres	63, 65, 66, 70, 72, 426
angulatus, Cyrtograpsus	13, 260, 261, 263, 434, 445
angusta, Cymopolia	184, 186, 210, 212
Sesarma	314
angustatus, Nautilograpsus	254
angustipes, Sesarma	286, 287, 303, 308, 311
Sesarma (Holometopus)	13, 311, 437
angustum, Sesarma	13, 286, 287, 313
Sesarme (Holometopus)	314, 437
angustus, Palicus	210
Annelids	62, 76, 132, 137, 138, 152, 166

	Page.
annulipes, Gelasimus	404, 405, 417
Anomura	11
ansoni, Leptograpsus	234
Aratus	226, 322
pisonii	13, 322, 323, 438
Aratv pinima	323
Arcotheres	62
palaensis	62
arctica, Saxicava	106
arenaria, Mya	76, 86, 87, 103, 148
Ocypoda	368, 369
arenarius, Cancer	367
Ocypode	368
arenata, Liosoma	144, 150
Arenicola cristata	159, 160
arenicola, Liosoma	144, 150
Molpadia	144, 150
Areograpsus	275
jamaicensis	275, 277
aristatus, Lithodomus	73, 74
Lithophaga	73, 74
armatum, Cardisoma	343
armatus, Gelasimus	378, 385, 387, 445
Ascidia	105
Ascidian, Black	94, 95
Ascidians	61, 87, 247, 250
Aspidograpsus	278
typicus	278
Atelecyclidae	14
atlantica, Prionoplax	12, 30, 425
Palicus	200
atlanticus, Prionoplax	30
audouinii, Cyclograpsus	329

B.

	Page.
bahamensis, Cymopolia	184, 185, 200, 432
Palicus	200
barbata, Frevillea	25, 26
Goneplax	25, 26, 27, 425
Pinnotheres	88
Sesarma	328
barbatus, Gnathochasmus	325, 328
Pinnotheres	63, 65, 88, 427
barbiger, Gecarcinus	352
Paraxanthus	352
barbimana, Sesarma	298
barbimanum, Sesarma	286
Sesarma (Sesarma)	298
barbimanus, Heterograpsus	266
barnharti, Pinnixa	130, 144, 149, 429
Bathyplax	16, 19
typhla	19, 425
typhlus	19
typhlus, var. oculiferus	19
Beach Crab	308
beaufortensis, Parapinnixa	107, 112

INDEX.

Bell-animalcules... 68
benedicti, Sesarma... 286, 319
   Sesarma (Holometopus)... 316, 437
bertheloti, Leptograpsus... 250
bidens, Chiromantes... 284
   Pachysoma... 284
   Sesarma... 284
bidentate, Sesarma... 295
bidentatum, Sesarma... 286
   Sesarma (Sesarme)... 295, 436
biolleyi, Sesarma... 13, 286, 287, 304
   Sesarma (Holometopus)... 306, 437
bipunctatum, Pinnotheres... 78
bipunctatus, Pinnotheres... 64, 66, 78, 445
Bird Stomachs... 313
bispinosa, Ocypode... 25
Bivalve Mollusks... 61, 87, 103, 104, 128
Black Crab... 352
   Land-crab... 355
blakei, Palicus... 188, 190, 191
Blue Crab... 401
   Land-crab... 352
borealis, Cardita... 86, 87
Boring Mollusk... 73
borradailei, Dissodactylus... 115, 121, 428
bosci, Euplax... 423
bouvieri, Parapinnixa... 12, 107, 111, 428
Brachygnatha... 14
Brachynotus... 273
   (Heterograpsus) jouyi... 273
   nudus... 268
   oregonensis... 270
Brachyrhyncha... 9, 14
Brachyura... 11, 13
Brackish-water Crab... 312
brasiliensis, Metopograpsus... 319
brevifrons, Gelasimus... 393
   Uca... 377, 393, 442
brevipes, Grapsus... 232
   Lycodes... 157
brevipollex, Pinnixa... 12, 130, 131, 169
Bromelia... 319
bromeliarum, Sesarma... 312, 313
Bufflehead... 172
Butter Clam... 103
Byssomia... 105
byssomiae, Fabia... 12, 101, 102, 105, 428
   Pinnotheres... 74, 105
Byssomya distorta... 106

C.

Caenocentrotus gibbosus... 176, 177
cailloti, Pelocarcinus... 364
californica, Mya... 87
   Pholas... 104
californicus, Stichopus... 174
californiensis, Eucrate... 42
   Pinnixa... 155, 157, 166
   Speocarcinus... 39, 42, 426
Callianassa... 39
Calling Crab... 374
calmani, Dissodactylus... 115, 125, 429
Calyptraea, species... 72
Cambarus... 189
Cancellus marinus minimus quadratus... 253

Cancer 1... 70
   arenarius... 367
   chabrus... 336
   cordatus... 341, 342, 347
   depressus... 332
   femoralis... 250
   glaberrimus... 254
   grapsus... 227
   (Grapsus) grapsus... 227
   marmoratus... 250, 267
   marmoreus... 250
   minutus... 253
   Pagurus, hirsutus, Americanus, pronus... 347
   palustris cuniculos sub terra agens... 378
   pinnophylax... 66
   pusillus... 254
   quadrata... 367
   quadratus... 284, 367
   rhomboides... 25
   ruricola... 237, 352
   squamosus... 332
   sulcatus, terrestris, sive montanus, Americanus... 352
   terrestris cuniculos sub terra agens... 352
   uca... 347
   uka... 381
   Uka una, Brasiliensis... 381
   variegatus... 234
   vca... 347, 381
   vocans... 367, 395, 400
   vocans major... 381
   vocator... 391
Cancridae... 14
canfieldi, Fabia... 12, 101, 102, 106, 428
Cangrejos Ajaes de Manglar... 347
   Ajaes Terrestres... 352
   Terrestres... 341
capensis, Plagusia... 336
Carcinoplacinae... 16, 17
Carcinoplax... 17
   dentatus... 18
Cardiosoma crassum... 345
   latimanus... 345
Cardisoma... 340, 346, 351, 354, 363
   armatum... 343
   carnifex... 343
   crassum... 13, 341, 345, 346, 439
   diurnum... 342
   guanhumi... 13, 341, 345, 424, 439
   quadrata... 342
Cardita borealis... 86, 87
carnifex, Cardisoma... 343
   Gecarcinus... 343
carolinensis, Speocarcinus... 12, 38, 39, 41, 426, 444
caronii, Cymopolia... 183
Catometopa... 9
Ceratophthalma... 366
   ceratophthalma... 366
   cursor... 366
ceratophthalma, Ceratophthalma... 366
   Ocypode... 366, 372
   Parocypoda... 366
Ceratoplax ciliata... 16
   sp... 16
chabrus, Cancer... 336
   Plagusia... 332, 336, 439

INDEX.

	Page.
Chaenostoma	423
orientale	423
chaetopterana, Pinnixa	130, 151, 158, 430
Chaetopterus	78, 132, 133
pergamentaceus	76, 151, 152, 153
Chama lazarus	66
macerophylla	66
Chanduya	240
Charitonetta albeola	172
Chasmagnathus	226, 329
convexus	329
granulata	329, 438
granulatus	329, 445
Chasmocarcinus	16, 54, 55
cylindricus	55, 59
latipes	12, 55, 57, 59
obliquus	55, 58, 426
typicus	12, 54, 55, 57, 59
Chasmophora	16, 37
macrophthalma	37
chilensis, Fabia	175, 176
Pinnaxodes	174, 175, 177, 431
Pinnotheres	175
chiragra, Sesarma	316, 317
Chiromantes	284, 285, 287
bidens	284
Ciccaba nigrolineata	405
Ciecie Ete	406
ciliata, Ceratoplax	16
Prionoplax	12, 30, 31
cimatodus, Gelasimus	387
cinerea, Sesarma	290, 300, 308
cinereum, Sesarma	13, 286, 287, 290, 303, 308
Sesarma (Holometopus)	300, 436
cinereus, Cyclograpsus	13, 325, 327, 328, 438
Grapsus	254, 300
cirripes, Cyrtograpsus	261, 262
Clam	144, 145, 147, 149, 174
Butter	103
Giant	144, 147, 148
Horse	145, 148
Long	76
Small	147
clavimana, Acanthopus	337
Clymenella	138
clypeatus, Planes	253, 254
coloradensis, Gelasimus	410
Uca	377, 410, 443
Columbus's Crab	253
Common Land-crab	355
concharum, Cryptophrys	62, 86
Pinnotheres	64, 66, 86, 88, 427
convexus, Chasmagnathus	329
copax, Modiola	73
cordata, Ocypode	341, 347
Ocypode (cardisoma)	342
Oedipleura	346, 348
Uca	346, 348
cordatus, Cancer	341, 342, 347
Ucides	13, 346, 347, 348, 350, 439, 444
corrugatus, Grapsus (Leptograpsus)	252
Pachygrapsus	241, 252, 445
Coryphaena	257
Corystidae	14
Crab, Beach	308
Black	352

	Page.
Crab, Blue	401
Brackish-water	312
Calling	374
Columbus's	253
Commensal	9, 61, 62, 67, 71, 85, 103, 147
Deaf-ear	398
Fiddler	10, 374, 375, 396, 401
Free-swimming	9, 61, 87
Fresh-water	295
Friendly	300
Ghost	10, 367
Gulf-weed	253
Land	10, 15, 339, 424
Larval	366, 368
Mangrove	237, 293, 294, 323
Marsh	284, 326
Mountain	352
Mud	9
Mussel	74
Oyster	10, 66, 70
Parasitic	9, 61
Pine	318
Plaster cast of	424
Red-jointed Fiddler	389
River	278
Rock	227
Round-fronted	9
Sand	10, 367, 401
Short-tailed	11
Spider	12
Spray	337
Square-fronted	9
Subfossil	362
Tree	237, 323
Turtle	253
Wood	300
crassatelloides, Pachydesma	105
crassimanus, Eucrate	52
Eucratopsis	52, 426, 444
Hemigrapsus	264
crassipes, Pachygrapsus	13, 240, 241, 244, 433
Pinnotheres	66
Sesarma	285, 294, 298
Sesarma (Sesarma)	294
crassum, Cardiosoma	345
Cardisoma	13, 341, 345, 346, 439
Sarmatium	321
crenulata, Lucapina	106, 174
Uca	13, 377, 378, 406, 409, 410, 443
crenulatus, Cyclograpsus	266
Gelasimus	409
Hemigrapsus	264, 266, 434
Lobograpsus	264, 266
Crepidula dilatata	72
crinitichelis, Dissodactylus	119
cristata, Arenicola	159, 160
Pinnixa	129, 131, 134, 136, 429
cristatipes, Cymopolia	184, 185, 186
Palicus	186
cruentata, Goniopsis	13, 236, 237, 240, 433
cruentatus, Goniograpsus	237
Goniopsis	239
Grapsus	237
Grapsus (Goniopsis)	237
Cryptophrys	62, 63
concharum	62, 86

65863—17——29

450 INDEX.

Cryptophrys pubescens... 87
Cucumber, Sea... 142, 150
curacaoense, Sesarma... 13, 285, 287, 294
    Sesarma (Sesarma)... 293, 435, 445
curacaoensis, Sesarma... 293
cursor, Ceratophthalma... 366
    Cymopolia... 184, 186, 215, 218, 219, 220, 432
    Palicus... 215
curvata, Sesarma... 321, 322
curvatum, Sarmatium... 321, 438
    Sesarma (Sarmatium)... 321
curvatus, Metagrapsus... 321
cyaneus, Planes... 254
Cyclograpsacea... 260, 283
Cyclograpsus... 226, 325
    audouinii... 329
    cinereus... 13, 325, 327, 328, 438
    crenulatus... 266
    gnatherion... 181
    integer... 13, 325, 326, 328, 438
    marmoratus... 267
    minutus... 328
    punctatus... 325, 327, 328, 438
Cyclometopa... 9
cylindrica, Pinnixa... 12, 128, 129, 131, 138, 139, 151 156, 159, 430
cylindricum, Pinnotheres... 159
cylindricus, Chasmocarcinus... 55, 59
Cymopolia... 6, 14, 183, 185
    acutifrons... 184, 186, 223
    affinis... 184, 185, 196, 432
    alternata... 12, 184, 185, 188, 191, 194, 196, 431
    angusta... 184, 186, 210, 212
    bahamensis... 184, 185, 200, 432
    caronii... 183
    cristatipes... 184, 185, 186
    cursor... 184, 186, 215, 218, 219, 220, 432
    dentata... 184, 185, 196, 202
    depressa... 184, 186, 212, 213
    dilatata... 215, 218
    faxoni... 12, 184, 185, 193, 194, 432
    floridana... 184, 186, 220, 431
    fragilis... 184, 186, 213, 215, 432
    gracilipes... 184, 186, 221, 223, 432
    gracilis... 184, 186, 218, 220, 432
    isthmia... 184, 185, 201, 432
    lucasii... 12, 184, 185, 193, 431
    obesa... 12, 184, 185, 204, 205, 207, 432
    rathbuni... 184, 185, 198, 200, 432
    sica... 184, 186, 208, 210, 211, 212, 431
    tuberculata... 12, 149, 184, 185, 207
    zonata... 12, 184, 185, 190, 192, 431, 432
Cymopoliidae... 10, 12, 15, 182
Cynthia... 82
Cyrtograpsus... 225, 260
    altimanus... 261, 262, 434
    angulatus... 13, 260, 261, 263, 434, 445
    cirripes... 261, 262
Cyrtoplax... 16, 45
    spinidentata... 45, 46, 426

D.

Deaf-ear Crab... 398
Decapoda... 11
declivifrons, Grapsus... 245

dentata, Cymopolia... 184, 185, 196, 202
dentatus, Carcinoplax... 18
    Palicus... 202
    Trizocarcinus... 17, 18, 425
depressa, Cymopolia... 184, 186, 212, 213
    Panoplax... 47, 49, 426
    Philyra... 331
    Plagusia... 331, 332, 334, 335, 438
    Zaops... 62, 63, 79
depressum, Pinnotheres... 79
depressus, Alpheus... 332
    Cancer... 332
    Gecarcinus... 355, 357
    Grapsus... 332
    Metopaulias... 317, 318, 438
    Palicus... 212
    Pinnotheres... 64, 65, 79, 427
    Plagusia... 332
digueti, Gecarcinus... 359, 360
dilatata. Crepidula... 72
    Cymopolia... 215, 218
diris, Grapsus... 254
Discoplax... 340
    longipes... 340
    rotundum... 340
Dissodactylus... 62, 114
    alcocki... 115, 124, 429
    borradailei... 115, 121, 428
    calmani... 115, 125, 429
    criniticheiis... 119
    encopei... 115, 119, 121, 428
    mellitae... 115, 117, 429
    nitidus... 115, 116, 428
    stebbingi... 115, 123, 429
distorta, Byssomya... 106
diurnum, Cardisoma... 342
Dorippidae... 10
Dromiacea... 14
Dromiidae... 10
dubius, Metopograpsus... 245

E

Echinaiachnius paima... 119
Echinoids... 61
Echinophilus... 114
    mellitae... 114, 117
Echiurus... 156
edulis, Mytilus... 76, 87, 103, 174
elata, Eucratoplax... 23
    Pilumnoplax... 21, 23, 425
emarginata, Encope... 119, 121
Encope emarginata... 119, 121
    michelini... 121
    sp... 119
encopei, Dissodactylus... 115, 119, 121, 428
Episesarma... 284, 288
Euchirograpsus... 225, 260, 281
    americanus... 282, 283, 435
    liguricus... 281
Eucrate californiensis... 42
    crassimanus... 52
Eucratoplax... 52
    elata... 23
    guttata... 52
    spinidentata... 46

INDEX. 451

Eucratopsis 16, 52
   crassimanus 52, 426, 444
   macrophthalma 37
   spinidentata 46
Euplax 366, 423
   bosci 423
   leptophthalma 424
   leptophthalmus 423, 424
Euryalidae 14
Eurychelus 375
   monilifer 375, 380
Euryplax 16, 34
   nitida 12, 34, 35, 37, 425
   polita 12, 34, 36
   politus 36
excellens, Gelasimus 387
   Gelasimus (Acanthoplax) 385
eximius, Grapsodius 259
eydouxi, Grapsus 241

F.

faba, Pinnixa 129, 142, 146, 147, 149, 166, 429
   Pinnothera 142
   Pinnotheres 142
Fabia 61, 69, 101, 175
   byssomiae 12, 101, 102, 105, 428
   canfieldi 12, 101, 102, 106, 428
   chilensis 175, 176
   lowei 101, 104, 428
   subquadrata 101, 102, 104, 428
faxoni, Cymopolia 12, 184, 185, 193, 194, 432
   Palicus 194
   Pinnixa 12, 129, 131, 133, 429
femoralis, Cancer 250
festae, Sesarma 286
   Sesarma (Holometopus) 313
   Uca 378, 420
Fiddler Crab 10, 374, 375, 396, 401
   Square-backed 300
fimbriata, Margaritophora 91, 93
floridana, Cymopolia 184, 186, 220, 431
   Pinnixa 12, 129, 131, 138, 141, 429
Flounder, Ocellated 158
fossor, Gecarcinus 347
   Ocypode 347
fragilis, Cymopolia 184, 186, 213, 215, 432
franciscana, Pinnixa 12, 130, 131, 161, 162, 430
Free-swimming Crabs 9, 61, 87
Fresh-water Crab 295
Frevillea 25
   barbata 25, 26
   quadridentata 32
   rosaea 27
   sigsbei 26
   tridentata 29
Friendly crab 300

G.

gaimardi, Plagusia 336
galapagensis, Uca 377, 403, 442
Gasteropod Mollusks 61, 62
gaudichaudii, Ocypoda 372, 373
   Ocypode 367, 373, 441
gayi, Leptograpsus 234
Gecarcinacea 339
Gecarcinidae 10, 13, 15, 339
Gecarciniens 339

Gecarcinus 340, 351, 362
   agricola 352
   barbiger 352
   carnifex 341
   depressus 355, 357
   digueti 359, 360
   fossor 347
   lagostoma 352, 361, 364, 440
   lagostomus 364
   lateralis 13, 352, 355, 358, 359, 440
   malpilensis 359, 361
   planatus 352, 359, 440
   quadratus 13, 351, 358, 440
   regius 352
   ruricola 13,
      237, 351, 352, 356, 358, 359, 360, 361, 440
   uca 347
Gecarcoidea 340, 362, 363
   lalandii 363, 364, 445
geddesi, Pinnotheres 63, 65, 70, 426
Gelasima maracoani 378
   pugilator 400
Gelasimus 10, 374
   (Acanthoplax) excellens 385
   affinis 391
   annulipes 404, 405, 417
   armatus 378, 385, 387, 445
   brevifrons 393
   cinnatodus 387
   coloradensis 410
   crenulatus 409
   excellens 387
   gibbosus 416, 417
   gracilis 409, 410
   grangeri 381
   heterocheles 381, 382
   heterophthalmus 384
   heteropleurus 385
   Hygomi 422
   insignis 386
   latimanus 422
   leptodactylus 420
   macrodactylus 404, 409
   maracoani 374, 378, 381
   minax 389
   mordax 391
   natalensis 378
   ornatus 386, 387, 445
   palustris 389, 395, 397
   panamensis 412
   perlatus 387
   platydactylus 381, 382, 383
   poeyi 420
   princeps 382
   pugilator 378, 395, 400
   pugnax 395
   rapax 397
   rectilatus 405
   speciosus 408
   stenodactylus 416, 420, 422
   styliferus 383
   subcylindricus 419
   tangeri 387
   vocans 395, 400
   vocans, var. A 395
   vocator 378, 391, 393, 395, 398, 400

452 INDEX.

	Page.
Geocarcinidae	339
Geograpsus	225, 231
lividus	13, 231, 232, 433
occidentalis	232
Gephyrean	156
Ghost Crab	10, 367
Giant Clam	144, 147, 148
gibbesi, Acanthopus	337
Perenon	337, 439
gibbosa, Uca	416, 420
gibbosus, Caenocentrotus	176, 177
Gelasimus	416, 417
gigantea, Ocypoda	342
giganteus, Saxidomus	145, 148
glaberimus, Pinnotheres	254
glaberrimus, Cancer	254
globosum, Pinnotheres	65
Glyptograpsus	225, 275
impressus	13, 275, 277, 435
jamaicensis	13, 275, 277, 435
spinipes	275
Glyptoplax	16, 48
pugnax	12, 48, 50, 52, 444
pusilla	50
smithii	12, 50, 51, 426, 444
gnatherion, Cyclograpsus	181
Gnathochasmus	325
barbatus	325, 328
gonagrus, Leptograpsus	241
Goneplacidae	9, 12, 14, 15
Goneplacinae	16, 24
Goneplat	25
Goneplax	16, 24, 25
barbata	25, 26, 27, 425
hirsuta	25, 28
hirsutus	28
maldivensis	29
rhomboides	25, 28, 29
rosaea	25, 27
sigsbei	25, 26, 425
tridentata	25, 29
Goniograpsus	236, 240
cruentatus	237
innotatus	244, 248
pulcher	239
simplex	244
varius	250
Goniopsis	225, 226, 236
cruentata	13, 236, 237, 240, 433
cruentatus	239
pulcher	239
pulchra	13, 236, 239, 433
ruricola	237
strigosus	231
Gonoplacés vigils	423
Gonoplacidae	15
Gonoplacinae	24
Gonoplax	25
maracoani	378
gracilipes, Cymopolia	184, 186, 221, 223, 432
Palicus	221
gracilis, Cymopolia	184, 186, 218, 220, 432
Gelasimus	409, 410
Metopograpsus	249
Pachygrapsus	241, 249, 433
Palicus	218, 221

	Page.
gracilis, Plagusia	332
granarius, Trichodactylus	266
grangeri, Gelasimus	381
granulata, Chasmagnathus	329, 438
Oediplax	44, 426
Pinnixa (Scleroplax)	171
Scleroplax	170, 171, 430
granulatus, Chasmagnathus	329, 445
Oediplax	44
Palicus	183
Trichodactylus	266
granulimanus, Speocarcinus	12, 39, 40, 41, 42, ,426
Grapsacea	226, 260
Grapsidae	10, 13, 15, 224, 225
Grapsinae	225, 226, 260
Grapsodius	225, 259
eximius	259
Grapsoidea	9
Grapsoidiens	224
Grapsus	9, 225, 226, 231, 232
altifrons	227, 230
brevipes	232
cinereus	254, 300
declivifrons	245
depressus	332
diris	254
eydouxi	241
(Goniopsis) cruentatus	237
(Goniopsis) pictus	227
grapsus	13, 227, 433
(Grapsus) pusillus	254
grapsus tenuicrustatus	229
guadalupensis	249
integer	326
(Leptograpsus) corrugatus	252
(Leptograpsus) rugulosus	249
lividus	232
longipes	237
maculatus	227
marmoratus	250
maurus	244
minutus	254
ornatus	227
pelagicus	254
pelii	237
personatus	234
pictus	227
planifrons	234, 235
strigilatus	234
strigosus	231
testudinum	254
transversus	244
variegatus	234
varius	250
webbi	227
grapsus, Cancer	227
Cancer (Grapsus)	227
Grapsus	13, 227, 433
Great Land-crab	341
Green Turtle	257
Grouper, Spotted	339
guadalupensis, Grapsus	249
Guanhumi	341
guanhumi, Cardisoma	13, 341, 345, 424, 439
guerini, Pinnateres	101
Pinnotheres	64, 65, 101

INDEX. 453

guerini, Sesarma............ 308, 311
Gulf-weed Crab............. 253
guttata, Eucratoplax............ 52
Gvanhumi............ 341

H.

haematocheir, Holometopus............ 284
    Sesarma............ 284
hanseni, Sesarma............ 286
    Sesarma (Holometopus)............ 315, 437
Hapalocarcinidae............ 15
helleri, Uca............ 377, 415, 443
Hemigrapsus............ 225, 264, 273
    affinis............ 264, 266, 434
    crassimanus............ 264
    crenulatus............ 264, 266, 434
    nudus............ 264, 267, 270, 434
    oregonensis............ 264, 270, 435
hemphilli, Pinnotheres............ 64, 65, 99, 428
hendersoni, Parapinnixa............ 12, 107, 109, 112, 428
heterocheles, Gelasimus............ 381, 382
heterochelos, Ocypoda............ 381
    Ocypode............ 378, 381
    Uca............ 13, 374, 376, 378, 381, 383, 441
Heterograpsus barbimanus............ 266
    marmoratus............ 268
    nudus............ 268
    sanguineus............ 266, 268
heterophthalmus, Gelasimus............ 384
heteropleura, Uca............ 376, 385, 445
heteropleurus, Gelasimus............ 385
hiatus, Pinnixa............ 130, 164, 167, 430
hillii, Orthograpsus............ 231, 232
hirsuta, Goneplax............ 25, 28
hirsutus, Goneplax............ 28
hirtimanus, Pinnotheres............ 64, 65, 101
hirtipes, Pinnaxodes............ 174, 175
holmesi, Pinnotheres............ 12, 63, 65, 66, 68, 426
Holograpsus............ 284
Holometopus............ 284, 286, 298
    haematocheir............ 284
Holothurians............ 128, 144, 149, 174
Homalaspis plana............ 352
Horse Clam............ 145, 148
    Mussels............ 76
humei, Hylaeocarcinus............ 363, 364
    Pelocarcinus............ 364
Hygomi, Gelasimus............ 422
Hylaeocarcinus............ 363
    humei............ 363, 364
    natalis............ 364

I.

iberingi, Sesarma (Holometopus) miersii............ 286, 304, 436
    Sesarma miersii............ 13, 286, 287
immaculata, Plagusia............ 332, 335, 439
impressus, Glyptograpsus............ 13, 275, 277, 435
inermis, Monolepis............ 363, 368
innotatus, Goniograpsus............ 244, 248
insignis, Acanthoplax............ 375, 385
    Gelasimus............ 386
    Uca............ 376, 385, 445
integer, Cyclograpsus............ 13, 325, 326, 328, 438
    Grapsus............ 326

intermedius, Limnocarcinus............ 363, 364
    Pachygrapsus............ 245
irradians, Pecten............ 76, 77, 78
Isopod Parasite............ 248
isthmia, Cymopolia............ 184, 185, 201, 432
isthmius, Pallens............ 201

J

jamaicensis, Areograpsus............ 275, 277
    Glyptograpsus............ 13, 275, 277, 435
jarvisi, Sesarma............ 286
    Sesarma (Sesarma)............ 296, 436
Jellyfishes............ 255
jouyi, Brachynotus (Heterograpsus)............ 273
    Tetragrapsus............ 273, 435
Juey............ 341

K.

Kaburi............ 347
Kellia laperousii............ 87
Key-hole Limpet............ 106, 174
    Urchin............ 119, 121
kuhlii, Ocypoda............ 372

L.

laevigata, Pinnixa............ 159
    Pinnotherelia............ 180, 181, 431
laevimanus, Pachygrapsus............ 245
laevis, Uca............ 350
lagostoma, Gecarcinus............ 352, 361, 364, 440
lagostomus, Gecarcinus............ 364
lalandei, Pelocarcinus............ 364
lalandii, Gecarcoidea............ 363, 364, 445
    Pelocarcinus............ 363
Lamellibranch Mollusks............ 62
Land-crab............ 10, 15, 339, 424
    Black............ 355
    Blue............ 352
    Common............ 355
    Great............ 341
    Mulatto............ 341
    White............ 341
laperousii, Kellia............ 87
Larval Crab............ 366, 368
lateralis, Gecarcinus............ 13, 352, 355, 358, 359, 440
    Ocypoda............ 355
latimana, Uca............ 422
latimanus, Cardiosoma............ 345
    Gelasimus............ 422
    Uca............ 378, 444
latipes, Chasmocarcinus............ 12, 55, 57, 59
lavis, Uca............ 348, 350
lazarus, Chama............ 66
Leidya............ 248
Leiolophus............ 337
    planissimus............ 337
leptodactyla, Uca............ 378, 420, 444
leptodactylus, Gelasimus............ 420
Leptograpsus............ 225, 234
    ansoni............ 234
    bertheloti............ 250
    gayi............ 234
    gonagrus............ 241
    marmoratus............ 250
    maurus............ 244

454   INDEX.

	Page
Leptograpsus rugulosus	245, 249
variegatus	234, 433
verreauxi	234
leptophthalma, Euplax	424
leptophthalmus, Euplax	423, 424
ligurlcus, Euchirograpsus	281
Limnocarcinus	363
intermedius	363, 364
Limpet, Key-hole	106, 174
linnaeana, Planes	254
Liosoma areuata	144, 150
arenicola	144, 150
lithodomi, Pinnotheres	64, 65, 66, 73
Lithodomus aristatus	73, 74
Lithophaga aristatus	73, 74
littoralis, Pinnixa	129, 142, 145, 429
lividus, Geograpsus	13, 231, 232, 433
Grapsus	232
Lobograpsus	264
crenulatus	264, 266
Lobworm	159, 160
loevigata, Pinnotherelia	181
Long Clam	76
longipes, Discoplax	340
Grapsus	237
Pinnixa	12, 129, 131, 137, 165
Tubicola	128, 137
lowei, Fabia	101, 104, 428
Raphonotus	104
Loxechinus albus	176, 177
Lucapina crenulata	106, 174
lucasii, Cymopolia	12, 184, 185, 193, 431
Palicus	193
Lycodes brevipes	157

M.

macrophylla, Chama	66
Macoma nasuta	148
macrocera, Ocypode	367
macrodactyla, Uca	404
macrodactylus, Gelasimus	404, 409
Uca	377, 404, 442
macrophthalma, Chasmophora	37
Eucratopsis	37
Macrophthalmidae	365
Macrophthalminae	365, 423
maculatum, Pinnotheres	74
maculatus, Grapsus	227
Pagurus	227
Pinnotheres	64, 65, 74, 427
Maestro-Sastre	382
magdalenense, Sesarma	286
Sesarma (Holometopus)	305, 436
major, Cancer vocans	381
Nautilograpsus	254
maldivensis, Goneplax	29
Malformation	409
malpilensis, Gecarcinus	359, 361
Mangrove Crab	237, 293, 294, 323
Maracoani	378
maracoani, Gelasima	378
Gelasimus	374, 378, 381
Gonoplax	378
Ocypode	378
Uca	13, 376, 378, 380, 381, 382, 441
marchei, Pelocarcinus	364

	Page
margarita, Pinnotheres	63, 64, 64, 91
Margaritophora fimbriata	91, 93
marinus, Planes	253, 258, 434
marmoratus, Cancer	250, 267
Cyclograpsus	267
Grapsus	250
Heterograpsus	268
Leptograpsus	250
Pachygrapsus	13, 241, 250, 434
marmoreus, Cancer	250
Marsh Crabs	284, 326
Master-Tailor	382
maurus, Grapsus	244
Leptograpsus	244
Pachygrapsus	13, 241, 244, 433
Megalops	11, 304, 365, 369
meinerti, Pinnaxodes	175, 177, 428
Mellita quinquesperforata	119
testudinata	119
mellitae, Dissodactylus	115, 117, 429
Echinophilus	114, 117
Metagrapsus	321
curvatus	321
pectinatus	321, 322
Metasesarma	226, 319
rousseauxi	319
rubripes	319, 437
Metopaulias	225, 317
depressus	317, 318, 438
Metopograpsus	brasiliensis	319
dubius	245
gracilis	249
minatus	245
michelini, Encope	121
Micropanope pusilla	50
miersii, Sesarma	286, 306, 308
Sesarma (Holometopus)	303, 304, 436
minax, Gelasimus	389
Uca	376, 389, 391, 441
Uca vocator, var	389
miniata, Sesarma	308, 310
minatus, Metopograpsus	245
minuta, Pinnixa	150
minutus, Cancer	253
Cyclograpsus	328
Grapsus	254
Nautilograpsus	253, 254
Pinnotheres	254
Planes	13, 253, 258, 434
Modiola copax	73
modiolus	76, 77, 104
tulipa	76, 78
modiolus, Modiola	76, 77, 104
Mollusk	72, 106
Bivalve	61, 87, 103, 104
Boring	73
Gasteropod	61, 62
Lamellibranch	62
Rock-boring	174
Molpadia arenicola	144, 150
monilifer, Eurychelus	375, 380
monilifera, Uca	13, 376, 378, 380, 441
monodactyla, Pinnixa	130, 131, 136
monodactylum, Pinnotheres	136
Monolepis	366
inermis	366, 368

INDEX. 455

	Page.
mordax, Gelasimus	391
Uca	13,376,391,395,441
mosen, Pinnotheres	64,65,94,95,427
Mountain Crab	352
Mud Crabs	9
mülleri, Sesarma	319
Mulatto Land-crab	341
Mulinia, sp	81
muliniarum, Pinnotheres	64,66,81,427
mülleril, Sesarma	298,319
muricata, Pinna	76,78
musica, Uca	377,417,444
Mussel	72,73,104
Crab	74
Mussel, Horse	76
Mya	145,147
arenaria	76,86,87,103,148
californica	87
Mytilus	77
algosus	72
edulis	76,87,103,174

N.

nasuta, Macoma	148
natalensis, Gelasimus	378
natalis, Hylaeocarcinus	364
Nautilograpsus	253
angustatus	254
major	254
minutus	253,254
smithii	254
nigrolineata, Ciccaba	405
nitida, Euryplax	12,34,35,37,425
Parapinnixa	12,107,111,113
Pinnixa	107,141
Pseudopinnixa	107
nitidus Dissodactylus	115,116,428
nudus, Brachynotus	268
Hemigrapsus	264,267,270,431
Heterograpsus	268
Pinnotheres	64,65,66,68,83
Pseudograpsus	267
nuttallii, Schizothaerus	144,145,148,149

O.

obesa, Cymopolia	12,184,185,204,205,207,432
obesus, Palicus	202,205
obliquus, Chasmocarcinus	55,58,426
occidentale, Sesarma	13,286,287
Sesarma (Holometopus)	299
occidentalis, Geograpsus	232
Ocypoda	372
Ocypoda kuhlii, var	372
Ocypode	13,367,372,373,441
Oedipleura	350
Pinnixa	12, 130,131,155,157,161,162,168,430
Sesarma	299
Ucides	13,347,350,439,440
octodentata, Pseudorhombila	43,426
oculiferus, Bathyplax typhlus, var	19
Ocypoda arenaria	368,369
gaudichaudii	372,373
gigantea	342
heterochelos	381
kuhlii	372

	Page.
Ocypoda kuhlii, var. occidentalis	372
lateralis	355
occidentalis	372
pugilator	400
rubra	352
urvillei	372
Ocypode	341,365,366
albicans	13,366,367,368,372,373,440,441
arenarius	368
bispinosa	25
(cardisoma) cordata	342
ceratophthalma	366,372
cordata	341,347
fossor	347
gaudichaudii	367,373,441
heterochelos	378,381
macrocera	367
maracoani	378
occidentalis	13,367,372,373,441
quadrata	367
reticulatus	290
rhombea	367
ruricola	341,342
(Sesarma) reticulatus	290
tourlourou	352
(uca) uca	347
urvillei	372
vca	347
Ocypodiacés ordinaires	366
Ocypodidae	10,13,15,365
Ocypodiens	365
Ocypodinae	365,366
Oediplax	16,44
granulata	44,426
granulatus	44
Oedipleura	346
cordata	346,348
occidentalis	350
oerstedi, Uca	377,414,443
ophioderma, Sesarma	286
Sesarma (Sesarma)	297
Opisthopus	128,172
transversus	172,173,430
orcutti, Pinnotheres	64,65,66,98,428
oregonensis, Brachynotus	270
Hemigrapsus	264,270,435
Pseudograpsus	270
orientale, Chaenostoma	423
orientalis, Plagusia	334,335
ornata, Amphitrite	152
ornatus, Gelasimus	386,387,445
Grapsus	227
Orthograpsus	226,231
hillii	231,232
Ostracotheres	63
politus	62,71
Ostrea parasitica	66
ostrearicola, Speocarcinus	39,41,426
ostrearius, Pinnotheres	70
ostreum, Pinnotheres	12,63,65,66,69,74,426
Owl Stomach	405
Oxyrhyncha	12
Oxystomata	13
Oyster	66,67,68,70,71,73,76,78,101
Oyster Beds	41,78
Crab	10,66,70
Oyster, Pearl	93

P.	Page.
Pachydesma crassatelloides	105
Pachygrapsus	225, 240, 253, 258
advena	245
corrugatus	241, 252, 445
crassipes	13, 240, 241, 244, 433
gracilis	241, 249, 433
intermedius	245
laevimanus	245
marmoratus	13, 241, 250, 434
maurus	13, 241, 244, 433
pubescens	13, 241, 252, 445
simplex	244
socius	245
transversus	13, 241, 244, 249, 433
Pachysoma	284
bidens	284
pacifica, Pholas	104
Pagurus	347
maculatus	227
palaensis, Arcotheres	62
Palicae	182
Palicidae	15, 182
Palicus	183
acutifrons	223
affinis	196
agassizi	196, 197
agassizii	196
alternatus	188, 196, 198
angustus	210
bahamensis	200
blakei	188, 190, 191
cristatipes	186
cursor	215
dentatus	202
depressus	212
faxoni	194
gracilipes	221
gracilis	218
granulatus	183
isthmius	201
lucasii	193
obesus	202, 205
rathbuni	198
sica	208
sicus	208
zonatus	190
palustris, Gelasimus	389, 395, 397
panamensis, Gelasimus	412
Pinnixa	131
Uca	377, 412, 443
Panopeus	47
Panoplax	16, 47, 48
depressa	47, 49, 426
Paphia	104
Parapinnixa	61, 107
affinis	12, 107, 111
beaufortensis	107, 112
bouvieri	12, 107, 111, 428
hendersoni	12, 107, 109, 112, 428
nitida	12, 107, 111, 113
Parasesarma	284
Parasite	245
Parasite, Isopod	248
parasitica, Ostrea	66
Paraxanthus barbiger	352
parma, Echinarachnius	119

	Page.
Parocypoda	366
ceratophthalma	366
patagoniensis, Pinnixa	129, 131, 135, 429
Pearl Oyster	93
Pecten irradians	76, 77, 78
tenuicostatus	76, 77
pectinatus, Metagrapsus	321, 322
pelagicus, Grapsus	254
pelii, Grapsus	237
Pelocarcinus	363
cailloti	364
humei	364
lalandei	364
lalandii	363
marchei	364
Percnon	226, 337
gibbesi	337, 439
planissimum	337
pergamentaceus, Chaetopterus	76, 151, 152, 153
Perisesarma	284, 287
perlatus, Gelasimus	387
personatus, Grapsus	234
Phallusia vermiformis	87
Philyra	331
depressa	331
Pholas	174
californica	104
pacifica	104
pica, Turbo	90
pictus, Grapsus	227
Grapsus (Goniopsis)	227
Piddock	174
pilosipes, Uca	348
Pilumnoplax	16, 21
americana	21
americanus	21
elata	21, 23, 425
sinclairi	21
sulcatifrons	21
Pine Crab	318
pinima, Aratv	323
Pinna muricata	76, 78
Pinnatores guerini	101
Pinnaxodes	128, 133, 174, 175
chilensis	174, 175, 177, 431
hirtipes	174, 175
meinerti	175, 177, 428
tomentosus	175, 178
Pinnixa	128, 170, 174, 179
affinis	12, 130, 131, 168
barnharti	130, 144, 149, 429
brevipollex	12, 130, 131, 169
californiensis	155, 157, 166
chaetopterana	130, 151, 158, 430
cristata	129, 131, 134, 136, 429
cylindrica	12, 128, 129, 131, 138, 139, 151, 156, 159, 430
faba	129, 142, 146, 147, 149, 166, 429
faxoni	12, 129, 131, 133, 429
floridana	12, 129, 131, 138, 141, 429
franciscana	12, 130, 131, 161, 162, 430
hiatus	130, 164, 167, 430
laevigata	159
littoralis	129, 142, 145, 429
longipes	12, 129, 131, 137, 165
minuta	159

	Page.		Page.
Pinnixa monodactyla	130, 131, 136	Pinnotheres reticulatus	64, 65, 66, 93, 427
nitida	107, 141	serrei	63, 65, 84, 427
occidentalis	12,	shoemakeri	64, 65, 95, 428
	130, 131, 155, 157, 161, 162, 168, 430	silvestrii	63, 65, 66, 91
panamensis	131	strombi	63, 65, 90, 427
patagoniensis	129, 131, 135, 429	taylori	64, 65, 66, 97, 427
retinens	130, 139, 431	transversalis	131, 154
sayana	12, 129, 131, 156, 430	Pinnotheridae	9, 12, 15, 61
schmitti	130, 162, 167, 430	Pinnotheridea	61
(Scleroplax) granulata	171	Pinnotherinae	61
species	156	pisonii, Aratus	12, 322, 323, 438
tomentosa	129, 141, 429	Sesarma	323
transversalis	12, 129, 131, 133, 134, 154, 429	Plagusia	226, 331, 337
tubicola	129, 165, 167, 430	capensis	336
tumida	149	chabrus	332, 336, 439
valdiviensis	130, 131, 154, 430	depressa	331, 332, 334, 335, 438
weymouthi	130, 166, 430	depressa tuberculata	332, 334, 335, 438
pinnophylax, Cancer	66	depressus	332
Pinnotheres	66	gaimardi	336
Pinnoteres silvestrii	91	gracilis	332
Pinnothera	62	immaculata	332, 335, 439
faba	142	orientalis	334, 335
Pinnotherelia	128, 180	sayi	332
laevigata	180, 181, 431	squamosa	332, 334
loevigata	181	tomentosa	336
Pinnothereliinae	127	tuberculata	334, 335
Pinnotherelinae	127	Plagusiacaea	331
Pinnotheres	10, 61, 62, 179	Plagusiinae	226, 331
angelica	72	Plagusinae	331
angelicus	63, 65, 66, 70, 72, 426	plana, Homalaspis	352
barbata	88	planatus, Gecarcinus	352, 359, 440
barbatus	63, 65, 88, 427	Planes	225, 253, 258
bipunctatum	78	clypeatus	253, 254
bipunctatus	64, 66, 78, 445	cyaneus	254
byssomiae	74, 105	linnaeana	254
chilensis	175	marinus	253, 258, 434
concharum	64, 66, 86, 88, 427	minutus	13, 253, 258, 434
crassipes	66	planifrons, Grapsus	234, 235
cylindricum	159	planissimum, Percnon	337
depressum	79	planissimus, Leiolophus	337
depressus	64, 65, 79, 427	Platychirograpsus	225, 278
faba	142	spectabilis	278, 281
geddesi	63, 65, 70, 426	typicus	278, 435
glaberimus	254	platydactyla, Uca	381, 382
globosum	65	platydactylus, Gelasimus	381, 382, 383
guerini	64, 65, 101	plicatus, Sesarma	284
hemphilli	64, 65, 99, 428	poeyi, Gelasimus	420
hirtimanus	64, 65, 101	polita, Euryplax	12, 34, 36
holmesi	12, 63, 65, 66, 68, 426	politus, Euryplax	36
lithodomi	64, 65, 66, 73	Ostracotheres	62, 71
maculatum	74	Pinnotheres	63, 65, 66, 71, 444
maculatus	64, 65, 74, 427	Portunidae	14
margarita	63, 64, 66, 91	Potamonidae	11, 14
minutus	254	princeps, Gelasimus	382
monodactylum	136	Uca	13, 376, 378, 382, 441, 445
moseri	64, 65, 94, 95, 427	Prionoplacinae	16, 29
mulinarum	64, 66, 81, 427	Prionoplax	16, 29
nudus	64, 65, 66, 68, 83	atlantica	12, 30, 425
orcutti	64, 65, 66, 98, 428	atlanticus	30
ostrearius	70	ciliata	12, 30, 31
ostreum	12, 63, 65, 66, 69, 74, 426	spinicarpus	29, 31
pinnophylax	66	Pseudograpsus nudus	267
politus	63, 65, 66, 71, 444	oregonensis	270
pubescens	63, 65, 66, 87	Pseudopinnixa	107
pugettensis	63, 65, 66, 82, 97, 427	nitida	107
pusillus	254	Pseudorhombila	16, 42

458 INDEX.

	Page.
Pseudorhombila octodentata	43, 426
quadridentata	42
Pseudorhombilinae	17
Ptenoplacidae	15
pubescens, Cryptophrys	87
Pachygrapsus	13, 241, 252, 445
Pinnotheres	63, 65, 66, 87
pugettensis, Pinnotheres	63, 65, 66, 82, 97, 427
pugilator, Gelasimus	378, 395, 400
Ocypoda	400
Uca	377, 396, 400, 412, 419, 442, 445
pugilis, Strombus	90
pugillator, Gelasima	400
pugnax, Gelasimus	395
Glyptoplax	12, 48, 50, 52, 444
Uca	377, 395, 396, 398, 400, 442
pulcher, Goniograpsus	239
Goniopsis	239
pulchra, Goniopsis	13, 236, 239, 433
punctatus, Cyclograpsus	325, 327, 328, 438
pusilla, Glyptoplax	50
Micropanope	50
pusillus, Cancer	254
Grapsus (Grapsus)	254
Pinnotheres	254

Q.

quadrata, Cancer	367
Cardisoma	342
Ocypode	367
Sesarma	284
quadratus, Cancer	284, 367
Gecarcinus	13, 351, 358, 440
quadridentata, Frevillea	32
Pseudorhombila	42
Tetraplax	32, 425
quinquesperforata, Mellita	119

R.

rapax, Gelasimus	397
Uca pugnax	377, 397, 398, 442
Raphonotus	101
lowei	104
subquadratus	101, 102, 106
rathbuni, Cymopolia	184, 185, 198, 200, 432
Palicus	198
recta, Sesarma	298, 316
rectilata, Uca	405
rectilatus, Gelasimus	405
Uca	376, 405, 406
rectum, Sesarma	286
Sesarma (Holometopus)	298, 436
Red-jointed Fiddler Crab	389
Red Tourlourou	352
regius, Gecarcinus	352
Reptantia	11
reticulata, Sesarma	290
reticulatum, Sesarma	13, 284, 285, 287, 292, 293
Sesarma (Sesarma)	290, 435
reticulatus, Ocypode	290
Ocypode (Sesarma)	290
Pinnotheres	64, 65, 66, 93, 427
retinens, Pinnixa	130, 139, 431
Retropluma	14, 15
Retroplumidae	15
Rhizocephalid	184

	Page.
rhizophorae, Sesarma	13, 285, 287
Sesarma (Sesarma)	294, 436
Rhizopidae	54
Rhizopinae	16, 54
rhombea, Ocypode	367
rhomboides, Cancer	25
Goneplax	25, 28, 29
ricordi, Sesarma	286, 308
Sesarma (Holometopus)	308, 437
River Crab	278
roberti, Sesarma	13, 286, 287, 312
Sesarma (Holometopus)	312, 437
Rock-boring Mollusk	174
Rock Crab	227
rosaea, Frevillea	27
Goneplax	25, 27
rotundum, Discoplax	340
rousseauxi, Metasesarma	319
rubra, Ocypoda	352
rubripes, Metasesarma	319, 437
Sesarma (Holometopus)	319
rugulosus, Grapsus (Leptograpsus)	249
Leptograpsus	245, 249
ruricola, Cancer	237, 352
Gecarcinus	13, 237, 351, 352, 356, 358, 359, 360, 361, 440
Goniopsis	237
Ocypode	341, 342
Sally Lightfoot	227, 229
Sand Crab	10, 367, 401
Dollar	119, 121
sanguineus, Heterograpsus	266, 268
Sargassum	256
Sarmatium	226, 321
crassum	321
curvatum	321, 438
Saxicava arctica	106
Saxidomus	148
giganteus	145, 148
sayana, Pinnixa	12, 129, 131, 156, 430
sayi, Plagusia	332
scabripes, Tetrias	179, 431
Scallops	76, 77
Schizothaerus	147
nuttallii	144, 145, 148, 149
schmitti, Pinnixa	130, 162, 167, 430
Scleroplax	128, 170
granulata	170, 171, 430
Sea Cucumber	149, 150
Squirt	95
Urchin	62, 79, 115, 119, 176
Serpula	244
Serpulid	184
serrei, Pinnotheres	63, 65, 84, 427
Sesarma	225, 284, 288, 298, 317, 319, 321, 322
aequatoriale	13, 285, 287
aequatorialis	292
africana	287
africanum	285
americana	311, 312
angusta	314
angustipes	286, 287, 303, 308, 311
angustum	13, 286, 287, 313
barbata	328

INDEX. 459

Sesarma barbimana... 298
   barbimanum... 286
   benedicti... 286, 319
   bidens... 284
   bidentata... 295
   bidentatum... 286
   biolleyi... 13, 286, 287, 304
   bromeliarum... 312, 313
   chiragra... 316, 317
   (Chiromantes) africanum... 287, 435
   cinerea... 290, 300, 308
   cinereum... 13, 286, 287, 290, 303, 308
   crassipes... 285, 294, 298
   curacaoense... 13, 285, 287, 294
   curacaoensis... 293
   curvata... 321, 322
   festae... 286
   guerini... 308, 311
   haematocheir... 284
   hanseni... 286
   (Holometopus) angustipes... 13, 311, 437
   (Holometopus) angustum... 314, 437
   (Holometopus) benedicti... 316, 437
   (Holometopus) biolleyi... 306, 437
   (Holometopus) cinereum... 300, 436
   (Holometopus) festae... 313
   (Holometopus) hanseni... 315, 437
   (Holometopus) magdalenense... 305, 436
   (Holometopus) miersii... 303, 304, 436
   (Holometopus) miersii ihoringi... 304, 436
   (Holometopus) occidentale... 299
   (Holometopus) rectum... 298, 436
   (Holometopus) ricordi... 308, 437
   (Holometopus) roberti... 312, 437
   (Holometopus) rubripes... 319
   (Holometopus) tampicense... 307, 437
   jarvisi... 286
   magdalenense... 286
   miersii... 286, 306, 308
   miersii iheringi... 13, 286, 287
   miniata... 308, 310
   mülleri... 319
   mulleri... 298, 319
   occidentale... 13, 286, 287
   occidentalis... 299
   ophioderma... 286
   (Perisesarma) africanum... 287
   pisonii... 323
   plicatus... 284
   quadrata... 284
   recta... 298, 316
   rectum... 286
   reticulata... 290
   reticulatum... 13, 284, 285, 287, 292, 293
   rhizophorae... 13, 285, 287
   ricordi... 286, 308
   roberti... 13, 286, 287, 312
   (Sarmatium) curvatum... 321
   (Sesarma) aequatoriale... 292
   (Sesarma) barbimanum... 298
   (Sesarma) bidentatum... 295, 436
   (Sesarma) crassipes... 294
   (Sesarma) curacaoense... 293, 435, 445
   (Sesarma) jarvisi... 296, 436
   (Sesarma) ophioderma... 297
   (Sesarma) reticulatum... 290, 435

Sesarma (Sesarma) rhizophorae... 294, 436
   (Sesarma) sulcatum... 289, 435
   (Sesarma) verleyi... 288, 435
   stimpsonii... 303, 308
   subintegra... 287
   sulcata... 289
   sulcatum... 285, 299
   tampicense... 286
   tetragonum... 284
   trapezium... 320
   verleyi... 285
   violacea... 321
Sesarmacea... 283
Sesarminae... 225, 283
shoemakeri, Pinnotheres... 64, 65, 95, 428
Short-tailed Crabs... 11
sica, Cymopolia... 184, 186, 208, 210, 211, 212, 431
   Palicus... 208
sicus, Palicus... 208
sigsbei, Frevillea... 26
   Goneplax... 25, 26, 425
silvestrii, Pinnoteres... 91
   Pinnotheres... 63, 65, 66, 91
simplex, Goniograpsus... 244
   Pachygrapsus... 244
sinclairi, Pilumnoplax... 21
Small Clam... 147
smithii, Glyptoplax... 12, 50, 51, 426, 444
   Nautilograpsus... 254
socius, Pachygrapsus... 245
speciosa, Uca... 13, 377, 378, 408, 409, 442
speciosus, Gelasimus... 408
spectabilis, Platychirograpsus... 278, 281
Speocarcinus... 16, 38
   californiensis... 39, 42, 426
   carolinensis... 12, 38, 39, 41, 426, 444
   granulimanus... 12, 39, 40, 41, 42, 426
   ostrearicola... 39, 41, 426
Spider Crabs... 12
spinicarpa, Uca... 377, 411, 443
spinicarpus, Prionoplax... 29, 31
spinidentata, Cyrtoplax... 45, 46, 426
   Eucratoplax... 46
   Eucratopsis... 46
spinipes, Glyptograpsus... 275
Spondylus... 74
Sponges... 247, 250, 255
Spotted Grouper... 339
Spray Crab... 337
squamosa, Plagusia... 332, 334
squamosus, Cancer... 332
Square-backed Fiddler... 300
Squilla... 39
Starfish... 68
stebbingi, Dissodactylus... 115, 123, 429
stenodactyla, Uca... 408, 409, 416, 417, 420
stenodactylus, Gelasimus... 416, 420, 422
   Uca... 13, 378, 416, 417, 418, 419, 420, 421, 443
Stichopus californicus... 174
stimpsonii, Sesarma... 303, 308
strigilatus, Grapsus... 234
strigosus, Goniopsis... 231
   Grapsus... 231
strombi, Pinnotheres... 63, 65, 90, 427
Strombus... 85

INDEX.

Strombus pugilis ... 90
stylifera, Uca ... 376, 383, 385, 441
   Uca platydactyla, var. ... 384, 385
styliferus, Gelasimus ... 383
subcylindrica, Uca ... 13, 378, 419, 444, 445
subcylindricus, Gelasimus ... 419
subintegra, Sesarma ... 287
subquadrata, Fabia ... 101, 102, 104, 428
subquadratus, Raphonotus ... 101, 102, 106
sulcata, Sesarma ... 289
sulcatifrons, Pilumnoplax ... 21
sulcatum, Sesarma ... 285, 299
   Sesarma (Sesarma) ... 289, 435

T.

tampicense, Sesarma ... 286
   Sesarma (Holometopus) ... 307, 437
tangeri, Gelasimus ... 387
   Uca ... 376, 387, 441
tangieri, Uca ... 387
Tapes ... 104, 145, 149
taylori, Pinnotheres ... 64, 65, 66, 97, 427
tenuicostatus, Pecten ... 76, 77
tenuicrustatus, Grapsus grapsus ... 229
testudinata, Mellita ... 119
testudinum, Grapsus ... 254
tetragonum, Sesarma ... 284
Tetragrapsus ... 225, 273
   jouyi ... 273, 435
Tetraplax ... 16, 32
   quadridentata ... 32, 425
Tetrias ... 128, 179
   scabripes ... 179, 431
thayeri, Uca ... 376, 406, 442
Tiliandsia ... 319
tomentosa, Pinnixa ... 129, 141, 429
   Plagusia ... 336
tomentosus, Pinnaxodes ... 175, 178
Tourlourou ... 341
   Red ... 352
tourlourou, Ocypode ... 352
transversalis, Pinnixa .. 12, 129, 131, 133, 134, 154, 429
   Pinnotheres ... 131, 154
transversus, Grapsus ... 244
   Opisthopus ... 172, 173, 430
   Pachygrapsus ... 13, 241, 244, 249, 433
trapezium, Sesarma ... 320
Tree Crab ... 237, 323
Trichodactylus granarius ... 266
   granulatus ... 266
tridentata, Frevillea ... 29
   Goneplax ... 25, 29
Trizocarcinus ... 16, 17
   dentatus ... 17, 18, 425
tuberculata, Cymopolia ... 12, 149, 184, 185, 207
   Plagusia ... 334, 335
   Plagusia depressa ... 332, 334, 335, 438
Tubicola ... 128
   longipes ... 128, 137
tubicola, Pinnixa ... 129, 165, 167, 430
tulipa, Modiola ... 76, 78
tumida, Pinnixa ... 149
Tunicates ... 62, 83, 98
Turbo pica ... 90
Turtle ... 255
   Crab ... 253

Turtle, Green ... 257
typhla, Bathyplax ... 19, 425
typhlus, Bathyplax ... 19
typicus, Aspidograpsus ... 278
   Chasmocarcinus ... 12, 54, 55, 57, 59
   Platychirograpsus ... 278, 435

U.

Uca ... 10, 341, 346, 366, 374
   amazonensis ... 391
   brevifrons ... 377, 393, 442
   coloradensis ... 377, 410, 443
   cordata ... 346, 348
   crenulata ... 13, 377, 378, 406, 409, 410, 443
   testae ... 378, 420
   galapagensis ... 377, 403, 442
   gibbosa ... 416, 420
   holleri ... 377, 415, 443
   heterochelos ... 13, 374, 376, 378, 381, 383, 441
   heteropleura ... 376, 385, 445
   insignis ... 376, 385, 445
   laevis ... 350
   latimana ... 422
   latimanus ... 378, 444
   lavis ... 348, 350
   leptodactyla ... 378, 420, 444
   macrodactyla ... 404
   macrodactylus ... 377, 404, 442
   maracoani ... 13, 376, 378, 380, 381, 382, 441
   minax ... 376, 389, 391, 441
   monilifera ... 13, 376, 378, 380, 441
   mordax ... 13, 376, 391, 395, 441
   musica ... 377, 417, 444
   oerstedi ... 377, 414, 443
   panamensis ... 377, 412, 443
   pilosipes ... 348
   platydactyla ... 381, 382
   platydactyla, var. stylifera ... 384, 385
   princeps ... 13, 376, 378, 382, 441, 445
   pugilator ... 377, 396, 400, 412, 419, 442, 445
   pugnax ... 377, 395, 396, 398, 400, 442
   pugnax rapax ... 377, 397, 398, 442
   rectilata ... 405
   rectilatus ... 376, 405, 406
   speciosa ... 13, 377, 378, 408, 409, 442
   spinicarpa ... 377, 411, 443
   stenodactyla ... 408, 409, 416, 417, 420
   stenodactylus ... 13, 378, 416, 417, 418, 419, 420, 421, 443
   stylifera ... 376, 383, 385, 441
   subcylindrica ... 13, 378, 419, 444, 445
   tangeri ... 376, 387, 441
   tangieri ... 387
   thayeri ... 376, 406, 442
   uca ... 346, 347
   una ... 347, 350, 374, 381
   uruguayensis ... 377, 413, 443
   vocator ... 378, 391, 393, 395, 398, 409
   vocator, var. minax ... 389
uca, Cancer ... 347
   Gecarcinus ... 347
   Ocypode (uca) ... 347
   Uca ... 346, 347
Ucides ... 340, 346
   cordatus ... 13, 346, 347, 348, 350, 439, 444
   occidentalis ... 13, 347, 350, 439, 440

INDEX. 461

	Page.
uka, Cancer	381
una, Uca	347, 350, 374, 381
Urchins, Sea	62, 79, 115, 119, 121, 176
uruguayensis, Uca	377, 413, 443
urvillei, Ocypoda	372
Ocypode	372

V.

valdiviensis, Pinnixa	130, 131, 154, 430
var. A, Gelasimus vocans	395
variegatus, Cancer	234
Grapsus	234
Leptograpsus	234, 433
varius, Goniograpsus	250
Grapsus	250
Varunacea	260
Varuninae	225, 260
Uca vna,	347
vca, Cancer	347, 381
Ocypode	347
Velella	257
verleyi, Sesarma	285
Sesarma (Sesarma)	288, 435
vermiformis, Phallusia	87
verreauxi, Leptograpsus	234
Vesicularia	153

	Page.
Violacea, Sesarma	321
vna, Uca	347
vocans, Cancer	367, 395, 400
Gelasimus	395, 400
vocator, Cancer	391
Gelasimus	378, 391, 393, 395, 398, 409
Uca	378, 391, 393, 395, 398, 409

W.

webbi, Grapsus	227
weymouthi, Pinnixa	130, 166, 430
White Land-crab	341
Wood Crab	300
Worm Tubes	61, 128, 133, 137, 152, 166
Worms	39, 128, 138, 159, 160

X.

Xanthidae	9, 14

Z.

Zaops	62
depressa	62, 63, 79
Zoea	11, 78, 304
zonata, Cymopolia	12, 184, 185, 190, 192, 431, 432
zonatus, Palicus	190
Zoothamnium	68

○